The ARRL
Extra Class License Manual
For Ham Radio

Twelfth Edition

By Ward Silver, NØAX

Contributing Editor: Mark Wilson, K1RO

Production Staff: Jodi Morin, **KA1JPA,** Assistant Production Supervisor, Layout
Michelle Bloom, **WB1ENT,** Production Supervisor, Layout
Maty Weinberg, **KB1EIB,** Editorial Assistant
David Pingree, **N1NAS,** Senior Technical Illustrator

Copyright © 2020 by

The American Radio Relay League, Inc

Copyright secured under the Pan-American Convention

All rights reserved. No part of this work may be reproduced in any form except by written permission of the publisher.
All rights of translation are reserved.

Printed in USA

Quedan reservados todos los derechos

ISBN: 978-1-62595-131-1

Twelfth Edition
First Printing

This book may be used for Extra license exams given beginning July 1, 2020. *QST* and the ARRL website (**www.arrl.org**) will have news about any rules changes affecting the Extra class license or any of the material in this book.

We strive to produce books without errors. Sometimes mistakes do occur, however. When we become aware of problems in our books (other than obvious typographical errors), we post corrections on the ARRL website. If you think you have found an error, please check **www.arrl.org/extra-class-license-manual** for corrections. If you don't find a correction there, please let us know by sending an e-mail to **pubsfdbk@arrl.org**.

The ARRL Extra Class License Manual   ON THE WEB

www.arrl.org/extra-class-license-manual

Visit *The ARRL Extra Class License Manual* home online for additional resources.

Contents

Foreword

ARRL Membership Benefits

What is Amateur Radio?

When to Expect New Books

Online Review and Practice Exams

About the ARRL

1	**Introduction**		
	1.1	The Extra Class License and Amateur Radio	1-1
	1.2	Extra Class Overview	1-3
	1.3	The Volunteer Testing Process	1-5
	1.4	How to Use This Book	1-9
2	**Operating Practices**		
	2.1	General Operating	2-1
	2.2	Amateur Satellites	2-7
3	**Rules and Regulations**		
	3.1	Operating Standards	3-1
	3.2	Station Restrictions	3-7
	3.3	Station Control	3-9
	3.4	Amateur Satellite Service	3-12
	3.5	Volunteer Examiner Program	3-14
	3.6	Miscellaneous Rules	3-18
4	**Electrical Principles**		
	4.1	Radio Mathematics	4-1
	4.2	Electrical and Magnetic Fields	4-4
	4.3	Principles of Circuits	4-6
5	**Components and Building Blocks**		
	5.1	Semiconductor Devices	5-1
	5.2	Optoelectronics	5-13
	5.3	Digital Logic	5-18

6	Radio Circuits and Systems	
	6.1 Amplifiers	6-1
	6.2 Signal Processing	6-12
	6.3 Digital Signal Processing (DSP) and Software Defined Radio (SDR)	6-25
	6.4 Filters and Impedance Matching	6-32
	6.5 Power Supplies	6-41

7	Radio Measurements and Performance	
	7.1 Test Equipment	7-1
	7.2 Receiver Performance	7-12
	7.3 Interference and Noise	7-21

8	Modulation, Protocols, and Modes	
	8.1 Modulation Systems	8-1
	8.2 Digital Protocols and Modes	8-5
	8.3 Amateur Television	8-17

9	Antennas and Feed Lines	
	9.1 Basics of Antennas	9-1
	9.2 Practical Antennas	9-10
	9.3 Antenna Systems	9-22
	9.4 Transmission Lines	9-28
	9.5 Antenna Design	9-40

10	Topics in Radio Propagation	
	10.1 Electromagnetic Waves	10-1
	10.2 Solar Effects	10-4
	10.3 HF Propagation	10-5
	10.4 VHF/UHF/Microwave Propagation	10-10

11	Safety	
	11.1 Hazardous Materials	11-1
	11.2 RF Exposure	11-3
	11.3 Grounding and Bonding	11-8

12	Glossary	

13	Extra Question Pool	
	Extra Class (Element 4) Syllabus	13-1
	Subelement E1 — Commission's Rules	13-4
	Subelement E2 — Operating Procedures	13-17
	Subelement E3 — Radio Wave Propagation	13-26
	Subelement E4 — Amateur Practices	13-32
	Subelement E5 — Electrical Principles	13-42
	Subelement E6 — Circuit Components	13-52
	Subelement E7 — Practical Circuits	13-64
	Subelement E8 — Signals and Emissions	13-82
	Subelement E9 — Antennas and Transmission Lines	13-89
	Subelement E0 — Safety	13-106

Advertising

Index

Foreword

Welcome to the twelfth edition of *The ARRL Extra Class License Manual*. You are holding the key to your final step up the amateur radio license ladder! With full access to the entire amateur radio frequency spectrum, you will be permitted to operate using every privilege granted to amateurs by the Federal Communications Commission.

With your increased privileges comes the challenge of increased responsibility to fulfill the Basis and Purpose of the Amateur Service as stated in Part 97.1 of the FCC's rules and regulations:
- Engaging in public service
- Advancing the radio art
- Enhancing your technical and operating skills
- Providing trained operators and technicians
- Enhancing international goodwill

That last point is significant in that even in this age of instant worldwide connectivity — no other group has access to direct personal communication without any intervening networks or equipment. As an Extra class licensee, you'll be able to take full advantage of those communication opportunities.

The ARRL Extra Class License Manual may seem huge, but it covers every one of the 622 questions. Each topic is addressed in sufficient detail that you can learn the "why" and "how" behind each answer. This will help you retain the information after you pass the exam and you will get more benefit and enjoyment from your upgraded license.

The book includes numerous examples to help you become comfortable with the necessary calculations. Graphics are included to help you visualize the concepts and explanations. If you would like a more concise study guide, *ARRL's Extra Q&A* is a companion to this book, presenting each question and a short explanation of the correct answer.

As with the license manuals for the Technician and General licenses, this manual organizes the material into a natural progression of topics. Each topic is followed by a list of questions from the exam on that subject. This makes the material easier to learn, remember, and use. Along with the printed manuals, additional resources are provided on the ARRL website at **www.arrl.org/extra-class-license-manual**. These web pages list supplemental references, such as a math tutorial, and links to resources you can use to go beyond the exam questions.

Of course, you aren't just studying to pass a license exam and we aren't satisfied just to help you pass. We want you to enjoy amateur radio to its fullest and that's why the ARRL provides opportunities for continued education, experimentation, and growth through technical and operating training resources, in both print and electronic form.

Be sure to take advantage of the technical and operating aids in the many books and supplies that make up our "Radio Amateur's Library." Make the Technology section of the ARRL's website (**www.arrl.org/technology**) a frequent stop on your web travels. If you're not yet an ARRL member, these are great reasons to join! Check *QST* each month for new material or browse the ARRL Publications Catalog on-line at **www.arrl.org/arrl-store**. You can also request a printed catalog or place an order by phone, 888-277-5289; by fax, 860-594-0303; by e-mail, **pubsales@arrl.org**

This twelfth edition of *The ARRL Extra Class License Manual* builds on the excellent material developed by authors and editors, many ARRL staff members, and readers of the earlier editions. You can help make this manual better, too. After you've used the book to prepare for your exam, e-mail your suggestions (including any corrections you think need to be made) to us at **pubsfdbk@arrl.org** or use the Feedback Form at the back of this book and mail the form. Your comments are welcome!

Upgrading your license is only the beginning of your adventure – you'll have access to the complete palette of the amateur experience with every one of ham radio's tool available. You can use these for personal enjoyment as well as apply them for the benefit of other amateurs and the public.

Thanks for making the decision to upgrade and reach the highest level of achievement in amateur radio. You won't regret it — good luck!

Ward Silver, NØAX
ARRL Contributing Editor
Newington, CT
March 2020

Get more from your Extra Class License with ARRL Membership

Membership in ARRL offers unique opportunities to advance and share your knowledge of amateur radio. For over 100 years, advancing the art, science, and enjoyment of amateur radio has been our mission. Your membership helps to ensure that new generations of hams continue to reap the benefits of the amateur radio community.

Here are just a few of the benefits you will receive with your annual membership. For a complete list visit, arrl.org/membership.

KNOWLEDGE

ARRL offers you a wealth of knowledge to advance your skills with lifelong learning courses, local clubs where you can meet and share ideas, and publications to help you keep up with the latest information from the world of ham radio.

ADVOCACY

ARRL is a strong national voice for preserving and protecting access to Amateur Radio Service frequencies.

SERVICES

From free FCC license renewals, to our Technical Information Service that answers calls and emails about your operating and technical concerns, ARRL offers a range of member services.

RESOURCES

Digital resources including e-mail forwarding, product review archives, e-newsletters, and more.

PUBLICATIONS

Members receive digital access to all four ARRL monthly and bimonthly publications – *QST*, the membership journal of ARRL; *On the Air*, an introduction to the world of amateur radio; *QEX*, which covers topics related to amateur radio and radio communications experimentation, and *National Contest Journal* (*NCJ*), covering radio contesting.

Two Easy Ways to Join

CALL
Member Services toll free at **1-888-277-5289**

ONLINE
Go to our secure website at **arrl.org/join**

What is Amateur Radio?

Perhaps you've just picked up this book in the library or from a bookstore shelf and are wondering what this amateur radio business is all about. Maybe you have a friend or relative who is a "ham" and you're interested in becoming one, as well. In that case, a short explanation is in order.

Amateur radio or "ham radio" is one of the longest-lived wireless activities. Amateur experimenters were operating right along with Marconi in the early part of the 20th century. They have helped advance the state-of-the-art in radio, television and dozens of other communications services since then, right up to the present day. There are more than 700,000 amateur radio operators or "hams" in the United States alone and several million more around the world!

Amateur radio in the United States is a formal *communications service*, administered by the Federal Communications Commission or FCC. Created officially in its present form in 1934, the amateur Service is intended to foster electronics and radio experimentation, provide emergency backup communications, encourage private citizens to train and practice operating, and even spread the goodwill of person-to-person contact over the airwaves.

Who Is a Ham and What Do Hams Do?

Anyone can be a ham — there are no age limits or physical requirements that prevent anyone from passing their license exam and getting on the air. Kids as young as 6 years old have passed the basic exam, and there are hams out there over the age of 100. You probably fall somewhere in the middle of that range.

Once you get on the air and start meeting other hams, you'll find a wide range of capabilities and interests. Of course, there are many technically skilled hams who work as engineers, scientists or technicians. But just as many don't have a deep technical background. You're just as likely to encounter writers, public safety personnel, students, farmers, truck drivers — anyone with an interest in personal communications over the radio.

The activities of amateur radio are incredibly varied. Amateurs who hold the Technician class license — the usual first license for hams in the US — communicate primarily with local and regional amateurs using relay stations called repeaters. Known as "Techs," they sharpen their skills of operating while portable and mobile, often joining emergency communications teams. They may instead focus on the burgeoning wireless data networks assembled and used by hams around the world. Techs can make use of the growing number of amateur radio satellites, built and launched by hams along with the commercial "birds." Technicians transmit their own television signals, push the limits of signal propagation through the atmosphere and experiment with microwaves. Hams hold most of the world records for long-distance communication on microwave frequencies, in fact!

John, W1RT operates this multi-band "rover" during VHF+ contests, driving to hilltops around New England such as from Mohawk Mountain where he is shown giving the antennas a little "hands on" adjustment.

Sisters Autumn and Hannah operate K5LBJ during the bi-annual School Club Roundup competition that features the radio clubs at schools across the country.

Brian Mileshosky, N5ZGT, during a NM5FD Field Day operation, showing his son, Landon, and daughter, Audrey, how to make and log contacts.

Ham radio can go right along with you on vacations and outdoor activities. Bill Paul, KD6JIU, gets on the air from his kayak, using a multiband antenna of his own design and construction.

Hams who advance or *upgrade* to General and then to Extra class are granted additional privileges with each step to use the frequencies usually associated with shortwave operation. This is the traditional amateur radio you probably encountered in movies or books. On these frequencies, signals can travel worldwide and so amateurs can make direct contact with foreign hams. No internet, phone systems, or data networks are required. It's just you, your radio, and the ionosphere — the upper layers of the Earth's atmosphere!

Many hams use voice, Morse code, computer data modes and even image transmissions to communicate. All of these signals are mixed together where hams operate, making the experience of tuning a radio receiver through the crowded bands an interesting experience.

One thing common to all hams is that all of their operation is noncommercial, especially the volunteers who provide emergency communications. Hams pursue their hobby purely for personal enjoyment and to advance their skills, taking satisfaction from providing services to their fellow citizens. This is especially valuable after natural disasters such as hurricanes and earthquakes when commercial systems are knocked out for a while. Amateur operators rush in to provide backup communication for hours, days, weeks or even months until the regular systems are restored. All this from a little study and a simple exam!

Want to Find Out More?

If you'd like to find out more about amateur radio in general, there is lots of information available on the internet. A good place to start is on the American Radio Relay League's (ARRL) ham radio introduction page at **www.arrl.org/what-is-ham-radio**. Books like *Ham Radio for Dummies* will help you fill in the blanks as you learn more.

Along with books and internet pages, there is no better way to learn about ham radio than to meet your local amateur operators. It is quite likely that no matter where you live in the United States, there is a ham radio club in your area — perhaps several! The ARRL provides a club lookup web page at **www.arrl.org/find-a-club** where you can find a club just by entering your Zip code or state. Carrying on the tradition of mutual assistance, many clubs make helping newcomers to ham radio a part of their charter.

If this sounds like hams are confident that you'll find their activities interesting, you're right! Amateur radio is much more than just talking on a radio, as you'll find out. It's an opportunity to dive into the fascinating world of radio communications, electronics, and computers as deeply as you wish to go. Welcome!

Arie, PA3A was one of four Dutch hams who participated in a Mercy Ships project in Sierra Leone. During their free time, they operated as 9L5MS on the HF bands.

Participating in a "radiosport" competition is a great way to build up your radio skills. The W2GD team specializes in 160 meter operation as shown here during the ARRL 160 Meter Contest.

When to Expect New Books

A Question Pool Committee (QPC) consisting of representatives from the various Volunteer Examiner Coordinators (VECs) prepares the license question pools. The QPC establishes a schedule for revising and implementing new question pools. The current question pool revision schedule is as follows:

Question Pool	Current Study Guides	Valid Through
Technician (Element 2)	*The ARRL Ham Radio License Manual,* 4th edition *ARRL's Tech Q&A,* 7th Edition	June 30, 2022
General (Element 3)	*The ARRL General Class License Manual,* 9th edition *ARRL's General Q&A,* 6th Edition	June 30, 2023
Amateur Extra (Element 4)	*The ARRL Extra Class License Manual,* 12th Edition *ARRL's Extra Q&A,* 5th Edition	June 30, 2024

As new question pools are released, ARRL will produce new study materials before the effective date of the new pools. Until then, the current question pools will remain in use, and current ARRL study materials, including this book, will help you prepare for your exam.

As the new question pool schedules are confirmed, the information will be published in *QST* and on the ARRL website at **www.arrl.org**.

Online Review and Practice Exams

Use this book with the *ARRL Exam Review for Ham Radio* to review material you are learning chapter-by-chapter. Take randomly generated practice exams using questions from the actual examination question pool. You won't have any surprises on exam day! Go to **www.arrl.org/examreview**.

About the ARRL

The seed for amateur radio was planted in the 1890s, when Guglielmo Marconi began his experiments in wireless telegraphy. Soon he was joined by dozens, then hundreds, of others who were enthusiastic about sending and receiving messages through the air — some with a commercial interest, but others solely out of a love for this new communications medium. The United States government began licensing amateur radio operators in 1912.

By 1914, there were thousands of amateur radio operators — hams — in the United States. Hiram Percy Maxim, a leading Hartford, Connecticut inventor and industrialist, saw the need for an organization to unify this fledgling group of radio experimenters. In May 1914, he founded the American Radio Relay League (ARRL) to meet that need.

ARRL is the national association for amateur radio in the US. Today, with approximately 150,000 members, ARRL numbers within its ranks the vast majority of active radio amateurs in the nation and has a proud history of achievement as the standard-bearer in amateur affairs. ARRL's underpinnings as amateur radio's witness, partner, and forum are defined by five pillars: Public Service, Advocacy, Education, Technology, and Membership. ARRL is also International Secretariat for the International Amateur Radio Union, which is made up of similar societies in 150 countries around the world.

ARRL's Mission Statement: To advance the art, science, and enjoyment of amateur radio.
ARRL's Vision Statement: As the national association for amateur radio in the United States, ARRL:

- Supports the awareness and growth of amateur radio worldwide;
- Advocates for meaningful access to radio spectrum;
- Strives for every member to get involved, get active, and get on the air;
- Encourages radio experimentation and, through its members, advances radio technology and education; and
- Organizes and trains volunteers to serve their communities by providing public service and emergency communications.

At ARRL Headquarters in Newington Connecticut, the staff helps serve the needs of members. ARRL publishes the monthly journal *QST*, and the bimonthly magazine *On the Air,* including interactive digital versions of both, as well as newsletters and many publications covering all aspects of amateur radio. Its Headquarters station, W1AW, transmits bulletins of interest to radio amateurs and Morse code practice sessions. ARRL also coordinates an extensive field organization, which includes volunteers who provide technical information and other support services as well as communications for public service activities. In addition, ARRL represents US radio amateurs to the Federal Communications Commission and other government agencies in the US and abroad.

Membership in ARRL means more than receiving *QST* each month. In addition to the services already described, ARRL offers membership services on a personal level, such as the Technical Information Service, where members can get answers — by phone, email, or the ARRL website — to all their technical and operating questions.

A bona fide interest in amateur radio is the only essential qualification of membership; an amateur radio license is not a prerequisite, although full voting membership is granted only to licensed radio amateurs in the US. Full ARRL membership gives you a voice in how the affairs of the organization are governed. ARRL policy is set by a Board of Directors (one from each of 15 Divisions). Each year, one-third of the ARRL Board of Directors stands for election by the full members they represent. The day-to-day operation of ARRL HQ is managed by a Chief Executive Officer and his/her staff.

Join ARRL Today! No matter what aspect of amateur radio attracts you, ARRL membership is relevant and important. There would be no amateur radio as we know it today were it not for ARRL. We would be happy to welcome you as a member! Join online at **www.arrl.org/join**. For more information about ARRL and answers to any questions you may have about amateur radio, write or call:

ARRL — The national association for Amateur Radio®
225 Main Street
Newington, CT 06111-1494
Tel: 860-594-0200
FAX: 860-594-0259
email: **hq@arrl.org**
www.arrl.org

Prospective new radio amateurs call (toll-free):
800-32-NEW HAM (800-326-3942)
You can also contact ARRL via email at **newham@arrl.org**
or check out the ARRL website at **www.arrl.org**

Chapter 1

Introduction

In this chapter, you'll learn about:
- Added frequencies and activities enjoyed by Extra licensees
- Reasons to upgrade from General or Advanced
- Requirements and study materials for the Extra exam
- How to prepare for your exam
- How to find an exam session
- Where to find helpful resources

Welcome to *The ARRL Extra Class License Manual*! You're about to begin the final chapter of your amateur radio license studies. By earning both a Technician and a General class license, you've learned a tremendous amount about the technology of radio and the operating practices that make it useful and effective. By upgrading to Extra, you'll complete that journey, with full access to everything amateur radio has to offer, joining more than 150,000 other "Extras."

1.1 The Extra Class License and Amateur Radio

Just about every ham thinks about obtaining the Extra class "ticket" at one point or another in their ham career — and why not? Extra class licensees have complete access to all frequencies available to the Amateur Service. There's also the good feeling of knowing that you've demonstrated broad and useful knowledge of technology, operating practices, and the FCC rules and regulations. Having an Extra class license doesn't mean you know everything there is to know — quite the contrary — but you will be better prepared to learn and grow within the service.

Table 1.1
Exam Elements Needed to Qualify for an Extra Class License

Current License	Exam Requirements	Study Materials
None or Novice	Technician (Element 2)	*The ARRL Ham Radio License Manual* or *ARRL's Tech Q&A*
	General (Element 3)	*The ARRL General Class License Manual* or *ARRL's General Q&A*
	Amateur Extra (Element 4)	*The ARRL Extra Class License Manual* or *ARRL's Extra Q&A*
Technician (issued on or after March 21, 1987)*	General (Element 3)	*The ARRL General Class License Manual* or *ARRL's General Q&A*
	Amateur Extra (Element 4)	*The ARRL Extra Class License Manual* or *ARRL's Extra Q&A*
General or Advanced	Amateur Extra (Element 4)	*The ARRL Extra Class License Manual* or *ARRL's Extra Q&A*

*Individuals who qualified for the Technician license before March 21, 1987, will be able to receive credit for Element 3 (General class) by providing documentary proof to a Volunteer Examiner Coordinator.

If you have been hesitant to study for the exam because you feel the theory is too difficult or your math and electronics background is rusty, take heart. Study patiently and make use of the available resources for Extra class students — you *will* succeed in passing the exam. It may take more than one attempt, but this is hardly unusual. You can continue to study and try again later. You might also try just taking a different version of the exam at the very same test session. Sooner or later, you'll be able to add "/AE" to your call sign on the air! The key is to make the personal commitment to passing the exam and be willing to study.

Table 1.1 shows what you need to qualify for Extra class, depending on your current license. Remember how you felt when you started studying for your Technician license and again when you hit the upgrade trail to General? Now you know that it's just a matter of persistent study and practice to obtain that coveted Extra class "ticket"! Congratulations for taking the first step — let's get started!

About Amateur Radio

If you're thumbing through this book wondering what amateur radio is all about, welcome to a unique and valuable hobby! More than 760,000 people in the United States and several million around the world have a license to operate on the radio wavelengths allocated to amateurs. You'll find them on the traditional shortwave bands sending signals around the world, just as they have for more than a century. Using portable and handheld equipment, they communicate with local and regional friends, too, even using amateur-built satellites!

Amateurs or "hams" are experimenters and innovators. Many design and build their own equipment and antenna systems. Hams have created novel and useful hybrids of computer and internet technology along with the traditional radio. It is possible today to use a tiny handheld radio to access local systems connected to the internet that relay the signals to a similarly-equipped ham halfway around the world. Along with the traditional Morse code and voice signals, hams are continually inventing ways to send digital data over radio. In fact, you may have used a technology adapted from an amateur invention!

It's not necessary to be an electrical engineer to be a ham! There are hams from all walks of life. During times of emergency, hams step in to assist by adapting their communications systems to the situation at hand. The ham spirit of volunteerism and "can do" helps relief and public safety agencies as well as private citizens around the world. If you can learn to operate a radio and follow some simple "rules of the road," you can participate. All it takes is an interest in radio and a willingness to learn and to help others.

Amateur radio is regulated by the Federal Communications Commission (FCC) and amateurs must be licensed to use the public airwaves that are allocated to their use. To get the license, prospective amateurs must pass a simple question-and-answer exam. In the United States, there are three license classes: Technician (entry-level), General and Amateur Extra (top-level). Each successive license grants additional privileges. This book is the study guide for the Extra class license exam.

Amateur radio has been around for quite a while — since before World War I. It's still a vital service today, at the forefront of public service communications, technical and operating innovation, and spreading goodwill around the world — one contact at a time. To learn more about amateur radio, log on to **www.arrl.org/what-is-ham-radio** for a guided tour of a truly fascinating hobby!

Many hams find it satisfying to give something back to their community. One way is to become involved in emergency communications by your local Amateur Radio Emergency Service (ARES®) group or becoming a NOAA SKYWARN volunteer.

1-2   Chapter 1

Morse Code — Very Much Alive!

Part of amateur radio since its beginnings, Morse code has been part of the rich amateur tradition for more than 100 years and many hams still use it extensively. At one time, Morse code proficiency at 20 words per minute (WPM) was required for an Extra class license. That requirement was lowered to a one-speed-fits-all 5 WPM and then dropped entirely in February 2007. Morse is likely to remain part of the amateur experience, however.

Morse (also known as CW, for *continuous wave* transmission) remains popular for solid reasons. It's easy to build CW transmitters and receivers. There is no more power-efficient mode of communications that is "copied" by the human ear. The extensive set of abbreviations, many of which have been adapted to text messaging, allow amateurs to communicate a great deal of information even if they don't share a common language.

Extra class licensees have access to exclusive CW segments at the bottom of the 80, 40, 20, and 15 meter bands. These are prime territory for contesting and DXing, a powerful incentive to learn the code or improve your code speed.

To help you learn Morse, the ARRL has a complete set of resources listed on its web page at **www.arrl.org/learning-morse-code**. Computer software and on-the-air code practice sessions are available for personal training and practice. Organizations such as CWops (**www.cwops.org**) and FISTS (**www.fists.org**) — an operator's style of sending is referred to as his or her "fist" — help hams learn Morse code and improve their skills.

CW operation is particularly effective when operating as a portable station with temporary antennas. Author Ward Silver, NØAX, took his mini-paddle along when activating Emerald Island, North Carolina, for the Islands On The Air (IOTA) awards program.

1.2 Extra Class Overview

Most of this book's readers will have already earned their General license, whether they have been a ham for quite a while or are new to the hobby. Some may have passed their Advanced class exam years ago (Advanced licenses are no longer issued but they can be renewed). No matter which type you might be, you're to be commended for making the effort to pass the Extra class exam!

UPGRADING TO EXTRA

As you begin your studies, remember that you've already overcome the hurdles of passing not just one, but at least two license exams! The Extra class exam questions are certainly more difficult, but you already know all about the testing procedure and the basics of ham radio. You can approach the process of upgrading with confidence!

The Extra class licensee has access to additional spectrum on four of the most popular amateur HF bands. Located at the lower edges of the CW and phone segments of the 80, 40, 20, and 15 meter bands, the "Extra sub-bands" are prime frequencies for DXing and contest operating. At other times, the less-crowded frequencies make operating more enjoyable. The improved technical understanding of upgrading to Extra will make you a more knowledgeable and skilled operator, too.

Once you've earned Extra class status, you'll be able to administer license exam sessions for any license class! Extra class VEs are needed to give both General and Extra class exams

Books to Help You Learn

As you study the material for the licensing exam, you will have lots of other questions about the how and why of amateur radio. The following references, available from your local bookstore or the ARRL (**www.arrl.org/shop**), will help fill in the blanks and give you a broader picture of the hobby. Make sure you have the latest editions that will have current information on rules, regulations, and fast-changing technology.

• ***Ham Radio for Dummies*** by Ward Silver, N0AX. Written for both new hams and for experienced hams interested in new activities, this book supplements the information in study guides with an informal, friendly approach to the hobby.

• ***ARRL Operating Manual***. With in-depth chapters on the most popular ham radio activities, this is your guide to nets, award programs, DXing, and more. It even includes a healthy set of references.

• ***Understanding Basic Electronics*** by Walter Banzhaf, WB1ANE. Students who want more technical background about electronics should take a look at this book. It covers the fundamentals of electricity and electronics that are the foundation of all radio.

• ***Basic Radio*** by Joel Hallas, W1ZR. This book goes beyond electronic circuits to explain how radios are designed and perform. It covers the key building blocks of receivers, transmitters, antennas, and propagation.

• ***ARRL Handbook***. This is the grandfather of all amateur radio references and belongs on the shelf of hams. Almost any topic you can think of in amateur radio technology is represented here. A new edition is published each year.

• ***ARRL Antenna Book***. After the radio itself, all radio communication depends on antennas. This book provides information on every common type of amateur antenna, as well as feed lines and related topics. It's filled with practical construction tips and techniques to help you build your own antennas.

Barb Kingery, AE7AQ and Rob Kingery, AE7AP, took portable equipment along on their backpacking trip to Glacier National Park, activating the Continental Divide for the Summits On the Air (SOTA) programs.

— an excellent opportunity to repay some of the assistance you've received. Technician and General class hams may turn to you for guidance, too. If you have a desire to teach other hams, as an Extra there's no limit to what you can do.

CALL SIGNS

A call sign is a very personal identification and many hams keep theirs for a lifetime. Will you *have* to change calls when you receive your Extra class license? No — just as you could keep your call sign when you upgraded to General, it is the same when you upgrade to Extra. However, you might want your on-the-air identity to reflect your new license status! Extra class licensees can choose vanity call signs from the coveted "one-by-two" or 1 × 2 series of call signs, such as K1EA or N6TR, or a 2 × 1 such as NN1N or KX9X. Extras are also allocated the set of 2 × 2 calls that begin with the letter combinations AA through AK, such as AB1FM. If you would like to know more about what calls

Amateurs played a significant role in the 2019 Seattle Earthquake and Tsunami Drill. The photo shows Seattle ACS (Auxiliary Communication Service) Lake City Hub radio operators Susanna Cunningham, WB7CON, Jeff Rodgers, KF7QLG, and Sandy Motzer, KI7KYS. They were one of several teams operating in the Seattle EOC (Emergency Operations Center) and at several locations around the city.

are available, it's fun to browse through helpful websites such as **www.radioqth.net** or **www.ae7q.com**. Pick your favorite and put it in a prominent location as an incentive to keep studying!

FOCUS ON HF AND ADVANCED MODES

Where the Technician and General class licenses introduced you to whole new areas of radio and electronics, the Extra class license is more focused. You'll be expected to build on what you learned for the first two licenses. If you still have your study guides for those licenses handy, they may provide useful background material. The sidebar "Books to Help You Learn" lists several additional reference books that provide more than enough information for the Extra class student. Here are some examples of topics you'll be studying:

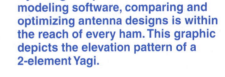

By using inexpensive antenna modeling software, comparing and optimizing antenna designs is within the reach of every ham. This graphic depicts the elevation pattern of a 2-element Yagi.

- The different types of station control
- AC impedance, resonance, and filters
- Contest and DX operating procedures
- Digital modes and operating procedures
- Test instruments and measurements
- Semiconductor devices and common RF circuits
- Special topics on antennas, feed lines, and propagation

Not every ham uses every mode and frequency, of course. By learning about this wider range of ideas, it helps hams to make better choices for regular operating. You will become aware of just how wide and deep ham radio really is. Better yet, the introduction of these new ideas may just get you interested in giving them a try!

1.3 The Volunteer Testing Process

The procedure for upgrading to Extra class is identical to the one you followed for other license exams. You must attend an exam session administered by Volunteer Examiners (VEs) accredited by a Volunteer Examiner Coordinator (VEC) and pass a 50-question written exam consisting of questions drawn from the Element 4 Question Pool. (You'll also have to have passed the elements for the Technician and General class licenses.) This is an increase of 15 questions from the General and Technician 35-question exams (see **Table 1.2**).

Table 1.2
Amateur License Class Examinations

License Class	Elements Required	Number of Questions
Technician	2 (Written)	35 (passing is 26 correct)
General	3 (Written)	35 (passing is 26 correct)
Extra	4 (Written)	50 (passing is 37 correct)

When you're ready, you'll need to find a test session. If you're in a licensing class, the instructor will help you find and register for a session. Otherwise, you can find a test session by using the ARRL's web page for finding exams, **www.arrl.org/exam**. If you can register for the test session in advance, do so. Other sessions, such as those at hamfests or conventions, are usually available to anyone who shows up, also known as *walk-ins*. You may have to wait for an available space though, so go early!

Bring the *original* of your current FCC-issued license and a photocopy to send with the application. You'll need a photo ID, such as a driver's license, passport, or employer's identity card. Two forms of identification are required if no photo ID is available. Know your Social Security Number (SSN) or FCC-issued Federal Registration Number (FRN). You can bring pencils or pens, blank scratch paper, and a calculator with the memory erased but any kind of electronic devices or phones that can access the internet are prohibited. For a complete list of items to bring or that are prohibited on exam day, please see the ARRL's web page **www.arrl.org/what-to-bring-to-an-exam-session**. (If you have a disability and need these devices to take the exam, contact the session sponsor ahead of time as described in the sidebar later in this chapter.)

Once you're signed in, you'll need to fill out a copy of the National Conference of Volunteer Examiner Coordinator's NCVEC Quick Form 605 (**Figure 1.1**). This is an application for a new or upgraded license. It is only used:
• At test sessions
• For a VEC to process a license renewal
• For a VEC to process a license change
• For a VEC to process a new club license or a club license change

Do not use an NCVEC Quick Form 605 for any kind of application directly to the FCC — it will be rejected. Use a regular FCC 605-Main Form. After filling out the form, pay the test fee and get ready. Check the ARRL VEC website **www.arrl.org/arrl-vec-exam-fees** or the website of the VEC certifying the exam for the current amount.

You will be given a question booklet and an answer sheet. Be sure to read the instructions, fill in all the necessary information and sign your name wherever it's required. Check to be sure your booklet has 50 questions and be sure to mark the answer in the correct space for each question.

If you've already passed the General class exam, it might be useful (and calming) to review. You don't have to answer the questions in order — skip the hard ones and go back to them. If you read the answers carefully, you'll probably find that you can eliminate one or more "distracters." Of the remaining answers, only

Volunteer Examiners will orchestrate every aspect of "Exam Day" and will let you know how you did right away. After you pass the exam, they'll issue you a Certificate of Successful Completion of Examination (CSCE), which allows you to start using your new privileges right away. Just be sure to sign "temporary AE" after your call sign until your upgrade appears in the official FCC online database or your new license shows up in the mail.

NCVEC QUICK-FORM 605 APPLICATION
AMATEUR OPERATOR/PRIMARY STATION LICENSE

SECTION 1 - TO BE COMPLETED BY APPLICANT
PLEASE PRINT **LEGIBLY!**

PRINT LAST NAME	SUFFIX (Jr., Sr.)	FIRST NAME	M.I.	STATION CALL SIGN (IF ANY)
GRIMALDI		AMANDA		KB1KJC

MAILING ADDRESS (Number and Street or P.O. Box)
225 MAIN ST.

CITY	STATE CODE	ZIP CODE (5 or 9 Numbers)	FEDERAL REGISTRATION NUMBER (FRN) - IF NONE, THEN SOCIAL SECURITY NUMBER (SSN)
NEWINGTON	CT	06111	0005189337

DAYTIME TELEPHONE NUMBER (Include Area Code)	E-MAIL ADDRESS (MANDATORY TO RECEIVE LICENSE NOTIFICATION EMAIL FROM FCC)
860-594-0200	myemail@arrl.org

Basic Qualification Question: *ANSWER REQUIRED IN ORDER TO PROCESS YOUR APPLICATION*

Has the Applicant or any party to this application, or any party directly or indirectly controlling the Applicant, ever been convicted of a felony by any state or federal court? ☐ YES ☒ NO

If "YES", see "FCC BASIC QUALIFICATION QUESTION INSTRUCTIONS AND PROCEDURES" on the back of this form.

I HEREBY APPLY FOR (Make an X in the appropriate box(es)):

☐ **EXAMINATION** for a new license grant

☒ **EXAMINATION** for **upgrade** of my license class

☐ **CHANGE** my **name** on my license to my new name
Former Name: _____
(Last name) (Suffix) (First name) (MI)

☐ **CHANGE** my mailing address to **above** address

☐ **CHANGE** my station **call sign** systematically
Applicant's Initials: To confirm _____

☐ **RENEWAL** of my license grant
Exp. Date: _____

Do you have another license application on file with the FCC which has not been acted upon?	PURPOSE OF OTHER APPLICATION	PENDING FILE NUMBER (FOR VEC USE ONLY)

I certify that:
- I waive any claim to the use of any particular frequency regardless of prior use by license or otherwise;
- All statements and attachments are true, complete and correct to the best of my knowledge and belief and are made in good faith;
- I am not a representative of a foreign government;
- I am not subject to a denial of Federal benefits pursuant to Section 5301 of the Anti-Drug Abuse Act of 1988, 21 U.S.C. § 862;
- The construction of my station will NOT be an action which is likely to have a significant environmental effect (See 47 CFR Sections 1.1301-1.1319 and Section 97.13(a));
- I have read and WILL COMPLY with Section 97.13(c) of the Commission's Rules regarding RADIOFREQUENCY (RF) RADIATION SAFETY and the amateur service section of OST/OET Bulletin Number 65.

Signature of Applicant:
X _Amanda_ Date Signed: 1/17/2020

SECTION 2 - TO BE COMPLETED BY ALL ADMINISTERING VEs

Applicant is qualified for operator license class:

☐ **NO NEW LICENSE OR UPGRADE WAS EARNED**

☐ **TECHNICIAN** Element 2

☐ **GENERAL** Elements 2 and 3

☒ **AMATEUR EXTRA** Elements 2, 3 and 4

DATE OF EXAMINATION SESSION
01-17-2020

EXAMINATION SESSION LOCATION
NEWINGTON, CT

VEC ORGANIZATION
ARRL

VEC RECEIPT DATE

I CERTIFY THAT I HAVE COMPLIED WITH THE ADMINISTERING VE REQUIRMENTS IN PART 97 OF THE COMMISSION'S RULES AND WITH THE INSTRUCTIONS PROVIDED BY THE COORDINATING VEC AND THE FCC.

1st VEs NAME (Print First, MI, Last, Suffix)	VEs STATION CALL SIGN	VEs SIGNATURE (Must match name)	DATE SIGNED
Perry Green	WY1O	Perry Green	01-17-20
2nd VEs NAME (Print First, MI, Last, Suffix) Steven Ewald	VEs STATION CALL SIGN WV1X	VEs SIGNATURE Steven Ewald	DATE SIGNED 01-17-20
3rd VEs NAME (Print First, MI, Last, Suffix) Rose-Anne Lawrence	VEs STATION CALL SIGN KB1DMW	VEs SIGNATURE Rose-Anne Lawrence	DATE SIGNED 01/17/20

DO NOT SEND THIS FORM TO FCC — THIS IS NOT AN FCC FORM.
IF THIS FORM IS SENT TO FCC, FCC WILL RETURN IT TO YOU WITHOUT ACTION.

NCVEC FORM 605 - March 2018
FOR VE/VEC USE ONLY - Page 1

Figure 1.1 — This sample NCVEC Quick Form 605 shows how your form will look after you have completed your upgrade to Extra.

Special Testing Procedures

The FCC allows Volunteer Examiners (VEs) to use a range of procedures to accommodate applicants with various disabilities. If this applies to you, you'll still have to pass the test, but special exam procedures can be applied. Contact your local VE or the Volunteer Examiner Coordinator (VEC) responsible for the test session you'll be attending. You may also contact the ARRL VEC Office at 225 Main St, Newington CT 06111-1494 or by phone at 860-594-0200, or via e-mail to **vec@arrl.org**. Ask for more information about special examination procedures.

one will be the best. If you can't decide on the correct answer, go ahead and guess. There is no penalty for an incorrect guess. When you're done, go back and check your answers and double-check your arithmetic — there's no rush!

Once you've answered all 50 questions, the VEs will grade and verify your test results. Assuming you've passed (your last amateur exam ever!) you'll fill out a *Certificate of Successful Completion of Examination* (CSCE) and the VE team will complete your NCVEC FCC Form 605. The exam organizers will submit your results to the FCC while you keep the CSCE as evidence that you've passed your Extra test.

You'll be more than ready to start using those new privileges as soon as you get home! When you give your call sign, append "/AE" (on CW or digital modes) or "temporary AE" (on phone). As soon as your license class is upgraded in the FCC's database of licensees, typically a week or two later, you can stop adding the suffix. The CSCE is good for 365 days or until receiving your paper license.

If you don't pass, don't be discouraged! You might be able to take another version of the test right then and there if the session organizers can accommodate you. Even if you decide to try again later, you now know just how the test session feels — you'll be more relaxed and ready next time. The Extra class sub-bands are full of hams who took their Extra test more than once before passing. You'll be in good company!

FCC AND ARRL VEC LICENSING RESOURCES

After you pass your exam, the examiners will file all of the necessary paperwork so that your license will be granted by the Federal Communications Commission (FCC). Within two weeks (often sooner) you will be able see your new license status — and your new call sign if you requested one — in the FCC's database via the ARRL website. The FCC has gone "paperless" and no longer routinely mails paper licenses, but you can still request a copy — see **www.arrl.org/obtain-license-copy**. The ARRL VEC can also process license renewals and modifications for you as described at **www.arrl.org/call-sign-renewals-or-changes**.

When you initially passed your Technician exam, you may have applied for your FCC Federal Registration Number (FRN) or may have been issued one automatically by FCC from the information gathered from your form. This allows you to access the information for any FCC licenses you may have and to request modifications to them. These functions are available via the FCC's Universal Licensing System (ULS) website (**www.fcc.gov/wireless/systems-utilities/universal-licensing-system**). Complete instructions for using the site are available at **www.arrl.org/universal-licensing-system**. When accessing the ULS, your FRN will allow you to watch the database for your license upgrade!

1.4 How to Use This Book

Designed to help the student really learn the material, the topics in this study guide build on one another. The first sections cover operating techniques and FCC regulations, while later chapters progress from simple electronics through radio signals. The book concludes with antennas, propagation, and safety topics. Each section includes the exact questions that could be on the exam, so you'll have a chance to test your understanding as soon as you complete each topic. Online references and sources supplement the information in the study guide, so you can get extra help on a topic or just read for interest. You'll have a lot of resources on your side during your studies!

This study guide will provide the necessary background and explanation for the answers to the exam questions. By learning this material, you will go beyond just learning the answers. You'll understand the fundamentals behind them — and this makes it easier to learn, remember, and use what you know. This book also contains many useful facts and figures that you can use in your station and on the air.

If you are taking a licensing class, help your instructors by letting them know about areas in which you need help. They want you to learn as thoroughly and quickly as possible, so don't hold back with your questions. Similarly, if you find the material particularly clear or helpful, tell them that, too, so it can be used in the next class!

Just before the Question Pool section of this book you'll find a large glossary of radio terminology. The Question Pool contains the complete set of exam questions and answers and it is followed by the Index and an advertising section with displays from some of amateur radio's best-known vendors of equipment and supplies.

This QRP (low power) wattmeter built by Fred Piering, WD9HNU, combines a home-built printed-circuit board (PCB) with an analog meter and LED level meter to display power and SWR (standing wave ratio) in the antenna feed line.

College clubs are great ways for hams to get licenses and learn about communications technology. Left to right are student members of the Ivy Tech Community College (Indiana) club Quinton Ritter, KD9MVK, Augustine Busch, KD9MVN, Jahrael Shabzultvorn, KD9MVJ, Karla Jimenez-Vasquez, KD9MVM, and Austin Owens, KD9MVL.

Introduction   1-9

WHAT WE ASSUME ABOUT YOU

You don't have to be a technical guru or an expert operator to upgrade to Extra class! The topics you will encounter build from the basic science of radio and electricity that you mastered for Technician and General. The math is a little more involved than for General class topics and tutorials are available at the *ARRL Extra Class License Manual* website, **www.arrl.org/extra-class-license-manual**. You should have a calculator capable of doing logarithms and trigonometric functions. You'll also be allowed to use it during the license exam, of course.

ADVANCED STUDENTS

If you have some background in radio, perhaps as a communications tech or radio operator, you may be able to short-circuit some of the sections. To find out, locate the shaded boxes in the text listing the exam questions for each topic. Turn to the Question Pool and if you can answer the questions correctly, move to the next topic in the text. It's common for technically-minded students to need help with the rules and regulations, while students with an operating background tend to need more help with the technical material. Regardless of your previous knowledge and experience, be sure that you can answer the questions because they will certainly be on the test!

SELF-STUDY OR CLASSROOM STUDENTS

The ARRL Extra Class License Manual can be used either by an individual student studying on his or her own, or as part of a licensing class taught by an instructor. If you're part of a class, the instructor will guide you through the book, section by section. The solo student can move at any pace and in any convenient order. You'll find that having a friend to study with makes learning the material more fun as you help each other over the rough spots.

Don't hesitate to ask for help! Your instructor can provide information on anything you find difficult. Classroom students may find asking their fellow students to be helpful. If you're studying on your own, there are resources for you, too! If you can't find the answer in the book or at the website, e-mail your question to the ARRL's New Ham Desk, **newham@arrl.org**. You may not be a new ham, but your question will be routed to the appropriate person. The ARRL's experts will answer directly or connect you with another ham who can answer your questions.

USING THE QUESTION POOL

The Element 4 Question Pool is divided into 10 subelements, E0 through E9. Each subelement is further divided into sections that focus on specific topics. For example, E1 on the Commission's Rules has six sections, E1A through E1F. Each question is then numbered E1Axx, with "xx" representing the number of the question in the section. The questions are given as lists at the beginning of each section of the text. The questions are also listed with the text covering the topic. Sometimes questions from several question pool sections may be discussed in a combined section of text. Material that addresses a question is indicated in the text by the question number in bold inside square brackets, such as [**E1A01**].

As you complete each topic be sure to review each of the exam questions highlighted in the shaded text boxes. This will tell you which areas need a little more study time. When you understand the answer to each of the questions, move on. Resist the temptation to just memorize the answers. Doing so leaves you without the real understanding that will make your new Extra class privileges enjoyable and useful. *The ARRL Extra Class License Manual* covers every one of the exam questions, so you can be sure you're ready at exam time.

When using the Question Pool section, cover the answers at the edge of the page to be sure you really do understand the question. Each question is accompanied by a cross-reference back to the page on which that topic is discussed. If you don't completely understand the question or answer, please go back and review that material. The ARRL's condensed guide *Extra Q&A* also provides short explanations for each one of the exam questions.

1-10   Chapter 1

ONLINE REVIEW AND PRACTICE EXAMS

Use this book with ARRL Exam Review for Ham Radio to review chapter-by-chapter. Take randomly-generated practice exams using questions from the actual examination question pool. You won't have any surprises on exam day! Go to **www.arrl.org/exam-review**.

EXTRA CLASS LICENSE MANUAL WEB PAGE

The ARRL also maintains a special web page for Extra class students at **www.arrl.org/extra-class-license-manual**. It provides or links to helpful supplements and clarifications to the material in the book. The useful and interesting online references listed there put you one click away from additional information on many topics.

FOR INSTRUCTORS

If you're an instructor, note that this edition of the study guide has the same organization as the previous edition. Topics are presented in a sequence intended to be easier for the student to learn. For example, the section on Radio Signals and Equipment comes after the

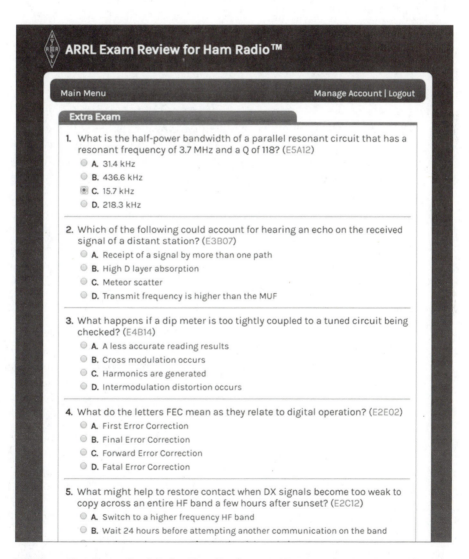

ARRL's online Exam Review for Ham Radio uses the question pool to construct an exam with the same number and variety of questions that you'll encounter on exam day. To find out more about ARRL's online practice exams, visit www.arrl.org/exam-review.

Want More Information?

Looking for more information about Extra class instruction in your area? Are you ready to take the Extra class exam? Do you need a list of ham radio clubs, instructors or examiners in your local area? The following web pages are very helpful in finding the local resources you need to successfully pass your Extra class exam:

- **www.arrl.org** — the ARRL's home page, it features news and links to other ARRL resources
- **www.arrl.org/extra-class-license-manual** — the website that supports this book
- **www.arrl.org/find-a-club** — a search page to find ARRL-Affiliated clubs
- **www.arrl.org/exam** — the ARRL VEC exam session search page
- **www.arrl.org/technical-information-service** — the ARRL's Technical Information Service is an excellent resource

Components and Circuits section. Because the Extra class exam topics are more sophisticated than for General class, it's more important to develop the context and background for each topic.

The ARRL has also created supporting material for instructors such as graphics files, handouts, and a detailed topics list. Check **www.arrl.org/resources-for-license-instruction** for support materials.

CONVENTIONS AND RESOURCES

Throughout your studies keep a sharp eye out for words in *italics*. These words are important so be sure you understand them. Many are included in the extensive Glossary in the back of the book. Another thing to look for are the addresses or URLs for web resources in **bold**, such as **www.arrl.org/extra-class-license-manual**. By browsing these web pages while you're studying, you will accelerate and broaden your understanding.

Question numbers in square brackets, such as [E1A01], indicate material that addresses a specific question. This will help you review or examine specific topics.

Throughout the book, there are many short sidebars that present topics related to the subject you're studying. These sidebars may just tell an interesting story or they might tackle a subject that needs separate space in the book. The information in sidebars supports the information you're studying.

TIME TO GET STARTED

By following these instructions and carefully studying the material in this book, soon you'll be joining the rest of the Extra class licensees! Each of us at the ARRL Headquarters and every ARRL member looks forward to the day when you join the fun. 73 (best regards) and good luck!

Extra Class (Element 4) Syllabus
Valid July 1, 2020 through June 30, 2024

SUBELEMENT E1 — COMMISSION'S RULES
[6 Exam Questions — 6 Groups] 75 Questions

E1A — Operating Standards: frequency privileges; automatic message forwarding; stations aboard ships or aircraft; power restriction on 630 and 2200 meter bands

E1B — Station restrictions and special operations: restrictions on station location; general operating restrictions; spurious emissions; antenna structure restrictions; RACES operations

E1C — Definitions and restrictions pertaining to local, automatic and remote control operation; IARP and CEPT licenses; emission and bandwidth standards

E1D — Amateur space and Earth stations; telemetry and telecommand rules; identification of balloon transmissions; one-way communications

E1E — Volunteer examiner program: definitions; qualifications; preparation and administration of exams; accreditation; question pools; documentation requirements

E1F — Miscellaneous rules: external RF power amplifiers; prohibited communications; spread spectrum; auxiliary stations; Canadian amateurs operating in the U.S.; special temporary authority; control operator of an auxiliary station

SUBELEMENT E2 — OPERATING PROCEDURES
[5 Exam Questions — 5 Groups] 61 Questions

E2A — Amateur radio in space: amateur satellites; orbital mechanics; frequencies and modes; satellite hardware; satellite operations

E2B — Television practices: fast scan television standards and techniques; slow scan television standards and techniques

E2C — Operating methods: contest and DX operating; remote operation techniques; Cabrillo format; QSLing; RF network connected systems

E2D — Operating methods: VHF and UHF digital modes and procedures; APRS; EME procedures; meteor scatter procedures

E2E — Operating methods: operating HF digital modes

SUBELEMENT E3 — RADIO WAVE PROPAGATION
[3 Exam Questions — 3 Groups] 40 Questions

E3A — Electromagnetic waves; Earth-Moon-Earth communications; meteor scatter; microwave tropospheric and scatter propagation; aurora propagation; ionospheric propagation changes over the day; circular polarization

E3B — Transequatorial propagation; long-path; ssordinary and extraordinary waves; chordal hop; sporadic E mechanisms

E3C — Radio horizon; ground wave; propagation prediction techniques and modeling; effects of space weather parameters on propagation

Introduction   1-13

SUBELEMENT E4 — AMATEUR PRACTICES
[5 Exam Questions — 5 Groups] 60 Questions

E4A — Test equipment: analog and digital instruments; spectrum analyzers; antenna analyzers; oscilloscopes; RF measurements; computer-aided measurements

E4B — Measurement technique and limitations: instrument accuracy and performance limitations; probes; techniques to minimize errors; measurement of Q; instrument calibration; S parameters; vector network analyzers

E4C — Receiver performance characteristics: phase noise, noise floor, image rejection, MDS, signal-to-noise ratio, noise figure, reciprocal mixing; selectivity; effects of SDR receiver non-linearity; use of attenuators at low frequencies

E4D — Receiver performance characteristics: blocking dynamic range; intermodulation and cross-modulation interference; third-order intercept; desensitization; preselector

E4E — Noise suppression and interference: system noise; electrical appliance noise; line noise; locating noise sources; DSP noise reduction; noise blankers; grounding for signals; common mode currents

SUBELEMENT E5 — ELECTRICAL PRINCIPLES
[4 Exam Questions — 4 Groups] 55 Questions

E5A — Resonance and Q: characteristics of resonant circuits: series and parallel resonance; definitions and effects of Q; half-power bandwidth; phase relationships in reactive circuits

E5B — Time constants and phase relationships: RL and RC time constants; phase angle in reactive circuits and components; admittance and susceptance

E5C — Coordinate systems and phasors in electronics: rectangular coordinates; polar coordinates; phasors

E5D — AC and RF energy in real circuits: skin effect; electromagnetic fields; reactive power; power factor; electrical length of conductors at UHF and microwave frequencies; microstrip

SUBELEMENT E6 — CIRCUIT COMPONENTS
[6 Exam Questions — 6 Groups] 70 Questions

E6A — Semiconductor materials and devices: semiconductor materials; germanium, silicon, P-type, N-type; transistor types: NPN, PNP, junction, field-effect transistors: enhancement mode; depletion mode; MOS; CMOS; N-channel; P-channel

E6B — Diodes

E6C — Digital ICs: Families of digital ICs; gates; Programmable Logic Devices (PLDs)

E6D — Toroidal and Solenoidal Inductors: permeability, core material, selecting, winding; transformers; piezoelectric devices

E6E — Analog ICs: MMICs, IC packaging characteristics

E6F — Electro-optical technology: photoconductivity; photovoltaic devices; optical sensors and encoders; optical isolation

SUBELEMENT E7 — PRACTICAL CIRCUITS
[8 Exam Questions — 8 Groups] 108 Questions

E7A — Digital circuits: digital circuit principles and logic circuits; classes of logic elements; positive and negative logic; frequency dividers; truth tables

E7B — Amplifiers: Class of operation; vacuum tube and solid-state circuits; distortion and intermodulation; spurious and parasitic suppression; microwave amplifiers; switching-type amplifiers

E7C — Filters and matching networks: types of networks; types of filters; filter applications; filter characteristics; impedance matching; DSP filtering

E7D — Power supplies and voltage regulators; Solar array charge controllers

E7E — Modulation and demodulation: reactance, phase and balanced modulators; detectors; mixer stages

E7F — DSP filtering and other operations; software defined radio fundamentals; DSP modulation and demodulation

E7G — Active filters and op-amp circuits: active audio filters; characteristics; basic circuit design; operational amplifiers

E7H — Oscillators and signal sources: types of oscillators; synthesizers and phase-locked loops; direct digital synthesizers; stabilizing thermal drift; microphonics; high-accuracy oscillators

SUBELEMENT E8 — SIGNALS AND EMISSIONS
[4 Exam Questions — 4 Groups] 45 Questions

E8A — AC waveforms: sine, square, and irregular waveforms; AC measurements; average power and PEP of RF signals; Fourier analysis; analog to digital conversion: digital to analog conversion; advantages of digital communications

E8B — Modulation and demodulation: modulation methods; modulation index and deviation ratio; frequency and time division multiplexing; Orthogonal Frequency Division Multiplexing

E8C — Digital signals: digital communication modes; information rate vs. bandwidth; error correction

E8D — Keying defects and overmodulation of digital signals; digital codes; spread spectrum

Introduction   1-15

SUBELEMENT E9 — ANTENNAS AND TRANSMISSION LINES
[8 Exam Questions — 8 Groups] 96 Questions

E9A — Basic Antenna parameters: radiation resistance, gain, beamwidth, efficiency; effective radiated power

E9B — Antenna patterns and designs: E and H plane patterns; gain as a function of pattern; antenna modeling

E9C — Practical wire antennas; folded dipoles; phased arrays; effects of ground near antennas

E9D — Yagi antennas; parabolic reflectors; circular polarization; loading coils; top loading; feed point impedance of electrically short antennas; antenna Q; RF grounding

E9E — Matching: matching antennas to feed lines; phasing lines; power dividers

E9F — Transmission lines: characteristics of open and shorted feed lines; coax versus open-wire; velocity factor; electrical length; coaxial cable dielectrics

E9G — The Smith chart

E9H — Receiving Antennas: radio direction finding antennas; Beverage antennas; specialized receiving antennas; long-wire receiving antennas

SUBELEMENT E0 — SAFETY
[1 exam question — 1 group] 11 Questions

E0A — Safety: RF radiation hazards; hazardous materials; grounding

Chapter 2

Operating Practices

In this chapter, you'll learn about:
- **Frequencies available to Extra class licensees**
- **DX and contest operating**
- **Using a remote station**
- **Satellite orbits and signals**
- **Satellite transponders, frequencies and modes**

Congratulations, you're going to be an Extra and no amateur frequency will be denied to your transmissions! You may have picked up quite a bit of operating experience as a Technician and General. If so, you can probably just scan this section for the items on which you need to brush up. If you have spent most of your operating time on VHF+ frequencies, or if you're wondering if the new HF frequencies are really that special, read on. This chapter also covers amateur satellite operations.

2.1 General Operating

EXTRA CLASS HF FREQUENCIES

The small additional portions of the HF spectrum available only to Extra class licensees don't look like much on the allocation charts in **Figure 2.1**, but they are prime real estate for the HF operator.

Why are these small slices of spectrum worth upgrading for? Traditional operating practice is for DX contacts to take place closer to the bottom edge of the bands. The higher ends of the bands are where most domestic QSOs and net operations take place. You'll find that nearly all serious DXers and contest operators hold Extra class licenses because that's where the contacts are most likely to be made! Even without DX contacts going on, the Extra class segments are less crowded.

Frequency Selection

As an Extra class operator, you will be expected to accommodate all sorts of band conditions. Be flexible. Know how to complete a scheduled contact, even if the planned frequency is busy. Always have a "plan B" of an alternate frequency or time. Use the resources available to all hams for planning your activities. For example, there are plenty of "contest calendars" on the web, such as those at **www.arrl.org/contests** or **www.contestcalendar.com**. Take a look to see if there is a major contest before making a schedule. Special event stations (**www.arrl.org/special-event-stations**) and DXpeditions may appear on your favorite frequency for a few days, as well. If you're a net control station or net manager, be sure that your net has an alternate frequency to accommodate a busy band or poor conditions — practice switching to Plan B so that you'll be ready when you really do need to use it!

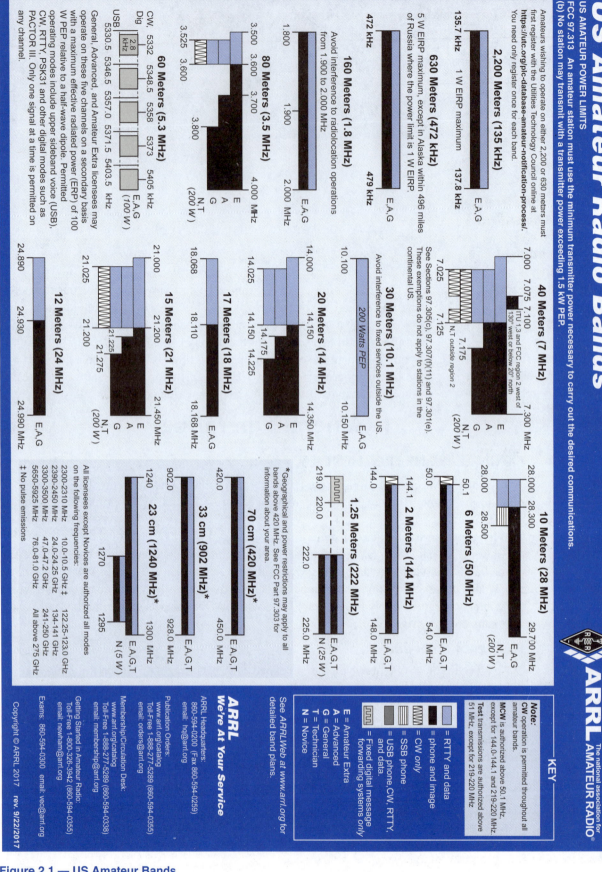

Figure 2.1 — US Amateur Bands

DXING

E2C05 — What is the function of a DX QSL Manager?

E2C08 — Which of the following contacts may be confirmed through the U.S. QSL bureau system?

E2C10 — Why might a DX station state that they are listening on another frequency?

E2C11 — How should you generally identify your station when attempting to contact a DX station during a contest or in a pileup?

Why chase DX? Why chase those elusive distant contacts through noisy bands and unruly pileups? DXing exemplifies a big part of what ham radio is about: the continuous improvement of equipment, antennas, propagation knowledge and operating skills. The quest for "a little more distance" or that special propagation opportunity drives a lot of technical and operating advances.

Your definition of DX will probably depend upon the bands on which you are operating. For HF operators, DX usually means any stations outside of your own country. On the VHF and UHF bands, however, DX may mean stations more than 50 or 100 miles away — beyond your radio horizon. DX extends into space, too — amateurs have been bouncing signals off the Moon to work nearly halfway around the world on microwave frequencies!

Once you've made some DX contacts, perhaps you'll want to chase an award or two. To confirm your contacts, you can use an on-line service such as the ARRL's online Logbook of The World (LoTW — **www.arrl.org/logbook-of-the-world**) or exchange a traditional paper QSL card. The ARRL's QSL Service (**www.arrl.org/qsl-service**) handles both US-to-DX and DX-to-US QSLs, but not US-to-US cards because there would be too many and for which regular mail is available. **[E2C08]** While mailing the card directly is quick and using the QSL bureau system is inexpensive, many DX stations use the services of a *QSL manager* who confirms contacts and sends out responding QSL cards for a DX station. **[E2C05]** Some DXpeditions also use online systems, such as the Online QSL Request System (OQRS — **www.m0urx.com/oqrs**) which also accepts contributions to help defray expenses. These are good compromises for active DXers.

DX Windows and Watering Holes

DX activity most common toward the low end of the bands, concentrated around calling frequencies and *DX windows*. **Table 2.1** shows some of the "watering holes" used by DXers on both HF and VHF/UHF.

Resources for DXing

This section of the *Extra Class License Manual* only touches on the basics of DX operating. To get the most out of your station and your operating time, there are many resources to turn to:

• *The ARRL Operating Manual* —this reference book has a detailed section on DXing discussing everything from DXing basics to maps and QSLing directions.

• *The Complete DXer*, by Bob Locher W9KNI — now in its third edition, this easy-to-read book provides valuable guidance with real-life stories and discussions.

• *The ARRL DX Bulletin* (**www.arrl.org/w1aw-bulletins-archive-dx**), the *OPDX Bulletin* (**www.papays.com/opdx.html**), and *Daily DX* (**www.dailydx.com**) are e-mailed to subscribers around the world. The DX World website (**dx-world.net**) also carries news and stories of interest to DXers.

Table 2.1
DX Windows and "Watering Holes"

Band	Frequency (MHz)
160 meters*	1.830 – 1.835
80 meters	3.505
75 meters†	3.795 – 3.800
40 meters	7.005 (CW)
20 meters	14.005 and 14.020 (CW)
	14.190 – 14.200 (Phone)
15 meters	21.005 and 21.020 (CW)
	21.195 and 21.295 (Phone)
10 meters	28.495 (Phone)
6 meters‡	50.100 – 50.130
	50.125 (US calling frequency)
2 meters‡	144.200

*This window is being phased out as allocations continue to align around the world.

†The recent change in US amateur allocations and continued realignment of allocations make this window less important.

‡Operation begins around the calling frequency, then moves to adjacent channels.

Operating Practices **2-3**

A DX window is a narrow range of frequencies in which QSOs take place between countries or continents that may not share frequency allocations or that have very narrow amateur bands. For example, 160 meters once had many different frequency allocations around the world because of conflicts with radionavigation systems. The 1830 to 1835 kHz window was a narrow slice of the band common to many different countries. As these conflicts disappear and allocations become worldwide, so will the need for a DX window.

Pileup Productivity

You've tuned across a bunch of stations giving their call signs repeatedly. If there is a pause after a station gives his call and a signal report, then many stations call, you've almost certainly found a DX pileup! Listen for someone giving the DX station's call during a contact, such as "T77C from W1JR, you're five-nine" on phone or "T77C DE W1JR 5NN" on CW. While the temptation is great to jump in there and call, don't! You have to be able to *hear* the DX station *before* you start calling. Otherwise, you are just causing QRM.

If all of the stations are on one frequency or close to one frequency, the DX station is probably working simplex. If the stations are spread out over a few kilohertz, the DX station is probably working *split*. Look for the DX station down a few kilohertz or more if the DX station is operating on a frequency unavailable to you, perhaps outside the US band entirely. This practice separates the signals of the calling stations from the DX station, reducing interference and improving efficiency. **[E2C10]**

Listen to the stations that get through — how are they operating? Are they from your area? Is the DX station staying on one frequency or tuning around for callers? Follow that pattern. In general, give your full call sign once or twice (using standard phonetics on phone), then pause to listen for the DX station. **[E2C11]** Remember that a little bit of listening will pay big dividends — if you are transmitting you *can't* be listening!

DXing Propagation

The DXer soon learns the truth of the old adage, "You can't work 'em if you can't hear 'em!" That makes understanding propagation crucial to DXing success. During the years that this edition of the *Extra Class License Manual* is in print (2020 to 2024), the sun will begin its next sunspot cycle. This will be a period of rapid changes in HF conditions, shifting the DXer's attention to the higher-frequency bands. Regardless, it's important to understand the basic variations of long-distance propagation that occur hour-to-hour, day-to-day, and season-to-season.

You can maximize your operating enjoyment by paying attention to band conditions. This will help you decide when to change bands or look for signals from different regions. It is especially important to notice openings and change bands during a period of low sunspot activity because the higher-frequency bands may close entirely after dark, with no DX stations to be heard.

For information about solar conditions, visit websites such as **www.spaceweather.com, www.swpc.noaa.gov/communities/radio-communications**, and **www.hfradio.org**. You will quickly learn what to watch for as ionospheric and solar conditions change. By listening to worldwide on-the-air beacons, such as those that are part of the International Beacon Project sponsored by Northern California DX Foundation and International Amateur Radio Union (**www.ncdxf.org**), you can correlate your expectations with actual behavior. The Reverse Beacon Network (**www.reversebeacon.net**) and websites such as DX Maps (**www.dxmaps.com**) can show you what's happening in real time!

Subscribe to ARRL propagation bulletins via your member information web page and read the columns on propagation in magazines and on websites. Supplement your tuning with DX spotting information from around the world to give you an idea about propagation in other areas.

2-4   Chapter 2

Use propagation prediction software, but remember it is statistical in nature and actual conditions may vary dramatically from the predictions. The software is only as good as its models of the Earth's geomagnetic field, so it may not predict unusual openings.

CONTESTING

E2C02 — Which of the following best describes the term "self-spotting" in connection with HF contest operating?

E2C03 — From which of the following bands is amateur radio contesting generally excluded?

E2C06 — During a VHF/UHF contest, in which band segment would you expect to find the highest level of SSB or CW activity?

E2C07 — What is the Cabrillo format?

What is a radio contest, anyway? Contests, also known as *radiosport*, are on-the-air operating events, usually held on weekends, in which operators try to make as many contacts as possible within defined time limits and according to a detailed set of rules. Contests provide a competitive outlet, enable quick additions to your state or DXCC totals, and offer a level of excitement that's hard to imagine until you've tried it!

Why contest? While each contest has its own particular purpose and operating rules, the main purpose for all amateur radio contests is to enhance communication and operating skills. When you optimize your station for best operating efficiency "in the heat of battle" and learn to pull out those weak stations to make the last available contacts, you are honing useful skills and building a station that can make a big difference in public service or emergency operating. The best way to maintain both is to use them on a regular basis. Contests provide a fun way to keep a keen edge on your equipment capabilities and your operating skills.

Every contest contact includes its own particular set of information (called an *exchange*) that participants send to each other. Most exchanges include call signs, signal report, contact sequence number (called a *serial number*), or location. You can find out the complete rules for contests using WA7BNM's Contest Calendar (**www.contestcalendar.com**), *QST*'s Contest Corral and the ARRL website (**www.arrl.org/contests**). These list contests for each month, showing the sponsor's website and the contest's exchange. Check the rules on the sponsor's website to see what you need to do to participate. Even if you're not a contester, these are good resources to let you know when the bands will be busy!

Contest activity is permitted by the FCC regulations on any frequency available to the station licensee that complies with the mode subband divisions. HF contest activity, like DXing with which it shares many traits, is usually concentrated toward the lowest frequencies of a phone or CW band, expanding upward according to activity levels. Digital contest

---

Resources for Contesting

There is much more to contesting than these few paragraphs, of course. The goal of this license study guide is to make you aware of basic contest principles in accordance with your new licensing privileges. If you'd like to know more about contesting, try some of these resources:

• *The ARRL Operating Manual* — this reference book has a detailed section on contest operating and equipment.

• *Amateur Radio Contesting for Beginners* by Doug Grant, K1DG — published by the ARRL, this book focuses on practical advice for those new to contesting

• *National Contest Journal* (**www.ncjweb.com**) — published by the ARRL, this bimonthly magazine includes in-depth articles on contesting, interviews, technical projects and product reviews, and scores of *NCJ* contests. Print subscriptions are available, and the magazine is available in digital form to all ARRL members.

• *ARRL Contest Update* (**www.arrl.org/the-arrl-contest-update**) is a biweekly e-mail newsletter covering contesting news, upcoming contests and log due dates, technical tips, and news about contest results. It's free to ARRL members via the member information page!

• *cq-contest* and *vhf-contesting* are two of the largest e-mail reflectors about contesting. Subscriptions are free via **www.contesting.com**.

Operating Practices **2-5**

activity centers around the digital calling frequencies for that mode. By general agreement, contesting does not take place on the 60, 30, 17 and 12 meter bands, giving non-contest operators some room during busy weekend events. **[E2C03]**

VHF and UHF contests, sometimes called VHF+ contests, are conducted in the "weak signal" areas at the low end of the VHF, UHF and microwave bands. Most activity is close to the calling frequencies for CW and SSB activity. **[E2C06]**

Submitting a Contest Log

You are encouraged to make contacts whether you intend to submit a log to the contest sponsors or not — no entry is required. But even if you've just entered the contest casually, making a handful of QSOs, go ahead and submit a log. The sponsor will appreciate your efforts and it allows them to more accurately gauge contest activity and score the results.

The standard method of submitting a log is as a *Cabrillo-formatted* computer file via an online log submission service or e-mail. You can submit a log for ARRL contests at **contest-log-submission.arrl.org**. The Cabrillo format is a standard for organizing the information in a submitted contest log so the sponsor can check and score the QSOs. **[E2C07]** You can read about the Cabrillo format at **wwrof.org/cabrillo**.

Cabrillo-formatted files consist of printable ASCII characters with the log information in fixed-position columns. Cabrillo also adds a number of standardized "header" lines that contain information about the log, such as the operator's name, call, and location, contest category, power, and so forth. **Figure 2.2** shows an example of a Cabrillo log file's header and QSO section.

```
START-OF-LOG: 3.0
LOCATION: CT
CALLSIGN: W1AW
CATEGORY-OPERATOR: SINGLE-OP
CATEGORY-BAND: ALL
CATEGORY-POWER: LOW
CATEGORY-MODE: SSB
CATEGORY-STATION: FIXED
CLAIMED-SCORE: 8184
CLUB: Popular Contest Club
CONTEST: ARRL-DX-SSB
CREATED-BY: Popular Contest Software
NAME: Hiram P. Maxim
ADDRESS: 225 Main St
ADDRESS-CITY: Newington
ADDRESS-STATE-PROVINCE: CT
ADDRESS-POSTALCODE: 06111
ADDRESS-COUNTRY: USA
OPERATORS: W1AW
SOAPBOX:
QSO: 14279 PH 2020-03-05 0230 W1AW        59 NH    KP2M     59 K
QSO: 14188 PH 2020-03-05 0232 W1AW        59 NH    KH7M     59 K
QSO: 14215 PH 2020-03-05 0233 W1AW        59 NH    CX1DZ    59 100
QSO: 14237 PH 2020-03-05 0234 W1AW        59 NH    PY2ZR    59 100
QSO: 14253 PH 2020-03-05 0235 W1AW        59 NH    P40L     59 K
QSO: 14300 PH 2020-03-05 0237 W1AW        59 NH    XQ3PC    59 1000
QSO: 14312 PH 2020-03-05 0238 W1AW        59 NH    TO66R    59 500
QSO: 14320 PH 2020-03-05 0238 W1AW        59 NH    CO2GG    59 500
END-OF-LOG:
```

Figure 2.2 —This is a sample of a Cabrillo-formatted contest log that can be submitted to the contest sponsors. The header section contains information about the entry and operator. Each QSO is listed on a separate line with all of the information in fixed columns for easy processing. After the last QSO the line "END-OF-LOG" concludes the log.

Using Spotting Networks

Many contests offer a *Single-Operator Assisted (SOA)* or *Single-Operator Unlimited (SOU)* category. This category usually has the same rules as the Single-Operator category with one major exception: The operator may use information from DX spotting networks such as **dxsummit.fi** and/or the Reverse Beacon Network (**www.reversebeacon.net**) that generate a stream of "spots" listing the call sign and frequency of stations operating in the contest.

Feel free to make use of the information from spotting networks, as long as it is permitted by the rules of the contest for your entry category. The only way in which spotting networks may *not* be used is to *self-spot*, that is announcing your own call sign and frequency on the spotting networks. **[E2C02]** This is because self-spotting is another form of CQing. HF contest rules generally require that soliciting contacts be done only on the bands of the contest and without intermediate stations or the internet.

REMOTE STATIONS

E2C01 — What indicator is required to be used by U.S.-licensed operators when operating a station via remote control and the remote transmitter is located in the U.S.?

Once rather unusual, the use of stations operated by remote control has become common. There are even stations you can use on a subscription basis. While this "changes the game," especially for award-chasing and contesting, using a remote station is allowing many hams who can't put up an effective station at home to get on (or stay on) the air. For most remote setups, all you need to operate them is an internet connection and a web browser. Hams have even operated while traveling on a commercial aircraft!

The ability to operate a station that is literally anywhere with the operator also literally anywhere brings with it some additional ethical and regulatory concerns. (See also the discussion on remote control in the Rules and Regulations chapter.) You must be licensed to transmit from that location and you must have permission to use that station.

You must identify your transmissions according to the rules for operation that apply at the transmitter if they are different from those that apply where the operator is located. If an FCC-licensed operator uses a remote station within the United States, no special identification beyond the FCC-issued call sign is required by the rules. **[E2C01]** However, if K1ABC operates a station in Brazil under remote control, the proper identification is PY/K1ABC (or something similar) to show the transmissions are coming from Brazil and not the United States.

2.2 Amateur Satellites

The spherical shape of the Earth and other factors limit terrestrial communication at VHF and UHF. Long-haul communication at VHF and UHF may require the use of higher effective radiated power (ERP) or may not be possible at all. The communication range of amateur stations is increased greatly by using repeaters, transponders, or store-and-forward equipment onboard satellites orbiting the Earth. There are several operational amateur satellites providing communications, plus the amateur equipment on board the International Space Station (ISS) and a number of experimental CubeSats. Most of the satellites can be accessed or used to relay signals with very modest equipment. (More information on using amateur satellites is available at **www.amsat.org**.)

UNDERSTANDING SATELLITE ORBITS

E2A01 — What is the direction of an ascending pass for an amateur satellite?

E2A06 — What are Keplerian elements?

E2A10 — What type of satellite appears to stay in one position in the sky?

E2A11 — What type of antenna can be used to minimize the effects of spin modulation and Faraday rotation?

Operating Practices **2-7**

Two factors affect a body in orbit around the Earth: forward motion (inertia) and gravitational attraction. Forward motion, the inertia of a body, tends to keep a body moving in a straight line in the direction it is moving at that instant. If the body is above the surface of the Earth, that straight line heads away into space. Gravity, on the other hand, tends to pull the body toward the Earth. When inertia and gravity are balanced, the object's path is a stable *orbit* around the Earth. One orbit is defined as one complete revolution about the Earth (the *orbital period*).

Johannes Kepler was the first to describe the mechanics of the orbits of the planets mathematically. His three laws of planetary motion, called "Kepler's Laws," also describe the lunar orbit and the orbits of artificial Earth satellites. **[E2A06]** Kepler's Laws can be expressed mathematically, and if you know the values of a set of measurements of the satellite orbit (called *Keplerian elements*), you can calculate the position of the satellite at any time. Keplerian elements for the current amateur satellites are available online from the AMSAT website, **www.amsat.org**.

Kepler's First Law tells us that all satellite orbits are shaped like an ellipse with the center of the Earth at one of the ellipse's focal points (**Figure 2.3** describes the geometry of an ellipse). The *eccentricity* of an ellipse (or an orbit) is equal to the distance from the center to one of the focal points divided by the semimajor axis. Notice that when the focal point is at the center, the eccentricity is 0, corresponding to a circle. The larger the eccentricity, the "thinner" the ellipse. The eccentricity of an elliptical orbit ranges between 0 and 1.

Kepler's Second Law is illustrated in **Figure 2.4**. The time required for a satellite to move in its orbit from point A to point B is the same as the time required to move from A′ to B′. What this means is that a satellite moves faster in its elliptical orbit when it is closer to the Earth, and slower when it is farther away. The area of section AOB is the same as the area of section A′OB′.

Kepler's Third Law tells us that the greater the average distance from the Earth, the longer it takes for a satellite to complete each orbit. The time required for a satellite to make a complete orbit around the Earth is called the *orbital period*. The low-flying amateur satellites and International Space Station (ISS) have nearly circular orbits with typical periods of approximately 90 minutes. Satellite AO-40 had a high, elliptical orbit and a period of more than 19 hours. If the orbit is high enough and circular, the satellite's orbit will be *geosynchronous* with a period equal to the time that it takes the Earth to rotate once about its axis. If the geosynchronous orbit is over the Earth's equator, the satellite appears to stay in the same position in the sky. This special type of orbit is called *geostationary*. **[E2A10]**

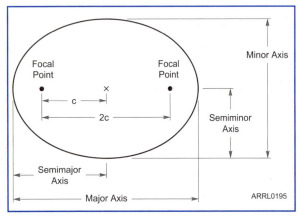

Figure 2.3 — This drawing illustrates the geometry of an ellipse. The ratio of distance c to the length of the semimajor axis is called the *eccentricity* of the ellipse. Eccentricity can vary from 0 (a circle) to 1 (a straight line).

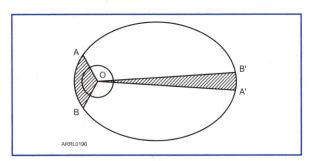

Figure 2.4 — A graphical representation of Kepler's second and third laws. The two shaded sections have equal areas. The time for the satellite to move along its orbit from A to B or from A' to B' is the same.

Orbital Definitions

Inclination is the angle of a satellite orbit with respect to Earth. Inclination is measured between the plane of the orbit and the plane of the equator (**Figure 2.5**). If a satellite is always over the equa-

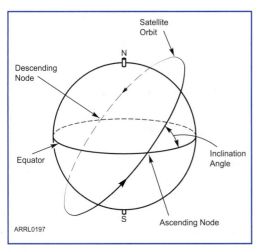

Figure 2.5 — An illustration of basic satellite orbit terminology.

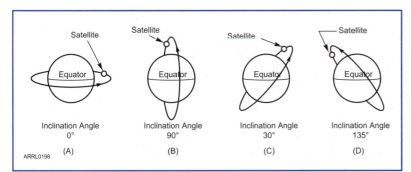

Figure 2.6 — Satellite orbit inclination angles are measured at an ascending node. The angle is measured from the equator to the orbital plane, on the eastern side of the ascending node.

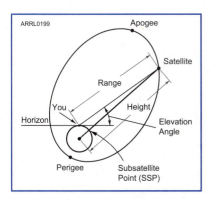

Figure 2.7 — A very elliptical orbit to show the difference between perigee and apogee. Most amateur satellites have orbits that are almost circular.

tor as it travels through its orbit, the orbit has an inclination of 0 degrees. If the orbit path takes the satellite over the poles, the inclination is 90 degrees. (If it goes over one pole, it will go over the other.) The inclination angle is always measured from the equator counterclockwise to the satellite path. **Figure 2.6** gives some examples of orbits with different inclinations.

A *node* is the point where a satellite's orbit crosses the equator. The *ascending node* is the point where the orbit crosses the equator when the satellite is traveling from south to north. Inclination is specified at the ascending node. *Equator crossing* (EQX) is usually specified in time (UTC) of crossing and in degrees west longitude. The *descending node* is the point where the orbit crosses the equator traveling from north to south. When the satellite is within range of your location, it is common to describe the pass as either an *ascending pass* (traveling south to north over your area) or a *descending pass* (traveling north to south). **[E2A01]**

The point of greatest height in a satellite orbit is called the *apogee* as shown in **Figure 2.7**. *Perigee* is the point of least height. Half the distance between the apogee and perigee is equal to the *semimajor axis* of the satellite orbit.

Faraday Rotation and Spin Modulation

The polarization of a radio signal passing through the ionosphere does not remain constant. A "horizontally polarized" signal leaving a satellite will not be horizontally polarized when it reaches Earth. The signal will seem to be changing polarization at a receiving station. This effect is called *Faraday rotation* and it is caused by the effect of the ionosphere on the signal passing through it. The best way to deal with Faraday rotation is to use circularly polarized antennas for transmitting and receiving.

Satellites are often stabilized by being spun like a gyroscope around an axis. This stabilizes the satellite and keeps it oriented in the same direction as it travels around the Earth. When the spacecraft's *spin axis* is not pointed directly at your ground station, you are likely to experience amplitude changes and possibly polarization changes resulting from the spacecraft rotation. This effect is called *spin modulation*. Using linear antennas (horizontal or vertical polarization) will deepen the spin-modulation fades to a point where they may become annoying. Circularly polarized antennas will minimize the effect, just as they do for Faraday rotation. **[E2A11]**

Operating Practices   2-9

SATELLITE OPERATION

E2A02 — Which of the following occurs when a satellite is using an inverting linear transponder?

E2A03 — How is the signal inverted by an inverting linear transponder?

E2A04 — What is meant by the term "mode" as applied to an amateur radio satellite?

E2A05 — What do the letters in a satellite's mode designator specify?

E2A07 — Which of the following types of signals can be relayed through a linear transponder?

E2A08 — Why should effective radiated power to a satellite that uses a linear transponder be limited?

E2A09 — What do the terms "L band" and "S band" specify regarding satellite communications?

E2A12 — What is the purpose of digital store-and-forward functions on an amateur radio satellite?

E2A13 — Which of the following techniques is normally used by low Earth orbiting digital satellites to relay messages around the world?

Of the many components of a satellite — batteries, solar panels, controllers — the most important to the amateur radio operator trying to use the satellite are the radio components. Satellites can have three different types of radio equipment onboard — repeaters, transponders, and store-and-forward systems.

Repeaters

A satellite-borne repeater operates in the same way as a terrestrial repeater. It receives FM voice signals on a single frequency or channel and retransmits what it receives on another channel. Satellite repeaters typically operate with their input and output frequencies on different bands (called a *cross-band repeater*) to allow them to dispense with the heavy and bulky cavity duplexers required for same-band operation. Otherwise, accessing a satellite repeater is just the same as a terrestrial repeater. With some satellites, you can even use low power, handheld transceivers with small beam antennas to make contacts. (See the AMSAT website for operational status and new satellites.)

Transponders

By convention, *transponder* is the name given to any linear translator that is installed in a satellite. In contrast to a repeater, a transponder's receive passband includes enough spectrum for many channels. An amateur satellite transponder does not use channels in the way that voice repeaters do. Received signals from an entire segment of a band are amplified, shifted to a new frequency range by a mixer and retransmitted by the transponder. See **Figure 2.8**.

The major hardware difference between a repeater and a transponder is signal detection. In a repeater, a received signal is converted to audio and then retransmitted. In a transponder, signals in the passband are shifted to an intermediate frequency (IF) for amplification and retransmission.

Operationally, the contrast is much greater. An FM voice repeater is a one-signal, one-mode-input and one-signal, one-mode-output device. A transponder can receive several signals at once and convert them to a new frequency range. Further, a transponder can be thought of as a multimode repeater. Whatever mode is received is retransmitted. **[E2A07]** The same transponder can simultaneously handle SSB and CW signals.

An *inverting linear transponder* inverts the *uplink* signals before retransmitting them on the *downlink*. This is done by using the mixer output's difference products instead of the sum. In other words, if you transmit to the satellite at the bottom of the uplink passband,

2-10   Chapter 2

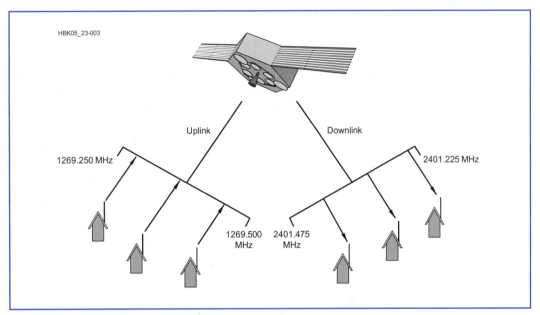

Figure 2.8 — A linear transponder acts much like a repeater, except that it relays an entire group of signals, not just one signal at a time. In this example, the satellite is receiving three signals on its 23 cm uplink passband and retransmitting them on its 13 cm downlink.

your signal position is reversed to the top of the downlink passband. In addition, if your uplink signal is upper sideband (USB), your downlink signal will be lower sideband (LSB). While this sounds confusing, there is a benefit — it counteracts the effects of Doppler shift on the received and transmitted signals. An increase in uplink frequency results in a corresponding decrease in downlink frequency. **[E2A02, E2A03]**

The use of a transponder rather than a channelized repeater allows more stations to use the satellite at one time. In fact, the number of different stations using a transponder at any one time is limited only by mutual interference, and the fact that the output power of the satellite is divided among the users. (On the Low Earth Orbit — LEO — satellites, the output power is a couple of watts. For higher-orbit, multi-band satellites, transponder outputs range from about 50 W on the 2.4 and 10 GHz bands to about 1 W on the 24 GHz band.) Because all users must share the power output, continuous-carrier modes such as FM and RTTY generally are not used through amateur satellites and all users should limit their transmitting ERP (effective radiated power) to allow as many stations as possible to use the transponder. **[E2A08]**

PACSATs

Packet radio store-and-forward systems in space, called *PACSATs*, provide an interesting mix of satellites and packet-radio technology. (See the Modulation, Protocols and Modes chapter for more about packet radio.) These small satellites function as packet bulletin board store-and-forward systems. A terrestrial station can send a message through a PACSAT by uploading it to the satellite for another station to download when the satellite is in view. **[E2A12, E2A13]** The most widely used store-and-forward satellite is the packet system on-board the International Space Station. Other short-lived "CubeSats" often provide temporary digital capabilities on an experimental basis. (See the "Satellite Status" page at **www.amsat.org** for more information on these satellites.)

Operating Practices

Satellite Operating Frequencies

Satellites used for two-way communication generally use one amateur band for the uplink and another for the downlink. Amateur satellites use a variety of uplink and downlink bands, ranging from HF through microwave frequencies. When describing satellite transponders, the input (uplink) band is given, followed by the corresponding output (downlink) band. For example, a 2 meter/70 cm transponder would have an input passband centered near 145 MHz and an output passband centered near 435 MHz.

Transponders usually are identified by *mode* — but not mode of transmission such as SSB or CW or RTTY. Mode has an entirely different meaning in this case. The operating mode of a satellite identifies the uplink and downlink frequency bands that the satellite is using. **[E2A04]**

Satellite operating modes are specified by letter designators that correspond to the frequency range of the uplink and downlink bands. For example, the letter V designates a satellite uplink or downlink in the 2 meter band and U designates uplinks and downlinks in the 70 cm band. So a satellite operating with an uplink on 70 cm and a downlink on 2 meters is in "Mode U/V." (Remember, the uplink band is always the first letter.) **Table 2.2** lists various satellite operating modes by their frequency bands and letter designations. **[E2A05, E2A09]**

An amateur satellite may operate on several of these modes, sometimes simultaneously. Some of the more flexible communications satellites can use various band combinations at different times. You will have to check the operating schedule for the particular satellite you plan to use.

Table 2.2
Satellite Operating Modes

Frequency Band	*Letter Designation*
HF	
15 and 10 meters (21-30 MHz)	H
VHF	
2 meters (144-146 MHz)	V
UHF	
70 cm (435-438 MHz)	U
23 cm (1.26-1.27 GHz)	L
13 cm (2.4-2.45 GHz)	S
5 cm (5.8 GHz)	C
3 cm (10.45 GHz)	X
1.5 cm (24 GHz)	K

Operating Example

Satellite Receive, First Letter *Satellite Transmit, Second Letter*	*Operating Frequency Range*
Mode U/V	435 – 438 MHz / 144 – 146 MHz
Mode V/U	144 – 146 MHz / 435 – 438 MHz
Mode L/U	1.26 – 1.27 GHz / 435 – 438 MHz
Mode V/H	144 – 146 MHz / 21 – 30 MHz
Mode H/S	21 – 30 MHz / 2.40 – 2.45 GHz
Mode L/S	1.26 – 1.27 GHz / 2.40 – 2.45 GHz
Mode L/X	1.26 – 1.27 GHz / 10.45 GHz
Mode C/X	5.8 GHz / 10.45 GHz

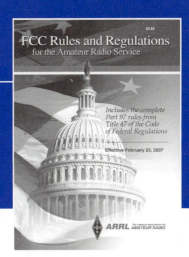

Chapter 3

Rules and Regulations

In this chapter, you'll learn about:
- **Operating regulations and frequency privileges**
- **Restrictions on station operation, location, and antennas**
- **Local, remote, and automatic station control**
- **Amateur-satellite service rules**
- **Rules for exams and volunteer examiners**
- **Auxiliary stations, power amplifier certification, spread spectrum requirements, and non-US operating agreements**

The Amateur Extra class license exam (Element 4) includes six questions about the FCC's Part 97 rules for the amateur service. These questions cover a wide range of topics, including operating standards, station restrictions and special operations, station control, the amateur-satellite service, the Volunteer Examiner program, and miscellaneous rules.

The *Extra Class License Manual* does not contain a complete listing of the Part 97 rules, but focuses on the specific rules on the exam. Every amateur should have an up-to-date copy of the complete Part 97 rules in their station or easily available from a website for reference. The ARRL's booklet *FCC Rules and Regulations for the Amateur Radio Service* contains the complete printed text of Part 97. The text of Part 97 is also available on the ARRL website at **www.arrl.org/part-97-text**.

3.1 Operating Standards

E1A01 — Which of the following carrier frequencies is illegal for LSB AFSK emissions on the 17 meter band RTTY and data segment of 18.068 to 18.110 MHz?

E1A02 — When using a transceiver that displays the carrier frequency of phone signals, which of the following displayed frequencies represents the lowest frequency at which a properly adjusted LSB emission will be totally within the band?

E1A03 — What is the maximum legal carrier frequency on the 20 meter band for transmitting USB AFSK digital signals having a 1 kHz bandwidth?

E1A04 — With your transceiver displaying the carrier frequency of phone signals, you hear a DX station calling CQ on 3.601 MHz LSB. Is it legal to return the call using lower sideband on the same frequency?

E1A05 — What is the maximum power output permitted on the 60 meter band?

E1A06 — Where must the carrier frequency of a CW signal be set to comply with FCC rules for 60 meter operation?

E1A07 — What is the maximum power permitted on the 2200 meter band?

E1A14 — Except in some parts of Alaska, what is the maximum power permitted on the 630 meter band?

E1C01 — What is the maximum bandwidth for a data emission on 60 meters?

E1C07 — At what level below a signal's mean power level is its bandwidth determined according to FCC rules?

E1C12 — On what portion of the 630 meter band are phone emissions permitted?

E1C13 — What notifications must be given before transmitting on the 630 meter or 2200 meter bands?

E1C14 — How long must an operator wait after filing a notification with the Utilities Technology Council (UTC) before operating on the 2200 meter or 630 meter band?

Having already passed your Technician and General class license exams, you're pretty familiar with the basic structure of the FCC rules. As an Extra class licensee, however, your growing amount of experience will expose you to a wider variety of operating situations and circumstances. While the exam can't cover every possible niche of the rules, it can touch on a few areas that you might encounter. The most obvious new experiences will be your new and exclusive HF operating privileges! The exam also covers the responsibilities of control operators when engaged in automatic message forwarding, operation aboard ships and planes, and operation under RACES rules.

FREQUENCY AND EMISSION PRIVILEGES

As an Extra class licensee, you will have all of the frequency privileges available in amateur radio. The privileges above 50 MHz are listed in Section 97.301(a) of the FCC rules and those below 30 MHz in §97.301(b). (Section is often represented by the § character when referencing a specific portion of the FCC rules and we'll use that convention in this manual.) **Table 3.1** contains the latest additions to our amateur bands. Any special frequency sharing requirements are stated in §97.303.

As an Extra class licensee you will have several exclusive privileges on segments of the 80, 75, 40, 20, and 15 meter bands. **Table 3.2** shows that you will have a total of 250 kHz of Extra class spectrum. While the exam does not contain questions about specific frequency privileges, it is still important to know what they are. (Figure 2.1, in Chapter 2, shows a handy chart of all US amateur privileges.) At the time this manual was prepared (early 2020), the

---

Table 3.1

New or Modified Amateur Frequency Privileges

Band	Operating Privileges	Frequency Privileges
2200 m	CW, RTTY, Data, Phone, Image	135.7-137.8 kHz
630 m[1]	CW, RTTY, Data, Phone, Image	472-479 kHz
60 m[2]	CW, RTTY, Data, Phone	5332, 5348, 5358.5, 5373, and 5405 kHz

Notes
1. RTTY, Data, Phone, and Image must use USB
2. The 5 allocated channels are 2.8 kHz wide, centered on these frequencies, §97.307(f)(14)

Table 3.2

Exclusive HF Privileges for Amateur Extra Class Operators

Band	Operating Privileges	Frequency Privileges
80 m	CW, RTTY, Data	3500-3525 kHz
75 m	CW, Phone, Image	3600-3700 kHz
40 m	CW, RTTY, Data	7000-7025 kHz
20 m	CW, RTTY, Data	14.000-14.025 MHz
	CW, Phone, Image	14.150-14.175 MHz
15 m	CW, RTTY, Data	21.000-21.025 MHz
	CW, Phone, Image	21.200-21.225 MHz

Table 3.3
How Mode Affects the Frequency the Transceiver Displays

Mode	Transmitted Carrier Frequency	Transmitted Signal Occupies	The Radio Displays
CW	14,040.00 kHz	14,039.75-14,040.25 kHz Assuming a 500 Hz-wide signal	14,040.0 kHz
LSB	14,200.00 kHz	14,199.70-14,197.00 kHz Assuming 300-3000 Hz audio	14,200.00 kHz
USB	14,200.00 kHz	14,200.30-14,203.00 kHz Assuming 300-3000 Hz audio	14,200.00 kHz
AM	14,200.00 kHz	14,197.00-14,203.00 kHz Assuming 300-3000 Hz audio	14,200.00 kHz
FM	29,600.00 kHz	29,603.00-29,597.00 kHz Assuming 300-3000 Hz audio	29,600.00 kHz

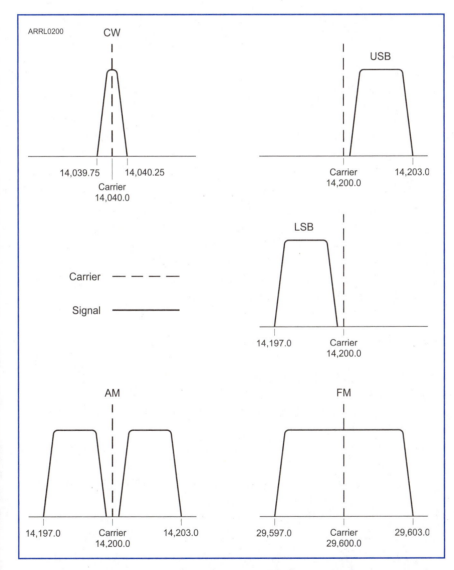

Figure 3.1 — The relationship between carrier frequency and the actual signal energy of typical emissions. Most radios are configured to show the carrier frequency of a signal, also referred to as the "dial frequency."

FCC was considering several proposals that would change emission privileges in the amateur bands. Rule changes will be published on the ARRL website, although the schedule for any announcements is unknown.

You must know where your transmitted signal is with respect to the displayed frequency on your transceiver! This is essential for both compliance with your license privileges and for operating convenience. What frequency is your radio displaying? This varies with operating mode, as shown in **Table 3.3** and **Figure 3.1**.

On CW, the radio displays the frequency of the transmitted signal. On AM, SSB, and FM the radio displays the carrier frequency of the signal. On AM and FM, the carrier frequency is in the middle of your transmitted signal. (The displayed frequency of some transceivers can be configured differently — check your radio's instruction manual.)

The bandwidth of a properly-adjusted amateur USB or LSB signal is from 2.5 to 3 kHz. That means the side-

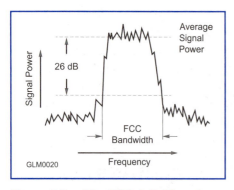

Figure 3.2 — The FCC definition of bandwidth is based on the signal's mean or average power. 26 dB is a power ratio of 400.

bands of an SSB phone signal extend up to 3 kHz from the carrier frequency as shown in Table 3.3. You must keep your sidebands entirely within the band so it's important to understand how close you can operate to the edge. On USB, it's prudent to set your carrier frequency no closer than 3 kHz below the band edge and on LSB no closer than 3 kHz above the band edge. [**E1A02**] If your carrier is closer than that to the band edge — say 14.349 MHz USB or 7.126 MHz LSB — your sidebands will be outside the band! [**E1A04**]

The same concerns apply to digital mode signals. Most digital signals today are transmitted using FSK or AFSK (audio frequency shift keying) techniques. (See the chapter Modulation, Protocols, and Modes for more information.) The signals are similar to USB or LSB signals as shown in Figure 3.1. You must make sure your sidebands remain within a band or band segment, just as when using a voice mode. For example, on 17 meters, when transmitting RTTY using LSB so that the mark and space sidebands are *below* the carrier frequency, setting your carrier frequency to 18.068 MHz, the lower band edge, would put the sidebands entirely outside the band. Similarly, for data segments inside a band if you are transmitting an USB AFSK signal with a bandwidth of 1 kHz, the closest you should set your carrier frequency to the edge of the data segment on 20 meters is 14.149 MHz. Any higher in frequency and the signal sidebands would extend into the phone segment that starts at 14.150 MHz. [**E1A01, E1A03**]

Remember that CW signals have sidebands, too, even though the signal is quite narrow compared to SSB. A typical CW signal has a bandwidth of 50 – 150 Hz, so setting the carrier frequency exactly on a band edge will result in your signal extending outside the band even if your displayed carrier frequency is exactly accurate.

Since signal bandwidth is determined by the signal's bandwidth, it's important to understand how the FCC measures bandwidth. (See §97.3(a)(8)) **Figure 3.2** shows a typical phone or data signal with most of its energy contained in a small range. The mean or average power of the signal in this range is used as a reference. The signal becomes weaker above and below the central range as the sidebands become less powerful. The signal's bandwidth is the difference between the frequencies at which the sideband strength is 26 dB (400 times) lower than the mean signal power. [**E1C07**] It doesn't matter whether the signal carrier frequency is centered in the range or to one side — it is only the average signal power that counts.

Special Restrictions

While most amateur bands are divided into Phone and CW/Data segments and have no other special restrictions on operating privileges, two bands are exceptions.

Amateurs may only transmit CW and data signals on the 30 meter band and are limited to 200 W output power — phone and image signals are not allowed.

Operation on the 60 meter band is restricted to certain channels and emission types [**E1A05, E1A06, E1C01**]:

• Amateurs may only use five 2.8 kHz-wide channels centered on 5332, 5348, 5358.5, 5373, and 5405 kHz.

• USB (upper sideband) voice is the only phone emission allowed. The carrier must be located 1.5 kHz below the center of the channel. This centers a properly-adjusted USB signal on the specified channel frequency.

• RTTY and data emissions are permitted as USB signals centered on the specified channel frequency (not just anywhere within the channel) as described above. The bandwidth of the transmitted signal may not exceed 2.8 kHz.

• CW is permitted with the carrier frequency set to the specified channel center frequency.

- Output power is limited to 100 W ERP (effective radiated power) relative to a dipole.
- Automatic control of RTTY and data emissions is not permitted.

These restrictions are necessary to coexist with other government fixed service stations. The 2015 World Radiocommunication Conference (WRC-15) in Geneva approved an allocation of 5351.5 – 5366.5 kHz to the Amateur Service on a secondary basis with a power limit of 15 W EIRP (*equivalent isotropic radiated power* — see the Antennas and Feed Lines chapter). As of early 2020, however, the FCC has not yet adopted the recommendations and made formal rules to allow amateurs access to this new band. Watch for announcements on the ARRL website when the new rules take effect.

On 630 meters, which is just below the AM broadcast band, amateurs are allowed a maximum power of 5 W EIRP. The power limit is even lower, 1 W EIRP, on the only amateur radio LF (low frequency) band, 2200 meters. Both of these bands require special attention to antenna system design and operating techniques for making contacts with low power. However, permitted emissions, including phone, may be transmitted anywhere in the band — good luck! [**E1A07, E1A14, E1C12**]

In order to operate on the 630 and 2200 meter bands, you must be sure you will not cause interference to power line control (PLC) systems that also operate in this frequency range. This requires informing the Utilities Technology Council (UTC) of your call sign and station coordinates. If after 30 days of notifying the UTC you have not received a notice that you are within 1 km of a PLC system, you may go ahead and operate. [**E1C13, E1C14**]

SPECIAL OPERATING RULES

E1A08 — If a station in a message forwarding system inadvertently forwards a message that is in violation of FCC rules, who is primarily accountable for the rules violation?

E1A09 — What action or actions you should you take if your digital message forwarding station inadvertently forwards a communication that violates FCC rules?

E1A10 — If an amateur station is installed aboard a ship or aircraft, what condition must be met before the station is operated?

E1A11 — Which of the following describes authorization or licensing required when operating an amateur station aboard a U.S.-registered vessel in international waters?

E1A13 — Who must be in physical control of the station apparatus of an amateur station aboard any vessel or craft that is documented or registered in the United States?

E1B09 — Which amateur stations may be operated under RACES rules?

E1B10 — What frequencies are authorized to an amateur station operating under RACES rules?

Automatic Message Forwarding

The FCC rules establish standards governing the operation of amateur radio stations. Within those standards, amateurs relish experimenting with methods of communication, such as types of modulation and digital protocols. Over the years, amateurs have developed or adapted technology to create radio-based data networks capable of automatically forwarding messages, such as packet radio and the Winlink system.

Amateurs have always been held accountable for any message transmitted from their stations. When all third-party communications were relayed by individual amateurs, such rules made sense and were easy to follow. With the advent of automatic message forwarding, requiring every individual station owner to personally screen every message was not feasible. In response, the rules now state that the FCC holds the originating station primarily responsible. [**E1A08**] The first station to forward an illegal message also bears some responsibility but other stations that automatically forward the message are not responsible. If you become

§97.219 — Message Forwarding System Rules

(a) Any amateur station may participate in a message forwarding system, subject to the privileges of the class of operator license held.

(b) For stations participating in a message forwarding system, the control operator of the station originating a message is primarily accountable for any violation of the rules in this Part contained in the message.

(c) Except as noted in paragraph (d) of this section, for stations participating in a message forwarding system, the control operators of forwarding stations that retransmit inadvertently communications that violate the rules in this Part are not accountable for the violative communications. They are, however, responsible for discontinuing such communications once they become aware of their presence.

(d) For stations participating in a message forwarding system, the control operator of the first forwarding station must:

(1) Authenticate the identity of the station from which it accepts communication on behalf of the system; or

(2) Accept accountability for any violation of the rules in this Part contained in messages it retransmits to the system.

§97.11 — Stations Aboard Ships or Aircraft

(a) The installation and operation of an amateur station on a ship or aircraft must be approved by the master of the ship or pilot in command of the aircraft.

(b) The station must be separate from and independent of all other radio apparatus installed on the ship or aircraft, except a common antenna may be shared with a voluntary ship radio installation. The station's transmissions must not cause interference to any other apparatus installed on the ship or aircraft.

(c) The station must not constitute a hazard to the safety of life or property. For a station aboard an aircraft, the apparatus shall not be operated while the aircraft is operating under Instrument Flight Rules, as defined by the FAA, unless the station has been found to comply with all applicable FAA rules.

aware that your station has inadvertently forwarded a communication that violates FCC rules, you should, of course, immediately discontinue forwarding that message to other stations. [E1A09]

RACES Operation

Any FCC-licensed amateur station may be operated under RACES (Radio Amateur Civil Emergency Service) rules, but to participate they must be certified by the responsible civil defense organization for the area served. [E1B09] RACES station operators do not receive any additional operator privileges because of their RACES registration. For example, a Technician class operator may only use Technician frequencies when serving as the control operator. Extra class operators must also follow the operator privileges granted by their license. In general, all amateur frequencies are available to stations participating in RACES operation. [E1B10] RACES stations may communicate with any RACES station as well as certain other stations as authorized by the responsible civil defense official. The complete RACES rules are covered in §97.407.

Stations Aboard Ships or Aircraft

You may have worked a maritime mobile station signing /MM or an aeronautical mobile with an /AM suffix. That sounds pretty exciting and there is absolutely no reason not to try it yourself except for one small caveat: If you want to operate on board a ship or aircraft you must have the radio installation approved by the master of the ship or the pilot in command of the aircraft. [E1A10] The sidebar "§97.11 — Stations Aboard Ships or Aircraft" lists the specific requirements for such an installation. You don't need any other special permit or permission from the FCC for such an operation, though. Your FCC amateur radio license is all that's needed. [E1A11]

You should also be aware that to operate from international waters (or international air space) from a vessel or plane registered in the US you must have an FCC-issued amateur license or reciprocal operating permit. [E1A13] This is because FCC rules apply to US-registered ships and aircraft. If the boat or plane travels into an area controlled by another country then you must have a license or permission from that country to operate. In international waters or air space, however, any amateur license is sufficient to permit you to operate. You must also obey the frequency restrictions that apply to the ITU Region in which you are operating.

3.2 Station Restrictions

E1B01 — Which of the following constitutes a spurious emission?

E1B03 — Within what distance must an amateur station protect an FCC monitoring facility from harmful interference?

E1B04 — What must be done before placing an amateur station within an officially designated wilderness area or wildlife preserve, or an area listed in the National Register of Historic Places?

E1B06 — Which of the following additional rules apply if you are installing an amateur station antenna at a site at or near a public use airport?

E1B07 — To what type of regulations does PRB-1 apply?

E1B08 — What limitations may the FCC place on an amateur station if its signal causes interference to domestic broadcast reception, assuming that the receivers involved are of good engineering design?

E1B11 — What does PRB-1 require of regulations affecting amateur radio?

E1C10 — What is the permitted mean power of any spurious emission relative to the mean power of the fundamental emission from a station transmitter or external RF amplifier installed after January 1, 2003 and transmitting on a frequency below 30 MHz?

In general, you can operate your amateur radio station anytime you want to. You can operate from nearly any location in the United States. Circumstances, however, may lead to some restrictions on your ability to operate.

OPERATING RESTRICTIONS

Under certain conditions spelled out in §97.121, the FCC may modify the terms of your amateur station license. These have to do with interference between a properly operating amateur station and users of other licensed services. Here's what §97.121(a) says about interference to broadcast signals:

"If the operation of an amateur station causes general interference to the reception of transmissions from stations operating in the domestic broadcast service when receivers of good engineering design, including adequate selectivity characteristics, are used to receive such transmissions, and this fact is made known to the amateur station licensee, the amateur station shall not be operated during the hours from 8 p.m. to 10:30 p.m., local time, and on Sunday for the additional period from 10:30 a.m. until 1 p.m., local time, upon the frequency or frequencies used when the interference is created."

Where the interference from the amateur station is causing a sufficient amount of interference, the FCC can impose limited quiet periods on the amateur station on the frequencies that cause interference. [**E1B08**] This is not a blanket injunction against interference to broadcast radio and TV signals. Note that the receiver must be "of good engineering design, including adequate selectivity characteristics." The majority of broadcast receivers, manufactured under stringent price constraints, omit key interference-rejection features, particularly filtering. The result is that quiet periods are rarely imposed by the FCC.

It is also required that the amateur station be operating properly without violating any rules, especially those regarding *spurious emissions*. §97.3(a)(42) defines a spurious emission as "an emission, on frequencies outside the necessary bandwidth of a transmission, the level of which may be reduced without affecting the information being transmitted." [**E1B01**] For stations installed in 2003 or later, spurious emissions must be at least 43 dB below the mean power of the fundamental signal. [**E1C10**] (See the previous section for a discussion of how the FCC defines signal bandwidth.)

It's important to realize that nearly *all* transmissions have some associated spurious emissions. The amateur may be unable to reduce the spurious emissions to zero; a station operat-

Rules and Regulations 3-7

ing completely legally may still transmit very low level spurious signals that are within the regulatory limits.

STATION LOCATION AND ANTENNA STRUCTURES

Unless restricted by private agreements such as deed restrictions and covenants, or by local zoning regulations, you are free to install and build whatever antennas and support structures you want — almost. There are certain conditions that may restrict the physical location of your amateur station and the height of the structures associated with it — your towers and antennas.

Restrictions on Location

If the land on which your station is located has environmental importance, or is significant in American history, architecture, or culture you may be required to take action as described in §97.13(a). For example, if your station will be located within the boundaries of an officially designated wilderness area, wildlife preserve, or an area listed in the National Register of Historic Places, you may be required to submit an Environmental Assessment to the FCC. [E1B04]

If your station will be located within 1 mile of an FCC monitoring facility, you must protect that facility from harmful interference. [E1B03] If you do cause interference to such a facility, the FCC Engineer in Charge may impose operating restrictions on your station.

Table 3.4 contains a list of these FCC facilities that must be protected from interference.

Restrictions on Antenna Structures

Amateurs may face considerable hurdles in attempting to build effective stations due to zoning limits, deed restrictions, and Homeowner Association (HOA) rules. The FCC recognized that amateurs needed some protection in order to be able to satisfy the Basis and Purpose of the amateur service stated in §97.1. This resulted in PRB-1, an FCC regulation which requires state and local or municipal zoning to make reasonable accommodations for amateur radio. [E1B07] PRB-1 by no means prevents any restrictions on antennas or stations, but it does require that regulations consider the needs of amateurs. [E1B11]

The FCC rules also place some limitations and restrictions on the construction of antennas for amateur stations. Without

---

§97.13 — Restrictions on Station Location

(a) Before placing an amateur station on land of environmental importance or that is significant in American history, architecture or culture, the licensee may be required to take certain actions prescribed by §§1.1305 – 1.1319 of this chapter.

(b) A station within 1600 m (1 mile) of an FCC monitoring facility must protect that facility from harmful interference. Failure to do so could result in imposition of operating restrictions upon the amateur station by a District Director pursuant to §97.121 of this Part. Geographical coordinates of the facilities that require protection are listed in §0.121(c) of this chapter.

---

§97.15 — Station Antenna Structures

(a) Owners of certain antenna structures more than 60.96 meters (200 feet) above ground level at the site or located near or at a public use airport must notify the Federal Aviation Administration and register with the Commission as required by Part 17 of this chapter.

(b) Except as otherwise provided herein, a station antenna structure may be erected at heights and dimensions sufficient to accommodate amateur service communications. [State and local regulation of a station antenna structure must not preclude amateur service communications. Rather, it must reasonably accommodate such communications and must constitute the minimum practicable regulation to accomplish the state or local authority's legitimate purpose. See PRB-1, 101 FCC 2d 952 (1985) for details.]

3-8   Chapter 3

Table 3.4
Protected FCC Locations

The following FCC field offices are listed in §0.121(b) and must be protected from interference:

Allegan, MI	Kingsville, TX
Belfast, ME	Laurel, MD
Canandaigua, NY	Livermore, CA
Douglas, AZ	Powder Springs, GA
Ferndale, WA	Santa Isabel, PR
Grand Island, NE	Vero Beach, FL
Kenai, AK	Waipahu, HI

prior FCC approval and notification to the Federal Aviation Administration (FAA), you may not build an antenna structure, including a tower or other support structure, higher than 200 feet. This applies even if your station is located in a valley or canyon. Unless it is approved, the antenna structure may not be more than 200 feet above the ground level at the station location. This includes all of the tower and antenna. The rule is needed to maintain aviation safety. Structures over 200 feet may be approved, but will be required to have warning markings and tower lighting.

If your antenna is located near a public-use airport, then further height limitations may apply. You must obtain approval from the FAA in such cases. [**E1B06**] The FAA rules limit antenna structure height based on the distance from the nearest active runways. If your antenna structure is no more than 20 feet above any natural or existing man-made structure then you do not need approval. See **www.fcc.gov/wireless/systems-utilities/antenna-structure-registration** for more information.

3.3 Station Control

E1C03 — How do the control operator responsibilities of a station under automatic control differ from one under local control?

E1C05 — When may an automatically controlled station originate third party communications?

E1C08 — What is the maximum permissible duration of a remotely controlled station's transmissions if its control link malfunctions?

E1D01 — What is the definition of telemetry?

E1D04 — Which of the following is required in the identification transmissions from a balloon-borne telemetry station?

E1D05 — What must be posted at the station location of a station being operated by telecommand on or within 50 km of the earth's surface?

E1D06 — What is the maximum permitted transmitter output power when operating a model craft by telecommand?

E1D12 — Which of the following amateur stations may transmit one-way communications?

The more experience you gain in amateur radio, the more varied and interesting the methods you'll find of putting together a functioning station. With the internet connected directly to the radio and all manner of "smart" interfaces that allow all the equipment to work in concert, the possibilities are literally endless. Even so, all of the different configurations can be reduced to one of three types of control: *local, remote,* and *automatic* as defined in the next few sections. FCC regulations must be followed under all methods of control.

What's important is that you, the *control operator*, understand the rules for each type of station. The FCC definition for a control operator [§97.3(a)(13)] is "an amateur operator designated by the licensee of a station to be responsible for the transmissions from that station to assure compliance with the FCC Rules." The control operator doesn't have to be the station owner and doesn't even have to be physically present at the transmitter, but someone must be responsible for all amateur transmissions, whether the equipment is directly supervised or not.

Rules and Regulations **3-9**

LOCAL CONTROL

The FCC defines *local control* [§97.3(a)(30)] as "the use of a control operator who directly manipulates the operating adjustments in the station to achieve compliance with the FCC Rules." Local control is the classic form of radio operation. If you are in your station, turning the VFO knob and pressing the PTT switch, that's local control. It doesn't matter whether the operator adjusts the equipment directly by hand or uses a computer to make changes or even uses a voice-activated speech system.

REMOTE CONTROL

Operating a station by remote control means that the *control point* is no longer at the radio — it's where the control operator is. The control point is wherever a control operator performs the station's control functions.

If you're not in direct contact with the radio, but are managing to operate it by means of some intermediary system, that's *remote control*. The intermediary system that allows you to operate the radio without being in direct contact with it — that's the *control link*.

Because it's possible that the control link could fail, you are expected to have some control backup systems that will keep the transmitter from being left on the air. If the control link malfunctions, §97.213 requires that backup control equipment should limit continuous transmissions to no more than three minutes. [**E1C08**]

It's important to be aware of the rules for remote control because more and more radio equipment is designed to support remote control. Many amateurs have constructed remote control stations that allow them to operate from antenna-restricted housing, for example. Another common example of remotely-controlled stations are the digital Winlink Express stations (**www.winlink.org**) that wait for a station to call them before responding. The Winlink station is considered to be remotely controlled by the calling operator.

TELECOMMAND AND ONE-WAY TRANSMISSIONS

The definition of *telecommand* is "one-way transmissions to initiate, modify, or terminate functions of a device at a distance." (§97.3(a)(44)) A station may be operated under remote control by telecommand. If the control link operates through the internet or by a cable or Wi-Fi from a separated front panel that is considered a "wireline" connection. If a radio link is used, however, an *auxiliary station* (§97.3(a)(7)) must be used. So that the control operator of the station can be identified, the following must be posted at the station being operated by telecommand if it is on or within 50 km of the Earth's surface [**E1D05**]:

- A photocopy of the station license
- A label with the name, address, and telephone number of the station licensee
- A label with the name, address, and telephone number of the control operator

Telecommand is commonly used for space stations as described in the next section and for model craft such as remotely piloted vehicles (RPVs, more commonly known as "drones"), model craft such as planes or boats, and for experimental platforms such as high-altitude balloons. The maximum power allowed for operating a model craft via telecommand is 1 W. [**E1D06**] In addition, a label with the station call sign and the licensee's name and address must be affixed to the station transmitter.

One-way transmissions such as for telecommand and telemetry are permitted for space stations, beacon stations, and the special telecommand stations that control space stations as defined in the next section on the amateur satellite service. [**E1D12**]

AUTOMATIC CONTROL

Does a human control operator have to be present at the control point to supervise every amateur transmission? No — repeaters are a very good example of stations operating under *automatic control* with no control operator present. Automatic control is defined in §97.3(a)(6) as "the use of devices and procedures for control of a station when it is transmitting so

Using Automatic Forwarding Systems

More and more hams are building, operating and using automated digital message forwarding systems. These systems require both a new on-the-air etiquette and a careful attention to the rules to ensure that the communication procedures and message content comply with FCC rules. As an Extra class licensee, you may become responsible as a control operator for an automated station or to lead a team using these systems. It's important that you understand and follow the rules that keep these systems compatible with amateur radio.

These systems use protocols to automate the process of calling, connecting, exchanging data, and disconnecting. The first such amateur system was packet radio, using the AX.25 protocol over VHF and UHF links. These protocols do a good job of getting the data from one station to another. Unfortunately, the protocols aren't built to recognize signals from other modes that may be operating on or near the same frequency.

This is a crucial difference between digital and analog communications modes in which the information is copied "by ear." It is easy for a human to recognize signals from other modes, even signals that may not be on the same frequency, just nearby. Digital systems have not yet developed that capability. To an HF digital protocol, CW or SSB voice signals are just an interfering tone or noise. A station using digital protocols is likely to react to them as interference or ignore them as it tries to connect with another digital station. This often disrupts the analog communications and is the reason why human supervision of digital stations is so important, particularly on HF where many modes share the same limited band space.

One of the most common sources of behavior-based conflict between modes is the use of semi-automated digital systems on HF. Amateurs have constructed "mailbox" stations that wait silently on a published frequency until called by another digital station using the same protocol. The mailbox station "will not transmit unless transmitted to" and so does not cause interference to signals from other modes. Unless, that is, another digital station calls in, causing the mailbox station to start the connection and transfer process. That makes it important for the operator of the calling station to listen carefully for other signals on the frequency.

If the operator does not listen, this allows the protocol controller to make the decisions about when to transmit. This can and does enable harmful interference to occur! Another cause of interference is the "hidden transmitter" problem caused by propagation in which the calling station can't hear the other stations on frequency but the mailbox station is heard by everyone. In this situation, even a human operator will not hear the other stations. Nevertheless, the best solution for everyone is for operators of stations attempting to connect to other digital stations to listen by ear first whenever operating on frequencies where other signals are likely to be present. It is not enough to watch for a BUSY light on a modem or controller — that light may only signify the presence of another recognizable digital signal.

Digital messages can also run afoul of content regulations when they are generated by non-amateurs. For example, it's fine to exchange e-mail messages about personal topics but not about work or financial matters. Remember that a non-ham sender is probably unaware of the restrictions on content with which amateurs must comply. Third-party regulations and agreements also apply — know the rules! Higher-speed digital systems may even support direct Internet access. Advertisements are commonplace on many web pages but are not allowed in amateur communications. For this reason alone, web browsing via Amateur Radio is not a good idea. It's important to follow the rules for our service even when they limit what we can do compared to online activities from home over non-amateur networks. It's important that Amateur Radio remain amateur in fact as well as spirit!

Rules and Regulations 3-11

Table 3.5

Allocations for Ground-Based Repeater Stations (§97.205)
29.5 – 29.7 MHz
51.0 – 54.0 MHz
144.5 – 145.5 MHz
146 – 148 MHz
222.15 – 225.0 MHz
420.0 – 431.0 MHz
433.0 – 435.0 MHz
438.0 – 450.0 MHz

Allocations for Automatically-Controlled Digital Stations (§97.221)
28.120-28.189 MHz
24.925-24.930 MHz
21.090-21.100 MHz
18.105-18.110 MHz
14.0950-14.0995 MHz
14.1005-14.112 MHz
10.140-10.150 MHz
7.100-7.105 MHz
3.585-3.600 MHz

that compliance with the FCC Rules is achieved without the control operator being present at a control point." [E1C03] The FCC limits the frequencies on which automatically-controlled stations may operate to make sure the amateur bands are used primarily by human operators. **Table 3.5** shows the frequencies on which ground-based stations may operate under automatic control either as repeaters or as digital stations.

Along with the familiar voice repeaters, auxiliary and beacon stations may be automatically controlled (discussed in the Miscellaneous Rules section later in this chapter). Beacon stations transmit continuously (limited to 100 W of output power) so that other amateurs can tell when propagation exists between their location and that of the beacon station.

There are special rules about the third-party messages and automatically controlled stations because of the power of message forwarding networks. (see the sidebar "Using Automatic Forwarding Systems") Third-party traffic is not limited to radiograms — it can be e-mail, digital files, or even keyboard-to-keyboard chat sessions if the content is transferred on behalf of someone who is not a licensed amateur. Automatically controlled stations may only relay third-party communications as RTTY or data emissions and are never allowed to originate the messages. [E1C05] These restrictions are in place to be sure amateur radio does not become an extension of the commercial data networks.

TELEMETRY

Amateur radio can be used to conduct experiments and make measurements. For example, a satellite might record the temperature, amount of solar radiation or other measurements and then transmit that information back to Earth. It is also important for the satellite operators to know the status of important parameters such as the state of battery charge, transmitter temperature, or other spacecraft conditions. High-altitude balloon-borne experiments and data logging stations on buoys or in remote sites also need to transmit similar information back to a host station.

When transmitted, this information is called *telemetry*, the general term for any one-way transmission of measurements to a receiver located at a distance from the measuring instrument. [E1D01] Telemetry transmissions can include any kind of data but must include the call sign of the transmitting station. [E1D04]

3.4 Amateur-Satellite Service

E1D02 — Which of the following may transmit special codes intended to obscure the meaning of messages?

E1D03 — What is a space telecommand station?

E1D07 — Which HF amateur bands have frequencies authorized for space stations?

E1D08 — Which VHF amateur bands have frequencies authorized for space stations?

E1D09 — Which UHF amateur bands have frequencies authorized for space stations?

E1D10 — Which amateur stations are eligible to be telecommand stations of space stations (subject to the privileges of the class of operator license held by the control operator of the station)?

E1D11 — Which amateur stations are eligible to operate as Earth stations?

Stations more than 50 km above the Earth's surface are called *space stations*. The term can be confusing since it is also commonly used to refer to the International Space Station (ISS). In the FCC rules (§97.207) "space stations" refers to *all* amateur stations located more than 50 km above the Earth's surface. Doubly confusing, the amateur equipment aboard the ISS is

3-12   Chapter 3

Table 3.6
Amateur-Satellite Service Definitions

The following definitions are included in section §97.3(a) of the FCC rules:

(3) *Amateur-satellite service.* A radiocommunication service using stations on Earth satellites for the same purpose as those of the amateur service.

(16) *Earth station.* An amateur station located on, or within 50 km of the Earth's surface intended for communications with space stations or with other Earth stations by means of one or more other objects in space.

(40) *Space station.* An amateur station located more than 50 km above the Earth's surface.

(43) *Telecommand.* A one-way transmission to initiate, modify, or terminate functions of a device at a distance.

(44) *Telecommand station.* An amateur station that transmits communications to initiate, modify, or terminate functions of a space station.

(45) *Telemetry.* A one-way transmission of measurements at a distance from the measuring instrument.

---

itself a space station aboard the International Space Station! Space stations operate according to the rules for the amateur satellite service. **Table 3.6** defines some of the terms used in the amateur-satellite service rules.

SPACE TELECOMMAND STATIONS

Since most space stations are not operated by amateurs under local control, amateurs must have some way to control the various functions of the satellite. A station that transmits communications to a satellite to initiate, modify, or terminate the various functions of a space station is a *space telecommand operation.* [**E1D03**] Stations that transmit these command communications are *telecommand stations.* Any amateur station that is designated by the space station licensee may serve as a telecommand station. §97.211 describes what telecommand stations are and what they may do.

Obviously, sending telecommand communications to a satellite should not be something any amateur can do. Unauthorized telecommand signals would likely disrupt or even damage the satellite. For this reason, the FCC allows telecommand stations to use special codes that are intended to obscure the meaning of telecommand messages. [**E1D02**] This is one of the few times an amateur may intentionally obscure the meaning of a message. Otherwise, anyone who copied the transmission could learn the control codes for the satellite.

SATELLITE LICENSING AND FREQUENCY PRIVILEGES

Any amateur radio station can also be a space station — assuming you can get it into space somehow! You'll need telecommand stations to control the satellite. Any amateur station may be a satellite telecommand station subject to the restrictions of the control operator's license class, of course. [**E1D10**] You just have to authorize them to do it and give them the telecommand information.

On the HF bands, a space station may only operate on the 17, 15, 12 and 10 meter bands and on portions of the 40 and 20 meter bands. [**E1D07**] Segments of 2 meters, 70 cm, 23 cm, 13 cm, and some microwave bands are also available for space station operation.
[**E1D08, E1D09**]

And then there are the satellite users who will be clamoring to "squirt your bird"! Once again, no special license is required — any amateur station can operate as an Earth station if the privileges of the license allow the operator to use the frequencies and modes on which the satellite operates. [**E1D11**] The only reason a satellite would not be usable by operators of every class of US license would be if it had an uplink frequency in either a General or Extra class subband on 40, 20, or 15 meters. There are no satellites currently active using those frequencies.

Rules and Regulations   **3-13**

3.5 Volunteer Examiner Program

E1E01 — For which types of out-of-pocket expenses do the Part 97 rules state that VEs and VECs may be reimbursed?

E1E02 — Who does Part 97 task with maintaining the pools of questions for all U.S. amateur license examinations?

E1E03 — What is a Volunteer Examiner Coordinator?

E1E04 — Which of the following best describes the Volunteer Examiner accreditation process?

E1E05 — What is the minimum passing score on all amateur operator license examinations?

E1E06 — Who is responsible for the proper conduct and necessary supervision during an amateur operator license examination session?

E1E07 — What should a VE do if a candidate fails to comply with the examiner's instructions during an amateur operator license examination?

E1E08 — To which of the following examinees may a VE not administer an examination?

E1E09 — What may be the penalty for a VE who fraudulently administers or certifies an examination?

E1E10 — What must the administering VEs do after the administration of a successful examination for an amateur operator license?

E1E11 — What must the VE team do if an examinee scores a passing grade on all examination elements needed for an upgrade or new license?

E1E12 — What must the VE team do with the application form if the examinee does not pass the exam?

By now, you've passed at least two amateur exams. You should be an expert on the Volunteer Examiner (VE) program! Just kidding, of course, but since you're about to pass your third and final exam (Right? Right!) why do you need to learn more about the program? Well, because as an Extra class licensee, you'll have the opportunity to administer those exams yourself! Extra class VEs are needed to give both General and Extra class exams, so this is a little advertisement for you to become a Volunteer Examiner and help others get their licenses.

The questions on the Extra class exam are designed to familiarize you with the FCC rules in Part 97, Section F that pertain to the "examiner side" of the exam. We'll start with some review of the program's structure.

THE VOLUNTEER EXAMINER COORDINATOR

A Volunteer Examiner Coordinator (VEC) is an organization that has entered into an agreement with the FCC to coordinate, prepare, and administer amateur license examinations. [**E1E03**] The organization must meet certain criteria before it can become a VEC. As described in §97.521, the organization must exist for the purpose of furthering the amateur service, but should be more than just a local radio club or group of hams. A VEC is expected to coordinate exams at least throughout an entire call district. The organization must also agree to coordinate exams for all classes of amateur operator license, and to ensure that anyone desiring an amateur license can register and take the exams without regard to race, sex, religion, or national origin.

THE VOLUNTEER EXAMINER

A VEC does not administer or grade the actual examinations. The VEC accredits licensed amateur radio operators — the Volunteer Examiners (VEs) — to administer exams. Each VEC is responsible for recruiting and training Volunteer Examiners to administer amateur

ARRL VEC, 225 Main Street, Newington, CT 06111
Phone: 1-860-594-0300   web: arrl.org/volunteer-examiners

VOLUNTEER EXAMINER APPLICATION FORM

PLEASE Type or Print Clearly in Ink   (check one)
☐ General
☐ Advanced
Call Sign: _____   ☐ Extra   License Expiration Date: _____

Name: _____
         (first,   MI,   last)

Mailing address (street or POB): _____
_____

City: _____ State: _____ ZIP: _____ Country: _____

Day phone: (____) _____ Night phone: (____) _____ Email address: _____

Has your FCC license ever been suspended or revoked? .... ☐ YES ☐ NO
Have you ever been disaccredited by another VEC? .... ☐ YES ☐ NO
If yes, which VEC(s) and when? _____
Do you have a call sign change (or Vanity call sign) pending with the FCC? ☐ YES ☐ NO
Do you have any Form 605 application pending with the FCC?...................... ☐ YES ☐ NO
Who can we contact to reach you, if you cannot be reached? _____
                                                                  (name)   (phone)

For Instant Accreditation, have you participated as a VE in another VEC program
and is your accreditation in that program current? .... ☐ YES ☐ NO
If yes, which VEC coordinated the test session? (enter VEC name here) _____
You MUST attach a copy of your credentials from that VEC to this form as proof.

CERTIFICATION

*By signing this Application Form, I certify that to the best of my knowledge
that the above information AND the following statements are true:*

1) I am at least 18 years of age.
2) I agree to comply with the FCC Part 97 Amateur Radio Service Rules, especially Subpart F (§97.509).
3) I agree to comply with examination procedures established by the ARRL as Volunteer Examiner Coordinator.
4) I understand that the ARRL as my coordinating VEC, or I as an accredited ARRL VE, may terminate this relationship at any time, with or without any reason or cause.
5) I understand that violation of the FCC Rules or willful noncompliance with the VEC will result in the loss of my VE accreditation, and could result in loss of my Amateur Radio operator or station licenses, or both.
6) I understand that, even though I may be accredited as a VE, if I am not able or competent to perform certain VE functions required for any particular examination, I should not administer that examination (§97.525).

_____   _____ _____
            (signature)   (call sign)   (date)

Look over your form for completeness, make sure it is signed and then send it or fax it to the ARRL VEC.
If *instant accreditation* is sought, you MUST indicate which VEC program you served as an administering VE and attach a copy of your other VEC credentials to this application. Otherwise your application *must include your completed open-book review.*

ARRL VEC -- VE APPLICATION 07/2018

Figure 3.3 — This is the ARRL VEC Application Form. It is available online in the *ARRL Volunteer Examiner Manual* (www.arrl.org/ve-manual). The *Manual* contains all you need to know to become an ARRL VEC-accredited Volunteer Examiner.

examinations under their program. The Volunteer Examiners determine where and when examinations for amateur operator licenses will be administered.

ACCREDITATION

When a VEC accredits a Volunteer Examiner, it is certifying that the amateur is qualified to perform all the duties of a VE as required by §97.509 and §97.525. The accreditation process is simply the steps that each VEC takes to ensure their VEs meet all the FCC requirements to serve in the Volunteer Examiner program. [**E1E04**] Each VEC has its own accreditation process. A VEC has the responsibility to refuse to accredit a person as a VE if the VEC determines that the person's integrity or honesty could compromise amateur license exams.

The ARRL VEC coordinates exams in all regions of the US, and would be pleased to have you apply for accreditation. You do not have to be an ARRL member to serve as an ARRL VE. In fact, one of the requirements of VECs is that they not demand membership in any organization as a prerequisite to serving as a VE!

If you are at least 18 years of age and hold at least a General class license, you meet the basic FCC requirements to be a VE. In addition, you must never have had your amateur license suspended or revoked. **Figure 3.3** shows the application form to become an ARRL VE.

EXAM PREPARATION

Coordinating amateur exams involves a bit more responsibility than simply recruiting amateurs to administer the exams. (§97.519 states the requirements for coordinating an exam session.) A VEC coordinates the preparation and administration of exams. Some VECs actually prepare the exams and provide their examiners with the necessary test forms, while others require their VEs to prepare their own exams or purchase exams from a qualified supplier. After the test is completed, the VEC must collect the application documents (NCVEC Form 605) and test results. After reviewing the materials to ensure accuracy, the VEC must forward the documentation to the FCC for applicants that qualify for a new license or a license upgrade.

All of the VECs must cooperate in the development and maintenance of the questions used on the exams. (§97.523) The set of all the questions available to be asked on an exam is called the *question pool*. [**E1E02**] A Question Pool Committee (QPC) works regularly to update the questions for each exam element. Exams are made up of questions selected from the question pool.

Volunteer Examiners may prepare written exams for all classes of amateur radio operator license. Section 97.507 of the FCC rules gives detailed instructions about who may prepare the various examination elements. You must hold a General, Advanced, or Amateur Extra license to prepare an Element 2 written exam for the Technician class license. Only Advanced and Amateur Extra licensees may prepare the Element 3 exam (General) and you must hold an Amateur Extra license to prepare an Element 4 exam (Amateur Extra).

If the VEC or a qualified supplier prepares the exams, they must still use amateurs with the proper license class to prepare the exams. In every case, the exams are prepared by selecting questions from the appropriate question pool.

EXAM SESSION ADMINISTRATION

Extra class exams must be administered by VEs holding Extra class licenses. To administer a General exam, you must hold an Advanced or Amateur Extra license. To administer a Technician license exam, you must hold a General, Advanced, or Amateur Extra license. VEs are prohibited from administering exams to close relatives as defined by the FCC. [**E1E08**] The requirements for VEs administering an exam are stated in §97.509.

Before actually beginning to administer an examination, the VEs should determine what

3-16   Chapter 3

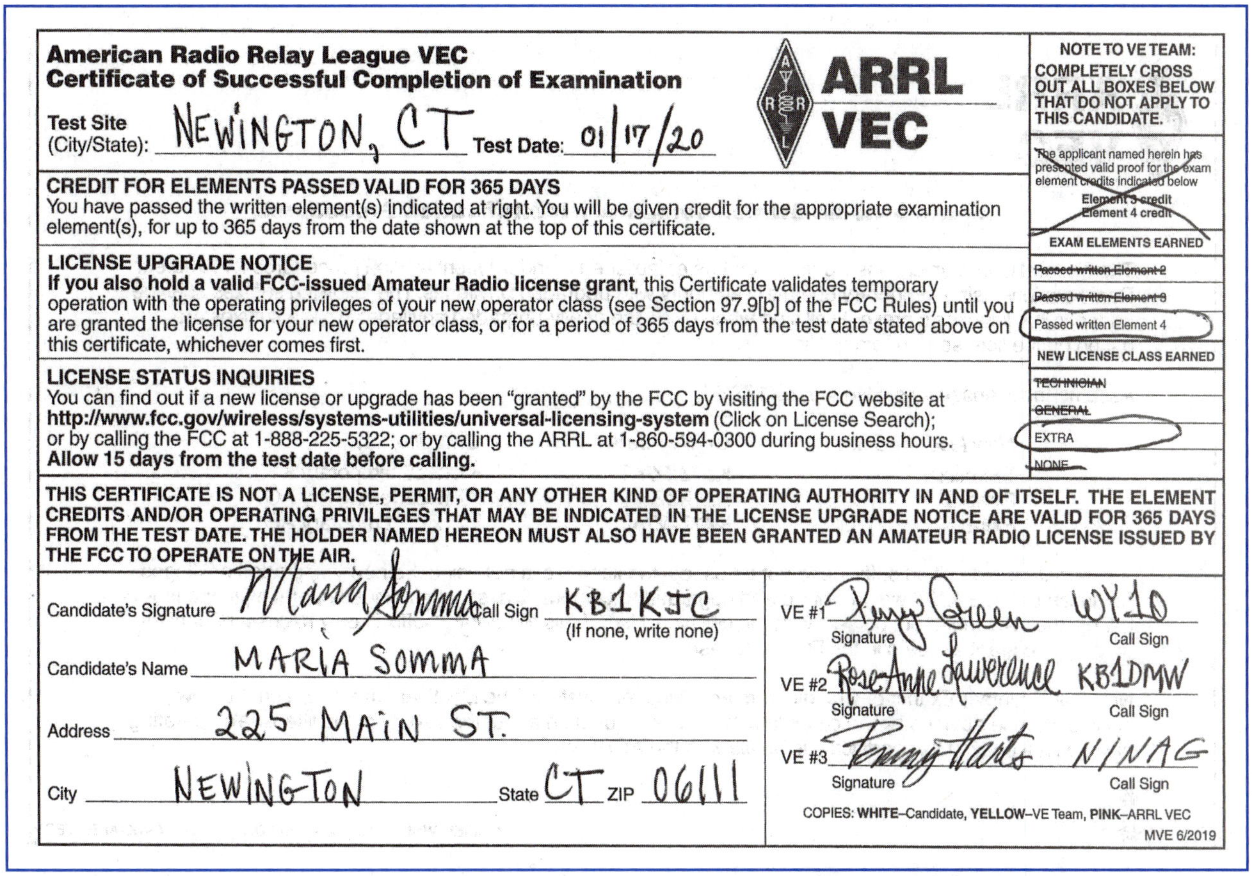

American Radio Relay League VEC
Certificate of Successful Completion of Examination

ARRL VEC

Test Site (City/State): NEWINGTON, CT Test Date: 01/17/20

NOTE TO VE TEAM: COMPLETELY CROSS OUT ALL BOXES BELOW THAT DO NOT APPLY TO THIS CANDIDATE.

~~The applicant named herein has presented valid proof for the exam element credits indicated below~~
~~Element 3 credit~~
~~Element 4 credit~~

CREDIT FOR ELEMENTS PASSED VALID FOR 365 DAYS
You have passed the written element(s) indicated at right. You will be given credit for the appropriate examination element(s), for up to 365 days from the date shown at the top of this certificate.

EXAM ELEMENTS EARNED
~~Passed written Element 2~~
~~Passed written Element 3~~
Passed written Element 4

LICENSE UPGRADE NOTICE
If you also hold a valid FCC-issued Amateur Radio license grant, this Certificate validates temporary operation with the operating privileges of your new operator class (see Section 97.9[b] of the FCC Rules) until you are granted the license for your new operator class, or for a period of 365 days from the test date stated above on this certificate, whichever comes first.

NEW LICENSE CLASS EARNED
~~TECHNICIAN~~
~~GENERAL~~
EXTRA
~~NONE~~

LICENSE STATUS INQUIRIES
You can find out if a new license or upgrade has been "granted" by the FCC by visiting the FCC website at **http://www.fcc.gov/wireless/systems-utilities/universal-licensing-system** (Click on License Search); or by calling the FCC at 1-888-225-5322; or by calling the ARRL at 1-860-594-0300 during business hours. **Allow 15 days from the test date before calling.**

THIS CERTIFICATE IS NOT A LICENSE, PERMIT, OR ANY OTHER KIND OF OPERATING AUTHORITY IN AND OF ITSELF. THE ELEMENT CREDITS AND/OR OPERATING PRIVILEGES THAT MAY BE INDICATED IN THE LICENSE UPGRADE NOTICE ARE VALID FOR 365 DAYS FROM THE TEST DATE. THE HOLDER NAMED HEREON MUST ALSO HAVE BEEN GRANTED AN AMATEUR RADIO LICENSE ISSUED BY THE FCC TO OPERATE ON THE AIR.

Candidate's Signature: Maria Somma Call Sign: KB1KJC (If none, write none)
Candidate's Name: MARIA SOMMA
Address: 225 MAIN ST.
City: NEWINGTON State: CT ZIP: 06111

VE #1 Signature: Perry Green Call Sign: WY1O
VE #2 Signature: Rose-Anne Lawrence Call Sign: KB1DMW
VE #3 Signature: Penny Harts Call Sign: N1NAG

COPIES: **WHITE**–Candidate, **YELLOW**–VE Team, **PINK**–ARRL VEC
MVE 6/2019

Figure 3.4 — The Certificate of Successful Completion of Examination (CSCE) documents that the holder has passed one or more amateur license exam elements. If you upgrade your license by passing an exam, the CSCE allows you to start using your new privileges immediately.

exam credit, if any, the candidates should be given as described in §97.505. For example, any candidates who already hold an amateur operator license must receive credit for having passed all of the exam elements necessary for that class of license. In addition, any candidate who presents a valid Certificate of Successful Completion of Examination (CSCE) must be given credit for each exam element that the CSCE indicates the examinee has passed. The combination of element credits and exam elements passed at the current exam session will determine if a candidate qualifies for a higher class of license.

As a voluntary service, VEs and VECs may not charge a fee to administer exams or receive any type of payment for the services they provide. Neither the VEC nor the VEs should have to bear the costs of administering exams out of their own pockets, however. FCC rule §97.527 provides a means for those being examined to reimburse the VEs and VEC for certain costs involved with the program. These costs include actual out-of-pocket expenses involved with preparing, processing, and administering license exams. [**E1E01**]

During the Exam

All three VEs are responsible for supervising the exam session and must be present during the entire exam session, observing the candidates to ensure that the session is conducted properly. [**E1E06**]

During the exam session, the candidates must follow all instructions given by the Volunteer Examiners. If any candidate fails to comply with a VE's instructions during an exam, the VE team should immediately terminate that candidate's exam. [**E1E07**]

When the candidates have completed their exams, the VEs must collect the test papers and grade them immediately. A score of 74% is the minimum to pass the exam. [**E1E05**] They then notify the candidates whether they passed or failed the exam. If any candidates did not pass all the exam elements needed to complete their license upgrade, then the examiners must return their applications to those candidates and inform them of the grades. [**E1E12**]

After grading the exams of those candidates who do pass the exam, the entire VE team must certify their qualifications for new licenses and that they have complied with the VE requirements on their application forms and issue each a CSCE (**Figure 3.4**) for their upgrade. [**E1E11**]

Maintaining control of the exam session and conducting it properly is key to the success of the amateur VE program. If the FCC determines that a VE has fraudulently administered or certified an exam, that VE can lose their amateur station license and have their operator privileges suspended. [**E1E09**] Such problems are extremely rare because of the high integrity of the amateur volunteer licensing program.

After the Exam

After an exam session, the VE Team must submit the application forms and test papers for all the candidates who passed to the coordinating VEC. [**E1E10**] They must do this within 10 days of the test session. This is to ensure that the VEC can review the paperwork and forward the information to the FCC in a timely fashion.

3.6 Miscellaneous Rules

The following sections cover topics of narrow interest. As an Extra class licensee, you'll be expected to know about lesser-visited areas of the FCC rules and have a more complete knowledge of important rules.

AUXILIARY STATIONS

E1F10 — Who may be the control operator of an auxiliary station?

Auxiliary stations [§97.3(a)(7)] are amateur stations, other than in a message forwarding system, that transmit communications point-to-point within a system of cooperating amateur stations. Amateurs are allowed to use auxiliary stations to provide point-to-point communications and control links between a remotely controlled station and its control point. Repeater systems may use point-to-point links to relay audio and control signals from one repeater to all other repeaters in the system. Control operators of auxiliary stations must hold a Technician, General, Advanced, or Extra class license. [**E1F10**]

You might set up a mobile rig as a cross-band repeater to act as an auxiliary station relaying your signals to and from a nearby low-power handheld transceiver. §97.201(b) limits auxiliary station transmissions to the 2 meter and shorter wavelength bands, excluding the 144.0 – 144.5 MHz, 145.8 – 146.0 MHz, 219 – 220 MHz, 222.0 – 222.15 MHz, 431 – 433 MHz, and 435 – 438 MHz segments.

EXTERNAL POWER AMPLIFIERS

E1F03 — Under what circumstances may a dealer sell an external RF power amplifier capable of operation below 144 MHz if it has not been granted FCC certification?

E1F11 — Which of the following best describes one of the standards that must be met by an external RF power amplifier if it is to qualify for a grant of FCC certification?

RF power amplifiers capable of operating on frequencies below 144 MHz may require FCC certification. Sections 97.315 and 97.317 describe the conditions under which certifi-

3-18 Chapter 3

cation is required, and set out the standards to be met for certification. Many of these rules apply to manufacturers of amplifiers or kits, but several points are important for individual amateurs. Amateurs may build their own amplifiers or modify amplifiers for use in an amateur radio station without concern for the certification rules. An unlicensed person may not build or modify any amplifier capable of operating below 144 MHz without a grant of FCC certification.

To receive a grant of certification, an amplifier must satisfy the spurious emission standards specified in §97.307(d) or (e) when operated at full power output or 1500 W, whichever is less. [**E1F11**] In addition, the amplifier must meet the spurious emission standards when it is placed in the "standby" or "off" position but is still connected to the transmitter. The amplifier must not be capable of amplifying the input signal by more than 15 dB, and it must exhibit no amplification between 26 and 28 MHz. (This is to prevent the amplifier from being used illegally on Citizen's Band frequencies.)

A manufacturer must obtain a separate grant of certification for each amplifier model. The FCC maintains a database of all certificated amplifier models, and an amplifier must be on that list before it can be marketed or sold for use in the US amateur service. Dealers may also sell non-certificated amplifiers if they were purchased in used condition and resold to another amateur for use in their station. [**E1F03**] FCC certification may be denied if the amplifier can be used in a telecommunication service other than the amateur service, or if it can be easily modified to operate between 26 and 28 MHz.

70 CM BAND RESTRICTIONS

E1B12 — **What must the control operator of a repeater operating in the 70 cm band do if a radiolocation system experiences interference from that repeater?**

E1F04 — **Which of the following geographic descriptions approximately describes "Line A"?**

E1F05 — **Amateur stations may not transmit in which of the following frequency segments if they are located in the contiguous 48 states and north of Line A?**

The 420 to 430 MHz band segment is allocated to the fixed and mobile services in the international allocations table on a primary basis worldwide. Canada has allocated this band segment to its fixed and mobile services, so US amateurs along the Canadian border are not permitted to transmit on these frequencies. An imaginary line, called Line A, runs roughly parallel to and south of the US-Canadian border. [**E1F04**] US stations north of this line may not transmit on the 420 to 430 MHz band. [**E1F05**] Section 97.3(a)(29) gives an exact definition of Line A. See **Figure 3.5**.

The FCC has also allocated portions of the 421 – 430 MHz band to the Land Mobile Service within a 50-mile radius centered on Buffalo, New York; Detroit, Michigan; and Cleveland, Ohio. Amateurs in these areas must not cause harmful interference to the Land Mobile or government radiolocation users.

Increased use of the 70 cm band for military radar systems places restrictions on amateur repeater systems, which are a secondary allocation in that band. This requires that the amateur station cease operation or make changes to the repeater to mitigate the interference. [**E1B12**]

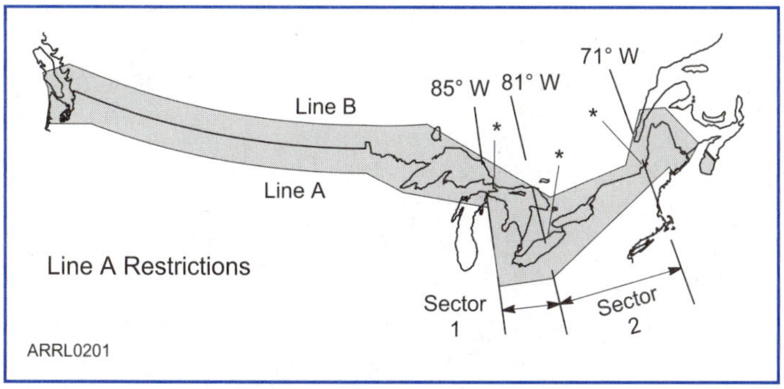

Figure 3.5 — Line A runs parallel to, and just south of, the US/Canadian border.

NATIONAL QUIET ZONES

E1B05 — What is the National Radio Quiet Zone?

Certain other restrictions may apply if your station is within specific geographical regions. For example, there is an area in Maryland, West Virginia and Virginia surrounding the National Radio Astronomy Observatory. This area is known as the *National Radio Quiet Zone*. [**E1B05**] The NRQZ serves to protect the interests of the National Radio Astronomy Observatory in Green Bank, West Virginia, and also Naval Research Laboratory at Sugar Grove, West Virginia.

If you plan to install an automatically controlled beacon station within the NRQZ, you will have to obtain permission from the National Radio Astronomy Observatory. §97.203 lists the details of this rule and the address to contact.

BUSINESS, PAYMENT, AND REBROADCASTING

E1C02 — Which of the following types of communications may be transmitted to amateur stations in foreign countries?
E1F07 — When may an amateur station send a message to a business?
E1F08 — Which of the following amateur station communications are prohibited?

Business communications rules for amateur radio are based on the principle that no transmissions are permitted in which you or your employer have a pecuniary (monetary) interest. This is, after all, amateur radio, and there are plenty of radio services available for commercial activities. That prohibition is the same for voice and digital messages, including those sent and received via message forwarding networks.

However, your own personal activities don't count as "business" communications. For example, it's perfectly okay for you to use ham radio to talk to your spouse about doing some shopping or to confer about what to pick up at the store. You can even send a message to a business over the air, to order something for example, as long as you don't do it regularly and as part of your normal income-making activities [**E1F07**]

Note that when you are contacting stations in other countries, communications are limited to remarks of a personal nature or incidental to amateur radio. No exception to the non-business rule is made for communications on behalf of nonprofit organizations. [§97.117] [**E1C02**]

Although it seems like a useful service, rebroadcasting propagation or weather forecasts on a regular basis is not permitted. [**E1F08**] The reason is that stations are already authorized to broadcast this information and it would be an unnecessary use of the amateur allocations. It is OK to relay an occasional forecast in support of amateur activities.

SPREAD SPECTRUM OPERATION

E1F01 — On what frequencies are spread spectrum transmissions permitted?
E1F09 — Which of the following conditions apply when transmitting spread spectrum emissions?

Spread spectrum (SS) communication is a modulation technique that spreads the signal over a wide bandwidth. The FCC refers to this as *bandwidth-expansion modulation*. If you want to learn more about these fascinating modes you can find the details in *The ARRL Handbook*.

The idea behind this communications mode is to spread a little power over a wide bandwidth to minimize interference rather than concentrating a lot of power in a narrow bandwidth. The FCC rules allow US amateurs to communicate using SS with other amateurs in areas regulated by the FCC and with other stations in countries that permit SS communica-

3-20   Chapter 3

tions. Amateurs are permitted to use SS as long as it does not cause harmful interference to other stations using authorized emissions. In addition, the SS transmission must not be used to obscure the meaning on any communication. [**E1F09**] Operation using spread spectrum techniques is restricted to frequencies above 222 MHz. [**E1F01**]

NON-US OPERATING AGREEMENTS

E1C04 — What is meant by IARP?

E1C06 — Which of the following is required in order to operate in accordance with CEPT rules in foreign countries where permitted?

E1C11 — Which of the following operating arrangements allows an FCC-licensed U.S. citizen to operate in many European countries, and alien amateurs from many European countries to operate in the U.S.?

E1F02 — What privileges are authorized in the U.S. to persons holding an amateur service license granted by the government of Canada?

There are three basic types of agreements that allow amateurs licensed in one country to operate from the territory of another country.

• European Conference of Postal and Telecommunications Administrations (CEPT) radio-amateur license — allows US amateurs to travel to and operate from most European countries and their overseas territories without obtaining an additional license or permit. (This does not automatically confer permission to enter restricted areas. You must also carry with you a copy of FCC Public Notice DA 16-1048.) Amateurs from countries that participate in CEPT may also operate in the US. [**E1C06, E1C11**] The CEPT treaty does not automatically allow remote-control operation across international borders. Be sure you know the rules for transmitting in the host country!

• International Amateur Radio Permit (IARP) — For operation in certain countries of Central and South America, the IARP allows US amateurs to operate without seeking a special license or permit to enter and operate from that country. [**E1C04**]

• ITU Reciprocal Permit — a reciprocal agreement between the US and a country that does not participate in either CEPT or IARP agreements.

The complete rules and procedures for obtaining permission to use your license elsewhere in the world are available on the ARRL's International Operating web page at **www.arrl.org/international-operating** and **www.arrl.org/international-regulatory**.

Amateurs who are citizens of a foreign country and are operating in the US under the terms of an operating agreement between the US and the other country will have the same privileges as a US Amateur Extra class licensee if they hold a full-privilege license from their country. This includes Canadians with an Advanced Qualification as well as operators with a Class 1 license issued by the European Conference of Posts and Telecommunications (CEPT) or a Class 1 International Amateur Radio Permit (IARP). Canadian amateurs operating in the US may operate to the extent of their Canadian license privileges, but not in excess of the US Extra class privileges. For example, Canadian amateurs operating in the US may not transmit SSB below 14.150 MHz even though they may do so from home. [**E1F02**]

SPECIAL TEMPORARY AUTHORITY

E1F06 — Under what circumstances might the FCC issue a Special Temporary Authority (STA) to an amateur station?

Occasionally, a new method or technique of communicating comes along that isn't covered by the FCC rules or would be in violation of the rules. If sufficiently good reasons are provided to the FCC, a *Special Temporary Authority* or STA may be granted to provide for experimental amateur communications. [**E1F06**] For example, STAs were granted to allow amateurs to experiment with spread spectrum communications before that mode was

Rules and Regulations   3-21

generally permitted on amateur frequencies. An STA was also granted allowing amateurs to experiment with communications to evaluate propagation and interference on 630 meters and 2200 meters.

STAs are temporary, lasting long enough for experiments to be performed and information accumulated. For particularly ambitious projects the STA might last for six months or longer. During this time, the STA does not grant amateurs the exclusive use of a frequency, nor does it waive all rules — only the ones explicitly covered in the STA. STAs can also be terminated by the FCC at any time if the operation is found to be causing interference, for example. STAs may result in changes to the FCC rules, but they are not permanent waivers of any rule.

Chapter 4

Electrical Principles

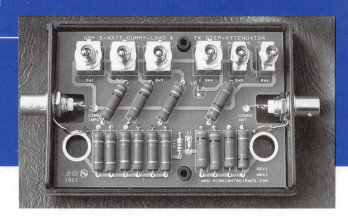

In this chapter, you'll learn about:
- **Rectangular and polar coordinates**
- **Complex numbers**
- **Electric, magnetic, and electromagnetic fields**
- **Time constants**
- **Impedance and admittance**
- **Reactive power and power factor**
- **Resonant circuits, Q, and bandwidth**
- **Inductors with magnetic cores**

The Amateur Extra class license exam will include questions about electrical principles from question pool Subelement 5 (four groups of questions) and Subelement 6 (six groups of questions). This chapter begins with some background in the way we use math to talk about electrical signals. Once prepared, we move on to the relationship between voltage and current in real circuits, impedance, resonance, and even magnetic materials used in inductors.

It would be easy to make this book very, very large by trying to teach fundamental subjects from the ground up. To keep focused, you'll be given references on where to find supplemental information on math and electronics. The *Extra Class License Manual* (ECLM) web page (**www.arrl.org/extra-class-license-manual**) is a good place to start looking, and it should be added to the bookmarks in your internet browser. Let's get going!

4.1 Radio Mathematics

You learned about frequency, phase, impedance, and reactance to pass your Technician and General exams. For the Extra exam, we go a little farther and learn how to work with those values in this initial section of the chapter. By learning these techniques, you'll be able to work with any combination of resistance (R), inductance (L), and capacitance (C). The very same techniques are used to describe signals in later sections.

RECTANGULAR AND POLAR COORDINATES

We can't actually see the electrons flowing in a circuit, or look at voltage or impedances, so we use math equations or graphs to describe what's happening. Graphs are drawings of what equations describe with symbols — they're both saying the same thing. The way in which mathematical quantities are positioned on the graph is called the *coordinate system*. *Coordinate* is another name for a numeric scale that divides a graph into regular units. The location of every point on the graph is described by a set of *coordinates*.

The two most common coordinate systems used in radio are the *rectangular-coordinate* system shown in **Figure 4.1** (sometimes called *Cartesian coordinates*) and the *polar-coordinate* system shown in **Figure 4.2**. (Additional information on coordinate systems is available on the Math Supplement page listed on the ECLM website.)

The horizontal line through the center of a rectangular coordinate graph is the *X axis*.

Electrical Principles 4-1

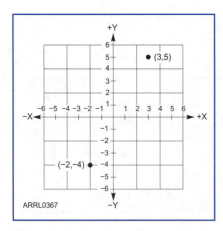

Figure 4.1 — Rectangular-coordinate graphs use a pair of axes at right angles to each other, each calibrated in numeric units. Any point on the resulting grid can be expressed in terms of its horizontal (X) and vertical (Y) values, called coordinates.

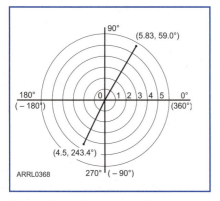

Figure 4.2 — Polar-coordinate graphs use a radius from the origin and an angle from the 0° axis to specify the location of a point. Thus, the location of any point can be specified in terms of a radius and an angle.

The vertical line through the center of the graph is the *Y axis*. Every point on a rectangular coordinate graph has two coordinates that identify its location, X and Y, also written as (X,Y). Every different pair of coordinate values describes a different point on the graph. The point at which the two axes cross — where the numeric values on both axes are zero — is called the *origin*, written as (0,0).

Coordinates with positive values of X and Y lie to the right of and above the origin, respectively. In Figure 4.1, the point with coordinates of (3,5) is located 3 units to the right of the origin along the X axis and 5 units above the origin along the Y axis. Another point at (−2,−4) is found 2 units to the left of the origin along the X axis and 4 units below the origin along the Y axis. Don't confuse the "X" that refers to position along the X-axis with the "X" that refers to reactance.

In the polar-coordinate system, points on the graph are also described by a pair of numeric values called *polar coordinates*. In this case we use a length, or *radius*, measured from the origin, and an *angle* from 0° to 360° measured counterclockwise from the 0° line as shown in Figure 4.2. The symbol r is used for the radius and θ for the angle. A number in polar coordinates is written $r\angle\theta$. So the two points described in the last paragraph could also be written as (5.83, ∠59.0°) and (4.5, ∠243.4°) and are drawn as polar coordinates in Figure 4.2. Unlike maps, the mathematical convention for the 0° direction is to the right and angle increases counterclockwise.

A negative angle essentially means, "turn the other way." With positive angles measured counterclockwise from the 0° axis, the polar coordinates of the point at lower left in Figure 4.2 would be (4.5, ∠−116.6°). When you encounter a negative value for the angle, it means to measure the angle clockwise from 0°. For example, −270° is equivalent to 90°; −90° is equivalent to 270°; 0° and −360° are equivalent; and +180° and −180° are equivalent. An angle can also be specified in *radians* (1 radian = 360 / 2π = 57.3 degrees) but all angles are in degrees in this book and on the exam.

In electronics, it's common to use both the rectangular and polar-coordinate systems when dealing with impedance problems. The examples in the next few sections of this book should help you become familiar with these coordinate systems and the techniques for changing between them.

COMPLEX COORDINATES

So far in your radio career, you've dealt exclusively with *real numbers* such as π (pi), 5 Ω, 2.5 mH, or 53.2 MHz. In solving equations that describe phase and angles, however, you will encounter numbers that contain the square root of minus one ($\sqrt{-1}$). Numbers based on $\sqrt{-1}$ are called *imaginary numbers* to distinguish them from the real numbers. For convenience, $\sqrt{-1}$ is represented as j in electronics. For example, $2j$, $0.1j$, $7j/4$, and $457.6j$ are all imaginary numbers. (Mathematicians use i for the same purpose, but i is used to represent current in electronics.) j also has another interesting property that you'll use: $1/j = -j$. Imaginary numbers are used when describing phase, rotation, or waveforms that change with time.

Real and imaginary numbers can be combined by using addition or subtraction. Combining real and imaginary numbers creates a hybrid called a *complex number*, such as $1 + j$ or $6 - j7$. (The convention in complex numbers is for j to be first in the imaginary part of the

Working With Polar and Rectangular Coordinates

Complex numbers representing electrical quantities can be expressed in either rectangular form $(a + jb)$ or polar form $(r \angle \theta)$. Adding complex numbers is easiest in rectangular form:

$(a + jb) + (c + jd) = (a + c) + j(b + d)$

Multiplying and dividing complex numbers is easiest in polar form:

$a\angle\theta_1 \times b\angle\theta_2 = (a \times b)\,(\theta_1 + \theta_2)$

and

$$\frac{a\angle\theta_1}{b\angle\theta_2} = \left(\frac{a}{b}\right) \angle(\theta_1 - \theta_2)$$

Converting from one form to another is useful in some kinds of calculations. For example, to calculate the value of two complex impedances in parallel you use the formula

$$Z = \frac{Z_1 Z_2}{Z_1 + Z_2}$$

To calculate the numerator $(Z_1 Z_2)$ you would write the impedances in polar form. To calculate the denominator $(Z_1 + Z_2)$ you would write the impedances in rectangular form. So you need to be able to convert back and forth from one form to the other. There is a good explanation of this process, with examples, on the **www.intmath.com/Complex-numbers/4_Polar-form.php** web page.

Here is the short procedure you can save for reference:

To convert from rectangular $(a + jb)$ to polar form $(r \angle \theta)$:

$$r = \sqrt{(a^2 + b^2)}$$

$$\theta = \tan^{-1}\left(\frac{b}{a}\right)$$

To convert from polar to rectangular form:

$a = r \cos \theta$

$b = r \sin \theta$

Many calculators have polar-rectangular conversion functions built-in and they are worth learning how to use. Be sure that your calculator is set to the angle units you prefer, radians or degrees. All calculations in this book and on the exam use units of degrees for angles.

Example

Convert $3 \angle 30°$ to rectangular form:

$a = 3 \cos 30° = 3\,(0.866) = 2.6$

$b = 3 \sin 30° = 3\,(0.5) = 1.5$

$3 \angle 30° = 2.6 + j1.5$

Example

Convert $0.8 + j0.6$ to polar form:

$$r = \sqrt{(0.8^2 + 0.6^2)} = 1$$

$$\theta = \tan^{-1}\left(\frac{0.6}{0.8}\right) = 36.8°$$

$0.8 + j0.6 = 1 \angle 36.8°$

Electrical Principles

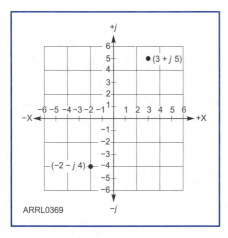

Figure 4.3 — The Y axis of a complex-coordinate graph represents the imaginary portion of complex numbers. This graph shows the same numbers as in Figure 4.1, graphed as complex numbers.

number.) These numbers come in very handy in radio, describing impedances, relationships between voltage and current, and many other phenomena.

If the complex number is broken up into its real and imaginary parts, those two numbers can also be used as coordinates on a graph using *complex coordinates*. This is a special type of rectangular-coordinate graph that is also referred to as the *complex plane*. By convention, the X axis coordinates represent the real number portion of the complex number and the Y axis coordinates represent the imaginary portion. For example, the complex number 6 – j7 would have the same location as the point (6,–7) on a rectangular-coordinate graph. **Figure 4.3** shows the same points as Figure 4.1, but now they are representing the complex numbers 3 + j5 and –2 – j4, respectively.

4.2 Electrical and Magnetic Fields

Electrical and magnetic energy are invisible — you can't detect them with any of your senses. All you can do is observe their effects such as when a resistor gets hot, a motor spins, or an electromagnet picks up iron or steel. The energy exists as a *field* — a region of space in which energy is stored and through which electrical and magnetic forces act. The metric system's basic unit for electrical and magnetic energy is the *joule* (pronounced "jewel") and abbreviated J. (For serious reading on the nature of fields, see **en.wikipedia.org/wiki/Electric_field** and **en.wikipedia.org/wiki/Magnetic_field**.)

You are already quite familiar with fields in the form of gravity. You are being pulled down toward the Earth as you read this (even astronaut-hams up there studying for their Extra exam) because you are in the Earth's *gravitational field*. Because your body has mass it interacts with the gravitational field in such a way that the Earth attracts you. (You have your own gravitational field, too, but many orders of magnitude smaller than that of the Earth.) Think of a bathroom scale as a "gravitational voltmeter" that instead of reading "volts," reads "pounds." The heavier something is, the stronger the Earth attracts it. Weight is the same as force. (Metric scales provide readings in kilograms, a unit of mass. To do so, the scales assume a standard strength for gravity in order to convert weight (a force) to an equivalent mass in kilograms.)

This field makes you do work, such as when you climb stairs. Work has a precise definition when it comes to fields: Work equals force times distance moved in the direction of the field's force. For example, let's say you pick up a mass — a stone that weighs 1 pound — and lift it to a shelf 10 feet above where it previously lay. How much work did you do? You moved a weight of 1 pound a distance of 10 feet against the attraction of the field, so you have done 10 foot-pounds of work. (It doesn't count if you move the stone sideways instead of vertically.)

What did that work accomplish? You stored gravitational energy in the stone equal to the amount of work that you performed! This stored energy is called *potential energy*, whether gravitational, electrical or magnetic. You could store the same amount of gravitational energy by lifting a 10-pound stone 1 foot or by lifting a stone that weighs 1/10th of a pound 100 feet. If you drop the stone (or it falls off the shelf), the same amount of potential energy is converted back to *kinetic energy* as the stone moves toward the Earth in the gravitational field.

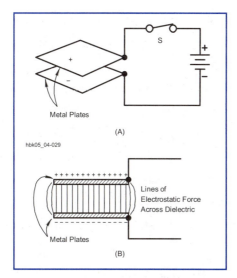

Figure 4.4 — A capacitor shows how an electric field is creating by causing an imbalance of electric charge between its electrodes. The direction of the field is from the positive to negative electrode.

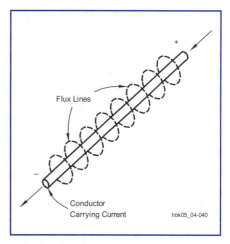

Figure 4.5 — The magnetic field around the current flowing in a wire. The field's orientation is determined by the direction of current.

In electronics we are interested in two types of fields: *electric fields* and *magnetic fields*. Electric fields can be detected as voltage differences between two points. The electric field's analog to gravitational mass is electric charge. Every electric charge has its own electric field, just as every mass has its own gravitational field. The more charged a body is, the "heavier" it is in terms of an electric field. Just as a body with mass feels a force to move in a gravitational field, so does an electric charge in an electric field.

Electrical energy is stored by moving electrical charges apart so that there is a voltage between them. **Figure 4.4** shows the electric field in a capacitor that is created by the imbalance of positive and negative charge. If the field does not change with time, it is called an *electrostatic field*. By convention, the direction of an electric field is assigned to be from positive charge to negative charge.

Magnetic energy is detected by its effects on moving electrical charges or current. Magnetic energy is stored through the motion of electric charge (current) creating a magnetic field. **Figure 4.5** shows the magnetic field around the current flowing through a wire. Magnetic fields that don't change with time, such as from a stationary permanent magnet, are called *magnetostatic fields*. The direction of the magnetic field is discussed in the following section.

Electric or magnetic potential energy is released by allowing the charges to move in the field. For example, electric energy is released when current flows from the charged-up capacitor in Figure 4.4. Magnetic energy stored by current flowing in an inductor is released when the current is allowed to change, such as when a relay or motor is turned off.

If the electric or magnetic fields are changing, they produce an *electromagnetic wave* that travels through space. Radio waves are electromagnetic, or EM, waves. They are described in more detail in the chapter on Topics in Radio Propagation.

Electrical Principles

4.3 Principles of Circuits

This section of the book covers the fundamentals of how electrical circuits work. We'll get into the relationship between voltage and current when inductance and capacitance are involved — that's when things get interesting! Understanding these basic ideas leads you directly to resonance, tuned circuits, Q, and all sorts of great radio know-how.

Remember that this book doesn't attempt to be an electronic textbook — there are plenty of good references available for that job. If you find yourself missing some crucial background, step back and read through one of the references on the ECLM website.

RC AND RL TIME CONSTANTS

E5B01 — What is the term for the time required for the capacitor in an RC circuit to be charged to 63.2% of the applied voltage or to discharge to 36.8% of its initial voltage?

E5B04 — What is the time constant of a circuit having two 220 microfarad capacitors and two 1 megohm resistors, all in parallel?

E5D06 — In what direction is the magnetic field oriented about a conductor in relation to the direction of electron flow?

In electronic circuits, electric and magnetic energy are constantly being stored and released by various components. This exchange is observed as the relationship between voltage and current in components that store electrical and magnetic energy.

Electrical Energy Storage

Electrical energy can be stored in a capacitor by applying a dc voltage across its terminals as shown in Figure 4.4. **Figure 4.6** illustrates a simple circuit for charging and discharging a capacitor. Assuming no energy was already stored in the capacitor, there will be an instantaneous inrush of current as charge moves into one capacitor terminal and out of the other when S1 is moved to the A position. The only limit on current is the resistance of the voltage source, the wires connecting the capacitor and the capacitor's internal conducting electrodes (sometimes called "plates"). The capacitor builds up a voltage (indicating electric field strength) as one set of electrodes accumulates an excess of electrons and the other set loses an equal number. Eventually, the voltage at the capacitor terminals is equal to the source voltage and the current ceases to flow.

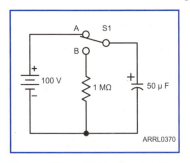

Figure 4.6 — A simple circuit for charging a capacitor (S1 in A position) and then discharging it through a resistor (S1 in B position).

If the voltage source is disconnected, the capacitor will remain charged at that voltage. The charge will stay on the capacitor electrodes as long as there is no path for the electrons to travel from one terminal to the other. The voltage across the capacitor is an indicator of how much electrical energy is stored in the electric field inside the capacitor. Since the charges (electrons) are no longer moving, the energy is stored in the electrostatic field between the electrodes. If a resistor is then connected across the capacitor terminals by moving S1 to the B position, the stored energy will be released as current flows through the resistor, dissipating the stored energy as heat.

Note that the discussion in this section deals with ideal components, such as resistors that have no stray capacitance or inductance associated with the leads or composition of the resistor itself. Ideal capacitors exhibit no losses and there is no resistance in the leads or capacitor electrodes. Ideal inductors are made of wire that has no resistance and there is no stray capacitance between turns. In practice there are no ideal components, so the behavior described here is modified a bit in real-life circuits. Even so, components can come pretty close to the ideal conditions. For example, a capacitor with very low leakage can hold a charge for days or even weeks.

Magnetic Energy Storage

When electrons flow, usually through a conductor, a magnetic field is produced. This magnetic field exists in the space around the conductor, and magnetic energy is stored in this space. The field can be detected by bringing a compass near a wire carrying dc current and watching the needle deflect. In fact, this is how the relationship between electrical current and magnetism was discovered!

Figure 4.7 illustrates the magnetic field around a wire connected to a battery. If the wire is wound into a coil so that the fields from adjacent turns add together, then a much stronger magnetic field can be produced. The direction of the magnetic field is a circle around the current at right angles to the current flow and can be determined by the *left-hand rule*. [**E5D06**] (See the sidebar "Magnetic Field Direction" for more about magnetic field polarity and direction.)

The use of the left-hand rule is required because we are referring to *electronic current* (the flow of electrons) and not to the more common *conventional current* (the flow of positive charge) that flows in the opposite direction. If electronic current is being discussed, use the left-hand rule and if conventional current, use the *right-hand rule*. The exam question refers to the flow of electrons and electronic current, so use the left-hand rule.

The strength of the magnetic field is proportional to the amount of current. Electrical energy from the voltage source is transferred to the magnetic field as the field is created. So we are storing energy by building up a magnetic field and that means work must be done against a voltage (also called an *electromagnetic force* or EMF) that is *induced* (created) in the circuit whenever the magnetic field (or current) is changing. The induced voltage is called a *back EMF* since it acts to oppose any change in the amount of current.

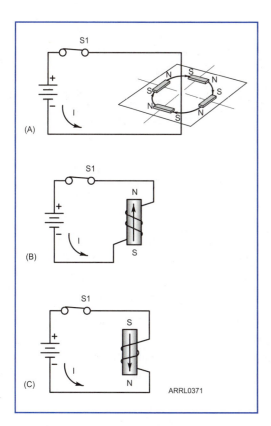

Figure 4.7 — A circuit showing the direction of a magnetic field around a straight wire (A) and in two coils wound in opposite directions (B,C). Note that electronic current (negative to positive) is used, not conventional current (positive to negative).

Magnetic Field Direction

How can you tell what direction a magnetic field is "pointing" or do they even point? Yes, they certainly do have a direction — think of the north and south poles of a magnet! For a permanent magnet, determining direction is pretty easy. All you have to do is observe the way the magnet wants to move when placed near another permanent magnet. Opposite poles (north-south and south-north) will attract each other.

In the case of magnetic fields created by current flowing, it's not so easy. Figure 4.7 shows how a magnetic field is oriented based on the direction of the electrons flowing in the circuit. (The figure shows *electronic current* — the true motion of the electrons from negative to positive. Most electronics uses *conventional current* that flows from positive to negative, the exact opposite of electronic current.) In order to find the direction of the field, the *left-hand rule* is used as described at **www.studyphysics.ca/30/Solenoidelectro.pdf**. You just have to know how the inductor is wound and in which direction the current is flowing.

Electrical Principles

Time Constant

If you connect a dc voltage source directly to a capacitor through a resistance, the higher the resistance's value, the longer it will take to charge the capacitor because the resistor limits current flow in the circuit. **Figure 4.8** shows an *RC circuit* (RC means resistor-capacitor) that alternately charges and discharges a capacitor (C) through a resistor (R). With the switch in position A, current through the resistor charges the capacitor to the battery voltage. When the switch is moved to position B, the capacitor returns its stored energy to the circuit as a current through the resistor. The amount of time it takes to charge the capacitor to a certain percentage of the applied voltage or discharge to a certain percentage of its maximum voltage is called the circuit's *time constant*.

The same general process applies to an *RL circuit* (RL means resistor-inductor) with an inductor in series with a resistor as in **Figure 4.9**. Instead of voltage increasing toward some maximum value when voltage is applied, it is the current through the inductor that increases. The time constant for this type of circuit is the amount of time it takes inductor current to increase or decrease to a specific percentage of its maximum or minimum value.

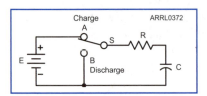

Figure 4.8 — The capacitor in this circuit charges and discharges through the resistor, illustrating the principle of an RC circuit's time constant.

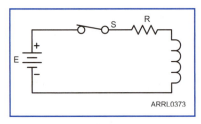

Figure 4.9 — The inductor current flowing through the resistor increases and decreases as voltage is applied, illustrating the principle of an RL circuit's time constant.

RC Circuit Time Constant Calculations

The time constant for a simple RC circuit as shown in Figure 4.8 is:

$$\tau = RC \qquad \text{(Equation 4.1)}$$

where:
τ is the Greek letter tau, used to represent the time constant.
R is the total circuit resistance in ohms.
C is the capacitance in farads.

Note that if R is in megohms and C is in microfarads, then τ is in seconds! Remember "megohms times microfarads equals seconds" and it will save you a lot of calculating time.

The capacitor charges and discharges according to an equation known as an *exponential curve*. **Figure 4.10** illustrates the charge and discharge curves, where the time axis is shown in terms of τ and the vertical axis is expressed as a percentage of the applied voltage. These graphs are true for any RC circuit.

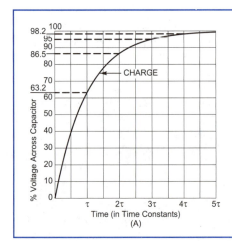
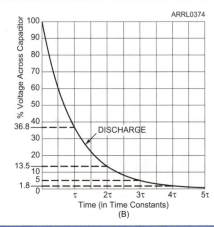

Figure 4.10 — The graph at A shows how the voltage across a capacitor rises with time when charged through a resistor as in Figure 4.6. Graph B shows how the voltage decreases with time as the capacitor is discharged through a resistor. From a practical standpoint, a capacitor is considered to be charged or discharged after a period of 5τ.

4-8 Chapter 4

If you know the time constant and the applied voltage, you can calculate the voltage on the capacitor at any instant of time. For a charging capacitor:

$$V(t) = E\,(1 - e^{-t/\tau}) \qquad\qquad \text{(Equation 4.2)}$$

where:

V(t) is the voltage across the capacitor at time t.
E is the applied voltage (the battery voltage in Figure 4.6)
t is the time in seconds since the capacitor began charging or discharging.
e is the base for natural logarithms, 2.718.
τ is the time constant for the circuit, in seconds.

If the capacitor is discharging from E volts, we have to write a slightly different equation:

$$V(t) = E\,(e^{-t/\tau}) \qquad\qquad \text{(Equation 4.3)}$$

These equations can be solved fairly easily with an inexpensive calculator that is able to work with natural logarithms (it will have a key labeled LN or LN X). In that case you could calculate the value for $e^{-t/\tau}$ as the inverse natural log of $-t/\tau$, written as $\ln^{-1}(-t/\tau)$. (For more information on natural logarithms and exponential functions, refer to the math supplement on the ECLM website.)

Actually, you do not have to know how to solve these equations if you are familiar with the results at a few important points. We'll show you how to use the solutions to the equations at these points as short cuts to most problems associated with time constants.

As shown on the graphs of Figure 4.10, it is common practice to think of charge or discharge time in terms of multiples of the circuit's time constant. If we select times of zero (starting time), one time constant (1τ), two time constants (2τ), and so on, then the exponential term in Equations 4.2 and 4.3 simplifies to e^0, e^{-1}, e^{-2}, e^{-3} and so forth. Then we can solve the equations for those values of time. Let's pick a value for battery voltage of E = 100 V so that the answers will be in the form of a percentage of any applied voltage.

V(0)	=	100 V $(1 - e^0)$	=	100 V $(1 - 1)$	=	0 V, or 0%
V(1τ)	=	100 V $(1 - e^{-1})$	=	100 V $(1 - 0.368)$	=	63.2 V, or 63.2%
V(2τ)	=	100 V $(1 - e^{-2})$	=	100 V $(1 - 0.135)$	=	86.5 V, or 86.5%
V(3τ)	=	100 V $(1 - e^{-3})$	=	100 V $(1 - 0.050)$	=	95.0 V, or 95%
V(4τ)	=	100 V $(1 - e^{-4})$	=	100 V $(1 - 0.018)$	=	98.2 V, or 98.2%
V(5τ)	=	100 V $(1 - e^{-5})$	=	100 V $(1 - 0.007)$	=	99.3 V, or 99.3%

After a time equal to five time constants has passed, the capacitor is charged to 99.3% of the applied voltage. This is fully charged for all practical purposes.

The equation used to calculate the capacitor voltage while it is discharging is slightly different from the one for charging. For values of time equal to multiples of the circuit time constant, the solutions to Equation 4.3 have a close relationship to those for Equation 4.2.

t = 0,	e^0	=	1,	so V(0)	=	100 V,	or 100%
t = 1τ,	e^{-1}	=	0.368,	so V(1τ)	=	36.8 V,	or 36.8%
t = 2τ,	e^{-2}	=	0.135,	so V(2τ)	=	13.5 V,	or 13.5%
t = 3τ,	e^{-3}	=	0.050,	so V(3τ)	=	5 V,	or 5%
t = 4τ,	e^{-4}	=	0.018,	so V(4τ)	=	1.8 V,	or 1.8%
t = 5τ,	e^{-5}	=	0.007,	so V(5τ)	=	0.7 V,	or 0.7%

Here we see that after a time equal to five time constants has passed, the capacitor has discharged to less than 1% of its initial value. This is fully discharged for all practical purposes.

From the calculations for a charging capacitor we can define the time constant of an RC circuit as the time it takes to charge the capacitor to 63.2% of the supply voltage. [E5B01] From the calculations of a discharging capacitor we can also define the time constant as the time it takes to discharge the capacitor to 36.8% of its initial voltage.

Electrical Principles 4-9

Another way to think of these results is that the discharge values are the complements of the charging values. Subtract either set of percentages from 100 and you will get the other set. You may also notice another relationship between the discharging values. If you take 36.8% (0.368) as the value for one time constant, then the discharged value is $0.368 \times 0.368 = 0.368^2 = 0.135$ after two time constants, $0.368^3 = 0.05$ after three time constants, $0.368^4 = 0.018$ after four time constants and $0.368^5 = 0.007$ after five time constants. You can change these values to percentages, or just remember that you have to multiply the decimal fraction times the applied voltage. If you subtract these decimal values from 1, you will get the values for the charging equation. In either case, by remembering the percentage 63.2% you can generate all of the other percentages without logarithms or exponentials!

In many cases, you will want to know how long it will take a capacitor to charge or discharge to some particular voltage. Probably the easiest way to handle such problems is to first calculate what percentage of the maximum voltage you are charging or discharging to. Then compare that value to the percentages listed for either charging or discharging the capacitor. Often you will be able to approximate the time as some whole number of time constants.

Suppose you have a 0.01-μF capacitor and a 2-MΩ resistor wired in parallel with a battery. The capacitor is charged to 20 V, and then the battery is removed. How long will it take for the capacitor to discharge through the resistor to reach a voltage of 7.36 V? First, calculate the percentage decrease in voltage:

$$\frac{7.36\ V}{20\ V} = 0.368 = 36.8\%$$

You should recognize this as the value for the discharge voltage after 1 time constant. Now calculate the time constant for the circuit using Equation 4.1.

$$\tau = RC = (2 \times 10^6\ \Omega) \times (0.01 \times 10^{-6}\ F) = 0.02\ \text{second}$$

It will take 0.02 second, or 20 milliseconds to discharge the capacitor to 7.36 V.

The circuit shown in Figure 4.8 is a series circuit. It is common to have a circuit with several resistors and capacitors connected either in series or parallel. If the components are wired in series we can still use Equation 4.1, but we must first combine all of the resistors into one equivalent resistor, and all of the capacitors into one equivalent capacitor. Then calculate the time constant using Equation 4.1, as before.

If the components are connected in parallel, there is an added complication when the circuit is charging, but for a discharging circuit you can still calculate a time constant. Again combine all of the resistors and all of the capacitors into equivalent values, and calculate the time constant using Equation 4.1. (If you have forgotten how to combine resistors and capacitors in series and parallel, review the appropriate sections of *The Ham Radio License Manual* or *The ARRL Handbook.*)

RC Circuit Examples

Let's look at an example of calculating the time constant for a circuit like the one in Figure 4.8, using values of 220 μF and 470 kΩ for C and R. To calculate the time constant, τ, multiply the R and C values, in ohms and farads.

$$\tau = RC = (470 \times 10^3\ \Omega) \times (220 \times 10^{-6}\ F) = 103.4\ \text{seconds}$$

You can calculate the time constant for any RC circuit in this manner.

If you have two 100-μF capacitors and two 470-kΩ resistors, all in series, first combine the resistor values into a single resistance and the capacitor values into a single capacitance.

4-10   Chapter 4

$$R_T \text{ (series)} = R_1 + R_2 = 470 \text{ k}\Omega + 470 \text{ k}\Omega = 940 \text{ k}\Omega = 940 \times 10^3 \text{ }\Omega$$

$$C_T \text{(series)} = \frac{1}{(1/C_1 + 1/C_2)} = \frac{1}{2/100} = 50 \text{ }\mu F = 50 \times 10^{-6} \text{ F}$$

Then the time constant is:

$$\tau = RC = (940 \times 10^3 \text{ }\Omega) \times (50 \times 10^{-6} \text{ F}) = 47 \text{ seconds}$$

Suppose you have two 220-μF capacitors and two 1-MΩ resistors all in parallel. Again, first combine the values into a single resistance and a single capacitance.

$$R_T \text{ (parallel)} = \frac{R_1 \times R_2}{R_1 + R_2} = \frac{1 \text{ M}\Omega \times 1 \text{ M}\Omega}{1 \text{ M}\Omega + 1 \text{ M}\Omega} = 0.5 \text{ M}\Omega = 5 \times 10^5 \text{ }\Omega$$

$$C_T \text{ (parallel)} = C_1 + C_2 = 220 \text{ }\mu F + 220 \text{ }\mu F = 440 \text{ }\mu F = 440 \times 10^{-6} \text{ F}$$

Then the time constant is:

$$\tau = RC = (5 \times 10^5 \text{ }\Omega) \times (440 \times 10^{-6} \text{ F}) = 220 \text{ seconds } [\textbf{E5B04}]$$

Suppose you have a 450-μF capacitor and a 1-MΩ resistor wired in parallel with a power supply. The capacitor is charged to 800 V, and then the power supply is removed. How long will it take for the capacitor to discharge to 294 V? First, calculate the percentage decrease in voltage:

$$\frac{294 \text{ V}}{800 \text{ V}} = 0.368 = 36.8\%$$

This is the value for the discharge voltage after one time constant. Now calculate the time constant for the circuit using Equation 4.1.

$$\tau = RC = (1 \times 10^6 \text{ }\Omega) \times (450 \times 10^{-6} \text{ F}) = 450 \text{ seconds}$$

Or you could have recalled "megohms times microfarads equals seconds" and made the calculation that way.

RL Circuit Time Constant Calculations

When resistance and inductance are connected in series there is a situation similar to what happens in an RC circuit. Figure 4.9 shows a circuit for storing magnetic energy in an inductor. When the switch is closed, a current will try to flow immediately. The instantaneous transition from no current to a value limited only by the voltage source and resistance represents a very large change in current and a back EMF is developed by the inductance. The back EMF is proportional to the rate of change of the current and its polarity is opposite to that of the applied voltage, meaning that it will oppose the change in current. The result is that the initial current is very small but increases quickly, gradually approaching the final current value given by Ohm's Law (I = E / R) as the back EMF decreases toward zero.

Figure 4.11 shows how the current through the inductor of Figure 4.9 increases as time passes. At any given instant, the back EMF will be equal to the difference between the voltage drop across the resistor and the battery voltage. You can see that when the switch is closed initially and there is no current, the back EMF is equal to the full battery voltage. Later on, the current will increase to a steady value and

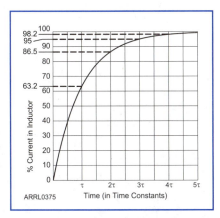

Figure 4.11 — This graph shows the current buildup in an RL circuit. Notice that the curve is identical to the voltage curve for a charging capacitor.

there will be no voltage drop across the inductor. The full battery voltage then appears across the resistor and the back EMF is zero. In practice, the current is essentially equal to the final value after 5 time constants. The curve looks just like the one we found for a charging capacitor.

The time constant depends on the circuit components, as for the RC circuit. For an RL circuit, the time constant is given by:

$$\tau = \frac{L}{R}$$
(Equation 4.4)

The equation for inductor current is another exponential curve, with an equation similar to Equation 4.2.

$$I(t) = \frac{E}{R}(1 - e^{-t/\tau})$$
(Equation 4.5)

where:

I(t) is the current in amperes at time t.
E is the applied voltage.
R is the circuit resistance in ohms.
t is the time in seconds after the switch is closed.
τ is the time constant for the circuit in seconds.

If we choose values of time equal to multiples of the circuit time constant, as we did for the RC circuit, then we will find that the current will build up to its maximum value in the same fashion as the voltage does when a capacitor is being charged. This time let's pick a value of 100 A for the maximum current, so that our results will again come out as a percentage of the maximum current for any RL circuit.

t = 0,	e^{-0}	=	1,	so I(0)	=	100 A (1 − 1)	=	0 A,	or	0%
t = 1τ,	e^{-1}	=	0.368,	so I(1τ)	=	100 A (1 − 0.368)	=	63.2 A,	or	63.2%
t = 2τ,	e^{-2}	=	0.135,	so I(2τ)	=	100 A (1 − 0.135)	=	86.5 A,	or	86.5%
t = 3τ,	e^{-3}	=	0.050,	so I(3τ)	=	100 A (1 − 0.050)	=	95.0 A,	or	95%
t = 4τ,	e^{-4}	=	0.018,	so I(4τ)	=	100 A (1 − 0.018)	=	98.2 A,	or	98.2%
t = 5τ,	e^{-5}	=	0.007,	so I(5τ)	=	100 A (1 − 0.007)	=	99.3 A,	or	99.3%

Notice that the current through the inductor will increase to 63.2% of the maximum value during 1 time constant. After 5 time constants, the current is within 1% of the maximum value.

PHASE ANGLE

E5B09 — What is the relationship between the AC current through a capacitor and the voltage across a capacitor?

E5B10 — What is the relationship between the AC current through an inductor and the voltage across an inductor?

Having learned that voltage and current don't rise and fall together in capacitors and inductors, you're ready to look at the situation when ac voltage is applied, instead of dc. The storage and release of electrical and magnetic energy is what creates the opposition to flow of ac current, the property of capacitors and inductors called *reactance*. (You learned about reactance when studying for your General class exam.)

To understand how ac voltage and current behave in inductors and capacitors, we'll take a look at the amplitudes of ac signals throughout the complete cycle. The relationship between the current and voltage waveforms at a specific instant is expressed as the phase of the waveforms. Phase essentially means time or the time interval between two events taking place. The event that occurs first is said to *lead* the second, while the trailing event *lags* the first.

4-12   Chapter 4

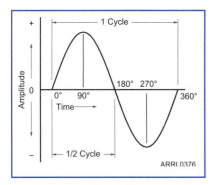

Figure 4.12 — An ac cycle is divided into 360° that are used as a measure of time or phase. Each degree corresponds to 1/360 of the cycle's period.

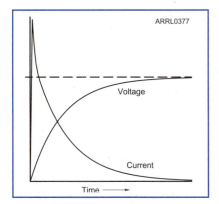

Figure 4.13 — This graph illustrates how the voltage across a capacitor changes as it charges with a dc voltage applied. The charging current is also shown.

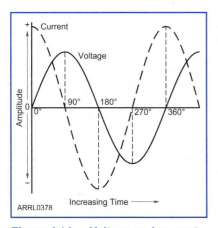

Figure 4.14 — Voltage and current phase relationships when an ac voltage is applied to a capacitor.

Since all ac cycles of the same frequency have the same *period* (or length in time), we can use the cycle's period as a basic unit of time. This makes phase measurement independent of the waveform frequency, relating only to position relative to the waveform's cycle. If two or more different frequencies are being considered, phase measurements are usually made with respect to the lowest frequency.

It is convenient to relate one complete cycle of the wave to a circle, and to divide the cycle into 360 degrees. **Figure 4.12** shows one complete cycle of a sine-wave voltage or current, with the wave broken into four quarters of 90° each. Each degree corresponds to 1/360 of the cycle's period. So a phase measurement is usually specified as an angle. Because we know the period of the waveform, we can convert degrees to time.

The *phase angle* between two waveforms is a measurement of the offset in time between similar points on each waveform — maximum-to-maximum, zero-crossing-to-zero-crossing, and so on. One of the waveforms is designated as the reference. A leading phase angle is positive and a lagging phase angle is negative.

AC Voltage-Current Relationship in Capacitors

Figure 4.13 shows the voltage across a capacitor as it charges and the charging current that flows into a capacitor with a dc voltage applied. As soon as a voltage is applied across an ideal capacitor, there is a sudden inrush of current as the capacitor begins to charge. That current tapers off as the capacitor is charged to the full value of applied voltage. By the time the applied voltage is reaching a maximum, the capacitor is also reaching full charge, and so the current into the capacitor goes to zero. A maximum amount of energy has been stored in the electric field of the capacitor at this point.

The situation is different when an ac voltage is applied because the applied voltage is not constant. **Figure 4.14** graphs the relative current and voltage amplitudes when an ac sine wave signal is applied. The scale does not represent specific current or voltage values. Here's what the graph of the two waveforms is telling us during intervals of one-quarter cycle of the voltage waveform:

0° to 90° — Voltage is zero, so no energy is stored in the capacitor. The applied voltage begins increasing and a large inrush of charging current occurs, just as is the case for an applied dc voltage. Current slows as more energy is stored in the capacitor.

90° to 180° — Applied voltage has reached a peak, so no additional charge flows into the capacitor and current flow stops — stored energy is at a maximum. As voltage begins to drop, that is the same as discharging the capacitor, so current reverses and energy is returned to the circuit.

180° to 270° — As the voltage reaches zero it is dropping at its fastest rate, so the discharge current in the reverse direction is at a maximum. Now the applied voltage is increasing again but with the opposite polarity. Energy is being stored in the capacitor again but with the voltage reversed. Charging current is now in the opposite direction, too, but decreases as more energy is stored in the capacitor.

270° to 360° — Once again, applied voltage has reached a peak but with reverse polarity. Charging current ceases as the voltage peaks and

Electrical Principles 4-13

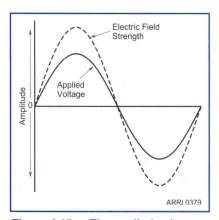

Figure 4.15 — The applied voltage, electric field strength, and stored energy in a capacitor are in phase.

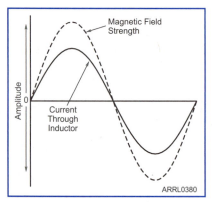

Figure 4.16 — The inductor current, magnetic field strength, and stored energy in an inductor are in phase.

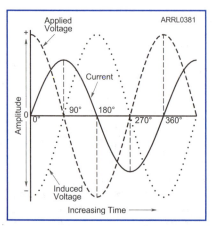

Figure 4.17 — Phase relationships between voltage and current when an alternating voltage is applied to an inductor.

begins to drop, repeating the situation between 180° and 270° but with the opposite polarity. When 360° arrives, voltage and current have the same relationship as at 0° and the cycle begins again.

Note that energy is stored in and discharged from the capacitor twice during each cycle — once with positive voltage across the capacitor and once with negative voltage. The current waveform describes electrons flowing in and out of the capacitor in response to the applied voltage. Energy storage is at a peak when voltage is maximum as shown in **Figure 4.15**. This occurs at 90° and 270° when current is zero. Note also that current reaches a peak 90° ahead of the voltage waveform. We say that the current through a capacitor leads the applied voltage by 90°. [E5B09] You could also say that the voltage applied to a capacitor lags the current through it by 90°. To help you remember this relationship, think of the word ICE. This will remind you that the current (I) comes before (leads) the voltage (E) in a capacitor (C). By convention, voltage is the reference waveform for phase angle so in a capacitor the phase angle is –90° (negative).

AC Voltage-Current Relationship in Inductors

The relationship between ac voltage and current in an inductor complements that in a capacitor. **Figure 4.16** shows that instead of stored energy being in phase with applied voltage, it is in phase with the inductor current. This causes the phase relationship between voltage and current to be reversed from that of the capacitor.

In the section on magnetic energy, you learned about back EMF. Back EMF is greatest when the magnetic field is changing the fastest. Furthermore, it is generated with a polarity that opposes the change in current or magnetic-field strength. So, when the current is crossing zero on the way to a positive peak, back EMF is at its greatest negative value. When the current is at the positive peak back EMF is zero and so on.

As before, let's examine the situation during each quarter cycle of the applied voltage waveform as shown in **Figure 4.17**. Along with applied voltage and inductor current, the back EMF waveform has been added. This will help explain the relationship between applied voltage and inductor current.

0° to 90° — Beginning at maximum applied voltage, the opposing induced voltage that resists changes in current flow is also at a maximum so current must increase slowly. As applied voltage falls, the change in current is also reduced and so induced voltage also decreases. As applied voltage reaches zero no additional current flows and induced voltage is zero. Stored energy is a maximum at this point.

90° to 180° — Applied voltage begins to increase in the reverse direction causing a reduction in current and stored energy. Induced voltage increases opposing the change in current. When applied voltage reaches a maximum with reverse polarity, current is now completely stopped and stored energy is zero.

180° to 270° — Reversed from the situation between 0° and 90°, current is now increasing in the opposite direction. Applied voltage is falling and so the rate of change of current is also falling, causing induced voltage to fall as well. As applied voltage reaches zero again, current and stored energy has reached a maximum.

270° to 360° — As between 90° and 180°, applied voltage is increasing in the opposite polarity to current, causing current to drop. The change in current also causes induced voltage to rise in opposition to the change in current. As applied voltage reaches a maximum, current and stored energy once again reach zero.

The phase relationship between applied ac voltage and current through an inductor is the opposite from their relationship in a capacitor. Current through an inductor *lags* the applied voltage by 90°. You could also say that the voltage applied to an inductor *leads* the current through it by 90°. [**E5B10**] A useful mnemonic for remembering these relationships is, "ELI the ICE man." The L and C represent the inductor and capacitor, and the E and I stand for voltage and current. Right away you can see that E (voltage) comes before (leads) I (current) in an inductor and that I comes before (leads) E in a capacitor. Using the same convention as for a capacitor, the phase angle in an inductor is 90° (positive).

Combining Reactance with Resistance

Up to this point we have studied the phase relationships between voltage and current only in inductors and capacitors. Actual circuits include resistance, either as a separate component or as part of the inductor or capacitor. This affects the phase angle between the voltage and current waveforms. The voltage across a resistor is in phase with the current through it, so if a circuit contains both resistance and reactance from either an inductance or capacitance, the resulting phase angle of current through all of the components will be less than 90°. The exact phase angle depends on the relative amounts of resistance and reactance in the circuit.

Revisiting reactance for a moment, reactance is defined as the opposition to ac current flow through an inductance or capacitance. A resistor opposes any type of current flow — ac or dc. You've just seen that inductors develop a back EMF that opposes changes in current flow which is the same thing as resisting changes in stored energy. Capacitors also resist changes in energy by opposing changes in voltage across them that would cause current to flow. This opposition to ac current flow is reactance. To combine reactances in series and parallel, use the same equations as when combining resistances.

Inductive reactance (X_L) increases with increasing frequency because as frequency goes up, so does the rate of change of the applied voltage and of inductor current. A higher rate of change increases the back EMF and thus the opposition to current flow. Similarly, higher inductance also increases inductive reactance. The equation for X_L is:

$$X_L = 2\pi f L$$

where:
  X_L is reactance in ohms.
  f is frequency in hertz.
  L is inductance in henrys.

Capacitive reactance increases with decreasing frequency because the longer cycle period means more current will flow, resulting in more energy change during each cycle. More energy change requires the voltage source to overcome wider swings in capacitor voltage and that has the same effect of opposing current flow. Lower capacitance also increases capacitive reactance. The equation for X_C is:

$$X_C = \frac{1}{2\pi f C}$$

where:
  X_C is reactance in ohms.
  f is frequency in hertz.
  C is capacitance in farads.

Electrical Principles 4-15

COMPLEX IMPEDANCE

E5C01 — Which of the following represents capacitive reactance in rectangular notation?

E5C02 — How are impedances described in polar coordinates?

E5C03 — Which of the following represents an inductive reactance in polar coordinates?

E5C04 — What coordinate system is often used to display the resistive, inductive, and/or capacitive reactance components of impedance?

E5C05 — What is the name of the diagram used to show the phase relationship between impedances at a given frequency?

E5C06 — What does the impedance 50–j25 represent?

E5C07 — Where is the impedance of a pure resistance plotted on rectangular coordinates?

E5C08 — What coordinate system is often used to display the phase angle of a circuit containing resistance, inductive and/or capacitive reactance?

E5C09 — When using rectangular coordinates to graph the impedance of a circuit, what do the axes represent?

When a circuit contains both resistance and reactance the combined effect of the two is called *impedance*, symbolized by the letter Z. Impedance is a more general term than either resistance or reactance and is given as a complex number to account for the phase difference between voltage and current.

Writing and Graphing Impedance and Phase Angle

Impedance values are written in rectangular form as $Z = R + jX$ where the value of reactance, X, can be positive (inductive, $+jX$) or negative (capacitive, $-jX$). For example, the impedance $50 - j25 \, \Omega$ consists of $50 \, \Omega$ of resistance and $25 \, \Omega$ of capacitive reactance. **[E5C01, E5C03, E5C06]**

Referring back to Figure 4.1, you can plot complex impedance and the associated phase angle using rectangular coordinates by using the horizontal axis for the value of R and the vertical axis for the value of X. **[E5C04, E5C09]** The *j* indicates that the reactance value (such as $j2$ or $-j3/2$) is a vertical distance along the Y axis.

When plotting impedances using complex coordinates, any point that falls on the horizontal axis from 0° to 180° is a *pure resistance* and has no reactive component. **[E5C07]** Any point that falls on the vertical axis from 90° to –90° (or 270°) is a *pure reactance* and has no resistive component.

Impedance can also be written in polar coordinates (see Figure 4.2) as $|Z|\angle\theta$, where $|Z|$ is the magnitude of the impedance and θ is its phase angle. **[E5C02]** Impedances in polar coordinates are plotted with the right side of the horizontal axis indicating 0°, the top half of the vertical axis indicating 90°, and so forth. **[E5C08]** Remember that polar coordinates and maps have different conventions for angles.

The polar coordinate representation of an impedance or admittance is also called *phasor notation* and the values are called a *phasor*. A *phasor diagram* shows impedance or admittance values plotted using polar coordinates. Remember that reactances change with frequency so a phasor diagram assumes the frequency is the same for all values. **[E5C05]** Phasors are a type of *vector* which is any quantity that has both a magnitude and a direction. In the case of phasors, the direction is the angular coordinate.

Either the rectangular or polar-coordinate system can be used to specify impedance. Choose rectangular coordinates to visualize the resistive and reactive parts. Choose polar coordinates to visualize the magnitude and the phase angle of the impedance.

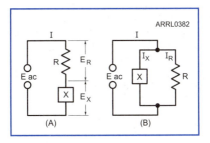

Figure 4.18 — Series and parallel circuits may contain resistance and reactance.

For More Information: Combining Resistance and Reactance

A circuit's reactance and resistance may be connected in series or in parallel, as shown in **Figure 4.18**. In these circuits, the reactance is shown as a box (X), to indicate that it can be either inductive or capacitive. In the series circuit shown at A, current is the same through both elements but with different voltages appearing across the resistance and reactance. In the parallel circuit shown at B, the same voltage is applied to both elements but different currents may flow in the two branches.

You can see that the phase relationship between current and voltage for the whole circuit can be anything between zero and ±90°. The phase angle depends on the relative amounts of resistance and reactance in the circuit.

The simple R + jX value assumes that all of the resistances and reactances have been combined into one equivalent value. If there is more than one resistor in the circuit, you must combine them to get one equivalent resistance value. Likewise, if there is more than one reactive element, they must be combined into one equivalent reactance. If there are several inductors and several capacitors, combine all the like elements and calculate the resulting capacitive and inductive reactance.

Capacitive and inductive reactances resist the flow of ac current in different ways and have opposite phase angles so that they cancel each other. This is reflected in the convention of capacitive reactance being treated as a negative value and inductive reactive as a positive value so that adding them together results in a smaller total reactance. If the two opposing reactances have equal values, the resulting cancellation means no reactance is present in the circuit at all!

Combining resistance and reactance is a little more complicated. When the resistance and reactance are in series, the two values can be combined in a relatively straightforward manner. The current is the same in all parts of the circuit ($I = I_R = I_X$), and the voltage is different across each component. Remembering that Z is a complex number, we can write an equation for the impedance in the form:

$$Z = \frac{E}{I} = \frac{E_R + jE_X}{I} = \frac{E_R}{I} + j\frac{E_X}{I}$$

(Equation 4.6)

This is really just Ohm's Law written for impedance instead of resistance, as we are used to seeing it. This equation also shows that we can consider the voltage and current associated with the resistive and reactive elements separately.

Can we say that Z = R + X? No! Because the phase angle of the current is different in the resistance (phase angle = 0°) and reactance (phase angle = ±90° as indicated by the j).

To find the actual impedance of the combined R and X, we need to take phase angle into account. This can be done graphically. Start by drawing the axes for a rectangular coordinate graph as shown in **Figure 4.19**. Resistance values correspond to the X axis and reactance values to the Y axis.

Start by assuming a current of 1 A flowing in the circuit of Figure 4.19A, so the voltage and impedance have the same numeric values of ohms and volts. (If I = 1 A, then numerically Z = E/1.)

Current is the same in both components so use the voltage across the

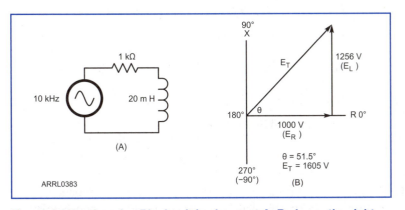

Figure 4.19 — A series RL circuit is shown at A. B shows the right triangle used to calculate the phase angle between the circuit current and voltage. The graph also shows the resulting impedance of the series RL circuit.

Electrical Principles 4-17

resistor as a reference and draw the voltage across the reactance in the direction of positive reactance on the graph. It is helpful to remember that the reason we label inductive reactance as + and capacitive reactance as – is because of this leading and lagging current-voltage relationship.

In the circuit of Figure 4.19, R = 1000 Ω so the voltage across it will be 1000 V. Draw this voltage as a line from the origin along the 0° axis (labeled "R 0°") to the point (1000,0). L = 20 mH and f = 10 kHz so,

$$X_L = 6.28 \times (20 \times 10^{-3}\,H) \times (10 \times 10^3\,Hz) = 1256\,\Omega.$$

Draw the voltage across $X_L = E_L = 1\,A \times 1256\,\Omega = 1256\,V$ as a line pointing straight up in the 90° direction. The line should be parallel to the reactance axis (labeled "90° X") from the end of the previous line to the point (1000,1256). Because the reactance is inductive, the phase angle between voltage and current is 90° and the voltage line extends upward from the X axis. If the reactance had been capacitive the phase angle would have been –90° and the voltage line would extend down from the X axis.

Complete the figure by drawing a line from the origin to the point (1000,1256). This represents the combination of voltages across 1000 Ω of resistance and 1256 Ω of inductive reactance. The complex impedance is $1000 + j1256\,\Omega$.

The right triangle you just created represents the solution to the problem. The length of the hypotenuse from the origin to (1000,1256) represents the magnitude of the voltage, $|E_T|$, across the combination of R and L for 1 A of current at 10 kHz. The angle θ is the phase angle between the voltage and the current.

The length of the hypotenuse and the angle can be calculated using trigonometry. (If you are unfamiliar with trigonometry, use the review references listed in the math supplement on the ECLM web page.)

$$|E_T| = \sqrt{1000^2 + 1256^2} = 1605\,V$$

$$\theta = \tan^{-1}\left(\frac{1256}{1000}\right) = 51.5°$$

If frequency or inductance increased, X_L would increase with the result that both E_T and θ would increase. If frequency or inductance decreased, X_L would also decrease, as would E_T and θ.

Completing the impedance calculation, remember that $Z = E_T / I$. We know I = 1 A and since it is the reference, it can be written as $1\angle0°$ A. We just determined that $E_T = 1605$ V with a phase angle of 51.5°, so E_T can be written as $1605\angle51.5°$ V. Therefore,

$$Z = \frac{E_T}{I} = \frac{1605\,\angle51.5°}{1\angle0°} = 1605\,\angle51.5°\,\Omega$$

If the reactance in the circuit had been capacitive (negative), the final impedance would have been $1605\,\angle{-51.5°}\,\Omega$.

ADMITTANCE AND SUSCEPTANCE

E5B02 — What letter is commonly used to represent susceptance?

E5B03 — How is impedance in polar form converted to an equivalent admittance?

E5B05 — What happens to the magnitude of a pure reactance when it is converted to a susceptance?

E5B06 — What is susceptance?

E5B07 — What is the phase angle between the voltage across and the current through a series RLC circuit if XC is 500 ohms, R is 1 kilohm, and XL is 250 ohms?

4-18   Chapter 4

E5B08 — What is the phase angle between the voltage across and the current through a series RLC circuit if XC is 100 ohms, R is 100 ohms, and XL is 75 ohms?

E5B11 — What is the phase angle between the voltage across and the current through a series RLC circuit if XC is 25 ohms, R is 100 ohms, and XL is 50 ohms?

E5B12 — What is admittance?

E5C10 — Which point on Figure E5-1 best represents the impedance of a series circuit consisting of a 400-ohm resistor and a 38-picofarad capacitor at 14 MHz?

E5C11 — Which point in Figure E5-1 best represents the impedance of a series circuit consisting of a 300-ohm resistor and an 18-microhenry inductor at 3.505 MHz?

E5C12 — Which point on Figure E5-1 best represents the impedance of a series circuit consisting of a 300-ohm resistor and a 19-picofarad capacitor at 21.200 MHz?

The reciprocal of impedance is *admittance (Y)* and the imaginary part of admittance is *susceptance (B)*. [**E5B02, E5B06, E5B12**] Admittance and susceptance values are measured in siemens (S).

Similar to impedance, admittance can be written in either rectangular or polar coordinates: $Y = G + jB$ or $|Y| \angle \theta$. Since impedance and admittance are reciprocals, converting one to the other is straightforward in polar form: $|Y| = 1/|Z|$

When taking the reciprocal of a phase angle in polar notation, the sign is changed from positive to negative or vice versa. So the admittance's phase angle is simply the negative of the impedance phase angle. Stated another way, $Y = 1/|Z| \angle -\theta$. Converting a pure reactance, $X \angle \theta$, to susceptance is very similar: $B = 1/|X| \angle -\theta$. [**E5B05**]

Calculating Impedances and Phase Angles

Let's get some practice working with impedances, admittances, susceptances, and with determining phase angles in simple circuits by using the following basic rules:

Rule 1: Impedances, resistances, and reactances in series add together

Rule 2: Admittance is the reciprocal of impedance ($Y = 1/Z$) and vice-versa. If only reactance is present in the impedance — a *pure reactance* — susceptance is the negative reciprocal of reactance ($B = -1/X$). For example, $Z = 1 / |Y| \angle \theta = |1/Y| \angle -\theta$ and $B = 1/|X| \angle -\theta$.

Rule 3: Admittances, conductances, and susceptances in parallel add together

Rule 4: Inductive and capacitive reactance in series cancel

Rule 5: $1/j = -j$

(See the sidebar, "Working With Polar and Rectangular Coordinates," presented earlier in this chapter for methods of working with complex numbers in rectangular and polar form.)

Note that when combining parallel elements, it is often easiest to convert them to conductances or susceptances, which can be added together directly. Then the combined element can be converted back to a resistance or reactance, if desired.

The following three examples practice the basic steps of converting between rectangular and polar forms and between admittance and impedance.

Example 4.1

Write the impedance $100 - j100 \, \Omega$ in polar form:

Step 1 — $r = \sqrt{100^2 + (100)^2} = 141$

Step 2 — $\theta = \tan^{-1}(-100/100) = -45°$

Step 3 — $Z = 141 \angle -45° \, \Omega$

Electrical Principles

and write the impedance 25∠60° in rectangular form:

Step 1 — R = |Z| cos 60° = 12.5 Ω

Step 2 — X = |Z| sin 60° = 21.7 Ω

Step 3 — Z = 12.5 + j21.7 Ω

Example 4.2

The first step in these examples needs more space from the question — see example 4.1 for how they should look. The steps and the calculations should all be equally spaced. Convert the admittance 7.09∠45° mS (millisiemens) to impedance in polar form.

Step 1 — Use rule 2 to find:

|Z| = 1 / 0.00709 = 141 Ω

θ = − (45°) = − 45°

Step 2 — Z = 141∠−45° Ω

Example 4.3

Convert the impedance 5∠−30° Ω to admittance in rectangular form. [E5B03]

Step 1 — Use rule 2 to find:

|Y| = 1 / 5 = 0.2 S
θ = − (−30°) = 30°

Step 2 — G = |Y| cos 30° = 0.17 S

Step 3 — B = |Y| sin 30° = 0.1 S

Step 4 — Y = 0.17 + j 0.1 S

Several of the following examples are found in Subelements E5B and E5C of the Extra class question pool. To apply the examples to the questions on the exam, remember that voltage is the reference for phase angle polarity so that if the phase angle is negative, voltage lags current.

Example 4.4

Using the circuit and diagram of **Figure 4.20**, pick the point on the graph in **Figure 4.21** that represents the impedance of a circuit consisting of a 300-Ω resistor in series with an 18-μH inductor at 3.505 MHz? [E5C11]

Step 1 — Calculate the inductor's reactance:
$X_L = 2\pi fL = 400$ Ω

Step 2 — Use rule 1 to add the resistance and reactance together:

Z = 300 + j400 Ω

Step 3 — Locate the point on the graph, 300 units along the X (horizontal) axis and +400 units on the Y (vertical) axis. This is Point 3.

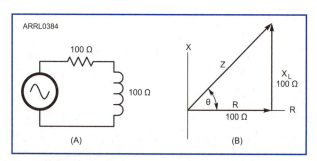

Figure 4.20 — A 100-Ω resistor is connected to an inductor that has a reactance of 100 Ω at some frequency. The impedance triangle shown at B shows the solution, given in the text.

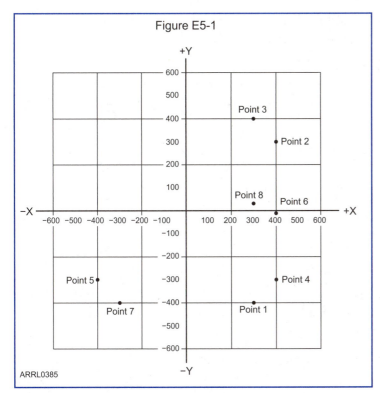

Figure 4.21 — This graph, Figure E5-1 from the Extra Class Question Pool, is used for several exam questions.

Example 4.5

Using the circuit and diagram of **Figure 4.22**, what is the impedance of the circuit consisting of a 100-Ω resistor in parallel with a capacitor that has $-j100$ Ω of reactance? Give the answer in rectangular form, polar form, and state the phase angle of the circuit. This type of calculation is not on the exam but now is a good opportunity to practice converting and working with impedance and admittance.

Step 1 — Use rules 2 and 5 to convert the impedances to admittances because so they can be added directly together:

$G = 1/R = 1/100 = 0.01$ S
$B_C = 1/X_C = 1/-j100 = j0.01$ S

Step 2 — Use rule 3 to add the admittances together:

$Y = 0.0141 \angle 45°$ S (shown in Figure 4.22B)

Step 3 — Use rule 2 to convert the admittance back to impedance:
$Z = 1/Y = (1/0.0141) \angle (0°-45°) = 71 \angle -45°$ Ω

Step 4 — The phase angle is equal to the angle of the impedance: $\theta = -45°$

Example 4.6

Using the graph in Figure 4.21, which point represents the impedance of a circuit consisting of a 400-Ω resistor in series with a 38-pF capacitor at 14 MHz? [**E5C10**]

Step 1 — Calculate the capacitor's reactance:

$X_C = \dfrac{1}{2\pi fC} = -300$ Ω (capacitive reactance is assigned a negative value)

Step 2 — Use rule 1 to add the resistance and reactance together:

$Z = 400 - j300$ Ω

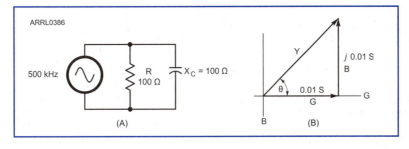

Figure 4.22 — Part A shows a 100-Ω resistor connected in parallel with a capacitor that has 100 Ω of reactance (X_C). Part B shows an "admittance triangle" to help you visualize the solution described in the text.

Electrical Principles

Step 3 — Locate the point on the graph, 400 units along the X (horizontal) axis and −300 units on the Y (vertical) axis. This is Point 4.

Example 4.7

Using the graph in Figure 4.21, which point represents the impedance of a circuit consisting of a 300-Ω resistor in series with a 19-pF capacitor at 21.200 MHz? [**E5C12**]

Step 1 — Calculate the capacitor's reactance:

$$X_C = \frac{1}{2\pi fC} = -400 \ \Omega \ \text{ (capacitive reactance is assigned a negative value)}$$

Step 2 — Use rule 1 to add the resistance and reactance together:

$$Z = 300 - j400 \ \Omega$$

Step 3 — Locate the point on the graph, 300 units along the X (horizontal) axis and −400 units on the Y (vertical) axis. This is Point 1.

Example 4.8

What is the phase angle between voltage and current in a series RLC circuit if X_C is 500 Ω, R is 1 kΩ, and X_L is 250 Ω? [**E5B07**]

Step 1 — Use rules 1 and 4 to add the resistance and reactances together:

$$Z = 1000 + j250 - j500 = 1000 - j250 \ \Omega$$

Step 2 — Convert Z to polar form:

$$r = \sqrt{(1000)^2 + (-250)^2} = 1031$$

$$\theta = \tan^{-1}(-250/1000) = -14°$$

$$Z = 1031\angle -14° \ \Omega$$

Step 3 — The phase angle is equal to the angle of the impedance: $\theta = -14°$. Since phase angle is from voltage to current, the negative angle indicates that voltage lags the current.

Example 4.9

What is the phase angle between voltage and current in a series RLC circuit if X_C is 100 Ω, R is 100 Ω, and X_L is 75 Ω? [**E5B08**]

Step 1 — Use rules 1 and 4 to add the resistance and reactances together:

$$Z = 100 + j75 - j100 = 100 - j25 \ \Omega$$

Step 2 — Convert Z to polar form:

$$r = \sqrt{100^2 + (-25)^2} = 103$$

$$\theta = \tan^{-1}(-25/100) = -14°$$

$$Z = 103\angle -14° \ \Omega$$

Step 3 — The phase angle is equal to the angle of the impedance: $\theta = -14°$. Since phase angle is from voltage to current, the negative angle indicates that voltage lags the current.

Compare this to the previous example and note that even though impedance changed by

Chapter 4

a factor of 10, phase angle was unchanged because the relative amounts of resistance and reactance are the same in both circuits.

Example 4.10

What is the phase angle between voltage and current in a series RLC circuit if X_C is 25 Ω, R is 100 Ω, and X_L is 50 Ω? [**E5B11**]

Step 1 — Use rules 1 and 4 to add the resistance and reactances together:

$$Z = 100 + j50 - j25 = 100 + j25 \ \Omega$$

Step 2 — Convert Z to polar form:

$$r = \sqrt{100^2 + (25)^2} = 103$$

$$\theta = \tan^{-1}(25/100) = 14°$$

$$Z = 103\angle 14° \ \Omega$$

Step 3 — The phase angle is equal to the angle of the impedance: $\theta = 14°$. Since phase angle is from voltage to current, the positive angle indicates that voltage leads the current.

REACTIVE POWER AND POWER FACTOR

E5D05 — What is the power factor of an RL circuit having a 30-degree phase angle between the voltage and the current?

E5D07 — How many watts are consumed in a circuit having a power factor of 0.71 if the apparent power is 500VA?

E5D08 — How many watts are consumed in a circuit having a power factor of 0.6 if the input is 200VAC at 5 amperes?

E5D09 — What happens to reactive power in an AC circuit that has both ideal inductors and ideal capacitors?

E5D10 — How can the true power be determined in an AC circuit where the voltage and current are out of phase?

E5D11 — What is the power factor of an RL circuit having a 60-degree phase angle between the voltage and the current?

E5D12 — How many watts are consumed in a circuit having a power factor of 0.2 if the input is 100 VAC at 4 amperes?

E5D13 — How many watts are consumed in a circuit consisting of a 100-ohm resistor in series with a 100-ohm inductive reactance drawing 1 ampere?

E5D14 — What is reactive power?

E5D15 — What is the power factor of an RL circuit having a 45-degree phase angle between the voltage and the current?

Power is the rate of doing work or using energy per unit of time. Going back to our example at the beginning of this chapter, if you did 10 foot-pounds of work in 5 seconds, then you provided power of 2 foot-pounds per second. If you generate 550 foot-pounds per second of power, you have generated 1 horsepower. So power is a way to express not only how much work you are doing (or how much energy is being stored); it also tells you how fast you are doing it. In the metric system, power is expressed in terms of the watt (W) — 1 watt means energy is being stored or work being done at the rate of 1 joule per second.

You learned earlier that when current increases through an inductor, energy is stored in the inductor's magnetic field. Energy is stored in the electric field of a capacitor when the volt-

Electrical Principles   4-23

age across it increases. That energy is returned to the circuit when the current through the inductor decreases or when the voltage across the capacitor decreases.

You also learned that the voltages across and currents in these components are 90° out of phase with each other. In one half of the cycle some energy is stored in the inductor or capacitor, and the same amount of energy returned on the next half cycle. A perfect capacitor or inductor does not dissipate or consume any energy, but current does flow in the circuit when a voltage is applied to it. If no energy is consumed in a perfect capacitor or inductor, then no work is done and no power is consumed.

Definition of Reactive Power

To pass the General class license exam, you learned that electrical power is equal to the RMS values of current multiplied by voltage:

$$P = I E \hspace{6em} \text{(Equation 4.7)}$$

There are certainly voltage and current present for the inductor and capacitor. Why is no power consumed? There is one catch in Equation 4.7 — it is only true when the current and voltage are in phase such as in a resistor where the phase angle is zero. The larger the phase angle, the smaller the amount of work done by the power source supplying the voltage and current. When the phase angle reaches ±90°, no work is being done at all and so the rate (or power) is equal to zero!

In a circuit's inductive or capacitive reactance, energy may be stored in and returned from the magnetic field in the inductor or the electric field in the capacitor but it will not be consumed as power. Only the resistive part of the circuit consumes and dissipates power as heat. [**E5D09**]

An ammeter and a voltmeter connected in an ac circuit to measure voltage across and current through an inductor or capacitor will both show non-zero values but multiplying them together does not give the true indication of the power being dissipated in the component. The meters do not account for the phase difference between voltage and current.

If you multiply the RMS values of voltage and current from these meters, you will get a quantity that is referred to as *apparent power* — a clue that multiplying RMS values of voltage and current doesn't always give the true picture! Apparent power is expressed in units of *volt-amperes* (VA) rather than watts. The apparent power in an inductor or capacitor is called *reactive power* or *nonproductive, wattless power*. [**E5D14**] Reactive power is expressed in volt-amperes-reactive (VAR). The apparent power in a resistor is called *real power* because voltage and current are in phase so that the power is dissipated as heat or causes work to be done.

Definition and Calculation of Power Factor

You can account for reactive power in a circuit by using phase angle to calculate the *power factor*. Power factor (abbreviated PF) relates the apparent power in a circuit to the real power. You can find the real power in a circuit containing resistance from Equations 4.8 and 4.9.

For a series circuit:

$$P = I^2 R \hspace{6em} \text{(Equation 4.8)}$$

where I is the RMS current.

For a parallel circuit:

$$P = E^2 / R \hspace{6em} \text{(Equation 4.9)}$$

where E is the RMS voltage.

4-24   Chapter 4

Both of these equations are easily derived by using Ohm's Law to solve for either voltage or current, ($E = I \times R$ and $I = E / R$) and replacing that term with the Ohm's Law equivalent.

One way to calculate power factor is simply to divide the real power by the apparent power:

$$PF = \frac{P_{REAL}}{P_{APPARENT}} \qquad \text{(Equation 4.10)}$$

If PF = 1, then the voltage and current are in phase and all of the apparent power is real power. If PF = 0, then the voltage and current are 90° out of phase and all of the apparent power is reactive power.

Figure 4.23 shows a series circuit containing a 75-Ω resistor and an inductor with an inductive reactance of 100 Ω at the signal frequency. The voltmeter reads 250 V RMS and the ammeter indicates a current of 2 A RMS. This is an apparent power of 250 V × 2 A = 500 VA. Use Equation 4.8 to calculate the power dissipated in the resistor:

$$P_{REAL} = I^2 R = (2 \text{ A})^2 \times 75 \text{ Ω} = 4 \text{ A}^2 \times 75 \text{ Ω} = 300 \text{ W}$$

Now by using Equation 4.10, calculate power factor:

$$PF = \frac{300 \text{ W}}{500 \text{ VA}} = 0.6$$

Another way to calculate the real power, if you know power factor, is given by:

$$P_{REAL} = P_{APPARENT} \times PF \qquad \text{(Equation 4.11)}$$

In our example,

$$P_{REAL} = 500 \text{ VA} \times 0.6 = 300 \text{ W } [\textbf{E5D10}]$$

Phase angle can also be used to calculate power factor and real power. You learned how to calculate the phase angle of either a series or a parallel circuit in the previous section. The power factor can be calculated from the phase angle by taking the cosine of the phase angle:

$$\text{Power factor} = \cos \theta \qquad \text{(Equation 4.12)}$$

where θ is the phase angle between voltage and current in the circuit. PF is positive whether the phase angle is positive or negative

You can see that for a circuit containing only resistance, where the voltage and current are in phase, the power factor is 1, and the real power is equal to the apparent power. For a circuit containing only capacitance or inductance in any combination, the power factor is 0 so there is no real power! For most practical circuits, which contain resistance, inductance and capacitance, and the phase angle is some value greater than or less than 0°, the power factor will be something less than one. In such a circuit, the real power will always be something less than the apparent power. This is an important point to remember.

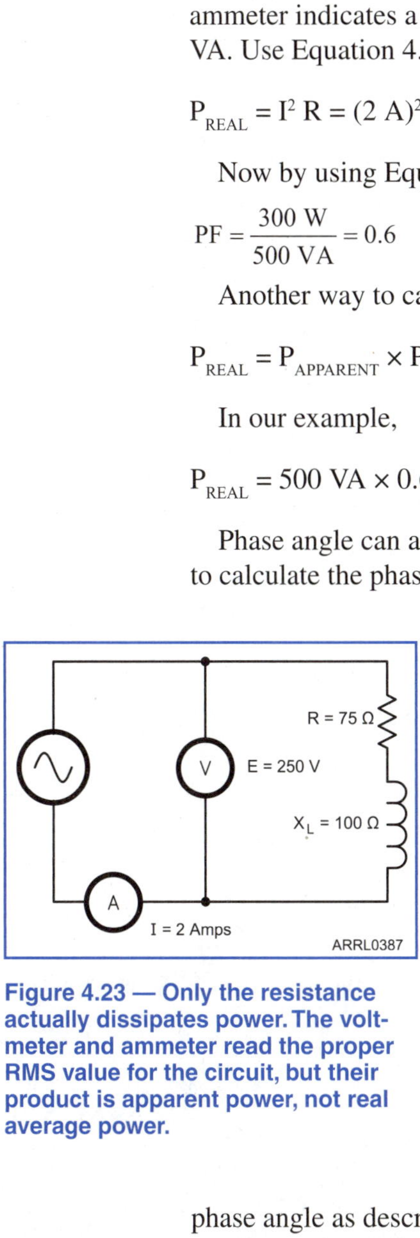

Let's try some sample problems assuming that you can find the phase angle as described in the previous section. (If you need review of the cosine function, use the math supplement on the *Extra Class License Manual* website, **www.arrl.org/extra-class-license-manual.**)

Figure 4.23 — Only the resistance actually dissipates power. The voltmeter and ammeter read the proper RMS value for the circuit, but their product is apparent power, not real average power.

Electrical Principles 4-25

Example 4.11

What is the power factor for an R-L circuit having a phase angle of 30°? 45°? 60°? Use Equation 4.12 to answer this question: [**E5D05, E5D11, E5D15**]

PF for phase angle of 30° = cos 30° = 0.866

PF for phase angle of 45° = cos 45° = 0.707

PF for phase angle of 60° = cos 60° = 0.500

Example 4.12

Suppose you have a circuit that draws 4 amperes of current when 100 V ac is applied. The power factor for this circuit is 0.2. What is the real power (how many watts are consumed) for this circuit? [**E5D12**]

Start by calculating apparent power using Equation 4.7.

$$P_{APPARENT} = 100 \text{ V} \times 4 \text{ A} = 400 \text{ VA}$$

Real power is then found using Equation 4.11:

$$P_{REAL} = 400 \text{ VA} \times 0.2 = 80 \text{ W}$$

Example 4.13

How much power is consumed in a circuit consisting of a 100-Ω resistor in series with a 100-Ω inductive reactance and drawing 1 ampere of current? [**E5D13**]

Because only the resistance consumes power:

$$P_{REAL} = I^2 R = (1 \text{ A})^2 \times 100 \text{ } \Omega = 100 \text{ W}$$

Example 4.14

How many watts are consumed in a circuit having a power factor of 0.6 if the input is 200 V ac at 5 amperes? [**E5D08**]

First, find apparent power using Equation 4.7:

$$P_{APPARENT} = I \text{ E} = 5 \text{ A} \times 200 \text{ V} = 1000 \text{ VA}$$

Then multiply by the power factor as in Equation 4.11:

$$P_{REAL} = P_{APPARENT} \times PF = 1000 \times 0.6 = 600 \text{ W}$$

Example 4.16

How many watts are consumed in a circuit having a power factor of 0.71 if the apparent power is 500 VA? [**E5D07**]

Use Equation 4.11 to find P_{REAL}:

$$P_{REAL} = P_{APPARENT} \times PF = 500 \times 0.71 = 355 \text{ W}$$

RESONANT CIRCUITS

E5A01 — What can cause the voltage across reactances in a series RLC circuit to be higher than the voltage applied to the entire circuit?

E5A02 — What is resonance in an LC or RLC circuit?

E5A03 — What is the magnitude of the impedance of a series RLC circuit at resonance?

E5A04 — What is the magnitude of the impedance of a parallel RLC circuit at resonance?

E5A06 — What is the magnitude of the circulating current within the components of a parallel LC circuit at resonance?

E5A07 — What is the magnitude of the current at the input of a parallel RLC circuit at resonance?

E5A08 — What is the phase relationship between the current through and the voltage across a series resonant circuit at resonance?

E5A14 — What is the resonant frequency of an RLC circuit if R is 22 ohms, L is 50 microhenries and C is 40 picofarads?

E5A16 — What is the resonant frequency of an RLC circuit if R is 33 ohms, L is 50 microhenries and C is 10 picofarads?

With all of the problems so far, we have used inductor and capacitor values that give different inductive and capacitive reactances. Have you wondered about what happens when both reactances are equal?

In a series circuit with an inductor and a capacitor, voltage leads the current by 90° in the inductor; in the capacitor, voltage lags the current by 90°. Since this is a series circuit the current through all of the components is the same. That means the voltages across the inductor and capacitor are 180° out of phase. Those voltages then cancel, leaving only the voltage across the resistance of the circuit which is in phase with the current.

In a parallel circuit containing inductance and capacitance, voltage is the same across both but it is the currents that are 180° out of phase. The current in the inductor lags the applied voltage by 90° and the current in the capacitor leads by 90°. The cancellation of the current leaves a parallel resistance as the only component in which current can flow and the remaining current is in phase with the voltage.

Whether the components are connected in series or parallel, we say the circuit is *resonant* or is at *resonance* when the inductive reactance value is the same as the capacitive reactance value. [E5A02] Remember that inductive reactance increases as frequency increases and that capacitive reactance decreases as frequency increases. The frequency at which the two are equal is the circuit's *resonant frequency*.

Calculation of Resonant Frequency

Figure 4.24 shows the intersection of two lines representing reactance vs frequency: the curved line, decreasing to the right, shows capacitive reactance $X_C = 1/2\pi f C$, and the straight line rising to the right shows inductive reactance $X_L = 2\pi f L$. At the frequency where the lines cross, $X_C = X_L$ and the circuit is resonant. The two lines cross at only one point — the resonant frequency of the circuit using those two components. Every combination of a capacitor and an inductor will be resonant at some frequency.

The scales of both axes in the figure are linear which results in the curved and straight lines. In most engineering manuals and

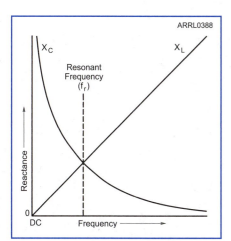

Figure 4.24 — A graph showing the relative change in inductive reactance and capacitive reactance as frequency increases. For any specific inductor-capacitor pair, there is only one frequency at which $X_L = X_C$, the resonant frequency, f_r.

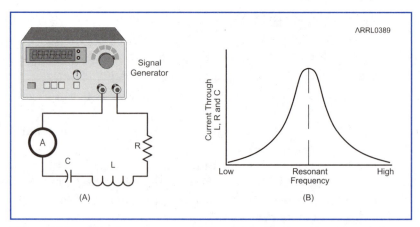

Figure 4.25 — A series-connected LC or RLC circuit presents a minimum value of resistance at the resonant frequency. Therefore, at resonance, the current passing through the circuits reaches a maximum.

texts, however, reactances are plotted on a graph with logarithmic axis scales where both types of reactance are plotted as straight lines. This makes it a lot easier to see where the resonant frequency is for a wide range of capacitance, inductance, and frequencies. You can see this kind of chart in the Electrical Fundamentals chapter of the *ARRL Handbook* or online at **www.rfcafe.com/references/electrical/frequency-reactance-nomograph.htm**.

Since resonance occurs when the reactances are equal, we can derive an equation to calculate the resonant frequency of any capacitor-inductor pair:

$$X_L = 2\pi f L$$

$$X_C = \frac{1}{2\pi f C}$$

Set $X_L = X_C$ at resonance, so:

$$2\pi f L = \frac{1}{2\pi f C} \text{ so}$$

$$(2\pi f)(2\pi f) = \frac{1}{LC} \text{ and}$$

$$4\pi^2 f^2 = \frac{1}{LC}$$

This leads to the formula for resonant frequency:

$$f_r = \frac{1}{2\pi\sqrt{LC}} \qquad \text{(Equation 4.13)}$$

Let's try calculating a resonant frequency. What frequency should the signal generator in **Figure 4.25** be tuned to for resonance if the resistor is 22 Ω, the coil is 50 µH, and the capacitor has a value of 40 pF? Probably the biggest stumbling block of these calculations will be remembering to convert the inductor value to henrys and the capacitor value to farads. After you have done that, use Equation 4.13 to calculate the resonant frequency. [E5A14]

50 µH = 50 × 10⁻⁶ H

40 pF = 40 × 10⁻¹² F

$$f_r = \frac{1}{6.28\sqrt{(50\times10^{-6})(40\times10^{-12})}} = 3.56\times10^6 = 3.56\,\text{MHz}$$

What would happen to f_r if the resistor value was changed to 47 Ω? Nothing! The value of the circuit's resistance does not affect the resonant frequency. This is because the resistor does not store electrical or magnetic energy.

Calculating the resonant frequency of a parallel circuit is exactly the same as for a series circuit. It does not matter if the L and C are in series or in parallel. For example, what is the resonant frequency of an RLC circuit if R is 33 Ω, L is 50 μH and C is 10 pF? [**E5A16**]

$$f_r = \frac{1}{6.28\sqrt{(50\times10^{-6})(10\times10^{-12})}} = 7.12\times10^6 = 7.12\,\text{MHz}$$

Using the same technique that we used to derive Equation 4.13, we can easily derive equations to calculate either the inductance or capacitance to resonate with a certain component at a specific frequency, f_r:

$$L = \frac{1}{(2\pi f_r)^2\, C} = \frac{1}{(2\pi)^2 (f_r)^2\, C} \qquad \text{if we know the value of C} \qquad \text{(Equation 4.14)}$$

and

$$C = \frac{1}{(2\pi f_r)^2\, L} = \frac{1}{(2\pi)^2 (f_r)^2\, L} \qquad \text{if we know the value of L} \qquad \text{(Equation 4.15)}$$

Let's try a couple of practical examples:

What value capacitor is needed to make a circuit that is resonant in the 80 meter band if you have a 20-μH inductor? Choose a frequency in the 80 meter band to work with, such as 3.6 MHz. Then convert to fundamental units: $f_r = 3.6 \times 10^6$ Hz and L = 20 × 10⁻⁶ H. Use Equation 4.15, since that one is written to find capacitance, the quantity we are looking for:

$$C = \frac{1}{(2\pi)^2\,(3.6\times10^6)^2\,(20\times10^{-6})}$$

$$C = \frac{1}{(39.48)(1.3\times10^{13})\,(20\times10^{-6})}$$

$$C = \frac{1}{1.03\times10^{10}} = 9.7\times10^{-11} = 97\times10^{-12} = 97\,\text{pF}$$

You can use a 100-pF capacitor. If you try solving this problem for both ends of the 80 meter band, you will find that you need a 103-pF capacitor at 3.5 MHz and a 79-pF unit at 4 MHz. So any capacitor value within this range will resonate in the 80 meter band with the 20-μH inductor.

Stored Energy in Resonant Circuits

Figure 4.25 shows a signal generator connected to a series RLC circuit. The signal generator produces a variable-frequency current through the circuit, which will cause a voltage to appear across each component. As discussed above, the voltages across the inductor and capacitor are always 180° out of phase.

When the signal generator produces an output signal at the resonant frequency of the circuit, the voltages across the inductor and capacitor are equal as well as out of phase. This means an equal amount of energy is stored in each component and is transferred between them on alternate half-cycles.

If the components have low amounts of resistive loss, the energy continually supplied by the signal generator will cause the voltages across the inductor and capacitor to build to levels several times larger than the voltage applied to the circuit! Similarly, the currents that flow back and forth between the inductor and capacitor to exchange the stored energy are

Electrical Principles   4-29

maximum at resonance. [**E5A01**] These currents are referred to as *circulating currents* because they flow back and forth as the energy circulates between the two reactances. (Storage of energy is also the reason why tuned circuits are sometimes called *tank circuits*.)

In analogy, consider pushing a playground swing. Even though the additional push on each swing is small, if friction is low the pushes can cause the amplitude of the swing's travel to be much larger than would be caused by any single push.

Impedance of Resonant Circuits Versus Frequency

With the voltages across the inductor and capacitor canceling each other the only impedance presented to the signal generator is that of the resistance, R. [**E5A03**] For perfect components and no resistance in the circuit, there would be nothing to restrict the current in the circuit. An ideal series-resonant circuit, then, "looks like" a short circuit to the signal generator. There is always some resistance in a circuit but if the total resistance is small, the current will be large according to Ohm's Law. The change in current with frequency is shown in Figure 4.25B. It reaches a maximum at the resonant frequency, f_r.

In a parallel-resonant circuit there are several current paths, but the same voltage is applied to the components. **Figure 4.26** shows a parallel LC circuit connected to a signal generator. The applied voltage causes current to flow in each of the three circuit branches. At resonance the current through the inductor will be 180° out of phase with the current in the capacitor and again they add up to zero.

As a result, the parallel resonant circuit has a high impedance and can appear to be an open circuit to the signal generator because the current from the signal generator is quite small. At resonance, the magnitude of the impedance of a circuit with a resistor, inductor and capacitor all connected in parallel will be approximately equal to the circuit resistance. [**E5A04, E5A07**]

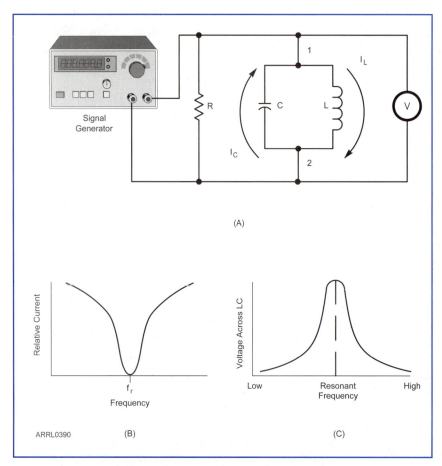

Figure 4.26B is a graph of the relative generator current. The current at the input of a parallel RLC circuit is a minimum at resonance. It is a mistake to assume, however, that because the generator current is small the current flowing through the capacitor and inductor is also small. As in a series resonant circuit, the energy being exchanged between the inductor and capacitor can build up to large values. At resonance, the circulating currents will be at maximum, limited only by resistive losses in the components. [**E5A06**] While the total current from the generator is small at resonance, the voltage measured across the tank reaches a maximum value at resonance. Figure 4.26C is a

Figure 4.26 — A parallel-connected LC or RLC circuit presents a very high resistance at the resonant frequency. Therefore, at resonance, the voltage across the circuit reaches a maximum.

graph of the voltage across the inductor and capacitor.

It is also interesting to consider the phase relationship between the voltage across a resonant circuit and the current through that circuit. Because the inductive reactance and the capacitive reactance are equal but opposite their effects cancel each other. The resulting current and voltage in a resonant circuit are in phase. This is true for both a series resonant circuit and a parallel resonant circuit. [**E5A08**]

Q OF COMPONENTS AND CIRCUITS

E4B08 — Which of the following can be used to measure the Q of a series-tuned circuit?
E5A05 — What is the result of increasing the Q of an impedance-matching circuit?
E5A09 — How is the Q of an RLC parallel resonant circuit calculated?
E5A10 — How is the Q of an RLC series resonant circuit calculated?
E5A11 — What is the half-power bandwidth of a resonant circuit that has a resonant frequency of 7.1 MHz and a Q of 150?
E5A12 — What is the half-power bandwidth of a resonant circuit that has a resonant frequency of 3.7 MHz and a Q of 118?
E5A13 — What is an effect of increasing Q in a series resonant circuit?
E5A15 — Which of the following increases Q for inductors and capacitors?

We have talked about ideal resistors, capacitors and inductors, and how they behave in ac circuits. We have shown that resistance in a circuit causes some departure from a circuit of ideal components by dissipating some of the stored energy. But how can we determine how close to the ideal a certain component comes? Or how much of an effect it will have on a designed circuit? We can calculate a value for inductors and capacitors that evaluates the relative merits of that component — the *quality factor* called *Q*. We can also assign a Q value to an entire circuit as a measure of how close to the ideal that circuit performs — at least in terms of its properties at resonance.

One definition of Q is the ratio of reactance to resistance. This is, in effect, the ratio of how much energy is stored to how much energy is dissipated. The lower the component's resistive losses, the higher the Q. [**E5A15**]

Figure 4.27 shows that a capacitor can be thought of as an ideal capacitor in series with a resistor and an inductor can be considered as an ideal inductor in series with a resistor. This parasitic resistance can't actually be separated from the inductor or capacitor, of course, but it acts just the same as if it were in series with an ideal, lossless component. The Q of a real inductor, L, is equal to the inductive reactance divided by the resistance and the Q of a real capacitor, C, is equal to the capacitive reactance divided by the resistance:

$$Q = \frac{X}{R} \qquad \text{(Equation 4.16A)}$$

If you want to know the Q of a circuit containing both parasitic resistance and actual resistors, both must be added together to find the value of R used in the equation. Since adding a resistor can only raise the total resistance, the Q of the circuit will always go down when resistance is added in series with an inductor or capacitor. There is no way to raise the Q of an inductor or capacitor except by building a component with less parasitic resistance.

In a resonant RLC circuit, the effect of resistance depends on whether the circuit is a parallel or series circuit. For a series circuit, increasing R represents increasing losses and lowers Q. In a parallel circuit, increasing R represents lower losses and raises Q. For that reason, the equations for Q in series and parallel resonant circuits are reciprocals:

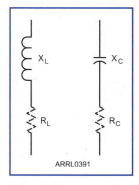

Figure 4.27 — A practical inductor can be considered as an ideal inductor in series with a resistor, and a practical capacitor can be considered as an ideal capacitor in series with a resistor.

Electrical Principles 4-31

$$Q_{SERIES} = \frac{1}{R}\sqrt{\frac{L}{C}} \text{ and } Q_{PARALLEL} = R\sqrt{\frac{C}{L}} \qquad \text{(Equation 4.16B)}$$

With a little algebra and knowing that $X_L = X_C$ at resonance, Q can be computed just knowing the reactance of either the inductor or capacitor at the resonant frequency [**E5A09, E5A10**]:

$$Q_{SERIES} = \frac{X}{R} \text{ and } Q_{PARALLEL} = \frac{R}{X} \qquad \text{(Equation 4.16C)}$$

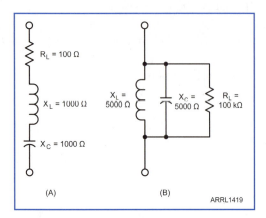

Figure 4.28 — A series-resonant circuit with a loss resistance of 100 Ω is shown at A and a parallel-resonant circuit with a loss resistance of 100 kΩ is shown at B. Q of the series circuit is 1000/100 = 10 and of the parallel circuit is 100,000/5000 = 20.

For example, in **Figure 4.28A** the Q of the circuit is 10 calculated from the 1000 Ω of reactance divided by the loss resistance of 100 Ω. In the parallel circuit of Figure 4.28B, the Q = 20, from 100 kΩ of resistance divided by 5000 Ω of reactance.

The relationship between Q of the resonant circuit and the circuit's internal voltages and circulating currents is now clear. The higher Q becomes, the higher the voltages and currents. [**E5A13**] In fact, for resonant circuits such as tuning networks for amplifiers and impedance matching, internal voltages and currents can become high enough to arc across tuning capacitors or melt soldered connections, even at modest power levels!

Another practical note, the internal resistance of a capacitor is usually much less than that for an inductor so we often ignore the resistance of a capacitor and consider only that associated with the inductor when computing Q of a resonant circuit. Stated another way, Q of the inductor is usually the limiting factor on Q of a resonant circuit.

Q and Resonant-Circuit Bandwidth

Bandwidth refers to the frequency range over which the circuit response in voltage or current is no more than 3 dB below the peak response. The −3 dB points are shown on **Figure 4.29**, and the bandwidths are indicated. (If you are not familiar with the use of decibels, see the math supplement on this book's web page.) Since this 3-dB decrease in signal represents the points where the circuit power is one half of the resonant power, the −3 dB points are also called *half-power points*. At these points, the voltage and current have been reduced to 0.707 times their peak values.

The frequencies at which the half-power points occur are f1 and f2; Δf is the difference between these two frequencies, and represents the *half-power (or 3-dB) bandwidth*. A circuit with a narrow bandwidth is said to be "sharp" and one with a wider bandwidth "broad." It is possible to calculate the bandwidth of a resonant circuit based on the circuit Q and the resonant frequency:

$$\Delta f = \frac{f_r}{Q} \qquad \text{(Equation 4.17)}$$

where:
Δf = the half-power bandwidth.
f_r = the resonant frequency of the circuit.
Q = the circuit Q

The higher the circuit Q, the smaller the bandwidth of a resonant circuit will be, whether it is a series or parallel circuit. [**E4B08**] Figure 4.29 shows the relative bandwidth of a circuit with two different Q values.

Let's calculate the half-power bandwidth of a parallel circuit that has a resonant frequency of 7.1 MHz and a Q of 150. The half-power bandwidth is found by Equation 4.17: [**E5A11**]

$$\Delta f = \frac{f_r}{Q} = \frac{7.1 \times 10^6}{150} = 47.3 \times 10^3 \text{ Hz} = 47.3 \text{ kHz}$$

To find the upper and lower half-power frequencies, subtract half the total bandwidth from the center frequency to get the lower half-power frequency and add half the bandwidth to get the upper half-power frequency. The response of this circuit will be at least half of the peak signal power for signals in the range 7.07635 to 7.12365 MHz.

Repeat the calculations for the following combination of resonant frequency and Q:

f_r = 3.7 MHz and Q = 118: Δf = 31.4 kHz [**E5A12**]

Impedance matching circuits that use inductances and capacitances also use circulating energy to transform one ratio of voltage to current (which is the definition of impedance) at the output to another at the input. Q of the components and of the circuit also affect how the circuit performs. As Q of such an impedance matching circuit increases, the internal voltages and currents increase and the bandwidth over which the impedance is matched decreases, just like that of a resonant circuit. [**E5A05**]

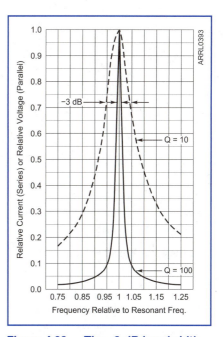

Figure 4.29 — The –3 dB bandwidth of two resonant circuits is shown. The circuit with the higher Q has a steeper response, and a narrower bandwidth. The vertical scale represents current for a series circuit and voltage for a parallel circuit.

COMPONENTS AT RF AND MICROWAVE FREQUENCIES

E5D01 — What is the result of skin effect?
E5D02 — Why is it important to keep lead lengths short for components used in circuits for VHF and above?
E5D04 — Why are short connections used at microwave frequencies?
E6D13 — What is the primary cause of inductor self-resonance?
E6E02 — Which of the following device packages is a through-hole type?
E6E09 — Which of the following component package types would be most suitable for use at frequencies above the HF range?
E6E10 — What advantage does surface-mount technology offer at RF compared to using through-hole components?
E6E11 — What is a characteristic of DIP packaging used for integrated circuits?
E6E12 — Why are DIP through-hole package ICs not typically used at UHF and higher frequencies?

Skin Effect and Q

As frequency increases, the electric and magnetic fields of signals do not penetrate as deeply into a conductor like a wire. At dc, the entire cross-section of the wire is used to carry currents. As the frequency increases, the effective area gets smaller and smaller as the current is confined closer and closer to the surface of the wire. [**E5D01**] This reduces the volume available to carry the electron flow and increases its effective resistance.

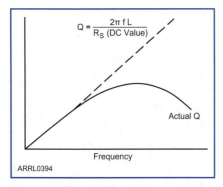

Figure 4.30 — At low frequencies, the Q of an inductor is proportional to frequency. At high frequencies, increased losses in the inductor cause Q to degrade from the expected value.

In the HF range, all current flows in the outer few thousandths of an inch of a conductor. At VHF and UHF, the depth is on the range of a few ten-thousandths of an inch. (This is why many VHF and UHF inductors are silver plated — to provide a low-resistance path for current.) In fact, at VHF and UHF, conductors could be made of metal-plated plastic without any ill effects!

Called *skin effect,* this is the major cause of why the parasitic resistance of inductors (due mainly to the resistance of the wire used to wind them) increases somewhat as the frequency increases. Because of the increasing reactance, inductor Q will increase with increasing frequency up to a point but then the parasitic resistance due to skin effect becomes greater and Q degrades as shown in **Figure 4.30**.

Self-Resonance

Because of *parasitic inter-turn capacitance* — very small capacitances that exist between the turns of an inductor — as shown in **Figure 4.31**, the inductor can become a *self-resonant* circuit at some sufficiently high frequency. [**E6D13**] Similar to a resonant circuit made of discrete components, the impedance of the inductor will peak at the self-resonant frequency and above the self-resonant frequency an inductor will appear capacitive! The amount of depends on the inductor's construction and there are several techniques used to control it. (Capacitors can also exhibit self-resonance.)

Self-resonance becomes critically important at VHF and UHF because the self-resonant frequency of many common components is at or below the frequency where the component will be used. In this case, special techniques can be used to construct components to operate at these frequencies by reducing the parasitic effects, or else the idea of lumped elements must be abandoned altogether in favor of microwave techniques such as stripline and waveguides.

Effects of Component Packaging at RF

A related effect is the *parasitic inductance* of the leads used to make contact to discrete components. Even straight wire has some inductance and while it is not a lot of inductance, it can be significant. For example, the #24 AWG wire typically used for the leads of discrete components has an inductance of about 24 nH per inch of length. In circuits operating at VHF and higher frequencies, including high-speed digital circuits, this inductive reactance

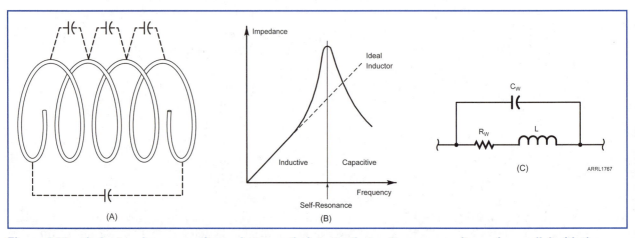

Figure 4.31 — Inductors have capacitance between their turns that acts as a capacitance in parallel with the inductance. The graph at B shows how distributed capacitance resonates with the inductance. The equivalent circuit of the inductor, including wire resistance, R_W, is shown at C.

can become significant and increases with frequency, making the circuit behave in unexpected (and usually unwanted) ways. Good design and construction practice at these frequencies is to minimize the effects of lead inductance by using surface-mount components or trimming the leads to be as short as possible. [**E5D02**]

As wavelength becomes shorter, the electrical length of component leads (and any connecting wire) causes phase shifts in the signal traveling along the lead. This phase shift can be very difficult to control and leads to oscillation and uneven frequency response at microwave frequencies. [**E5D04**]

Integrated circuits were a great advance in reducing lead length and increasing operating frequency. The popular DIP (dual in-line package) style of IC rapidly became a standard for analog and digital circuits, allowing operation into the low UHF range. It features two rows of pins spaced 0.1 inch apart, with the rows from 0.3 to 0.6 inch apart along the opposite sides of a rectangular plastic or ceramic body. [**E6E11**] Since the pins are inserted in holes in the printed-circuit board and extend through the board to be soldered on one or both sides of the board, the DIP package is an example of a *through-hole* component. [**E6E02**] Components such as resistors and capacitors with wire leads are also through-hole parts.

As device complexity and operating frequency increased, however, even the leads of compact DIP packages became too long. [**E6E12**] The solution was found in *surface-mount* components that don't have leads at all, just terminals on the side of the package. Surface-mount or SMT components are placed directly on a circuit board's exposed pads that are coated with solder paste. The entire assembly is heated until the solder paste melts and makes the permanent connection, holding the component to the board. SMT components can be as small as 1 millimeter on a side, which means the circuit can be constructed with shorter circuit-board traces. Because SMT components have less parasitic inductance and capacitance, they are usable well into the VHF, UHF, and microwave range. [**E6E09, E6E10**]

MAGNETIC CORES

E6D01 — **Why should core saturation of an impedance matching transformer be avoided?**

E6D04 — **Which materials are commonly used as a core in an inductor?**

E6D05 — **What is one reason for using ferrite cores rather than powdered iron in an inductor?**

E6D06 — **What core material property determines the inductance of an inductor?**

E6D07 — **What is current in the primary winding of a transformer called if no load is attached to the secondary?**

E6D08 — **What is one reason for using powdered-iron cores rather than ferrite cores in an inductor?**

E6D09 — **What devices are commonly used as VHF and UHF parasitic suppressors at the input and output terminals of a transistor HF amplifier?**

E6D10 — **What is a primary advantage of using a toroidal core instead of a solenoidal core in an inductor?**

E6D11 — **Which type of core material decreases inductance when inserted into a coil?**

E6D12 — **What is inductor saturation?**

As you've seen, inductors store magnetic energy, creating reactance. Inductors are usually visualized as the classic winding of wire from one end to the other of a round form — this winding shape is called *solenoidal* — giving rise to the common term "coil." An inductor's *core* is whatever material the wire is wound around, even air. (An inductor whose core consists of air is called *air-wound*.)

Solenoidal coils make great figures in books, but a winding of wire around a hollow form filled with nothing but air is a relatively inefficient way to store magnetic energy. A form made of magnetic material increases the storage of energy because it focuses the magnetic

Electrical Principles **4-35**

field created by the current in the surrounding winding. The stronger magnetic field increases the inductance of the inductor.

Inductance is determined by the number of turns of wire on the core and on the core material's *permeability*. [**E6D06**] Permeability is a measure of a magnetic field in the core compared to the strength of the field with a core of air. Cores with higher permeability have more inductance for the same number of turns on the core. In other words, if you make two inductors with 10 turns around different core materials, the core with a higher permeability will have more inductance.

Manufacturers offer a wide variety of materials, or *mixes*, to provide cores that will perform well over a desired frequency range. Powdered-iron cores combine fine iron particles with magnetically-inert binding materials. Combining materials such as nickel-zinc and manganese-zinc compounds with the iron produces ceramic *ferrite* cores. The chemical names for iron compounds are based on the Latin word for iron, *ferrum*, so this is how these materials get the name ferrite. Inductors with magnetic material cores are also called *ferromagnetic inductors*.

The choice of core materials for a particular inductor presents a compromise of features. Powdered-iron cores generally have better temperature stability and maintain their characteristics at higher currents. [**E6D08**] Ferrite cores generally have higher permeability values, however, so inductors made with ferrite cores require fewer turns to produce a given inductance value. [**E6D05**]

Some inductors are made to be adjustable by winding them on a form containing a movable, threaded core or *slug*. The core is adjusted with a screwdriver or tuning tool to move the core in and out of the coil. Inserting the core into the coil changes the coil's inductance. Ferrite cores, the most common, have a high relative permeability and increase inductance as the core is inserted. The low relative permeability of brass cores causes a reduction in inductance. [**E6D04, E6D11**]

Magnetic cores are also used for transformers that couple power from a primary winding to a secondary winding through the core. When using transformers of any sort, it is important to avoid exceeding the core's ability to store magnetic energy, an effect called *saturation*. When saturation occurs, the output waveform becomes distorted, generating harmonics and other distortion products. [**E6D01, E6D12**] A transformer's core will contain some magnetic energy from *magnetizing current* in the primary winding even if no load is connected to the secondary. [**E6D07**]

Core Shape — Toroids and Beads

The shape of an inductor's core also affects how its magnetic field is contained. For a solenoidal core, the magnetic field exists not only in the core, but in the space around the inductor. This allows the magnetic field to interact with, or *couple* to, other nearby conductors. This coupling often creates unwanted signal paths and interactions between components so external shields or other isolation methods must be used.

To reduce unwanted coupling, the donut-shaped *toroid* core is used. When wire is wound on such a core, a *toroidal* inductor is produced. Nearly all of a toroidal inductor's magnetic field is contained within the toroid core. [**E6D10**] Toroidal inductors are one of the most popular inductor types in RF circuits because they can be located close to each other on a circuit board with almost no interaction. See **Figure 4.32** for a photo of a variety of toroidal inductors.

Toroid cores are very useful for solving a variety of radio-frequency interference (RFI) problems. For example, you might select a type-43 mix ferrite core and wind several turns of a

Figure 4.32 — This photo shows a variety of inductors wound on toroid cores.

telephone wire or speaker leads through the core to produce a *common-mode choke*. Such a choke is designed to suppress any RF energy flowing in common on all of these wires. Audio signals flow through the choke unimpeded but the RF signals are blocked.

A *ferrite bead* is a very small core with a hole designed to slip over a component lead. These are often used as suppressors for VHF and UHF oscillations at the input and output terminals of HF and UHF amplifiers, for example. [**E6D09**] The use of ferrite beads as parasitic suppressors points out another interesting property of these core materials — their loss changes with frequency. Each mix has a different set of loss characteristics with frequency. While we normally want to select an inductor core material that will have low loss at a particular frequency or over a certain range, at times we want to select a core material that will have high loss to absorb or dissipate energy.

For More Information

Calculating the inductance of a particular toroidal inductor is simple. First, you must know the inductance index value for the particular core you will use. This value, known as A_L, is found in the manufacturer's data. For powdered-iron toroids, A_L values are given in microhenrys per 100 turns-squared. See *The ARRL Handbook* for more complete information about the different types of cores and their characteristics.

To calculate the inductance of a powdered-iron toroidal inductor when the number of turns and the core material are known, use Equation 4.18.

$$L \text{ (for powdered iron cores)} = \frac{A_L \times N^2}{10,000} \quad \text{(Equation 4.18)}$$

where:
L = inductance in µH.
A_L = inductance index, in µH per 100 turns-squared.
N = number of turns.

Often you want to know how many turns to wind on the core to produce an inductor with a specific value. In that case, solve Equation 4.18 for N.

$$N = 100\sqrt{\frac{L}{A_L}} \quad \text{(Equation 4.19)}$$

When winding wire on a toroid, keep in mind that if the wire simply passes through the center of the core, you have a 1-turn inductor as illustrated in **Figure 4.33**. Each time the wire passes through the center of the core it counts as another turn. A common error is to count one complete wrap around the core ring as one turn. That can produce a two-turn inductor, however.

The calculations for ferrite toroids are nearly identical but the A_L values are given in millihenrys per 1000 turns-squared instead of microhenrys per 100 turns-squared because

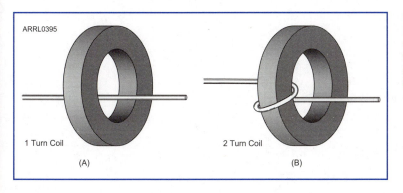

Figure 4.33 — Proper turns counting is important when you wind a toroidal inductor. Each pass through the center of the core must be counted. Part A shows a one-turn inductor and Part B shows a two-turn inductor.

the permeability of ferrite is higher. This requires a change of the constant in Equation 4.18 from 10,000 to 1,000,000. Use Equation 4.20 to calculate the inductance of a ferrite toroidal inductor.

$$L \text{ (for ferrite cores)} = \frac{A_L \times N^2}{1,000,000}$$

(Equation 4.20)

where:

L = inductance in mH.

A_L = inductance index, in mH per 1000 turns-squared.

N = number of turns.

Chapter 5

Components and Building Blocks

In this chapter, you'll learn about:
- How semiconductor devices are made
- Types of diodes and rectifiers
- Bipolar and field effect transistors and RF integrated circuits
- Digital logic basics and families
- Optoelectronics such as solar cells and optocouplers

The Extra class license exam presents basic questions about electronic components — diodes, transistors, ICs, and other devices. These are contained in question pool Subelement 6 (six groups of questions) and Subelement 7 (eight groups of questions). You won't have to become a circuit designer to answer the exam questions, but you'll be expected to know what types of devices are used in radio circuits and their important characteristics.

This chapter presents the fundamentals of how the devices operate. We'll start with semiconductor materials and work our way up to diodes, transistors, and integrated circuits. As a comprehensive explanation of these electronic building blocks is well beyond the scope of this book, turn to the *ARRL Handbook* or the references on the *Extra Class License Manual* website for more information.

5.1 Semiconductor Devices

E6A02 — Which of the following semiconductor materials contains excess free electrons?
E6A04 — What is the name given to an impurity atom that adds holes to a semiconductor crystal structure?

Before you can understand the operation of electronic circuits, you must know some basic information about the devices that make up those circuits. This section presents the information about semiconductors and other active devices you need to know to pass your Extra class license exam. You will find descriptions of several types of diodes and transistors, RF and digital integrated circuits (ICs), and various types of display and optoelectronic devices.

MATERIALS

Silicon (Si) and germanium (Ge) are the materials normally used to make semiconductor materials. (The element *silicon* [SIL-i-kahn] is not the same as the household lubricants and rubber-like sealers called *silicone* [sil-i-CONE]). Silicon has 14 protons and 14 electrons, while germanium has 32 of each. Silicon and germanium atoms both have four shareable or *valence* electrons in their outer layer of electrons. This arrangement allows these four electrons to be shared with other nearby atoms.

Atoms that arrange themselves into a regular pattern by sharing electrons form *crystals*. **Figure 5.1** shows silicon and

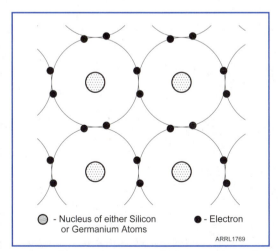

Figure 5.1 — Silicon and germanium atoms arrange themselves into a regular pattern — a crystal. Each atom in this crystal structure is sharing its outermost four electrons with other nearby atoms.

Components and Building Blocks   5-1

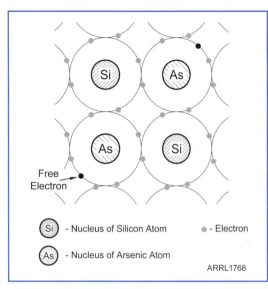

Figure 5.2 — Adding antimony or arsenic atoms to the silicon or germanium crystals results in an extra or free electron in the crystal structure, producing N-type semiconductor material. An arrow points to one of the free electrons.

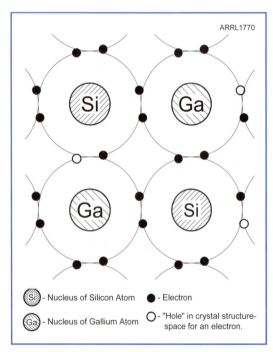

Figure 5.3 — Manufacturers can add gallium or indium atoms to the silicon or germanium crystals. These impurity atoms have only three electrons to share. This leaves a hole or space for another electron in the crystal structure, producing P-type semiconductor material.

germanium crystals. (Different kinds of atoms might arrange themselves into other patterns.) The crystals made by silicon or germanium atoms do not make good electrical conductors or insulators. That's why they are called *semiconductors*. Under the right conditions they can act as either conductors or insulators. Semiconductor materials also exhibit properties of both metallic and nonmetallic substances. Semiconductors are solid crystals. They are strong and not easily damaged by vibration or rough handling. We refer to electronic parts made with semiconductor materials as *solid-state devices*.

To control the electrical characteristics of semiconductor material, manufacturers add other atoms to these crystals through a carefully controlled process called *doping*. The atoms added in this way produce a material that is no longer pure silicon or pure germanium. We call the added atoms *impurities*. The impurities are generally chosen for their ability to alter the way in which electrons are shared within the crystal.

As an example, the manufacturer might add some atoms of arsenic (As) or antimony (Sb) to the silicon or germanium while making the crystals. Arsenic and antimony atoms each have five electrons to share — an extra shareable electron compared to the crystal of pure silicon. **Figure 5.2** shows how an atom with five electrons in its outer layer fits into the crystal structure. In such a case, there is an extra or *free* electron in the crystal and we call the semiconductor material made in this way *N-type* material. (This name comes from the extra free electrons in the crystal structure.) **[E6A02]**

The impurity atoms are electrically neutral, just as the silicon or germanium atoms are. The extra electrons are considered "free" because they are not so strongly shared with adjacent atoms and are freer to move within the crystal structure. Impurity atoms that create (donate) free electrons to the crystal structure are called *donor impurities*.

Now let's suppose the manufacturer adds some gallium or indium atoms instead of arsenic or antimony. Gallium (Ga) and indium (In) atoms only have three electrons that they can share with other nearby atoms. When there are gallium or indium atoms in the crystal there is an extra space where an electron could fit into the structure.

Figure 5.3 shows an example of a crystal structure with spaces where an electron could be present. We call this space for an electron a *hole*. The semiconductor material produced in this way is *P-type* material. Impurity atoms that produce extra holes for electrons in the crystal structure are called *acceptor impurities*. **[E6A04]**

JUNCTION DIODES

E6A03 — Why does a PN-junction diode not conduct current when reverse biased?
E6B07 — What is the failure mechanism when a junction diode fails due to excessive current?

The *junction diode*, also called the *PN-junction diode*, is made from two layers of semiconductor material joined together. One layer is made from P-type (positive) material. The other layer is made from N-type (negative) material. The name PN junction comes from the way the P and N layers are joined to form a semiconductor diode. **Figure 5.4** illustrates the basic concept of a junction diode.

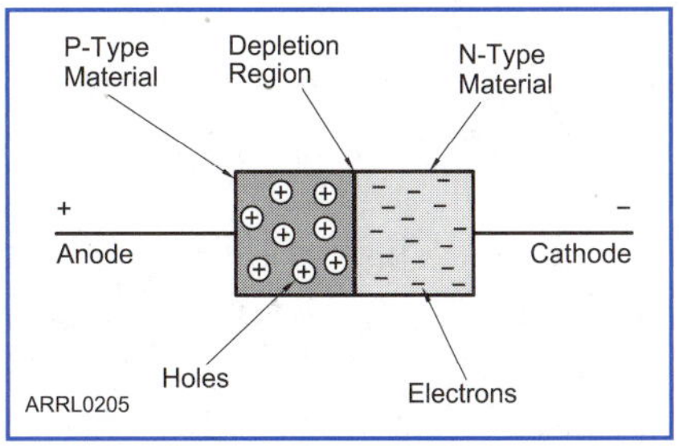

When no voltage is applied to a diode, the junction between the P-type and N-type material acts as a barrier that prevents carriers from flowing between the layers. This happens because the majority carriers (the electrons and holes) combine where the two types of material are in contact, leaving no carriers to support current flow unless a voltage is applied from an external source. This barrier to current flow is called the *depletion region*.

The P-type side of the diode is called the *anode*. The N-type side is called the *cathode*. When voltage is applied to a junction diode as shown at A in **Figure 5.5**, charge carriers flow across the barrier and the diode conducts. With the anode positive with respect to the cathode, electrons are attracted across the junction from the N-type material, through the P-type material, and on through the circuit to the positive battery terminal. Holes are attracted in the opposite direction by the negative voltage from the battery. Electrons are supplied to the cathode and removed from the anode by the wires connected to the battery. When the diode is connected in this manner it is said to be *forward biased*. *Conventional current* (which flows from positive to negative) in a diode flows from the anode to the cathode. The electrons flow in the opposite direction.

Figure 5.5B shows the schematic symbol for a diode, drawn as it would be used in the circuit instead of as the pictorial of semiconductor blocks used in part A. The arrow on the schematic symbol points in the direction of conventional current instead of electronic current which is the flow of the actual electrons.

If the battery polarity is reversed, as shown in Figure 5.5C, the excess electrons in the N-type material are attracted away from the junction toward the positive battery terminal. Similarly, the holes in the P-type material are attracted away from the junction toward the negative battery terminal. When this happens, electrons do not flow across the junction to the P-type material and the diode does not conduct. When the anode is connected to a negative voltage source and the cathode is connected to a positive voltage source, the device is said to be *reverse biased*. **[E6A03]**

The voltage required for carriers to move across the PN junction results in a *forward voltage* across the diode when it is conducting. For silicon diodes, forward voltage is approximately 0.6 to 0.7 V; it is 0.2 to 0.3 V for germanium diodes.

Junction diodes are used as rectifiers to allow current in one direction only. When an ac signal is applied to a diode, it will be forward biased and conduct during one half of the cycle, allowing current to

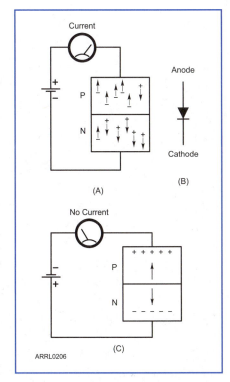

Figure 5.4 — A PN junction consists of P-type and N-type material separated by a thin depletion region in which the majority charge carriers are not present.

Figure 5.5 — At A, the PN junction is forward biased and conducting. B shows the schematic symbol used to represent a diode, oriented so that its internal structure is the same as in A. Conventional current flows in the direction indicated by the arrowhead in the symbol. At C, the PN junction is reverse biased, so it does not conduct.

Components and Building Blocks 5-3

flow to the load. During the other half of the cycle, the diode is reverse biased and current does not flow. The ac current becomes pulses of dc, always flowing in the same direction.

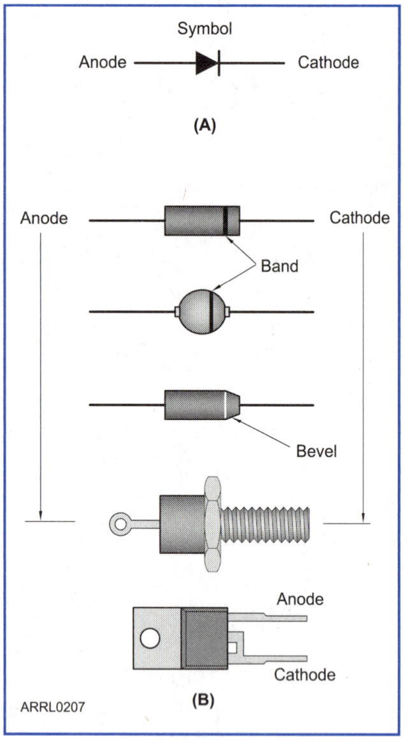

The maximum average forward current is the highest average current that can flow through the diode in the forward direction for a specified *maximum allowable junction temperature*. If allowed to get too hot, the diode will be damaged or destroyed. **[E6B07]**

Diodes designed to safely handle forward currents in excess of a few amps are packaged so they may be mounted on a heat sink. The heat sink helps the diode package dissipate heat more rapidly, keeping the diode junction temperature at a safe level. The metal case or tab of a power diode is usually electrically connected to one of the diode's layers so it must usually be insulated from ground.

Figure 5.6 shows some of the more common diode-case styles, as well as the general schematic symbol for a diode. The line, or spot, on a diode case indicates the cathode lead. Check the case or the manufacturer's data sheet for the correct connections.

SCHOTTKY BARRIER DIODES

E6B02 — What is an important characteristic of a Schottky diode as compared to an ordinary silicon diode when used as a power supply rectifier?

E6B06 — Which of the following is a common use of a Schottky diode?

E6B08 — Which of the following is a Schottky barrier diode?

E6B09 — What is a common use for point-contact diodes?

If a PN-junction's P-type material is replaced with a metal layer as in **Figure 5.7A** a *Schottky barrier* is created which has similar rectifying properties but with a lower forward voltage than an all-semiconductor junction. **[E6B08]** (Schottky was the physicist who developed this structure.) For example, the Schottky barrier diode's forward voltage is 0.2 to 0.5 V, compared to the 0.6 to 0.7 V for silicon PN-junction diodes. **[E6B02]** The lower forward voltage results in lower power dissipation than PN-junction diodes for the same amount of current, so Schottky diode rectifiers are widely used in power supply circuits. Figure 5.7B shows the schematic symbol for a Schottky diode.

Point-Contact Diodes

In a junction diode, the P and N layers are separated only by the junction, forming a capacitor: two charged plates separated by a thin dielectric. Although the internal capacitance of a PN-junction diode may be only a few picofarads, this capacitance can cause problems in RF circuits, especially at VHF and above. Junction diodes may be used from dc to the microwave region, but the *point-contact diode* has low internal capacitance that is specially designed for RF applications.

Figure 5.8 illustrates the internal structure of a point-contact diode. The point-contact diode has a much smaller surface area at the junction than does a PN-junction diode. When a point-contact diode is manufactured, the main portion of the device is made from N-type material and a thin aluminum wire, often called a *whisker*, is placed in contact with the semiconductor surface forming a Schottky

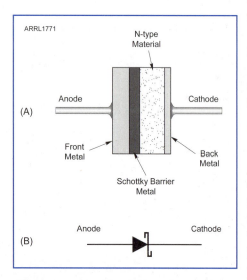

Figure 5.6 — The schematic symbol for a diode is shown at A. Typical diode packages are shown at B.

Figure 5.7 — The Schottky barrier diode substitutes a metal layer for the P-type material of a PN-junction. This results in a lower forward voltage drop than for a PN-junction diode. The schematic symbol for a Schottky barrier diode is shown at B.

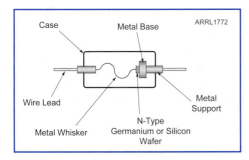

Figure 5.8 — The internal structure of a point-contact diode. The schematic symbol for point-contact diodes is the same as junction diodes.

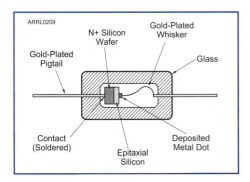

Figure 5.9 — This drawing represents the internal structure of a hot-carrier diode. The whisker contact is attached to a metal contact directly on the semiconductor material, improving mechanical and electrical performance over a point-contact diode.

barrier. The result is a diode that exhibits much less internal capacitance than PN-junction diodes, typically 1 pF or less. This means point-contact Schottky diodes are better suited for VHF and UHF applications than are PN-junction diodes. Point-contact Schottky diodes are generally used as UHF mixers and as RF detectors at VHF and below. **[E6B06, E6B09]**

Hot-Carrier Diodes

Another type of Schottky barrier diode with low internal capacitance and good high-frequency characteristics is the *hot-carrier diode*. ("Hot" refers to the diode's higher electron velocities compared to a PN-junction diode.) This device is very similar in construction to the point-contact diode but with an important difference depicted in **Figure 5.9**.

The whisker in a hot-carrier diode is physically attached to a metal dot deposited on the element. The hot-carrier diode is mechanically and electrically superior to the point-contact diode. Some of the advantages of the hot-carrier type are improved power-handling characteristics, lower contact resistance and improved immunity to burnout caused by transient noise pulses.

Hot-carrier diodes are often used in mixers and detectors at VHF and UHF. **[E6B06]** In this application, hot-carrier diodes are superior to point-contact diodes because they exhibit greater conversion efficiency and generate less noise.

ZENER DIODES

E6B01 — What is the most useful characteristic of a Zener diode?

Zener diodes (named for their inventor) are a special class of PN-junction diode used as voltage references and voltage regulators. As discussed earlier, leakage current rises as reverse voltage is applied to a diode. At first, this leakage current is very small and changes very little with increasing reverse voltage. There is a point, however, at which the leakage current rises suddenly. Beyond this point, the current increases very rapidly for a small increase in voltage; this is called the *avalanche point*. The *Zener voltage* is the voltage necessary to cause avalanche. Normal junction diodes would be destroyed if they were operated in this region, but Zener diodes are specially manufactured to safely withstand the avalanche current.

Since the current in the avalanche region can change over a wide range while the voltage stays practically constant, this kind of diode can be used as a voltage regulator. **[E6B01]** The voltage at which avalanche occurs can be controlled precisely in the manufacturing process. Zener diodes are calibrated in terms of avalanche voltage.

Zener diode voltage regulators, shown in **Figure 5.10**, provide a

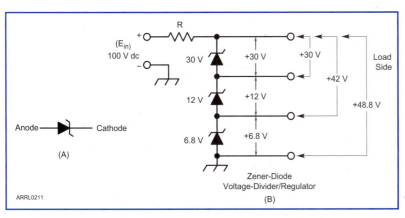

Figure 5.10 — The schematic symbol for a Zener diode is shown at A. B is an example of how Zener diodes are used as voltage regulators.

nearly constant dc output voltage, even though there may be large changes in load resistance or input voltage. As voltage references, they exhibit a stable voltage that remains constant over a wide temperature range.

Zener diodes are currently available with voltage ratings between 1.8 and 200 V. Their power ratings range from 250 mW to 50 W. They are packaged in the same case styles as junction diodes. Usually, Zener diodes rated for 10 W dissipation or more are made in stud- or tab-mount cases.

VARACTOR DIODES

E6B04 — What type of semiconductor device is designed for use as a voltage-controlled capacitor?

As mentioned above, junction diodes exhibit an appreciable internal capacitance. It is possible to change the internal capacitance of a diode by varying the amount of reverse bias applied to it, changing the separation of the carriers outside the depletion region. *Variable-capacitance diodes* and *varactor diodes* (variable reactance diodes) are designed to take advantage of this property, creating voltage-controlled capacitors. **[E6B04]** Varicap is a trade name for these diodes.

Varactors provide various capacitance ranges from a few picofarads to more than 100 pF. Each style has a specific minimum and maximum capacitance. The higher the maximum capacitance, the greater will be the minimum capacitance. A typical varactor can provide capacitance changes over a 10:1 range with bias voltages in the 0- to 100-V range.

Common schematic symbols for a varactor diode are given in **Figure 5.11**. These devices are used in frequency multipliers at power levels as great as 25 W, in remotely tuned circuits and in frequency modulator circuits.

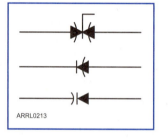

Figure 5.11 — These schematic symbols are commonly used to represent varactor diodes.

PIN DIODES

E6B05 — What characteristic of a PIN diode makes it useful as an RF switch?

E6B11 — What is used to control the attenuation of RF signals by a PIN diode?

A PIN (positive/intrinsic/negative) diode is formed by diffusing P-type and N-type layers onto opposite sides of an almost pure silicon layer, called the *I region* because conduction is carried out by the electrons *intrinsic* to a normal silicon crystal. **Figure 5.12** shows the three layers of the PIN diode. This layer is not "doped" with P-type or N-type charge carriers, as are the other layers. Any charge carriers found in this layer are a result of the natural properties of the pure semiconductor material. In the case of silicon, there are relatively few free charge carriers. PIN-diode characteristics are determined primarily by the thickness and area of the I region. The outside layers are designated P+ and N+ to indicate heavier than normal doping of these layers. PIN diodes are represented by the same schematic symbol as a PN-junction diode.

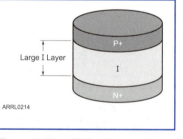

Figure 5.12 — This diagram illustrates the inner structure of a PIN diode. The top and bottom layers are labeled P+ and N+ to indicate very heavy levels of doping impurities are used.

PIN diodes respond to RF in three different ways depending on how they are biased:

• With reverse bias, the charge carriers move very slowly. Their slow response time causes the PIN diode to look like a resistor to RF currents, effectively blocking them so the PIN diode is cut-off and acts like an open circuit to RF.

• With zero bias, there are essentially no free charge carriers available to conduct so the PIN junction acts like a very small capacitor, often small enough that the diode can be considered an open circuit.

• With forward bias, the PIN diode acts like a resistance that decreases with increasing

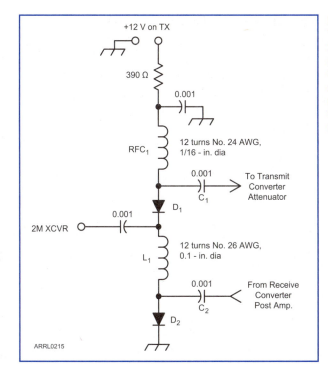

Figure 5.13 — PIN diodes may be used as RF switches. This schematic shows PIN diodes (D1, D2) switching a 2 meter transceiver between a transmit converter and a receive converter. Applying 12 V turns both diodes ON, shorting the receive converter input to ground and connecting the transmit converter to the 2 meter rig. When bias (12 V) is removed, the diodes are both OFF, disconnecting the transmit converter and reconnecting the receive converter.

bias current. Thus, the amount of resistance that a PIN diode exhibits to RF can be controlled by changing the amount of forward bias applied.

These characteristics allow the PIN diode to act as a switch or attenuator. **[E6B05, E6B11]** PIN diodes are faster, smaller, more rugged, and more reliable than relays or other electromechanical switching devices.

Figure 5.13 shows a circuit in which PIN diodes are used to build an RF switch. This diagram shows a transmit/receive switch for use between a 2 meter transceiver and a UHF or microwave transverter. With no bias, or with reverse bias applied to the diode, the PIN diode exhibits a high resistance to RF, so no signal will flow from the generator to the load. When forward bias is applied, the diode resistance will decrease, allowing the RF signal to pass. The amount of insertion loss (resistance to RF current) is determined primarily by the amount of forward bias applied; the greater the forward bias current, the lower the RF resistance.

LIGHT-EMITTING DIODES

E6B03 — What type of bias is required for an LED to emit light?
E6B10 — In Figure E6-2, what is the schematic symbol for a light-emitting diode?

Light-emitting diodes (*LEDs*) are designed to emit light when they are forward biased so that current passes through their PN junctions. **[E6B03]** As a free electron combines with a hole, it gives off light of a specific wavelength or color. LEDs are very efficient light sources.

The color of the LED depends on the material or combination of materials used for the junction. LEDs are available in many colors. By controlling the energy difference between electrons and holes, LED color can also be controlled. The intensity of the light given off is proportional to the amount of current. Red, green, and yellow LEDs are typically made from gallium arsenide, gallium phosphide, or a combination of these two materials. Blue LEDs use materials such as silicon carbide or zinc selenide. A white LED is really a blue LED with a yellowish phosphor coating on the inside of the package that glows when struck by blue light from the LED. The combination of blue light given off by the LED and yellow light from the phosphor appears white to the human eye.

LEDs are packaged in plastic cases or in metal cases with a transparent end. LEDs are useful as replacements for incandescent panel and indicator lamps. In this application they offer long life, low current drain, and small size. One of their most important electronic applications is as numeric displays in which arrays of tiny LEDs are arranged to provide illuminated segments that form numbers. The schematic symbol and a typical case style for the LED are shown in **Figure 5.14**. **[E6B10]**

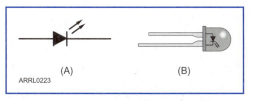

Figure 5.14 — The schematic symbol for an LED is shown at A. B is a drawing of a typical LED case style.

Components and Building Blocks 5-7

A typical red LED has a forward voltage of 1.6 V. Yellow and green LEDs have higher forward voltages (2 V for yellow and 4 V for green). The forward-bias current for a typical LED ranges between 10 and 20 mA for maximum brilliance. High-intensity LEDs used for lighting use much higher currents. As with other diodes, the current through an LED can be varied with series resistors. Varying the current through an LED will affect its intensity; the voltage across the LED, however, will remain fairly constant.

BIPOLAR TRANSISTORS

E6A06 — What is the beta of a bipolar junction transistor?
E6A07 — Which of the following indicates that a silicon NPN junction transistor is biased on?
E6A08 — What term indicates the frequency at which the grounded-base current gain of a transistor has decreased to 0.7 of the gain obtainable at 1 kHz?

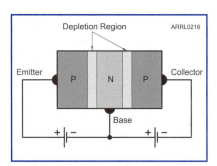

Figure 5.15 — A bipolar junction transistor consists of two layers of N- or P-type material sandwiching a layer of the opposite type of material. This drawing shows the internal structure of a PNP transistor.

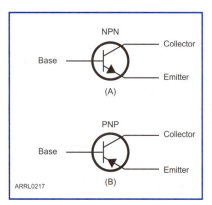

Figure 5.16 — The schematic symbol for an NPN transistor is shown at A and for a PNP transistor at B.

The *bipolar junction transistor* (BJT) is a type of three-terminal, PN-junction device able to use a small current to control a large current — in other words, amplify current. It is made of two layers of N- or P-type material sandwiching a thin layer of the opposite type of material between them as illustrated in **Figure 5.15**. If the outer layers are P-type material and the middle layer is N-type material, the device is called a *PNP transistor* because of the layer arrangement. If the outer layers are N-type material, the device is called an NPN transistor. A transistor is, in effect, two PN-junction diodes back-to-back. **Figure 5.16** shows the schematic symbols for PNP and NPN bipolar transistors. The three layers of the transistor sandwich are called the *emitter*, *base*, and *collector*. A diagram of the construction of a typical PNP transistor is given in Figure 5.15.

In an actual bipolar transistor, the base layer (in this case, N-type material) is much thinner than the outer layers. Just as in the PN-junction diode described in the previous section, a depletion region forms at each junction between the P- and N-type material. These depletion regions form a barrier to current flow until forward bias is applied across the junction between the base and emitter layers.

Forward-bias voltage across the emitter-base section of the sandwich causes electrons to flow through it from the base to the emitter. As the free electrons from the N-type material flow into the P-type material, holes from the P-type material flow the other way into the base. Some of the holes will combine with free electrons in the base, but because the base layer is so thin, most will move right on through into the P-type material of the collector.

As shown, the collector is connected to a negative voltage with respect to the base. Normally, reverse bias would prevent current from flowing across the base-collector junction. The collector, however, now contains an excess of holes because of those from the emitter that overshot the base. Since the voltage source connected to the collector produces a negative voltage, the holes from the emitter will be attracted to that power supply connection, creating current flow from the emitter to the collector.

A BJT is "biased on" when a forward voltage drop is present across the emitter-base junction and the collector-base junction is reverse-biased. For silicon transistors, the emitter-to-base "on" voltage is 0.6 to 0.7 V from the base-to-emitter for NPN and from the emitter-to-base for PNP transistors. **[E6A07]**

Bipolar junction transistors are used in a wide variety of applications, including ampli-

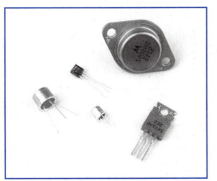

Figure 5.17 — Transistors are packaged in a wide variety of case styles, depending on the application in which they are intended to be used.

fiers (from very low level to very high power), switches, oscillators, and power supplies. They are used at all frequency ranges from dc through the UHF and microwave range. Transistors are packaged in a wide variety of case styles. Some of the more common case styles are depicted in **Figure 5.17**.

Transistor Characteristics

Because of the transistor's construction, the current through the collector will be considerably larger than that flowing through the base. When a transistor's base-emitter junction is forward biased, collector current increases in proportion to the amount of bias current applied. The ratio of collector current to base current is called the *current gain*, or *beta*. Beta is expressed by the Greek symbol β. **[E6A06]** It is calculated from the equation:

$$\beta = I_c / I_b \qquad \text{(Equation 5.1)}$$

where:
  I_c = collector current
  I_b = base current

For example, if a 1-mA base current results in a collector current of 100 mA the beta is 100. Typical betas for junction transistors range from as low as 10 to as high as several hundred. Manufacturers' data sheets specify a range of values for β. Individual transistors of a given type can have widely varying betas.

Another important transistor characteristic is *alpha*, expressed by the Greek letter α. Alpha is the ratio of collector current to emitter current, given by the equation:

$$\alpha = I_c / I_e \qquad \text{(Equation 5.2)}$$

where:
  I_c = collector current
  I_e = emitter current

The smaller the base current, the closer the collector current comes to being equal to that of the emitter and the closer alpha comes to being 1. For a junction transistor, alpha is usually between 0.92 and 0.98.

The transistor is *saturated* when further increases in base-emitter current do not increase the collector current, and the transistor is said to be *fully on* when the transistor is saturated. At the other end of the scale, when the transistor is reverse-biased, there is no current from the emitter to the collector and the transistor is at *cutoff*. When used to amplify a signal, a transistor operates between these two extremes. By operating at either cutoff or saturation, the transistor can be used as a switch.

Transistors have important frequency characteristics. The *alpha cutoff frequency* is the frequency at which the current gain of a transistor decreases to 0.707 times its gain at 1 kHz. Alpha cutoff frequency is considered to be the practical upper frequency limit of a transistor configured as a common-base amplifier. **[E6A08]**

Beta cutoff frequency is similar to alpha cutoff frequency, but it applies to transistors connected as common-emitter amplifiers. Beta cutoff frequency is the frequency at which the current gain of a transistor in the common-emitter configuration decreases to 0.707 times its gain at 1 kHz. (These amplifier configurations are explained in the Radio Circuits and Systems chapter's section on amplifier circuits.)

FIELD EFFECT TRANSISTORS

E6A05 — How does DC input impedance at the gate of a field-effect transistor compare with the DC input impedance of a bipolar transistor?
E6A09 — What is a depletion-mode FET?
E6A10 — In Figure E6-1, what is the schematic symbol for an N-channel dual-gate MOSFET?
E6A11 — In Figure E6-1, what is the schematic symbol for a P-channel junction FET?
E6A12 — Why do many MOSFET devices have internally connected Zener diodes on the gates?

Field-effect transistors (*FETs*) are given that name because the current through them is controlled by the effect of an electric field or voltage, as opposed to current as for the bipolar junction transistor. There are two types of field-effect transistors in common use today: the *junction FET* (*JFET*) and the *metal oxide semiconductor FET* (*MOSFET*). The basic characteristic of both FET types is a very high input impedance — typically 1 megohm or greater. This is considerably higher than the input impedance of a bipolar transistor. FETs are made in the same types of packages as bipolar transistors. Some different case styles are shown in **Figure 5.18**.

Figure 5.18 — FETs are packaged in cases much like those used for bipolar junction transistors.

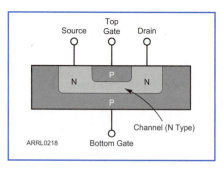

Figure 5.19 — The construction of a junction field-effect transistor (JFET). The top and bottom gate terminals are connected together outside the cross section shown here.

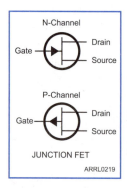

Figure 5.20 — The schematic symbol for an N-channel JFET has an arrow pointing toward the center line that represents the channel. The arrow points away from the channel for a P-channel JFET.

JFETs

The basic JFET construction is shown in **Figure 5.19**. The JFET can be thought of simply as a bar of semiconductor material that acts like a variable resistance. The terminal into which the charge carriers enter is called the *source*. The opposite terminal is called the *drain*. The terminals that control the resistance between source and drain are called *gates*. The material connecting the source and drain is called the *channel*. There are two types of JFET, named N-channel and P-channel for the type of material that forms the channel. The schematic symbols for the two JFET types are illustrated in **Figure 5.20**. **[E6A11]**

Two gate regions, made of the opposite type of semiconductor material used for the channel, are created on opposite sides of the JFET channel and connected together. When a reverse-bias voltage is applied from the top and bottom gate-channel junctions to the source, an electric field is set up across the channel. The electric field controls the normal electron flow through the channel. As the gate voltage changes, the electric field varies and that varies source-to-drain current. The gate terminal is always reverse biased, so very little current flows in the gate terminal and the JFET has very high input impedance — unlike the bipolar transistor which has much lower input impedance. **[E6A05]**

Because channel current is controlled by voltage on the gate, the gain of a FET is measured as *transconductance* (g_m), the ratio of output current to input voltage. Transconductance is measured in siemens (S), the inverse of ohms.

MOSFETs

The construction of a *metal-oxide semiconductor field-effect transistor* (*MOSFET*), sometimes called an *insulated gate field-effect transistor* (*IGFET*), and its schematic symbol are illustrated in **Figure 5.21**. In the MOSFET, the gate is insulated from the source/

5-10 Chapter 5

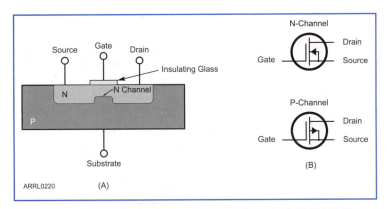
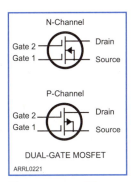

Figure 5.21 — Part A shows the construction of a MOSFET. The schematic symbols (B) for MOSFETs show that the gate terminal is not connected to the channel as in a JFET. As in the JFET symbols, the arrow's direction indicates the type of channel material. These are single-gate MOSFET symbols.

Figure 5.22 — The schematic symbols for N-channel and P-channel dual-gate MOSFETs. Notice that the direction of the arrows again indicates the type of channel material.

drain channel by a thin dielectric layer. Since there is very little current through this dielectric the input impedance is even higher than in the JFET — typically 10 megohms or greater. The schematic symbols for N-channel and P-channel dual-gate MOSFETs are shown in **Figure 5.22**. **[E6A10]** Some types of MOSFETs have two gates to which different voltages can be applied for special applications, such as mixers.

Nearly all the MOSFETs manufactured today have built-in gate-protective Zener diodes. Without this provision the gate insulation can be punctured easily by small static discharges. **[E6A12]** The protective diodes are connected between the gate (or gates) and the source lead of the FET. The diodes are generally not shown on the schematic symbol.

Enhancement and Depletion-Mode FETs

There are two types of field-effect transistors: *enhancement mode* and *depletion mode*. A depletion-mode device corresponds to Figure 5.19, where a channel exists without gate voltage applied. **[E6A09]** The gate of a depletion-mode device is reverse biased in operation. When the reverse bias is applied between the gate and source the channel is *depleted* of charge carriers and current decreases.

Enhancement-mode devices are constructed so there is no channel without voltage applied to the gate. The channel conducts current only when a gate-to-source voltage is applied that causes the channel to be able to conduct current. When the gate of an enhancement-mode device is forward biased, current begins to flow through the source/drain channel. The higher the forward bias on the gate the more current through the channel. JFETs cannot be used as enhancement-mode devices because if the gate is forward biased it will conduct like a forward-biased diode.

The gates of MOSFETs are insulated from the channel region, so they may be used as enhancement-mode devices. Both polarities may be applied to the gate without the gate becoming forward biased and conducting. Some MOSFETs are designed to be used without bias on the gate. The MOSFET operates in the enhancement mode when the gate is forward biased, and in the depletion mode when the gate is reverse biased.

Components and Building Blocks   5-11

RF INTEGRATED DEVICES

E5D03 — What is microstrip?

E6A01 — In what application is gallium arsenide used as a semiconductor material?

E6E01 — Why is gallium arsenide (GaAs) useful for semiconductor devices operating at UHF and higher frequencies?

E6E03 — Which of the following materials is likely to provide the highest frequency of operation when used in MMICs?

E6E04 — Which is the most common input and output impedance of circuits that use MMICs?

E6E05 — Which of the following noise figure values is typical of a low-noise UHF preamplifier?

E6E06 — What characteristics of the MMIC make it a popular choice for VHF through microwave circuits?

E6E07 — What type of transmission line is used for connections to MMICs?

E6E08 — How is power supplied to the most common type of MMIC?

Integrated circuits (ICs) make up most of the internal circuitry of modern electronics. If you open up a new transceiver or a computer you may be hard-pressed to find very many discrete transistors! One of the last types of electronics to convert to ICs has been VHF, UHF, and microwave circuits. Even though transistors in an IC are able to handle the high-speed signals, making an IC that would work properly in many different circuits is a tough challenge. Advances in circuit design have finally made it possible to use ICs at these high frequencies just as for lower frequencies. In fact, mobile telephones would be impossible to build without ICs that include UHF and microwave functions!

The most common RF IC used by amateurs is a *monolithic microwave integrated circuit* (*MMIC*). It's quite unlike most other ICs you've seen. Most MMICs are quite small, often called "pill packages" because they look like a small pill with four leads coming out of the device at 90° to each other. The most common MMIC has an input lead, an output lead, and two ground leads.

Wait a minute! Two ground leads? Where's the power lead? Many MMICs don't have a separate power lead — dc power to the internal electronics and the RF output from the MMIC both use the same lead! Power is supplied through a resistor and/or RF choke to the output lead. **[E6E08]** The typical dc operating voltage for an MMIC amplifier is 12 V. A small series *blocking capacitor* keeps the dc voltage from getting to any other circuits as shown in the schematic of **Figure 5.23**. MMICs use this method because of its simplicity and the extra ground lead helps ensure that the amplifier circuit operates properly over the entire frequency range.

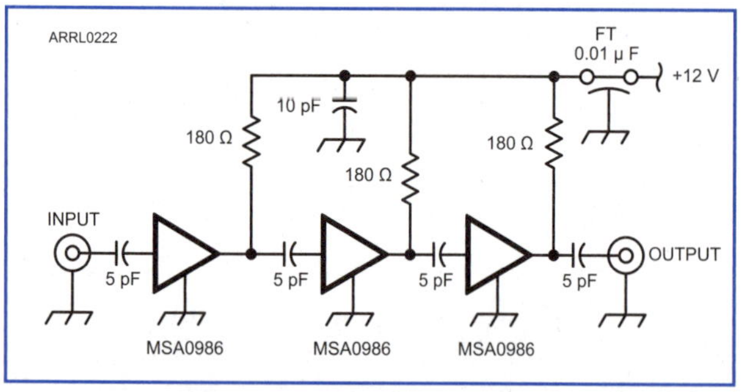

MMIC devices have well-controlled operating characteristics such as gain, noise figure and input/output impedance, requiring only a few external components for proper operation. **[E6E06]** As "building blocks," MMICs can greatly simplify an amplifier design for circuits at UHF and microwave frequencies because the circuits and the IC input and output impedances are all close to 50 Ω. **[E6E04]**

As a design example, a MAR-6 MMIC could be used to build a receive preamplifier for a 1296 MHz receiver with just a few external resistors and capacitors. This device provides 16 dB of gain for signals up to 2 GHz, with a noise figure around 3 dB. Many MMIC amplifier devices have noise figures in the

Figure 5.23 — The schematic diagram of the amplifier for UHF and microwave bands shown in Figure 5.24. Three MMIC devices provide a lot of gain without requiring a lot of complex circuitry. Operating power is supplied to the MMICs at their output pins with 5 pF capacitors coupling the RF signal to the next stage while blocking the dc voltage.

5-12   Chapter 5

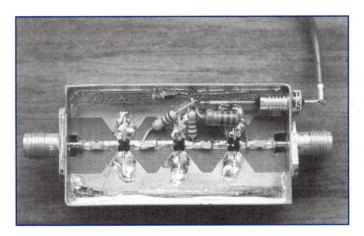

Figure 5.24 — This simple utility amplifier is suitable for low-level amplification on the amateur bands from 903 MHz through 5.7 GHz. The schematic is shown in Figure 5.23. The three MMICs are visible as round, black "pills" in a straight line between the input jack at left and the output jack at right. Short sections of microstrip transmission line connect the MMIC devices.

range of about 3.5 to 6 dB. A high-performance, low-noise UHF preamplifier can have a noise figure of 2 dB or less. [**E6E05**]

Above the VHF and low-UHF spectrum, the gain of silicon and germanium devices falls rapidly because the charge carriers don't move through those materials fast enough. Materials with higher *carrier mobility* are required. RF transistors and MMICs can operate well into the microwave range by using gallium arsenide (GaAs) and gallium nitride (GaN). [**E6A01, E6E01, E6E03**]

Circuits built using MMICs generally employ *microstrip construction techniques*. **Figure 5.24** is an example of microstrip techniques used to build an amplifier based on the schematic in Figure 5.23. Double-sided circuit board material is used, with one side forming a ground plane. Precisely sized traces over the ground plane form a 50-Ω transmission line. [**E5D03, E6E07**] MMICs and other components are soldered directly to these feed line sections. The amplifier module in Figure 5.24 includes just three MMICs, three resistors, four chip (surface-mount) capacitors and a feed-through capacitor to bring the supply voltage into the enclosure.

5.2 Optoelectronics

E6F01 — What absorbs the energy from light falling on a photovoltaic cell?
E6F02 — What happens to the conductivity of a photoconductive material when light shines on it?
E6F03 — What is the most common configuration of an optoisolator or optocoupler?
E6F04 — What is the photovoltaic effect?
E6F05 — Which describes an optical shaft encoder?
E6F06 — Which of these materials is most commonly used to create photoconductive devices?
E6F07 — What is a solid-state relay?
E6F08 — Why are optoisolators often used in conjunction with solid-state circuits when switching 120 VAC?
E6F09 — What is the efficiency of a photovoltaic cell?
E6F10 — What is the most common type of photovoltaic cell used for electrical power generation?
E6F11 — What is the approximate open-circuit voltage produced by a fully illuminated silicon photovoltaic cell?

Optics may not seem to have a lot to do with radio, but there are many components that are hybrids of optical and electronic functions. These are called *optoelectronics* and they make use of the optical properties of semiconductors to perform useful functions. The most commonly utilized optical properties are *photoconductivity*, in which light interacts with a semiconductor to change its conductivity and the *photovoltaic effect*, in which light causes current to flow.

PHOTOCONDUCTIVITY

To understand photoconductivity, we must start with the *photoelectric effect*. In simple terms, this refers to electrons being knocked loose from the atoms of a material when light shines on it. While a complete explanation of light's interaction with semiconductor material is beyond the scope of this book, we will simply describe some of the basic principles behind photoelectricity.

Let's revisit the basic structure of an atom as shown in **Figure 5.25**. The nucleus contains protons (positively charged particles) and neutrons (with no electrical charge). The number of protons in the nucleus determines the atom's element. Carbon has six protons, oxygen has eight and copper has 29, for example. The nucleus of the atom is surrounded by the same number of negatively charged electrons as there are protons in the nucleus so that an atom has zero net electrical charge.

The electrons surrounding the nucleus are found in specific energy levels, as shown in Figure 5.25. The increasing energy levels are shown as larger and larger spheres surrounding the nucleus. (While this picture is not really accurate, it will help you get the idea of the atomic structure.) For an electron to move to a different energy level, it must either gain or lose a certain amount of energy. One way that an electron can gain the required energy is by absorb-

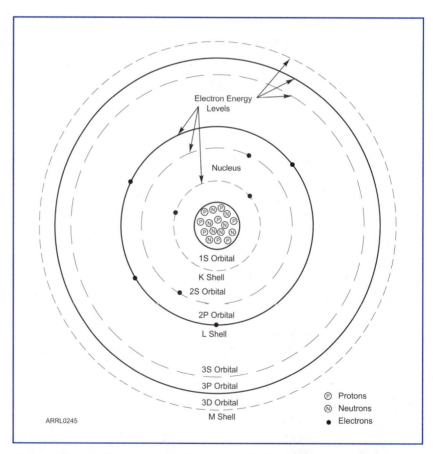

Figure 5.25 — In the structure of an atom, shells of electrons surrounding the nucleus have increasing energy levels with increasing distance from the nucleus. The photoelectric effect is caused by electrons in the outermost shell absorbing a photon of light with sufficient energy to cause them to leave the atom. As free electrons they can flow as current.

ing electromagnetic energy in the form of a photon of light. The electron absorbs the energy from the photon and jumps to a new energy level. An electron that has absorbed energy and jumped to a higher energy level is called *excited*.

If the light photon has enough energy, the electron can be freed completely from the atom. In a metallic conductor this *free electron* can now flow as an electric current. The current can then flow through a circuit connected to the material illuminated by the photons. This is the basis of the *photoelectric effect*.

The Photoconductive Effect

With this simple model of the atom in mind, it is easy to see that an electric current through a wire or other material depends on electrons being pulled away or knocked free from atoms. The rate of electrons moving past a certain point in the wire specifies the current. Every material presents some opposition to this flow of electrons, and that opposition is the *resistivity* of the material. If you include the length and cross-sectional area of a specific object or piece of wire, then you know the resistance of the object:

$$R = \rho l/A \qquad \text{(Equation 5.3)}$$

where:

ρ is the lower case Greek letter rho, representing the resistivity of the material.
l is the length of the object.
A is the cross-sectional area of the object.
R is the resistance.

Conductivity is the reciprocal of resistivity, and conductance is the reciprocal of resistance:

$$\sigma = 1/\rho \qquad \text{(Equation 5.4)}$$

where σ is the lower case Greek letter sigma, which represents conductivity and

$$G = 1/R \qquad \text{(Equation 5.5)}$$

where G is the conductance.

You just learned that according to the photoelectric effect electrons can be knocked loose from atoms when light strikes the surface of the material. With this principle in mind, you can see that those free electrons will make it easier for a current to flow through the material. But even if electrons are not knocked completely free of the atom, excited electrons in the higher-energy-level regions are more easily passed from one atom to another.

All of this discussion leads us to one simple fact: it is easier to produce a current when some of the electrons associated with an atom are excited. The conductivity of the material is increased and the resistivity is decreased. **[E6F02]** The total conductance of a piece of wire may increase and the resistance decrease when light shines on the surface. That is the nature of the *photoconductive effect*. Materials that respond to the photoconductive effect are said to exhibit *photoconductivity*.

The photoconductive effect is more pronounced and more important in crystalline semiconductor materials than in ordinary metal conductors. **[E6F06]** With a piece of copper wire, for example, the conductance is normally high so any slight increase because of light striking the wire surface will be almost unnoticeable. The conductivity of semiconductor crystals such as germanium, silicon, cadmium sulfide, cadmium selenide, gallium arsenide, lead sulfide and others is low when not illuminated but the increase in conductivity is significant when light shines on their surfaces. (This also means that the resistance decreases.)

Each material shows its biggest change in conductivity over a different range of light frequencies. For example, lead sulfide responds best to frequencies in the infrared region while cadmium sulfide and cadmium selenide are both commonly used in visible light detectors, such as are found in cameras.

Components and Building Blocks **5-15**

OPTOELECTRONIC COMPONENTS

Most semiconductor devices are sealed in plastic or metal cases so that no light will reach the semiconductor junction. Light will not affect the conductivity and hence the operating characteristics of such a transistor or diode. But if the case is made with a window to allow light to pass through and reach the junction, then the device characteristics will depend on how much light is shining on it. Such specially made devices have a number of important applications in amateur radio.

A *phototransistor* is a special device designed to allow light to reach the transistor junction. Light, then, acts as the control element for the transistor. In fact, in some phototransistors there is no base lead at all. In others, a base lead is provided, so you can control the output signal in the absence of light. You can also use the base lead to bias the transistor to respond to different light intensities. In general, the gain of the transistor is directly proportional to the amount of light shining on the transistor. A phototransistor can be used as a *photodetector* — a device that detects the presence of light.

Optocouplers and Optoisolators

An *optocoupler* or *optoisolator* is an LED and a phototransistor sharing a single IC package. [E6F03] Applying current to the LED causes it to emit light and the light from the LED causes the phototransistor to turn on. Because they use light instead of a direct electrical connection, optoisolators provide one of the safest ways to transfer signals between circuits using widely differing voltages.

Optoisolators have a very high impedance between the light source (input) and the phototransistor (output). There is no current between the input and output terminals.

The LEDs in most optocouplers are infrared emitters, although some operate in the visible-light portion of the electromagnetic spectrum. For this reason, they are often used when 120 V ac circuits are to be switched under the control of low-power digital signals. [E6F08]

Figure 5.26A shows the schematic diagram of a typical optocoupler. In this example, the phototransistor base lead is brought outside the package. The ratio of the output current to input current is called the *current transfer ratio* (CTR). As shown at B, a Darlington phototransistor can be used to improve the CTR of the device.

In an IC optoisolator, the light is transmitted from the LED to the phototransistor detector by means of a plastic light pipe or small gap between the two sections.

By combining an optocoupler with power transistors, the functions of an electromechanical relay can be implemented by solid-state components. The resulting *solid-state relay* (SSR) can operate much faster than an electromechanical relay and can be controlled directly by digital circuits. [E6F07]

A separate LED or infrared emitter and matching phototransistor detector can be separated by some small distance to use a reflective path or other external gap. In this case, changing the path length or blocking the light will change the transistor output. This can be used to detect an object passing between the detector and light source, for example.

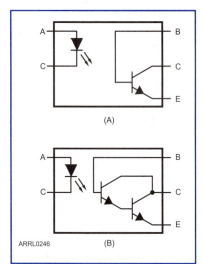

Figure 5.26 — Optocouplers consist of an LED that emits light and a phototransistor to detect light. The device shown at A uses a single-transistor detector while the unit shown at B uses a Darlington phototransistor to increase the current transfer ratio.

The Optical Shaft Encoder

An optoelectronic device widely used in radio equipment is an *optical shaft encoder*. It consists of an array of emitters and detectors. A plastic disc with a pattern of alternating clear and opaque radial bands rotates through a gap between the emitters and detectors as illustrated in **Figure 5.27**. By using an array of emitters and two detectors, a microprocessor can detect the rotation direction and speed of the wheel. [E6F05] Modern transceivers use a system

5-16   Chapter 5

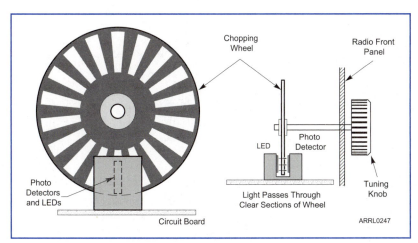

Figure 5.27 — An illustration of the operation of an optical shaft encoder, often used on the tuning and control knobs of transceivers.

like this to control the frequency of a synthesized VFO. To the operator, the tuning knob may feel like it is mechanically tuning the VFO, but there is no tuning capacitor or other mechanical linkage connected to the knob and light-chopping wheel. Inexpensive shaft encoders are often used for switches and selector controls, as well.

PHOTOVOLTAIC CELLS

The photoelectric effect can also be put to use to generate electrical energy as well as to change or control electrical properties. At a PN junction, such as in a diode, charge carriers create a depletion region as described earlier in this chapter. For charge carriers to cross the junction, voltage must be applied.

If a PN junction is exposed to light, photons will be absorbed by the electrons in the semiconductor material. **[E6F01]** If the photons have the correct energy, free electrons in the N-type material can be excited sufficiently to move across the depletion region into the P-type material. Effectively, this is the same as a hole moving the other way. As long as the junction is illuminated, electrons and holes can be made to flow across the junction, creating a voltage difference from one side of the junction to the other.

If a circuit is provided between the two sides of the junction, the voltage caused by the photons being absorbed by the electrons will cause a current to flow. This current represents the conversion of light energy from the photons to electrical energy carried by the electrons in the circuit. This is the *photovoltaic effect*. **[E6F04]** A PN-junction designed to absorb photons and create electrical energy is called a *photovoltaic* or *PV cell*. The cross section of a photovoltaic cell is shown in **Figure 5.28**.

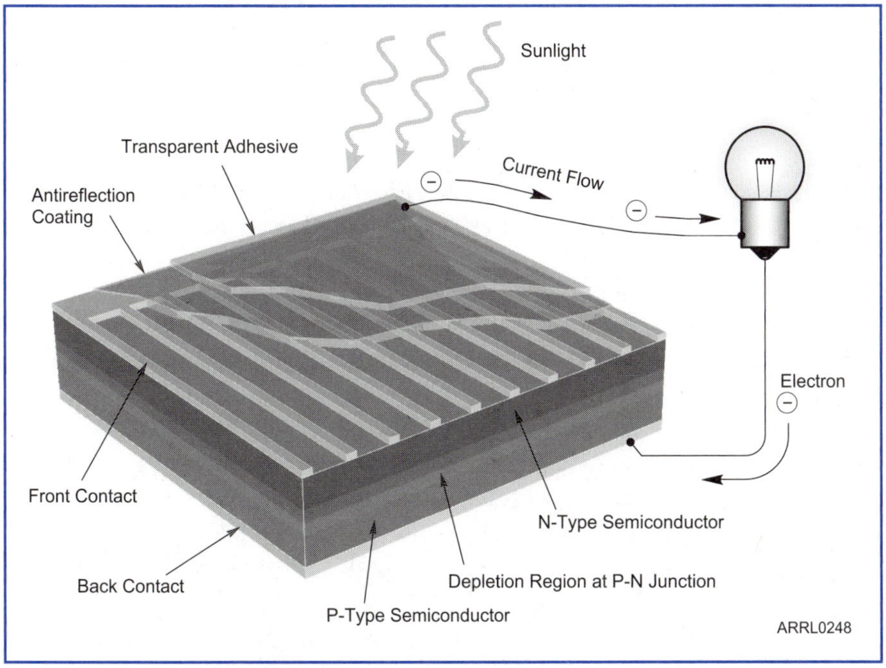

The voltage developed by a photovoltaic cell depends on the material from which it is made. For example, a fully-illuminated cell made of silicon, the most common material used for PV cells, develops an open-circuit voltage of approximately 0.5 V. **[E6F10, E6F11]** The amount of current such a cell can produce is determined by the degree of illumina-

Figure 5.28 — The construction of a photovoltaic cell. As long as the PN junction is illuminated, electrons and holes can absorb energy and flow across the junction causing a current to flow in the external circuit. This current represents the conversion of light energy from the photons to electrical energy carried by the electrons in the circuit.

Components and Building Blocks   5-17

tion and the *conversion efficiency* of the material — the relative fraction of light energy that is converted to electrical energy in the form of current. **[E6F09]** Almost any semiconductor material can be made into a photovoltaic cell, but the highest efficiency in cells made from a single material is currently found in cells made from gallium arsenide (GaAs).

Photovoltaic energy is a commercially viable option to generate electrical energy in large quantities. The photovoltaic or *solar* cell shown in Figure 5.28 is made from semiconductor material but other types of optically-active material such as mixtures of metal and semiconductor material, organic molecules, and nanomaterials may be used.

5.3 Digital Logic

E6C03 — What is tri-state logic?
E6C08 — In Figure E6-3, what is the schematic symbol for a NAND gate?
E6C10 — In Figure E6-3, what is the schematic symbol for a NOR gate?
E6C11 — In Figure E6-3, what is the schematic symbol for the NOT operation (inverter)?
E7A07 — What logical operation does a NAND gate perform?
E7A08 — What logical operation does an OR gate perform?
E7A09 — What logical operation is performed by an exclusive NOR gate?
E7A10 — What is a truth table?
E7A11 — What type of logic defines "1" as a high voltage?

Digital electronics is an important aspect of amateur radio. Everything from simple digital circuits to sophisticated microcomputer systems are used in modern amateur radio. Even simple equipment often includes a microprocessor. Applications in the radio realm include digital communications, code conversion, signal processing, station control, frequency synthesis, satellite telemetry, message handling, and other information-handling operations.

You've already been exposed to digital logic functions in your General license studies. For the Extra exam, we'll examine the fundamentals of digital logic and digital electronics. You'll be introduced to synchronous logic circuits and their applications. Are you ready — 1 or 0? I'll take that as a 1!

LOGIC BASICS

Boolean Algebra

The fundamental principle of digital electronics is that a signal can have only a finite number of discrete values or *states*. In *binary* digital systems signals may have two states, represented in base-2 arithmetic by the numerals 0 and 1. The binary states described as 0 and 1 may represent an OFF and ON condition or as space and mark in a communications transmission such as CW or RTTY. **Figure 5.29** illustrates a typical binary signal.

The simplest digital devices are switches and relays. Electronic digital systems, however, are created using *digital ICs* — integrated circuits that generate, detect, or in some way process digital signals. Whether switches or microprocessors, though, all digital systems use common mathematical principles known as *logic*. We'll start with the rules for combining different digital signals, called *combinational logic*. These rules are derived from the mathematics of binary numbers, called *Boolean algebra*, after its creator, George Boole.

In binary digital logic circuits each combination of inputs results in a specific output or combination of outputs. Except during transitions of the input and output signals (called *switching transitions*), the state of the output is determined by the simultaneous state(s) of the input signal(s). A combinational logic function has one and only one output state corresponding to each

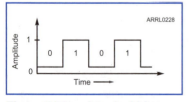

Figure 5.29 — A typical binary signal may have either of two signal levels, shown as 0 when the signal has the amplitude representing a logic value of 0 and 1 when the signal amplitude represents 1. Digital systems use many other combinations of amplitude and timing.

combination of input states. The output of a combinational logic circuit is determined entirely by the information at the circuit's inputs.

The individual circuits that perform the simplest mathematical functions are called *elements*. Combinational logic elements may perform arithmetic or logical operations. Regardless of their purpose, these operations are usually expressed in arithmetic terms. Digital circuits add, subtract, multiply, and divide but normally do it in binary form using two states that we represent with the numerals 0 and 1.

Binary digital circuit functions are represented by equations using Boolean algebra. The symbols and laws of Boolean algebra are somewhat different from those of ordinary algebra. The symbol for each logical function is shown here in the descriptions of the individual logical elements.

The logical function of a particular element may be described by listing all possible combinations of input and output values in a *truth table*. Such a list of all input combinations and their corresponding outputs characterize, or describe, the function of any digital device. **[E7A10]**

One-Input Elements

There are two logic elements that have only one input and one output: the *noninverting* buffer and the *inverter* or NOT circuit (**Figure 5.30**). **[E6C11]** The noninverting buffer simply passes the same state (0 or 1) from its input to its output. In an inverter or NOT circuit, a 1 at the input produces a 0 at the output, and vice versa. NOT indicates inversion, negation or complementation. Notice that the only difference between symbols for the noninverting buffer and the inverter is the small circle or triangle on the output lead. This is used to indicate inversion on any digital-logic circuit symbol. The Boolean algebra notation for NOT is a bar over the variable or expression.

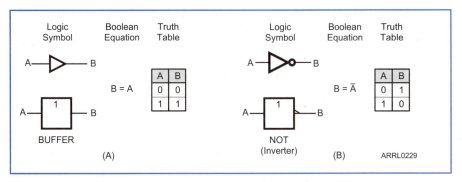

Figure 5.30 — Schematic symbols for a noninverting buffer (A) and inverter or NOT function (B) are shown. The distinctive (triangular) shape is used in ARRL publications. The square symbol is an alternate. The Boolean equation for the buffer and a truth table for the operation are also given.

The AND Operation

A *gate* is defined as a combinational logic element with two or more inputs and one output state that depends on the state of the inputs. Gates perform simple logical operations and can be combined to form complex switching functions. So as we talk about the logical operations used in Boolean algebra, you should keep in mind that each function is implemented by using a gate with the same name. For example, an AND gate implements the AND operation.

The AND operation results in a 1 only when all inputs or *operands* are 1. That is, if the inputs are called A and B, the output is 1 only if A and B are both 1. In Boolean notation, the logical operator AND is usually represented by a dot between the variables (•). The AND function may also be signified by no space between the variables. Both forms are shown in **Figure 5.31**, along with the schematic symbol for an AND gate.

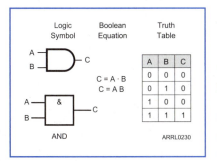

Figure 5.31 — Schematic symbols for a two-input AND gate are shown. The distinctive (round-nosed) shape is used in ARRL publications. The square symbols in this and following figures are an alternate. The Boolean equation and truth table for the operation are also given.

Components and Building Blocks 5-19

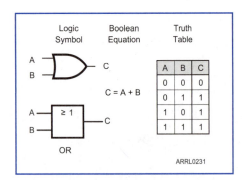

Figure 5.32 — Schematic symbols for a two-input OR gate are shown. The Boolean equation and truth table for the operation are also given.

The OR Operation

The OR operation results in a 1 at the output if any or all inputs are 1. In Boolean notation, the + symbol is used to indicate the OR function. The OR gate shown in **Figure 5.32** is sometimes called an INCLUSIVE OR. Study the truth table for the OR function in Figure 5.32. You should notice that the OR gate will have a 0 output only when all inputs are 0. [E7A08]

The NAND Operation

The NAND operation means NOT AND. A NAND gate (**Figure 5.33**) is an AND gate with an inverted output. A NAND gate produces a 0 at its output only when all inputs are 1. In Boolean notation, NAND is usually represented by a dot between the variables and a bar over the combination, as shown in Figure 5.33. [E6C08, E7A07]

The NOR Operation

The NOR operation means NOT OR. The truth table in **Figure 5.34** show that a NOR gate produces a 1 output only when all of the inputs are 0. [E6C10] In Boolean notation, the variables have a + symbol between them and a bar over the entire expression to indicate the NOR function.

The EXCLUSIVE NOR Operation

The EXCLUSIVE OR (XOR) operation results in an output of 1 only when one of the inputs is 1. If both inputs are 1 then the output is 0. The Boolean expression ⊕ represents the EXCLUSIVE OR function. Inverting the XOR function results in the EXCLUSIVE NOR (XNOR) operation. **Figure 5.35** shows the schematic symbol for an EXCLUSIVE NOR gate and its truth table. [E7A09]

Positive- and Negative-True Logic

Logic systems can be designed to use two types of *logic polarity*. *Positive* or *positive-true logic* uses the highest voltage level (HIGH) to represent binary 1 and the lowest level (LOW) to represent 0. If the opposite representation is used (HIGH = 0 and LOW = 1), that is *negative* or *negative-true logic*. In the element descriptions to follow, positive logic will be used. [E7A11]

Positive and negative logic symbols are compared in **Figure 5.36**. Small circles (*state indicators*) on the input side of a gate signify negative logic. The use of negative logic sometimes simplifies the Boolean algebra associated with logic circuits.

Consider a circuit having two inputs and one output, and suppose you desire a HIGH output

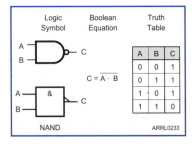

Figure 5.33 — Schematic symbols for a two-input NAND gate are shown. The Boolean equation and truth table for the operation are also given.

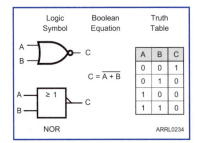

Figure 5.34 — Schematic symbols for a two-input NOR gate are shown. The Boolean equation and truth table for the operation are included.

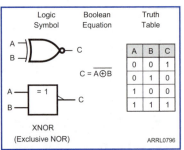

Figure 5.35 — Schematic symbols for a two-input EXCLUSIVE NOR (XNOR) gate are shown. The Boolean equation and truth table for the operation are included.

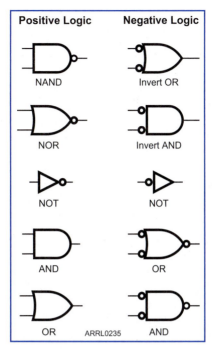

Figure 5.36 — A comparison of positive- and negative-true logic symbols for the common logic elements. The small circles at an input or output indicate inversion.

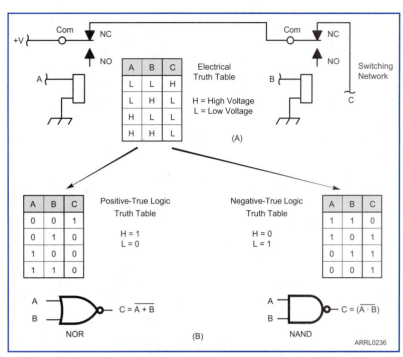

Figure 5.37 — The truth table at A describes a desired logic function. The two truth tables at B describe positive logic (NOR) and negative logic (NAND) implementations of that function.

only when both inputs are LOW. A search through the truth tables shows the NOR gate has the proper characteristics. The way the problem is posed (the words "only" and "both") suggests the AND (or NAND) function, however. A negative-logic NAND is functionally equivalent to a positive-logic NOR gate. The NAND symbol better expresses the circuit function in the application just described. **Figure 5.37** shows the implementation of a simple function as a NOR or NAND gate, depending on the logic convention chosen. Notice that the truth tables prove the circuits perform identical functions. You should verify this to be true by comparing the lists of input and output conditions.

Tri-State Logic

In digital circuits it is common for many ICs to be connected in parallel on a *data bus* or *address bus* to share data and addressing information. In this configuration, only one IC output may control the signals on the bus at a time and all other IC outputs must "stand by" by changing their outputs to act as a high impedance without attempting to drive the bus connection to a HIGH or LOW state. ICs with this ability are referred to as *tri-state logic* in which an output can be HIGH, LOW, or high-impedance. [E6C03]

SEQUENTIAL AND SYNCHRONOUS LOGIC

E7A01 — Which circuit is bistable?
E7A02 — What is the function of a decade counter?
E7A03 — Which of the following can divide the frequency of a pulse train by 2?
E7A04 — How many flip-flops are required to divide a signal frequency by 4?
E7A05 — Which of the following is a circuit that continuously alternates between two states without an external clock?
E7A06 — What is a characteristic of a monostable multivibrator?

The output state of a *sequential-logic* circuit is determined by both its present inputs and

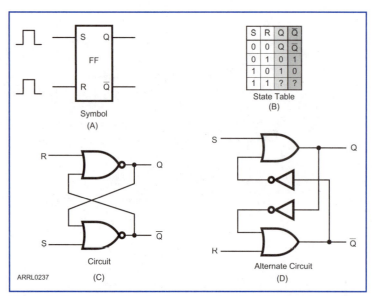

Figure 5.38 — A positive-logic, unclocked R-S flip-flop is used to illustrate the operation of flip-flops in general. Where Q and $\overline{Q}$ are shown in the state table, the previous output states are retained. A question mark (?) indicates an invalid state in which you cannot be sure what the output will be. C shows an R-S flip-flop made from two NOR gates. The circuit shown at D is another implementation using two OR gates and two inverters.

previous output states. The dependence on previous output states implies that the circuit must have some type of memory.

Flip-Flops

A *flip-flop* (also known as a *bistable multivibrator*) is a binary sequential-logic element with two stable states: the *set* state (1 state) and the *reset* state (0 state). The term *bistable* means that the circuit has two stable states and it can stay in either of them indefinitely. **[E7A01]** Thus, a flip-flop can store one bit of information. A flip-flop used to store information is sometimes called a *latch*. The schematic symbol for a flip-flop is a rectangle containing the letters FF, as shown in **Figure 5.38**. (These letters may be omitted if the function is obvious.)

Flip-flop inputs and outputs are normally identified by one or two letters. For example, the flip-flop in Figure 5.38 is an R-S type. The *state table* of Figure 5.38 shows that if S and R are both zero, the states of the Q and $\overline{Q}$ outputs are unchanged. The state table also shows that you can't be sure what the outputs (Q and $\overline{Q}$) will be if both inputs are high at the same time. There are normally two output signals that are complements of each other, designated Q and $\overline{Q}$ (read as Q NOT). If Q = 1 then $\overline{Q}$ = 0 and vice versa. See **Table 5.1** for a summary of the flip-flop output signal behavior.

Synchronous and Asynchronous Flip-Flops

The terms synchronous and asynchronous are used to characterize a flip-flop or individual inputs to an IC. In *synchronous flip-flops* (also called *clocked*, *clock-driven* or *gated* flip-flops), the output follows the input only at prescribed times determined by the clock input. *Asynchronous flip-flops* are sometimes called *unclocked* or *data-driven* flip-flops because the output can change whenever the inputs change. Asynchronous inputs are those that can affect the output state independently of the clock. Synchronous inputs affect the output state under control of the clock input.

Table 5.1
Flip-Flop Output Designations

Output	Action	Restrictions
Q (Set)	Normal output	Only two output states are possible: Q = 1 and Q = 0
$\overline{Q}$ (Reset)	Inverted output	Output states are the opposite of Q: $\overline{Q}$ = 0 and $\overline{Q}$ = 1

Notes
1) $\overline{Q}$ is the complement of Q.
2) The normal output is normally marked Q or unmarked.
3) The inverted output is normally marked $\overline{Q}$. If there is a 1 state at Q, there will be a 0 state at $\overline{Q}$.
4) Alternatively the inverted output may have a (negative) polarity indicated (a small right triangle on the outside of the flip-flop rectangle at the inverted output line). For lines with polarity indicators, be aware that a 1 state in negative logic is the same as a 0 state in positive logic. This is the convention followed by the International Electrotechnical Commission.

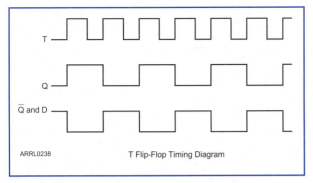

Figure 5.39 — The timing diagram for a flip-flop. Since the output changes at half the rate of the input signal, the flip-flop is acting as a divide-by-two counter.

Dynamic versus Static Inputs

Dynamic (edge-triggered) inputs can affect the outputs only when the clock changes state. This type of input is indicated on logic symbols by a small triangle (called a *dynamic indicator*) on the symbol where the input line is attached. Unless there is an inversion or *negation* indicator (a small circle or triangle outside the symbol), the 0-to-1 transition is the recognized transition. This is called *positive-edge triggering*. The negation indicator means that the input is *negative-edge triggered* and responds to 1-to-0 transitions.

Static (level-triggered) inputs are recognizable by the absence of the dynamic indicator on the logic symbol. Input states (1 or 0) of static inputs are what causes the flip-flop to act.

The timing diagram of **Figure 5.39** shows a flip-flop output changing state with each positive clock pulse. So if the output is 0 initially, it will change to a 1 on the leading edge of the first positive clock pulse and it will change back to 0 on the leading edge of the next positive clock pulse. All types of flip-flops can be configured or connected to work this way.

The flip-flop thus provides one complete output pulse for every *two* input pulses, dividing the input signal's frequency by two. **[E7A03]** Two such flip-flops could be connected sequentially to divide the input signal by four, and so on. **[E7A04]** (There is more about digital frequency divider circuits later in this section.)

One-Shot or Monostable Multivibrator

A *monostable multivibrator* (or *one-shot*) has one stable state and an unstable (or quasi-stable) state. The circuit can stay in the unstable state for a time determined by RC circuit components connected to the one-shot. When triggered, it switches to the unstable state and then returns after a set time to its original, stable state until triggered again. **[E7A06]** When the time constant has expired the one-shot reverts to its stable state until retriggered. Thus, the one-shot outputs a single pulse when triggered.

In **Figure 5.40**, the popular 555 timer IC is shown connected as a one-shot multivibrator. The action is started by a negative-going trigger pulse applied between the trigger input and ground. The trigger pulse causes the output (Q) to go positive until capacitor C charges to two-thirds of V_{CC} through resistor R. At the end of the timing period, the capacitor is quickly discharged to ground. The output remains at logic 1 for a time determined by:

$$T = 1.1\ RC \qquad \text{(Equation 5.6)}$$

where:
R is resistance in ohms.
C is capacitance in farads.
T is time in seconds.

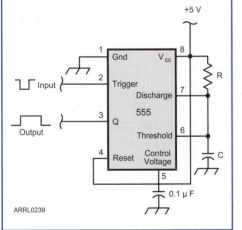

Figure 5.40 — A 555 timer IC can be connected to act as a one-shot multivibrator. See text for the formula to calculate values for R and C.

Astable Multivibrator

An *astable* or *free-running multivibrator* is a circuit that continuously switches between two unstable states. **[E7A05]**

An astable multivibrator circuit using the 555 timer IC is shown in **Figure 5.41**. Capacitor C_1 repeatedly charges to two-thirds V_{CC} through R_1 and R_2, and discharges to one-third V_{CC} through R_2. The ratio (R_1:R_2) sets the duty cycle. The frequency is determined by:

Components and Building Blocks 5-23

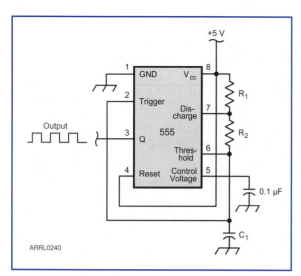

Figure 5.41 — A 555 timer IC can be connected as an astable multivibrator. See text for the formula to calculate values of R_1, R_2 and C_1.

$$f = \frac{1.46}{C_1[R_1 + 2 \times R_2]} \quad \text{(Equation 5.7)}$$

where:
R is resistance in ohms.
C is capacitance in farads.

Dividers and Counters

A *counter*, *divider*, or *divide-by-n* counter is a circuit composed of multiple flip-flops that produces an output pulse after a specified number (n) of input pulses have occurred. In a counter consisting of flip-flops connected in series, when the first stage changes state it affects the second stage and so on. Each input pulse toggles the counter circuit to the next state. The outputs from all of the flip-flops can form a composite output that forms a binary number representing the total pulse count.

A *ripple*, *ripple-carry*, or *asynchronous* counter passes the count from stage to stage; each stage is clocked by the preceding stage so that the change in the circuit's state "ripples" through the stages. In a *synchronous* counter, each stage is controlled by a common clock so that the outputs of all stages change at the same time.

Most counters have the ability to clear the count to 0. Some counters can also be preset to a desired count. Some counters may *count up* (increment) and some *count down* (decrement). Up/down counter ICs are able to count in either direction, depending on the status of a control input.

Internally, a *decade counter* IC has 10 output states. Some counters have a separate output pin for each of these 10 states while others have only one output connected to the last bit of the counter. The last flip-flop stage produces one output pulse for every 10 input pulses. [E7A02]

LOGIC FAMILIES

E6C04 — Which of the following is an advantage of BiCMOS logic?
E6C05 — What is an advantage of CMOS logic devices over TTL devices?
E6C06 — Why do CMOS digital integrated circuits have high immunity to noise on the input signal or power supply?
E6C07 — What best describes a pull-up or pull-down resistor?
E6C09 — What is a Programmable Logic Device (PLD)?

While there may be just one symbol for a NAND gate or a decade counter, there are lots of different types of digital circuits and components that can perform the necessary functions. Digital logic device manufacturers strive for consistency across their product line, creating a whole series of logic ICs with similar characteristics optimized for certain types of applications, such as low power consumption or high switching speed. All of the logic elements are available within that technology so that the circuit's signals are compatible with other similar ICs. These groups of similar ICs are called *families*. Within a logic family, all of the devices will have similar input and output signal constraints and will switch at approximately the same speed.

TTL Characteristics

Transistor-transistor logic (*TTL*) is one of the oldest bipolar logic families, so called because the gates are made entirely of bipolar transistors. Most TTL ICs are identified by 7400/5400 series numbers. For example, the 7490 is a decade counter IC. More modern families have much higher performance but the organization of logic functions and part numbering are very similar to TTL.

All of the logic elements described earlier in this section have TTL IC implementations. Some examples are the 7400 quad NAND gate, the 7432 quad OR gate and the 7408 quad AND gate. (The *quad* in these names refers to the fact that there are four individual gate circuits on the single IC chip.) Other examples of 7400 series ICs are the 7404 hex inverter, and the 7476 dual flip-flop. (*Hex* refers to the six inverters on a single IC.) The 7404 contains six separate inverters, each with one input and one output, in a single 14-pin package. A diagram of the 7404 is shown in **Figure 5.42**. The 7476 includes two J-K flip-flops on one IC.

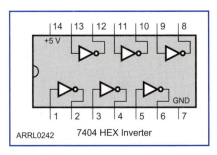

Figure 5.42 — This symbol is the schematic representation of a 7404 TTL hex inverter.

TTL ICs require a +5-V power supply. The supply voltage can vary between 4.7 and 5.3 V, but 5 V is optimum. There are also limits on the input-signal voltages. To ensure proper logic operation, a HIGH, or 1 input must be between 2 V and 5 V and a LOW, or 0 input must be no greater than 0.8 V. To prevent permanent damage to a TTL IC, HIGH inputs must be no greater than 5.5 V, and LOW inputs no more negative than –0.6 V. TTL HI outputs will fall somewhere between 2.4 V and 5.0 V, depending on the individual chip and load current. TTL LOW outputs will range from 0 V to 0.4 V. The ranges of input and output levels are shown in **Figure 5.43**. Note that the guaranteed output levels fall conveniently within the input limits. This ensures reliable operation when TTL ICs are interconnected.

TTL inputs that are left open, or allowed to "float," will cause the internal circuitry to assume a HIGH or 1 state, but operation may be unreliable. If an input should be HIGH, it is better to tie the input to the positive supply through a *pull-up resistor* (usually a 1 to 10-kΩ resistor). If an input must be kept LOW, it may be connected directly to the power supply return or common or a *pull-down resistor* may be used. In either case, the resistors ensure that the input is kept at a known logic level. Pull-up resistors are also used for *open-collector* outputs that depend on the external resistor to power the output transistor. **[E6C07]**

There are several variants of the TTL family that provide different characteristics and are identified by letters following the "74" in the part number. For example, a logic device number beginning with "74LS" is from the Low-power Schottky TTL family and a part number beginning with "74HC" is from the High-speed CMOS version of TTL logic. Within a family, it is almost always the case that parts with the same logic function will have the same pin connections. For example, all the inverters in the 7404, 74LS04, 74H04, 74S04, 74HC04, and so on families will have the same pin connections or *pinouts* as in Figure 5.42.

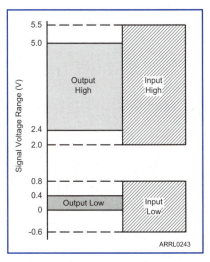

Figure 5.43 — The input and output signal-voltage ranges for the TTL family of logic devices.

CMOS Characteristics

Complementary metal-oxide semiconductor (*CMOS*) devices are composed of N-channel and P-channel FETs combined on the same substrate. Because both N and P-channel FETs can be combined on the same substrate, the circuitry can be placed in a smaller-sized area. This also helps reduce the cost of these ICs. CMOS logic has become the most widely used form of digital logic in the world because of its high switching speed, small size of the individual gates and other elements, and far lower power consumption than TTL. When a CMOS gate is not switching, it draws very little power, for example. **[E6C05]**

One of the most popular CMOS families is the parts carrying 4000-series part numbers.

Components and Building Blocks   **5-25**

For example, a 4001 IC is a quad, two-input NOR gate. The 4001 contains four separate NOR gates, each with two inputs and one output. Some other examples are the 4011 quad NAND gate, the 4081 quad AND gate, and the 4069 hex inverter.

Mentioned previously, the 74HC00-series of part numbers are pin-compatible with the 7400 TTL family, offering equivalent switching speed at much lower power. If you come across a device whose part number begins 74C or 74HC, you should be aware that a C in the part number probably indicates that it is a CMOS device.

The 4000-series of CMOS ICs (model numbers between 4000 and 4999) will operate over a much larger power-supply range than TTL ICs. The power-supply voltage can vary from 3 V to as much as 18 V. CMOS output voltages depend on the power-supply voltage. A HIGH output is generally within 0.1 V of the positive supply connection, and a LOW output is within 0.1 V of the negative supply connection (ground in most applications). For example, if you are operating CMOS gates from a 9 V battery, a logic 1 output will be somewhere between 8.9 and 9 V, and a logic 0 output will be between 0 and 0.1 V.

The switching threshold for CMOS inputs is approximately half the supply voltage. **Figure 5.44** shows these input and output voltage characteristics. The wide range of input voltages gives the CMOS family great immunity to noise, since noise spikes will generally not cause a transition in the input state. Even the TTL-compatible CMOS families have a slightly higher noise immunity because of their wider HIGH and LOW signal ranges. [E6C06]

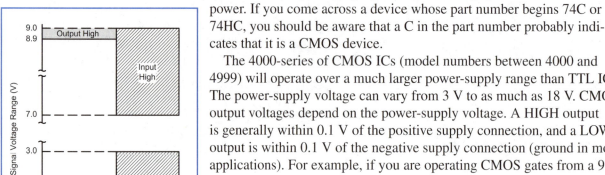

Figure 5.44 — This diagram shows 4000-series CMOS-device input and output signal-voltage ranges with a 9-V supply.

All CMOS ICs require special handling because of the thin layer of insulation between the gate and substrate of the MOS transistors. Even small static charges can cause this insulation to be punctured, destroying the gate. CMOS ICs should be stored with their pins pressed into special conductive foam. They should be installed in a socket, or else a soldering iron with a grounded tip should be used to solder them on a circuit board. Wear a grounded wrist strap when handling CMOS ICs to ensure that your body is at ground potential. Any static electricity discharge to or through the IC before it installed in a circuit may destroy it.

BiCMOS Logic

Because both bipolar and CMOS technology each have certain performance advantages, combining them in a single IC creates devices that can operate with the speed and low output impedance of bipolar transistors and the high input impedance and reduced power consumption typical of CMOS. This is referred to as BiCMOS technology. [E6C04] This allows ICs to combine analog functions, such as amplifiers and oscillators, with digital functions such as control and switching circuits.

Programmable Logic

Instead of creating complex logic functions from individual ICs, it is far more practical to use *programmable logic devices* or *PLDs*. PLDs are single ICs that consist of thousands of logic gates, sequential logic, switches, registers, and other complex functions up to and including microprocessors. [E6C09] Some PLDs are composed primarily of logic gates and are known as *programmable gate arrays* or *PGAs*. The desired circuit is designed by special software and a programmer device transfers the circuit design into the PLD where it is stored. The PLD is then dedicated to perform the functions programmed into it. Whether a PLD or PGA is used, it is possible to create extremely complex functions in the single IC that operates at very high speed. In fact, many software-defined radios (SDR) use PLDs to implement all filtering, modulation, and demodulation functions.

Chapter 6

Radio Circuits and Systems

In this chapter, you'll learn about:
- Amplifier characteristics and design
- Oscillators and frequency synthesis
- Mixers and modulators
- Detectors and demodulators
- Digital signal processing (DSP) and software defined radio (SDR)
- Filter types and characteristics
- Impedance matching
- Power supplies

You have studied dc and ac electronics principles and the basic properties of some modern solid-state components. Now you are ready for the subject of this chapter, how they are used in amateur radio. The circuits include amplifiers (both low- and high-power), signal processing (such as oscillators and modulators), filters, impedance matching circuits, and power supplies.

We can't provide a complete treatment of every subject, but the material will help you understand the circuits sufficiently to pass your Extra class exam. If you need more background or are interested in the details of the topic, consult the references listed on the *Extra Class License Manual* (ECLM) website (**www.arrl.org/extra-class-license-manual**) or the *ARRL Technical Information Service* web pages at **www.arrl.org/technical-information-service**. The *ARRL Handbook* covers all of these topics in depth, as well.

The topics of each exam question are explained, along with example calculations. Make sure you can answer the questions before going on. Ready for some circuits?

6.1 Amplifiers

When amateurs talk about amplifiers, the subject is often the piece of equipment that amplifies the output of a transceiver to several hundred watts or more. Far more numerous, however, are the much smaller amplifier circuits that increase the power of small signals in our radios and test instruments. Yet all of them have much in common.

In a piece of equipment where several amplifier circuits work together, each separate amplifier circuit is called a *stage*, just as a rocket has stages. A stage whose output signal is the input to another amplifier, particularly in a transmitter, is called a *driver*. The last amplifier in a piece of transmitting equipment is called the *final amplifier*, or simply the *final*. The circuit to which an amplifier delivers its output power is called a *load*. A load may be anything from another circuit to a dummy load to an antenna. Attaching a load to the output of an amplifier is called *loading*.

Amplifier Gain

The *gain* of an amplifier is the ratio of the output signal to the input signal. An amplifier's *voltage gain* is the ratio of its output and input voltages. *Current gain* is the ratio of output and input current, and *power gain* is the ratio of output and input power levels.

We often state the gain of a stage as a "voltage gain of 16" or a "power gain of 25," both simple ratios. But for very large ratios, such as an IF amplifier gain of 90 dB (1,000,000,000), it is easier to express and work with *decibels*. Decibels have been part of your license studies since the Technician exam and won't be covered again in this manual, but a primer on the dB is available on the ECLM website.

Input and Output Impedance

An amplifier's *input impedance* is the equivalent impedance that it presents to the preceding or driving stage. There is no single component that creates the input impedance. It is a combined effect of the components making up the circuit and the way the circuit is designed. Input impedance is measured as the ratio of input voltage to input current at the amplifier's input terminals. Input impedance almost always changes with frequency and may also change with the circuit's operating characteristics.

Output impedance is a bit more difficult to define. It is the equivalent impedance of a signal source representing the amplifier output. Low output impedance implies that the source can maintain a constant voltage over wide ranges of current. High output impedance sources maintain constant current while voltage may vary. Amplifiers that are intended to deliver significant output power generally have low output impedances, such as 50 Ω.

DISCRETE DEVICE AMPLIFIERS

E7B10 — In Figure E7-1, what is the purpose of R1 and R2?
E7B11 — In Figure E7-1, what is the purpose of R3?
E7B12 — What type of amplifier circuit is shown in Figure E7-1?
E7B13 — Which of the following describes an emitter follower (or common collector) amplifier?
E7B15 — What is one way to prevent thermal runaway in a bipolar transistor amplifier?
E7B18 — What is a characteristic of a grounded-grid amplifier?

The discussion in this section will be limited to amplifier circuits using bipolar transistors. The general techniques also apply to FETs and vacuum tubes and some topics associated with tube circuits are discussed later in this chapter. You will eventually want to learn about amplifiers using FETs and tubes, but for the purpose of helping you pass your Extra class exam we will concentrate on bipolar transistor circuits. To learn more about how FET and tube-type amplifiers function, we recommend that you turn to appropriate sections of *The ARRL Handbook*.

Basic Circuits

Amplifier circuits used with bipolar junction transistors (BJT) fall into one of three types, known as the *common-emitter, common-base* and *common-collector* circuits. "Common" means that the referenced transistor electrode — base, emitter, collector — serves as a reference terminal for both the input and output connections. The common terminal is usually circuit ground, as shown in the following circuits.

A bipolar transistor amplifier is essentially a current amplifier. Current in the base-emitter circuit controls larger currents in the collector-emitter circuit. To use the transistor as a voltage amplifier, the amplifier's output current flows through a resistive load and the resulting voltage, or change in voltage, is the amplifier's voltage signal output.

Bipolar transistor base-emitter junctions must be forward biased and the base-collector junctions reverse biased (see the Components and Building Blocks chapter) in order to act as current amplifiers. (Forward bias will be assumed when the word *bias* is used unless stated otherwise.) In circuits using an NPN transistor, the collector and base must be positive with respect to the emitter. Conversely, when using a PNP transistor, the base and collector must be negative with respect to the emitter. The required bias is provided by a power source that supplies the collector-to-emitter voltage and emitter-to-base voltage. These bias voltages cause two currents to flow: collector-emitter current and base-emitter current. The direction of current flow depends on the type of transistor used.

Either type of transistor, PNP or NPN, can be used with a negative- or positive-ground power supply. Correct bias must still be maintained, however. The combination of bias and collector-emitter current is called the circuit's *operating point*. The operating point with no input signal present is called the circuit's *quiescent* or *Q-point*. Field-effect transistor (FET) circuits have many similarities to BJT circuits but are not covered by the exam. See the *ARRL Handbook* for more information on FET circuit operation and design.

Common-Emitter and Common-Collector Circuits

Common-emitter amplifiers are the type of amplifier most often used, so we'll use that as our example of amplifier circuit operation as shown in **Figure 6.1**. You can recognize the common-emitter circuit by the value for resistance in the emitter circuit (R3 in Figure 6.1) being much smaller (or even absent) than that in the collector circuit (R4), or the emitter resistor being bypassed with a capacitor (C3 in Figure 6.1). [E7B12]

In bipolar transistors, the emitter current is in phase with the base current. As input voltage increases, so does base-emitter current which causes collector-emitter curent to increase. As more collector current flows, the voltage drop across R4 increases, lowering the voltage at the circuit's output. Thus, the input and output signals are out of phase as shown in Figure 6.1.

R1 and R2 form a *voltage divider* to create a bias voltage. This fixed dc voltage on the transistor base creates a stable operating point for the transistor. [E7B10] Determining the operating point is a compromise between amplifier circuit gain and transistor power dissipation.

As bias is increased, the collector current is higher but so is the temperature of the transistor's internal junctions. As junction temperature increases so does the gain of a BJT, causing collector-emitter current to increase even more. Even if the base-emitter bias is kept steady by R1 and R2, for large enough emitter currents or if the transistor gets hot enough, the mutually reinforcing conditions of increasing temperature and gain can result in *thermal runaway* in which the transistor junctions are overheated and destroyed. Some kind of negative feedback or *bias stabilization* is required to prevent this from happening.

One solution is to add resistance in the emitter circuit (R3) to create *degenerative emitter feedback* or *self-bias*. [E7B11] Here's how it works: As emitter current increases, so does the dc voltage across R3. This increasing voltage reduces the base-emitter forward bias established by R1 and R2, also reducing emitter current. The resulting balancing act stabilizes the transistor's operating point and prevents thermal runaway. [E7B15]

For a given load, the voltage gain of the common-emitter circuit is controlled by the ratio of R4 to R3. The values of R4 and R3 also set the value of collector current for a given bias voltage.

A variation of the common-emitter circuit leaves both R4 and R3 in place but connects C2 to ground meaning this circuit is a

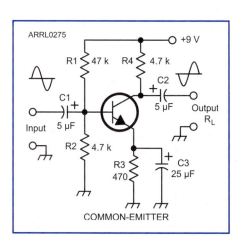

Figure 6.1 — The common-emitter amplifier circuit has a voltage divider bias source (R1 and R2), degenerative emitter feedback for bias stabilization (R3), and a resistive collector load (R4). C1 and C2 block dc but pass ac input and output signals. C3 acts as an emitter bypass to increase ac gain while maintaining stable behavior at dc.

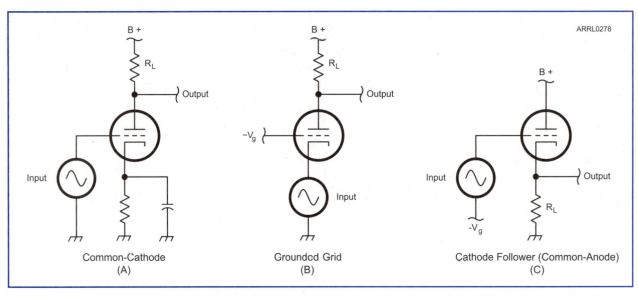

Figure 6.2 — Each of the common tube amplifier configurations — common-cathode (A), grounded-grid (B), common-anode (C) — has its analog in transistor amplifier circuits.

common-collector amplifier. The output signal is taken from the emitter. The emitter voltage is in-phase with or "follows" the base input voltage so this circuit is called an *emitter follower*. Since R3 is usually a low value, the output impedance of the emitter follower is also low, making it a good choice for driving low-impedance loads like coaxial cables. **[E7B13]** These circuits are discussed in more detail in the following section "For More Information."

Similarities of Vacuum Tube Circuits

Each of the bipolar transistor's or FET's electrodes — emitter or source; base or gate; collector or drain — has a comparable vacuum tube electrode. The analog of the emitter and source is the tube's *cathode*. The base and gate correspond to a tube's *grid* and the collector and drain correspond to the tube's *anode* or *plate*. Thus, each of the three transistor amplifier circuits discussed here has a corresponding vacuum tube amplifier circuit. **Figure 6.2** shows the three tube amplifier configurations. Like the transistor amplifiers, tube circuits have unique characteristics:

• *Common-cathode:* Input signal applied to the grid, relatively high input impedance and power gain, often requires neutralization at VHF, output is in-phase with the input signal.

• *Grounded-grid*: Input signal applied to the cathode, no current gain, low input impedance matches well to 50-Ω feed line, grounded grid reduces need for neutralization. **[E7B18]**

• *Common-anode* (*cathode follower*): Input signal applied to the grid, high input impedance, no voltage gain, output is in-phase with the input signal.

For More Information

The low reactance of C3 at audio frequencies *bypasses* R3 for ac signals which increases ac signal gain as explained below. C1 and C2 are coupling capacitors which allow the desired ac signals to pass into and out of the amplifier, while blocking the dc bias voltages. Their values are chosen to provide a low reactance at the signal frequency.

The input impedance of the common-emitter amplifier is fairly high — several kΩ is typical. The output impedance depends largely on the collector circuit resistance (R4). The common-emitter circuit has a lower *cutoff frequency* (the frequency at which current gain is reduced by half, or 3 dB) than does the common-base circuit, but it can provide the highest power gain of the three amplifier circuit configurations.

The common emitter amplifier's voltage gain (the letter A represents gain and A_V is voltage gain) is determined by the following ratio. The numerator is the parallel combination of an external load, R_L, connected between the output terminals and the sum of R3 and R4. ("//" means "the parallel combination of") The second resistance is the sum of R3 and the transistor's internal emitter resistance (26 mV/I_e) in the denominator.

$$A_V \text{ (dc gain)} = \frac{-[(R3 + R4)//R_L]}{\left(R3 + \dfrac{26}{I_e}\right)} \qquad \text{(Equation 6.1)}$$

where:

I_e is the emitter current in milliamperes
R_L is the output load.

The minus sign indicates that the input and output signals are out of phase — an increasing input voltage results in a decreasing output voltage. The quantity $26/I_e$ is also called the *dynamic emitter resistance* and is abbreviated r_e.

Equation 6.1 tells us that as R3 gets larger, A_V gets smaller. Wouldn't it be nice if we could get rid of R3 and have higher gain, but retain its stabilizing effect on bias? That is the function of C3 — at dc, R3 provides bias stabilization, while for ac signals, the low reactance of C3 bypasses R3 and changes the equation for voltage gain to:

$$A_V \text{ (ac gain)} = \frac{-(R4//R_L)}{r_e} \qquad \text{(Equation 6.2)}$$

Thus, the common emitter circuit has a different value of gain at dc than for ac.

For example, if in the circuit of Figure 6.1 the emitter current is 1.3 mA and R_L is omitted:

$$A_V \text{ (dc gain)} = \frac{-4700}{(470+20)} = -9.6$$

and

$$A_V \text{ (ac gain)} = \frac{-4700}{20} = -235$$

That's quite a difference in gain! To express this voltage-gain ratio in decibels, take the logarithm of 235 (the minus sign is ignored), and multiply by 20:

$$A_V \text{ (in dB)} = 20 \log (235) = 20 \times 2.37 = 47.4 \text{ dB}$$

In a more complete analysis of the circuit, the effects of C1 and C2 cannot be ignored at low frequencies. The increasing reactance of the coupling capacitors reduces gain at low frequencies until, at dc, input and output signals are completely blocked. The increasing reactance of C3 as frequency decreases also causes it to be less effective as a bypass so gain eventually drops to the dc value. As frequency increases, the transistor's current gain also decreases, reducing gain at high frequencies. For this reason, this value of A_V is said to be *mid-band gain*.

In **Figure 6.3**, the common-collector or emitter-follower circuit's input and output signals are in phase. C2 is a collector bypass capacitor. This amplifier also uses input- and output-coupling capacitors, C1 and C3. R3 acts to stabilize the circuit bias. The circuit's ac load consists of R3//R_L. The common-collector circuit can be recognized by the output voltage being taken from the emitter and the collector being kept at ac ground with a bypass capacitor.

Common-collector amplifiers are often used as *buffer* amplifiers that isolate low-power or sensitive stages from a heavy or varying output load. The circuit also performs *impedance*

Radio Circuits and Systems   6-5

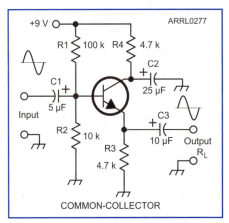

Figure 6.3 — The common-collector amplifier is also known as an emitter follower because the output voltage across the emitter resistor, R3, is in phase with and very nearly equal to that of the input signal.

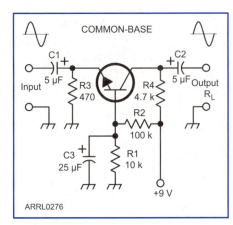

Figure 6.4 — The common-base amplifier is used when a low input impedance and high output impedance are needed. It is often used for RF preamplifiers and input circuits, since its input impedance is near 50 Ω. The circuit has approximately the same output and input current, but can have high values of voltage gain.

conversion. It has high input impedance and low output impedance. The input impedance is approximately equal to the load impedance connected to the output terminals, R_L, divided by $(1 - \alpha)$. Since α, the ratio of collector to emitter current, is close to 1, that means $(1 - \alpha)$ is a small number making the input impedance quite large. Having input impedance depend on load impedance is a disadvantage of this type of amplifier, especially if the load impedance varies with frequency.

Figure 6.4 shows the common-base amplifier circuit. It can be recognized as the common-emitter "on its side," with the input signal applied to the emitter instead of the base. When viewed from this perspective, the bias resistors, R1 and R2, have much the same configuration as in the common-emitter circuit, and that input and output coupling capacitors are also used. C3 bypasses the base to ground so that a steady dc forward bias current flows in the base-emitter circuit.

By using the transistor in this way the circuit has no current gain because the collector output current is equal to the emitter current less the small base current. This also means that the input and output signals are in-phase. The output impedance of the circuit is high so the collector current is almost independent of whatever R_L is connected to the output terminals. As a result, the common-base amplifier can have fairly large voltage gains:

$$A_V = \frac{R_L}{R3} \qquad \text{(Equation 6.3)}$$

Common-base amplifiers are used as an impedance converter when signals from a low-impedance source (such as a 50-Ω feed line) must drive a higher impedance load (such as another amplifier circuit's input). Common-base amplifiers are frequently used as receiver preamplifiers because of their high voltage gain.

OP AMP AMPLIFIERS

E7G01 — What is the typical output impedance of an op-amp?
E7G03 — What is the typical input impedance of an op-amp?
E7G04 — What is meant by the term "op-amp input offset voltage"?
E7G06 — What is the gain-bandwidth of an operational amplifier?

E7G07 — What magnitude of voltage gain can be expected from the circuit in Figure E7-3 when R1 is 10 ohms and RF is 470 ohms?

E7G08 — How does the gain of an ideal operational amplifier vary with frequency?

E7G09 — What will be the output voltage of the circuit shown in Figure E7-3 if R1 is 1000 ohms, RF is 10,000 ohms, and 0.23 volts DC is applied to the input?

E7G10 — What absolute voltage gain can be expected from the circuit in Figure E7-3 when R1 is 1800 ohms and RF is 68 kilohms?

E7G11 — What absolute voltage gain can be expected from the circuit in Figure E7-3 when R1 is 3300 ohms and RF is 47 kilohms?

E7G12 — What is an operational amplifier?

The *operational amplifier* or *op amp* is a high-gain, *direct-coupled*, *differential* amplifier that amplifies dc signals as well as ac signals. [**E7G12**] Direct-coupling means that the circuit's internal components and stages are connected directly together without blocking, coupling, or bypass capacitors, so that it works with dc and ac signals in exactly the same way.

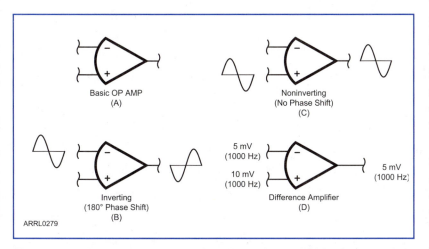

Figure 6.5 — Part A shows the basic schematic symbol for an operational amplifier (op amp). Parts B through D show how the output responds to signals at the op amp's inputs.

The input to a differential amplifier is the difference between two input signals. Op amps were originally used in analog computers for performing mathematical operations such as multiplying numbers and extracting square roots; hence the name operational amplifier.

Operational amplifiers have two inputs, one *inverting* and one *non-inverting*, as shown in **Figure 6.5**. Signals connected to the inverting (labeled –) and non-inverting (labeled +) inputs result in out-of-phase and in-phase output signals, respectively. Because it is a differential amplifier, the op amp amplifies the difference between the signals at its two inputs, regardless of the absolute voltage level at either input — it is only the difference that matters.

Op Amp Characteristics

A theoretically perfect (*ideal*) op amp would have the following characteristics: infinite input impedance, zero output impedance, infinite voltage gain that does not vary with frequency, and zero output when the input is zero. [**E7G01, E7G03, E7G08**] Because of this, the characteristics of op amp circuits are controlled by components external to the op amp itself. These criteria can be approached in a practical op amp as described in the following paragraphs.

The voltage gain of a practical op amp without feedback (*open-loop gain*) is often as high as 120 dB (1,000,000). Op amps are rarely used as amplifiers in the open-loop configuration, however. Usually, some of the output is fed back to the inverting input, where it acts to reduce and stabilize the circuit gain. The more negative feedback that is applied, the more stable the amplifier circuit will be.

The open loop gain of a practical op amp decreases linearly with increasing frequency. The *gain-bandwidth* of an op amp is the frequency range over which the open-loop voltage gain is equal to or greater than 1 (0 dB). [**E7G06**]

The gain of the circuit with negative feedback is called the *closed-loop gain*. The higher

Radio Circuits and Systems   6-7

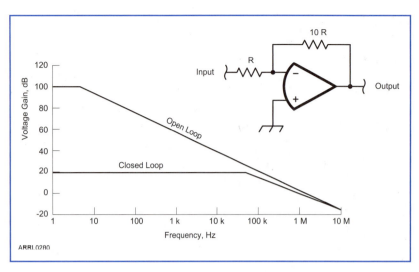

Figure 6.6 — The open-loop gain and closed-loop gain are shown as a function of frequency.

the open-loop gain, the more negative feedback that can be used and still have a useful amount of closed-loop gain. By connecting the op amp in a closed-loop circuit as shown in **Figure 6.6**, circuit gain remains constant over a wide frequency range.

If the input terminals of an op amp are shorted together, the output voltage should be zero. With most op amps there will be a small output voltage, however. This voltage offset results from imbalances between the op amp's input transistors. The op amp's *input-offset voltage* specifies the voltage between the amplifier inputs that will produce a zero output voltage, assuming the amplifier is in an open-loop circuit. [E7G04] Offset-voltages range from millivolts in consumer-grade devices down to nanovolts or microvolts in premium op amps.

Basic Amplifier Circuits

Op amps make excellent, low-distortion amplifiers. They can be used to make oscillators that generate sine, square, and even sawtooth waves. Used with negative feedback, their high input impedance and linear characteristics make them ideal for use as instrumentation amplifiers that amplify signals for precise measurements. There are many books, such as *IC Op-Amp Cookbook* by Walter Jung, that describe useful op-amp circuits.

The op amp's high gain amplifies the difference between the voltages between its inputs. Applying negative feedback causes the op amp to attempt to drive the input difference voltage to zero. The op amp's high input impedance allows current into or out of the inputs to be ignored. The usefulness of these two negative feedback concepts — input difference voltage driven to zero and no input current — will become apparent as we derive the gain for the simple *inverting* op amp circuit in **Figure 6.7**. (The circuit is inverting because the input and output signals are out of phase.)

The op amp's high gain forces the voltages at the inverting and non-inverting terminals to be approximately equal. Since the non-inverting input is connected to ground, the voltage at the inverting input will be forced to ground potential, no matter what the value of the circuit input and output voltages (as long as they are within the power supply range). Maintaining one input at ground potential without a direct ground connection is called a *virtual ground*.

Voltage gain for the inverting op-amp circuit in Figure 6.7 is determined solely by R1 (the input resistor) and R_F (the feedback resistor). In order to maintain the inverting input at ground potential, any input current $I_{IN} = V_{IN} / R1$ must be balanced by an equal and opposite feedback current $I_F = -V_{OUT} / R_F$, or:

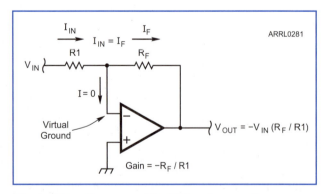

Figure 6.7 — Because of its high gain, negative feedback forces the op amp to keep the inverting and non-inverting inputs at nearly the same voltage. To do so requires balancing input current through R1 and feedback current in R_F, resulting in voltage gain of $-R_F/R1$.

$$\frac{V_{IN}}{R_1} = \frac{-V_{OUT}}{R_F}$$

so

$$A_V = \frac{V_{OUT}}{V_{IN}} = \frac{-R_F}{R_1} \qquad \text{(Equation 6.4)}$$

Op amp circuit gain is generally stated as a magnitude ($|A_V|$) and either as inverting or non-inverting. This dependence only on external components makes computing circuit gain easy as shown in the following examples:

Example 6.1

What is the voltage gain of the circuit in Figure 6.7 if R1 = 1800 Ω and R_F = 68 kΩ? **[E7G10]**

$$|A_V| = \frac{R_F}{R1} = \frac{68000}{1800} = 38$$

Example 6.2

What is the voltage gain of the circuit in Figure 6.7 if R1 = 10 Ω and R_F = 470 Ω? **[E7G07]**

$$|A_V| = \frac{R_F}{R1} = \frac{470}{10} = 47$$

Example 6.3

What is the voltage gain of the circuit in Figure 6.7 if R1 = 3300 Ω and R_F = 47 kΩ? **[E7G11]**

$$|A_V| = \frac{R_F}{R1} = \frac{47000}{3300} = 14$$

Example 6.4

What will be the output voltage of the circuit in Figure 6.7 if R1 = 1000 Ω and R_F = 10 kΩ and the input voltage = 0.23 V? **[E7G09]**

$$|A_V| = \frac{R_F}{R1} = \frac{10000}{1000} = 10$$

The circuit is inverting, so $V_{OUT} = -A_V V_{IN} = -10 \,(0.23) = -2.3$ V

COMPARATORS

E6C01 — What is the function of hysteresis in a comparator?
E6C02 — What happens when the level of a comparator's input signal crosses the threshold?

A voltage comparator is another special form of op amp circuit, shown in **Figure 6.8**. It has two analog signals as its inputs and its output is either high or low depending on whether the noninverting or inverting signal voltage is higher, respectively. Thus, it "compares" the input voltages.

A standard operational amplifier can act as a comparator by connecting the two input voltages to the noninverting and inverting inputs with no input or feedback resistors. If the voltage of the noninverting input is higher than that of the inverting input, the output voltage will be driven to the positive limit. If the inverting input is at a higher potential than the noninverting input, the output voltage will be driven to the negative limit. External resis-

Radio Circuits and Systems 6-9

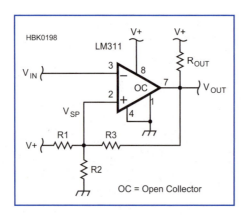

Figure 6.8 — A comparator circuit in which the output voltage is low when voltage at the inverting input is higher than the setpoint voltage, V_{SP}, at the non-inverting input. R3 creates hysteresis by allowing more current to flow through R1 when the comparator output is low, shifting the setpoint by a few millivolts.

tors generate a reference voltage, V_{SP}, called the *setpoint* or *threshold* to which the input signal is compared. Thus the comparator changes its output state depending on whether the unknown voltage is above or below the threshold. **[E6C02]**

Comparator circuits use *hysteresis* to prevent "chatter" — the output of the comparator switching rapidly back and forth when noise causes the input voltage to cross the setpoint threshold repeatedly. The unstable output can be confusing to the circuits acting on the comparator output. **[E6C01]**

Hysteresis is a form of positive feedback that "moves" the setpoint by a few millivolts in the direction opposite to that in which the input signal crossed the setpoint threshold. The output of the comparator is fed back to the positive input through resistor R3, adding or subtracting a small amount of current from the divider and shifting the setpoint.

CLASSES OF OPERATION

E7B01 — For what portion of the signal cycle does each active element in a push-pull Class AB amplifier conduct?

E7B02 — What is a Class D amplifier?

E7B03 — Which of the following components form the output of a class D amplifier circuit?

E7B04 — Where on the load line of a Class A common emitter amplifier would bias normally be set?

E7B06 — Which of the following amplifier types reduces even-order harmonics?

E7B07 — Which of the following is a likely result when a Class C amplifier is used to amplify a single-sideband phone signal?

E7B14 — Why are switching amplifiers more efficient than linear amplifiers?

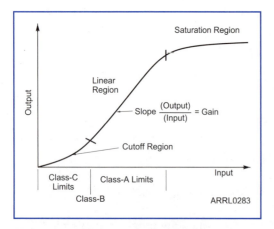

Figure 6.9 — This graph shows the three regions of amplifier operation: cutoff, linear and saturation. Each region of the curve corresponds to a different operating class.

Figure 6.9 is a graph showing a typical amplifier circuit's output versus input. As more input signal is applied, the output increases because of the amplifier's gain. When an amplifier transistor's collector or drain current becomes very small (cutoff) or very large (saturation), changes in the input signal cause less and less change in the current, creating distortion in the output signal.

The amplifier's operating point is controlled by the amount of bias. By changing the operating point the amplifier can be configured to operate closer to saturation or cutoff or somewhere in between. The effect of the operating point on the amplifier's linearity is called the amplifier's *operating class*.

The three basic operating classes for analog amplifier circuits are *Class A*, *Class B*, and *Class C*. The class of operation with characteristics intermediate between the A and B classes is called *Class AB* operation. Each class has advantages and

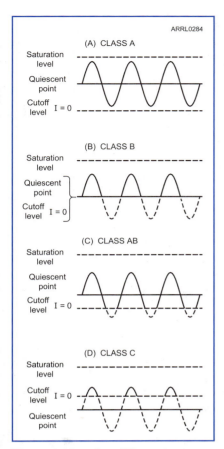

Figure 6.10 — Amplifier output waveforms for various classes of operation. All waveforms assume a sine-wave input signal.

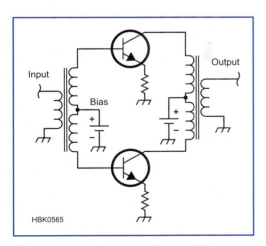

Figure 6.11 — A push-pull amplifier consisting of two transistors biased to operate in Class B at their cutoff points. Each transistor conducts for alternating halves of each cycle. The output transformer recombines the signals from each transistor to form a complete output signal.

disadvantages. **Figure 6.10** shows the output signal from amplifiers operating in Class A, B, AB and C.

Class A

In Class A operation, bias is adjusted so that the amplifier's operating point is halfway between saturation and cutoff regions as shown in Figure 6.10A. [**E7B04**] This keeps the amplifier in the linear region on the graph at all times. As long as the input signal is not too large, the output signal is a linear reproduction of the input with a minimal amount of distortion. The efficiency of a Class A amplifier is low (typically 25-30%) because there is always a significant amount of current being drawn from the power supply, even with no input signal.

Class B

Class B operation sets the bias at the cutoff level as shown in Figure 6.10B. In this case, the output signal only appears during one-half of the input sine wave. The advantage is increased efficiency with practical amplifiers often attaining 60% efficiency.

Class B amplifiers are often used at audio frequencies by connecting two tubes or transistors in a *push-pull circuit* as in **Figure 6.11**. This circuit is popular in RF power amplifiers. While one transistor is cut off, the other is conducting, so both halves of the signal waveform are present in the output. This reduces the amount of distortion in the output and will reduce even-order harmonics. [**E7B06**]

Class AB

For a Class AB amplifier, the drive level and bias are adjusted so that the operating point allows the amplifier to enter the cutoff region on signal peaks. Figure 6.10C shows the output signal for a Class AB amplifier. It operates for between 180 and 360 degrees of the signal cycle. This class of operation is often used for voice signals when a small amount of distortion is acceptable as the price of improved efficiency which is often more than 50%. [**E7B01**] Harmonics and other distortion products are reduced by using filters at the amplifier output.

Class C

For Class C amplification the bias is adjusted so that the operating point is in the cutoff region. The amplifier only conducts current during part of a half-cycle of the input signal, creating pulses at the signal frequency, as shown in Figure 6.10D. The result is that the operating efficiency can be quite high — up to 80% with proper design. Linearity is very poor so Class C amplifiers can only be used for CW and FM signals which do not require linear amplification. Using a Class C amplifier for SSB or digital signals would result in too much distortion and the output signal would occupy excessive bandwidth [**E7B07**] A tuned filter is required at the output of a Class C amplifier to reduce harmonics and other distortion products.

Switching or Switchmode Classes

Switching or *switchmode* operation goes beyond Class C with the transistor acting entirely as a switch — either completely saturated or completely cutoff nearly all of the time. This results in very low power dissipation by the transistor and efficiencies of more than 90% which is significantly higher than for linear amplifier classes. [**E7B14**]

There are many classes of switchmode amplifiers but all employ same the basic mechanism. For example, Class D amplifiers are used to amplify audio signals with the switching taking place at many times the highest audio frequency to be amplified. The switching action creates an output waveform that is a series of squared-off pulses very rich in harmonics. A low-pass filter at the amplifier output removes the harmonics while leaving signals in the desired range of frequencies unattenuated. [**E7B02, E7B03**]

DISTORTION AND INTERMODULATION

E7B16 — What is the effect of intermodulation products in a linear power amplifier?
E7B17 — Why are odd-order rather than even-order intermodulation distortion products of concern in linear power amplifiers?

The linearity of the amplifier stage is important because it describes how faithfully the input signal will be reproduced at the output. Any nonlinearity results in a distorted output. The Class A amplifier will have the least amount of distortion, while a Class C amplifier produces a severely distorted output. The tradeoff is of non-linearity for efficiency.

A consequence of nonlinearity is that the output waveform will contain harmonics of the input signal. Distortion causes a pure sine wave input signal to become a complex combination of sine waves at the output. To remove harmonics, RF amplifiers use a tuned output or *tank* circuit. The tank circuit gets its name from storing energy in the inductance and capacitance. Like a mechanical flywheel, the tank circuit smooths the pulses that occur from turning the amplifying device off for parts of each cycle, reducing harmonics.

This is especially useful if you are amplifying a pure sine-wave signal, such as for CW, and want to take advantage of the increased efficiency offered by a Class C amplifier. The tuned circuit will reduce the unwanted harmonics generated by a nonlinear amplifier stage.

If the input signal consists of sine waves of more than one frequency, such as a voice signal, the nonlinearity of the amplifier will also create *intermodulation products* or *intermod*. Intermodulation products are created at the sum and difference of all of the harmonics of the input signals and so would be considered spurious signals if transmitted.

The severity of intermodulation distortion depends on the *order* of the products that are created, even or odd. (Intermodulation in receivers is explained in detail in the Radio Signals and Measurements chapter.) *Even-order products* result in spurious signals near harmonics of the input signal and *odd-order products* result in spurious signals near the frequencies of the input signals. [**E7B16**] The higher the harmonics that are combined, the weaker the product. This means that the lower odd-order intermodulation products, specifically third-order, are more likely to cause interference to signals near the desired transmit frequency. [**E7B17**]

6.2 Signal Processing

In this section, we begin to study basic that are useful in radio: oscillators, modulators and demodulators, detectors, mixers, phase-locked loops, and frequency synthesis. These generate or manipulate signals in support of getting information from one point to another via radio. The general name for these functions is *signal processing*. These functions can be performed by circuits or software. It is easiest to understand them as circuits first, however, and much equipment is circuit-based, so that will be our focus in this section. We begin with the source of nearly all signals — the oscillator.

OSCILLATOR CIRCUITS AND CHARACTERISTICS

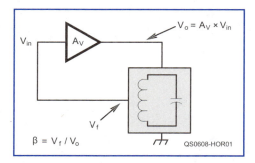

Figure 6.12 — An oscillator consists of an amplifier plus a feedback network, shown here as a parallel LC circuit. The LC circuit acts as a filter, restricting feedback to its resonant frequency.

To create an oscillator, we need three things: an amplifier with gain at the desired frequency, a circuit that provides positive feedback from the output of the amplifier to its input, and a filter that restricts the feedback to the desired frequency. These are connected in a *feedback loop* as shown in a block diagram by **Figure 6.12**.

The feedback loop is designed so that at the frequency of interest, the product of amplifier gain, A_V, and feedback ratio, β, is equal to or greater than 1. $A_V\beta$ is also known as *loop gain*, the total gain experienced by a signal all the way through the amplifier, the feedback circuit (the LC circuit in Figure 6.12), and back to the input. (The feedback ratio, β, is different than the current gain, β, of transistors.)

Signals of the right frequency will be amplified, a portion fed back to the amplifier input, amplified again, and so on, becoming self-sustaining and creating a steady output signal. If the phase difference is not exactly right, however, then the returned portion of the output signal becomes progressively farther and farther out of phase on each trip and won't build up a sustained signal.

The resonator serves two purposes. It provides the necessary phase shift at the desired frequency, and it acts as a filter for signals in the amplifier loop so that only the desired signals are amplified. (The dc power and bias connections are omitted for clarity.)

INSTABILITY AND PARASITIC OSCILLATION

E7B05 — What can be done to prevent unwanted oscillations in an RF power amplifier?

E7B08 — How can an RF power amplifier be neutralized?

The combination of positive feedback, gain, and filtering can cause any amplifier to become unstable whether the amplifier is a small circuit or a large RF power amplifier. The amplifier can oscillate steadily, in bursts, or generate noise. Even low-power amplifiers can experience instability resulting in noisy outputs, generating spurious signals, and interference to other services if oscillations are coupled back to an antenna.

Negative feedback can stabilize an RF amplifier as described in the following sections on *neutralization* and *parasitic suppression*. [**E7B05**] Care in terminating the amplifier input and output, attention to proper circuit layout, bonding of enclosures, and proper shielding of the input from the output can also prevent oscillation.

Neutralization

As we saw from the basic oscillator circuit, oscillation can occur when some of the output signal is fed back in phase with the input signal as positive feedback. As the output voltage increases so will the feedback signal. The re-amplified signal can then build up to the point where it is self-sustaining and the amplifier is now an oscillator.

To prevent the unwanted oscillations, it is necessary to

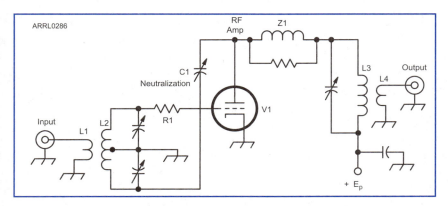

Figure 6.13 — Examples of neutralization and parasitic suppression techniques used with a tube-type RF amplifier.

Radio Circuits and Systems   6-13

cancel the positive feedback. This is done by providing an alternate path back to the input for a portion of the output signal. The out-of-phase signal then cancels the in-phase signal and prevents oscillation. [**E7B08**]

One neutralization technique for vacuum-tube RF power amplifiers is shown in **Figure 6.13**. In this circuit the neutralization capacitor, C1, is adjusted to have the same value of reactance as the plate-to-grid *interelectrode capacitance* that is providing the feedback path causing the oscillation. By connecting C1 to the tuned input circuit, the phase shift results in the feedback signal having the opposite phase of the unwanted plate-to-grid feedback signal, canceling it.

Parasitic Oscillations

Oscillations can also occur in an amplifier on frequencies that have no relation to those intended to be amplified. Oscillations of this sort are called *parasitics* because they absorb power from the circuits in which they occur. Parasitics occur because of resonances that exist in the input or output circuits, enabling positive feedback to occur.

Parasitics are most likely to occur above the operating frequency as a result of stray capacitance and lead inductance along with interelectrode capacitance. In some cases it is possible to eliminate parasitics by changing lead lengths or the position of leads so as to change their capacitance and inductance and thus the resonant frequency.

An effective method of suppressing parasitics in HF vacuum tube amplifiers is to insert a parallel combination of a small inductor and resistor in series with the grid or plate lead. Such a *parasitic suppressor* is labeled Z1 in Figure 6.13. The coil's reactance is high enough at VHF/UHF that those signals must pass through the resistor while HF signals pass easily through the coil. The resistor value is chosen to load the VHF/UHF feedback path heavily enough to prevent oscillation. Values for the coil and resistor are usually found experimentally as different layouts require different suppressor values.

RF OSCILLATORS

E7H01 — What are three oscillator circuits used in amateur radio equipment?
E7H03 — How is positive feedback supplied in a Hartley oscillator?
E7H04 — How is positive feedback supplied in a Colpitts oscillator?
E7H05 — How is positive feedback supplied in a Pierce oscillator?
E7H06 — Which of the following oscillator circuits are commonly used in VFOs?
E7H13 — Which of the following is a technique for providing highly accurate and stable oscillators needed for microwave transmission and reception?

Back in the 1920s, radio engineers Hartley and Colpitts came up with the two circuits of **Figure 6.14A** and 6.14B that became popular in radio designs. In each, feedback is created by routing part of the emitter circuit through a voltage divider created by two reactances. The connection to the voltage divider is called a *tap* and such a circuit or component is said to be *tapped*. If the reactive divider is a pair of capacitors, it's a Colpitts oscillator. [**E7H04**] If the reactive divider is a pair of inductors or, more frequently, a single tapped inductor, the circuit is a Hartley oscillator. [**E7H03**] These same circuits are in wide use today at nearly 100 years of age! (You can remember which is which by thinking, "C is for capacitors and Colpitts" and "H is for henrys and Hartley.")

The Hartley and Colpitts oscillator circuits are very similar in behavior but their differences influence the designer's preferred choice. For example, the Hartley has a wider tuning range than the Colpitts and fewer components. The Colpitts, however, avoids the tapped inductor and has several popular variants with good stability — a Hartley oscillator is less stable than the Colpitts. Both circuits are *LC oscillators*, meaning their operating frequency is controlled by inductors (L) and capacitors (C).

If a quartz crystal is inserted in the feedback path to control the frequency of feedback, the

6-14   Chapter 6

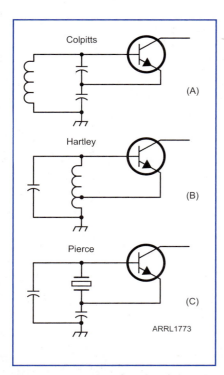

Figure 6.14 — The Colpitts (A) and Hartley (B) oscillators work on the same principle, but use different connections to the LC resonator to provide feedback. A quartz crystal is inserted in the feedback path at C to create the Pierce oscillator. Power and output connections are omitted for simplicity.

result is a Pierce crystal oscillator as shown in Figure 6.14C. [**E7H05**] The Pierce oscillator is the most stable of the three major oscillator circuits — the Colpitts, Hartley, and Pierce. [**E7H01**]

Even crystals aren't stable enough, however, for direct use at microwave frequencies where more advanced technology is required. The most easily available stable frequency reference is signals from the Global Positioning System (GPS) satellites. Each satellite contains a high-accuracy atomic frequency reference on board and many GPS receivers supply a frequency reference output synchronized to the GPS signal. A *GPS disciplined* oscillator is synchronized with the GPS signal. Other sources of high-accuracy frequencies are rubidium oscillators and temperature-stabilized high-Q dielectric resonators such as those found in laboratory test equipment. [**E7H13**]

Variable-Frequency Oscillators

While the quartz crystal oscillator has excellent frequency stability, amateurs need to be able to tune their radios over a frequency range. This requires a *variable-frequency oscillator* (VFO). VFOs are created by using a variable component in the oscillator's resonant circuit. The tradeoff is that the resulting frequency is not as stable as that of a crystal-controlled oscillator. Both Hartley and Colpitts oscillators can be used as VFOs. [**E7H06**] The usual technique for adjustable LC oscillators is to use a Colpitts oscillator in which an adjustable tuning capacitor is placed in parallel with the inductor. Numerous variations on this scheme can be found in the technical references on this book's website.

CRYSTALS FOR OSCILLATORS

E6D02 — What is the equivalent circuit of a quartz crystal?
E6D03 — Which of the following is an aspect of the piezoelectric effect?
E7H12 — Which of the following must be done to ensure that a crystal oscillator provides the frequency specified by the crystal manufacturer?

Quartz is a natural *piezoelectric* material with the ability to change mechanical energy (such as pressure or deformation) into an electrical potential (voltage) and vice versa. This property is known as the piezoelectric effect. [**E6D03**]

Quartz crystals can be sliced into plates with resonant vibration frequencies ranging from a few thousand hertz to tens of megahertz depending on the dimensions of the plate. What makes the crystal resonator valuable is that it has an extremely high Q, ranging from a minimum of about 20,000 to as high as 1,000,000. The high Q means that the frequency of vibration is very stable and precise. That makes quartz an excellent material to be used at radio frequencies because of its excellent temperature stability and its mechanical ruggedness.

The mechanical properties of a quartz crystal resonator, or simply crystal, are very similar to the electrical properties of a tuned circuit. We therefore have an equivalent circuit for the crystal. The electrical coupling to the crystal is through holder plates or electrodes that sandwich the crystal between them. A small capacitor is formed by the two plates with the crystal plate as then dielectric between them. The crystal itself is equivalent to a series-resonant circuit and, together with the capacitance of the holder, forms the equivalent circuit shown in **Figure 6.15**. [**E6D02**]

The major advantage of a crystal used in an oscillator circuit is its frequency stability. In

Radio Circuits and Systems 6-15

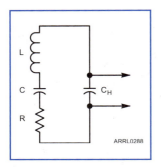

Figure 6.15 — The electrical equivalent circuit of a quartz crystal in a holder. L, C and R are the electrical equivalents of the crystal's mechanical properties and C_H is the capacitance of the holder plates, with the crystal serving as the dielectric.

an LC oscillator, the spacing of coil turns can change with vibration and the plates of a variable capacitor can move. On the other hand, the frequency of a crystal is much less apt to change with thermal or mechanical changes. By controlling the angle at which the crystal plate is cut across the plane of the quartz structure, manufacturers are able to control the crystal's temperature coefficient and other parameters. If better frequency stability is required, the crystal can also be placed in a crystal oven that maintains a constant temperature.

Another way to affect the frequency at which a crystal oscillator operates is by adding capacitance in parallel with the crystal. This changes the resonant frequency of the tuned circuit formed by the crystal and its associated components. In fact, crystals are quite sensitive to the effects of external capacitance. The crystal manufacturer will specify what capacitance must be placed in parallel with the crystal in order for it to resonate at the intended frequency. [**E7H12**]

MICROPHONICS AND THERMAL DRIFT

E7H02 — What is a microphonic?
E7H07 — How can an oscillator's microphonic responses be reduced?
E7H08 — Which of the following components can be used to reduce thermal drift in crystal oscillators?

The stability of an oscillator depends not only on the electronic circuit but the environment in which the oscillator operates. Any physical change in the oscillator will also change the feedback path, including the frequency at which the phase shift is just right to sustain oscillation. There are two primary sources of change — mechanical vibration and changes in temperature.

Mechanical vibration can affect an oscillator in many ways. The most obvious is that it moves or vibrates the components, changing the stray capacitance between them or even changing their electrical value a small amount. For a crystal oscillator relying on the crystal's piezoelectric properties, mechanical vibration can affect the crystal, too.

The oscillator can be surprisingly sensitive to vibrations. For example, if you tune in the signal from an oscillator and tap on its enclosure, you'll likely hear the frequency change a little bit. This response to mechanical vibration is called a *microphonic response* because the oscillator is acting as a kind of microphone, converting sound waves (the vibrations) into an electrical signal (the change in frequency). [**E7H02**] *Microphonics* can be quite annoying and radio designers take great pains to mechanically isolate oscillators from vibration by padding, shock absorbers, and careful component selection and layout. [**E7H07**]

The other challenge all oscillators face is *thermal drift*. Unless all of the oscillator's components — including internal semiconductor junctions — can be maintained at a constant temperature, the oscillator's frequency will shift with the temperature as component values change. As with microphonics, oscillator designers take care to minimize the oscillator's sensitivity to changing temperatures. One way to do this is by using components such as capacitors with NP0 (negative-positive zero) temperature coefficients. [**E7H08**] Another way is to heat the oscillator circuit and insulate it from the surrounding environment. Crystal ovens can be used to maintain a crystal at a constant temperature, as well.

FREQUENCY SYNTHESIS

Modern radios do not use continuously tunable oscillator circuits to control signal frequency. Instead, a technique called *frequency synthesis* is used to create signals with precisely controlled frequencies that vary in small steps of 100 Hz or less. The primary method of frequency synthesis used in commercial HF radios is *direct digital synthesizers* (*DDS*). *Phase-locked loop* (*PLL*) synthesizers were once universal in commercial radio equipment but have been largely replaced by DDS which requires less analog circuitry and are easier to integrate into digital ICs.

DIRECT DIGITAL SYNTHESIZERS (DDS)

E7H09 — What type of frequency synthesizer circuit uses a phase accumulator, lookup table, digital to analog converter, and a low-pass anti-alias filter?

E7H10 — What information is contained in the lookup table of a direct digital frequency synthesizer?

E7H11 — What are the major spectral impurity components of direct digital synthesizers?

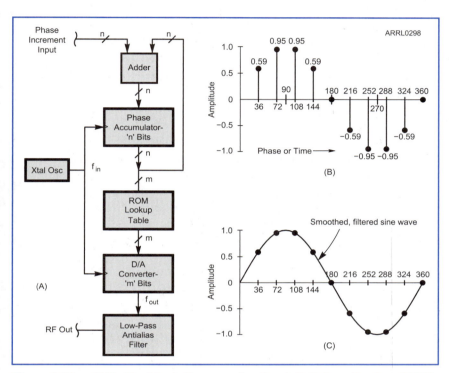

Figure 6.16 — (A) is the block diagram of a direct digital synthesizer (DDS). (B) shows the amplitude values found in the ROM lookup table for a particular sine wave being generated. (C) shows the smoothed output signal from the DDS, after it goes through the low-pass anti-alias filter.

Figure 6.16 shows the block diagram of a direct digital synthesizer. This type of synthesizer is based on the concept that we can define a sine wave by specifying a series of amplitude values spaced at equal phase angles. The frequency of the sine wave is then determined by the *sampling rate* at which the synthesizer steps through successive values.

The crystal oscillator sets the sampling rate for the amplitude values. The *phase increment* input to the adder block sets the number of samples for one cycle. The oscillator *clock* signal tells the *phase accumulator* to read the data from the adder and then increment the adder value by the phase increment. The phase accumulator value varies between 0 and 360, corresponding to one complete cycle of the output waveform. **[E7H09]** The *ROM lookup table* contains the waveform's amplitude values at each angle represented by the phase accumulator. A digital-to-analog converter (DAC) changes the digital values from the lookup table to an analog output voltage creating the waveform. **[E7H10]**

The major spectral impurity components produced by a direct digital synthesizer are *spurs* (unwanted spurious signals) at specific discrete frequencies determined by the clock signal frequency and other digital components of the DDS synthesizer. Careful design can place those spurs outside of the amateur bands. **[E7H11]**

For More Information

In a DDS, the larger the phase increment at each step, the higher the frequency of the output signal will be. For example, suppose our synthesizer uses a 10-kHz crystal oscillator. This means there will be one sample every 0.1 ms. If the phase increment is set to 36°, there will be 10 samples in each cycle: 0°, 36°, 72°, 108°, 144°, 180°, 216°, 252°, 288° and 324°. The next sample, at 360° starts the second cycle. The total time for these 10 samples is 1 ms, which means the sine wave defined by these samples has a frequency of 1 kHz. Figure 6.16B shows a representation of the sine values found in the lookup table for these phase angles. If the phase increment is changed to 72° there will be five samples per cycle. Each cycle will take 0.5 ms, so the frequency of this new signal is 2 kHz.

PHASE-LOCKED LOOPS (PLL)

E7H14 — What is a phase-locked loop circuit?
E7H15 — Which of these functions can be performed by a phase-locked loop?

In a phase-locked loop, the frequency of a variable oscillator is continuously compared to the phase of a stable, fixed-frequency reference oscillator. If the variable oscillator's frequency is too high, its phase will begin to lead that of the reference. This phase difference is used to decrease the frequency of the variable oscillator until it is back in phase with the reference oscillator. Too low a frequency causes an increasing phase lag, which is used to increase the frequency of the oscillator. In this way, the variable oscillator is *phase-locked* to the reference so that their frequencies are kept exactly equal.

To be used as a tunable oscillator, however, the frequency of the variable oscillator must be able to change yet still remain under control of the reference. **Figure 6.17** shows how a PLL works. The *phase detector* outputs a voltage corresponding to the phase difference between the oscillators. This voltage is passed through a low-pass *loop filter*, amplified by a *loop amplifier*, and used to control the frequency of a *voltage-controlled oscillator* (VCO). The combination of phase-detector, filter and VCO create an electronic "servo" loop. Using the reference oscillator as the frequency input to the servo loop creates a *phase-locked loop*. **[E7H14]**

If the loop is "in lock," meaning that the frequency of the VCO is under control, the amount of frequency variation over which the loop can maintain control of the VCO is called the loop's *lock range*. If the loop is not in lock and the divided-down frequencies of the VCO and reference are gradually brought closer to each other the loop will "capture" the VCO at some point. The difference between the maximum and minimum frequencies for which this occurs is the loop's *capture range*. The characteristics of the loop filter and amplifier determine the stability and tuning speed of the PLL. The PLL is continuously adjusting the output signal frequency with respect to the reference oscillator, creating variations in the phase of the signal. These variations create a broadband *phase noise* that is the main spectral impurity produced by PLLs and a primary reason why PLLs have been replaced by DDS oscillators.

Along with frequency synthesis, a PLL can also be used to perform both FM modulation

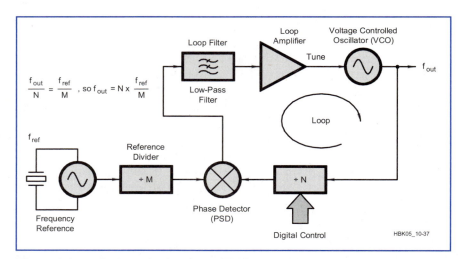

Figure 6.17 — A phase-locked loop (PLL) synthesizer acts to keep the divided-down signal from its voltage-controlled oscillator (VCO) phase-locked to the divided-down signal from its reference oscillator. By changing M and N, fine tuning steps in the VCO frequency can be made with the same frequency stability of the reference oscillator.

and demodulation. [E7H15] If the modulating signal is added to the VCO control signal, the output is a direct FM signal. If a divided-down FM signal is input to the phase detector instead of a VCO output, the output of the phase detector will be a replica of the modulating signal.

MIXERS

E7E08 — What are the principal frequencies that appear at the output of a mixer circuit?

E7E09 — What occurs when an excessive amount of signal energy reaches a mixer circuit?

Mixer circuits are used to change the frequency of a signal. In a superheterodyne receiver, this means converting the received signal to the *intermediate frequency* (IF) so it can be amplified and filtered more efficiently. In this way, the receiver can be optimized for the best signal-handling characteristics such as linearity and selectivity without the need to retune many circuit elements every time you change the received frequency. **Figure 6.18** shows the symbol for a mixer.

Mixers are also used to change the frequency of a signal as it progresses through a transmitter. The principles of operation are much the same for mixers, detectors, and modulators, so the discussion of how mixers work will prepare you to better understand the remaining topics.

If the mixing process is performed by devices that can amplify the signals, then the circuit is called an *active mixer*. If the mixer uses *passive* components that cannot amplify a signal to perform the mixing function, it is called a *passive mixer*. This also results in *conversion loss* caused by losses in the passive components. **Figure 6.19** shows the schematic of a passive, *double-balanced mixer* (DBM) that uses four diodes in a ring (similar to, but different from a full-wave rectifier) to multiply the RF and LO signals together.

When two sine waves are combined in a *nonlinear* circuit (one whose output is not a scaled replica of its input) such as a mixer, the output signal is a complex waveform that has principal components at the frequencies of the two original signals and two *product* signals. The product signals are sine waves whose frequencies are the sum and difference of the frequencies of the two original signals. [E7E08] Also included are higher orders of combinations of the harmonics from the input signals, although these are usually weaker than the primary sum and difference frequencies and are ignored in this discussion. The signal of varying frequency (in a receiver, the desired input signal) is usually referred to as the *RF signal*. The signal generated in the radio to mix with the RF signal is called the *local oscillator* (*LO*), and the resulting output signal is called

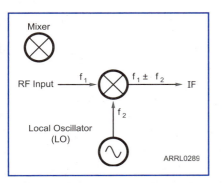

Figure 6.18 — The mixer combines the two input signals (RF and LO). This produces the mixing products that compose the IF signal at the output.

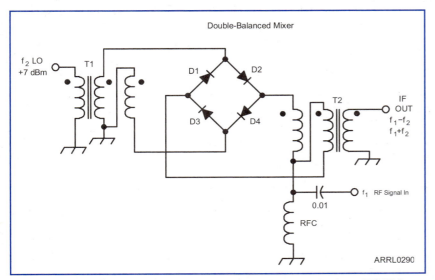

Figure 6.19 — The popular diode-ring, double-balanced mixer (DBM) is an example of a passive mixer circuit.

the *IF signal*. In a superheterodyne receiver, the LO frequency is varied so that one of the mixer IF output products is at the frequency for which the IF circuits are designed.

One of the products can be selected from the output combination by using a filter. Of course, the better the filter, the lower the level of the unwanted products or the two input signals in the resulting output signal. By using *balanced mixer* techniques the mixer circuit provides isolation of the various signal connections or *ports* so that the RF, LO, and IF signals will not appear at any other port. This prevents the two input signals (RF and LO) from reaching the output. In that case the filter needs to remove only the unwanted mixer output products.

The mixer stages in a high-performance receiver must be given careful consideration because they have a great impact on the ability of the receiver to perform properly in the presence of strong signals. (See the section Receiver Performance in the chapter on Radio Signals and Measurements.) The RF signal should be amplified only enough to overcome mixer losses. Otherwise, strong signals will overload the mixer circuit. This causes desensitization and the higher order combinations of frequencies will appear as highly undesirable spurious signals or *intermodulation distortion* (*IMD*) products at or near the IF frequency, interfering with the desired signal. [**E7E09**] A mixer should be able to handle strong signals, called *strong-signal performance*, without generating spurious signals.

MODULATORS

E7E07 — What is meant by the term "baseband" in radio communications?

Modulation is the process of adding information to an unmodulated radio-frequency (RF) signal, also known as a *carrier*. The modulating voice or data is called the *baseband signal* or *baseband information*. Baseband is the term for the frequency range occupied by a message signal prior to modulation. [**E7E07**] The baseband or message signal can be audio for voice modes or it can be a data signal for digital modes. Circuits that perform the modulation process are called *modulators*.

Any aspect of the carrier signal can be varied to add the information. Varying the amplitude of the signal is *amplitude modulation* (*AM*). Varying the phase or frequency of the carrier are both forms of *angle modulation*, referring to the phase angle of the signal. The two main types of angle modulation are *phase modulation* (*PM*) and *frequency modulation* (*FM*).

AMPLITUDE MODULATION AND SINGLE SIDEBAND

An AM signal is actually a composite signal containing three individual RF signals. Along with the RF carrier, the remaining two signals are the *sidebands*: one at a frequency that is the sum of the carrier and message signal frequencies (the *upper sideband* or *USB*), the other at their difference (the *lower sideband* or *LSB*). **Figure 6.20** shows the spectrum of an AM signal generated by modulating a 10 MHz carrier with a 1 kHz sine wave audio signal.

SSB: The Filter Method

E7E04 — What is one way a single-sideband phone signal can be generated?

In an AM signal, both sidebands contain essentially the same information — the amplitude and frequencies of the modulating signal. Only one sideband needs to be transmitted to reconstruct the original information, and so the unwanted sideband can be discarded along with the carrier which carries no information.

The first step of generating SSB by the *filter method* is to combine

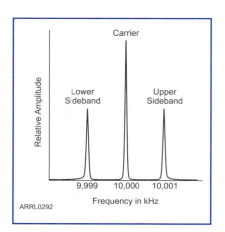

Figure 6.20 — The result of amplitude modulating a 10 MHz carrier with a 1 kHz sine wave shows the upper sideband (USB) at 10 MHz + 1 kHz and the lower sideband (LSB) at 10 MHz − 1 kHz.

the baseband signal with an RF carrier in a *balanced modulator*. A balanced modulator is a type of mixer with output products that are the sum and difference of the message and carrier frequencies — the sidebands — and the carrier signal is canceled. The resulting signal consists only of the USB and LSB signals and is called *double-sideband, suppressed-carrier* or DSB-SC. Finally, a narrow bandpass filter is used to remove one of the sidebands along with any remaining carrier signal. [**E7E04**] This leaves only the single sideband signal as shown in **Figure 6.21**.

Figure 6.22 shows the block diagram of a basic SSB transmitter. Once the SSB signal has been generated, a mixer can shift its frequency to whatever amateur band we want. It is then amplified by the linear amplifier to useful power levels.

For More Information

We can also create the SSB signal directly by manipulating the phase of the message and RF signals. This is called the *quadrature method* and the block diagram of such a system is shown in **Figure 6.23**. The quadrature method is examined in this chapter's section on DSP and SDR.

The audio and carrier signals are each split into equal components with a 90° phase difference (called *quadrature*) and applied to individual balanced modulators. When the DSB-SC outputs of the modulators are added together in the RF combiner, one sideband is reinforced and the other is canceled. The desired sideband, USB or

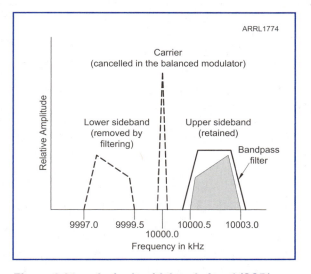

Figure 6.21 — A single-sideband signal (SSB) can be created by using a balanced modulator that eliminates the carrier followed by a narrow bandpass filter to remove the unwanted sideband.

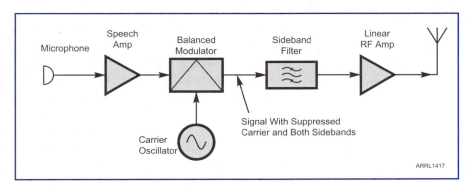

Figure 6.22 — A block diagram showing the filter method of generating and transmitting an SSB signal.

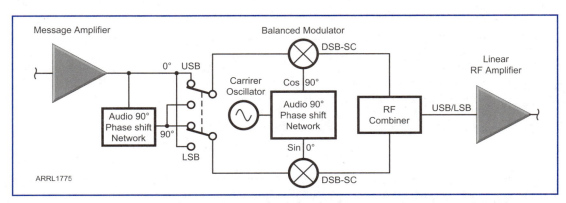

Figure 6.23 — The block diagram of a quadrature SSB generator. By combining the independent DSB signals from the balanced modulators, one sideband is canceled and an SSB signal results.

Radio Circuits and Systems

LSB, can be selected by either switching the message inputs to the balanced modulators (as shown) or by switching the RF inputs to the balanced modulators.

FREQUENCY AND PHASE MODULATION

E7E01 — Which of the following can be used to generate FM phone emissions?
E7E02 — What is the function of a reactance modulator?

Most methods of producing FM fall into two general categories: *direct FM* and *indirect FM*. As you might expect, each has its advantages and disadvantages. Let's look at the direct FM method first.

The only way to produce a true FM signal with no phase modulation is with a *reactance modulator* acting on an oscillator as in **Figure 6.24**. [**E7E01**] If the modulating signal controls the variable reactance, the result is a direct FM signal.

Varying phase also varies a signal's frequency, so from a practical view, *indirect FM* is the same as PM. The same type of reactance modulator circuit can be used as a *phase modulator* to vary reactance in a tuned RF amplifier circuit and thus vary the phase of the current. This produces *phase modulation* (*PM*).

Figure 6.24 shows an example of a reactance modulator creating direct FM as described below. The audio input signal causes the capacitance of the varactor diode to change and in turn, that changes the LC ratio of the oscillator's resonant circuit and its frequency. The modulator's sensitivity (measured in Hz/V) depends on the varactor diode's change of capacitance per volt of input signal. A variable inductance could also be used but is a less practical choice. [**E7E02**]

For practical reactance modulators, the modulated oscillator is usually operated on a relatively low frequency so the carrier frequency is very stable. Frequency multipliers then increase the signal's frequency to the final desired output frequency as in **Figure 6.25**. It is important to note that when the frequency is multiplied so is the frequency deviation. The amount of deviation produced by the modulator must be adjusted carefully to give the proper deviation at the final output frequency.

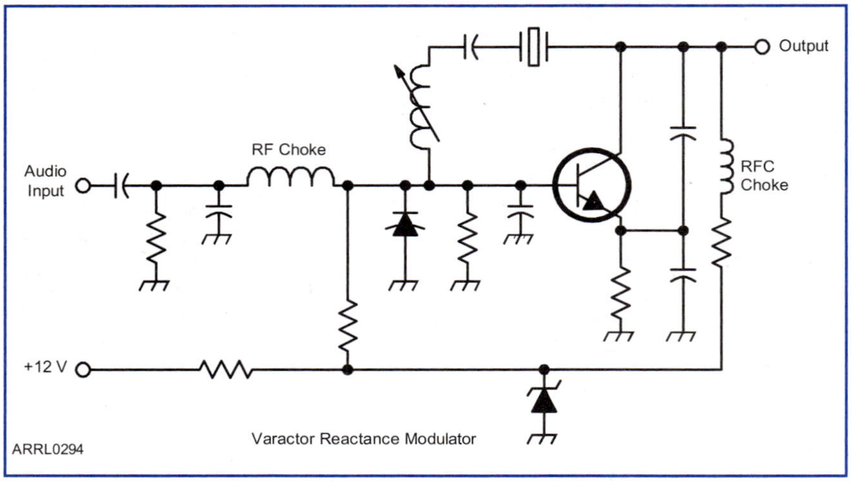

Figure 6.24 — The changing capacitive reactance of the varactor diode connected to the series-LC-crystal circuit of this oscillator changes the output frequency under control of the audio input signal, producing direct FM modulation.

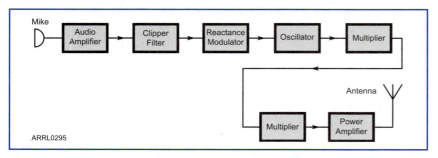

Figure 6.25 — The block diagram of a typical direct-FM transmitter. The amount of frequency deviation produced by the reactance modulator is multiplied along with the carrier frequency by the multiplier stages.

PRE-EMPHASIS AND DE-EMPHASIS

E7E05 — What circuit is added to an FM transmitter to boost the higher audio frequencies?
E7E06 — Why is de-emphasis commonly used in FM communications receivers?

Frequency deviation increases with the modulating audio frequency in PM. (Higher audio frequencies produce greater frequency deviation.) Therefore, it is necessary to use a low-pass filter on the modulating audio to attenuate frequencies above 3000 Hz before modulation takes place. This prevents the generation of unwanted sidebands far from the carrier frequency that cause interference on nearby channels.

In a direct FM system, frequency deviation does not increase with modulating audio frequency so no low-pass filter is needed. However, the lower energy of higher-frequency speech components means that the recovered signal is susceptible to high-frequency noise. To reduce hiss and high-frequency noise in the receiver, an audio circuit called a *pre-emphasis network* is added to a direct FM modulator. Pre-emphasis applied to an FM transmitter gives the deviation characteristics of PM. [**E7E05**]

The reverse process of applying a low-pass filter, called *de-emphasis*, is used at the receiver to restore the audio spectrum to its original relative proportions and reduce high-frequency noise. A transmitter that uses PM does not need a pre-emphasis network. Thus, an FM receiver with a de-emphasis network is compatible with both PM signals and FM signals using pre-emphasis. [**E7E06**]

DETECTORS AND DEMODULATORS

E7E03 — What is a frequency discriminator stage in a FM receiver?
E7E10 — How does a diode envelope detector function?
E7E11 — Which type of detector circuit is used for demodulating SSB signals?

Detectors and *demodulators* have much the same job — to recover the modulating information from a modulated RF signal. A detector circuit extracts the information directly from the signal. A demodulator reverses the modulation process to recover the information. Detectors tend to be simpler circuits than demodulators, but the recovered signal is generally not as accurate a replica of the original modulating signal as when a demodulator is used. Each has a place in radio communications.

Detectors

The simplest type of detector, used in the very first radio receivers, is the *diode detector*. It works by rectifying, then filtering, the received RF signal. [**E7E10**] A complete, simple receiver is shown in **Figure 6.26**. This circuit only works for strong AM signals so it is not used very much today except for experimentation. It does serve as a good starting point to understand detector operation, however. In early

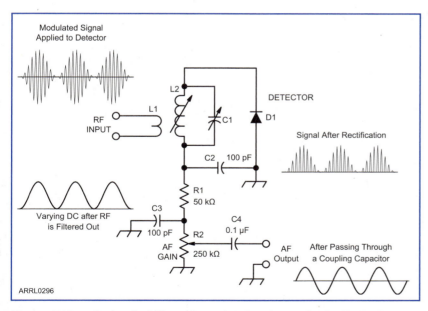

Figure 6.26 — A simple AM receiver circuit using a single diode detector. L2 and C1 are tuned to the desired receive frequency. D1 rectifies the signal, creating a waveform from which the envelope can be easily transformed into an audio signal.

crystal radio sets, a steel "cat's whisker" pressing on a lead crystal created a diode (an early type of Schottky barrier diode) to rectify the signal. Sensitive headphones then recovered the audio signal.

Product Detectors

A *product detector* is a type of mixer that follows the IF stages in a superheterodyne receiver. It combines the IF output signal with the output from a *beat-frequency oscillator* (*BFO*). The BFO frequency is chosen so that one of the sum-and-difference output products is at audio frequencies. Product detectors are used for SSB, CW, and RTTY reception. [**E7E11**] The BFO is named for the audio frequency difference or "beat" between its output and the IF output.

For example, if the receiver's IF is 455 kHz and the operator prefers listening to a CW signal with a 700 Hz tone, the BFO could be set to 455.7 kHz, creating sum-and-difference products at 700 Hz and 910.7 kHz. An audio filter then removes the higher frequency component. The BFO frequency could also be set to 454.3 kHz, achieving the same result. The same process can also be used on SSB and RTTY signals.

Detecting FM Signals

The most common FM detector is the *frequency-discriminator* circuit of **Figure 6.27**. It uses a transformer tuned to the receiver's IF to detect FM signals. [**E7E03**] The primary signal is introduced to the secondary winding's center tap through a capacitor. For an unmodulated input signal, the resulting voltages on either side of the secondary's center tap will cancel. But when the signal frequency changes, there is a phase shift in the two output voltages that varies at the audio frequency of the modulating signal. The two voltages are rectified by a pair of diodes, and the resulting difference in output voltage becomes the audio signal.

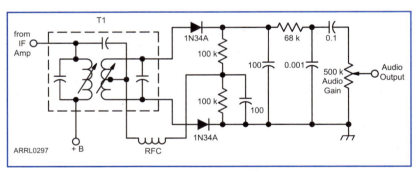

Figure 6.27 — A typical frequency-discriminator circuit used for FM detection. A modulated signal results in an imbalanced output from the transformer that is rectified and turned into audio by the diodes and RC filters.

6.3 Digital Signal Processing (DSP) and Software Defined Radio (SDR)

Digital signal processing (*DSP*) is an integral part of modern electronics, revolutionizing the industry just as the transistor and microprocessor did. The capabilities of DSP have improved so dramatically that nearly all of the functions once implemented in a transceiver by analog circuits are now performed as mathematical functions inside a digital processor. This new kind of radio is a *software defined radio* (*SDR*). All of an SDR-type radio's primary functions are controlled by software that can be changed to suit the needs of the radio's user. SDR has become the dominant technology in amateur radio transceivers, but there will always be analog radios and circuits.

To explain the basics of DSP and SDR for reference in subsequent sections of this book, we now present edited portions of material originally written for the *ARRL Handbook*. If you are interested in learning more about DSP and SDR, it is strongly suggested that you read the full *Handbook* chapter or study one of the numerous books and articles on the subject.

DIGITAL SIGNAL PROCESSING (DSP)
Sequential Sampling

DSP is about rapidly measuring analog signals, recording the measurements as a series of numbers, processing those numbers, then converting the new sequence back to analog signals. How we process the numbers depends on which of many possible functions we are performing.

The process of generating a sequence of numbers that represent periodic measurements of a continuous analog waveform is called *sequential sampling*. Each number in the sequence is a single measurement of the instantaneous amplitude of the waveform at a sampling time. When we make the measurements continually at regular intervals, the result is a sequence of numbers representing the amplitude of the signal at evenly spaced times. This process is illustrated in **Figure 6.28** showing how an analog signal is converted to digital form.

Note that the frequency of the sine wave being sampled is much lower than the sampling frequency, f_s. In other words, we are taking many samples during each cycle of the sine wave. The sampled waveform does not contain information about what the analog signal did between samples, but it still roughly resembles the sine wave. Were we to feed the analog sine wave into a spectrum analyzer, we would see a single signal component at the sine wave's frequency. Obviously, the spectrum of the sampled waveform is not the same, since it is a stepwise representation consisting of discrete steps at each value before jumping to the next value.

The sampled signal's spectrum can be predicted and interpreted as shown in

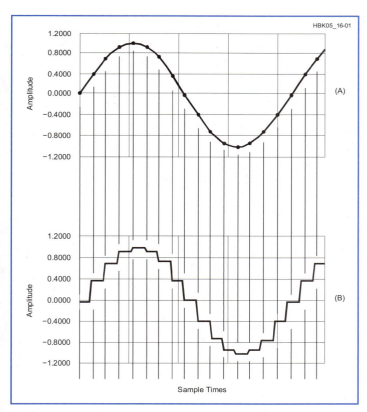

Figure 6.28 — Sine wave of frequency much less than the sampling frequency (A). The sampled sine wave (B).

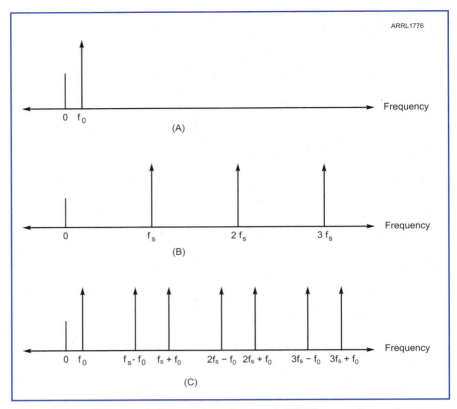

Figure 6.29 — Spectrum of an analog sine wave (A). The spectrum of a sampling function (B). The spectrum of the sampled sine wave (C).

Figure 6.29. The analog sine wave's spectrum is shown in Figure 6.29A, above the spectrum of the sampling function in Figure 6.29B. The sampled signal is just the product of the two signals as shown in Figure 6.29C.

The sampling process is equivalent to a mixing process — both perform a multiplication of the two input signals. Note that the sampled spectrum repeats at intervals of f_s. These repetitions are called *aliases* and are as real as the fundamental in the sampled signal data. Each contains all the information necessary to fully describe the original signal. In general, we are only interested in the fundamental, but let's see what happens when the sampling frequency is less than that of the analog input.

Sine Wave, Alias Sine Wave

E7F05 — How frequently must an analog signal be sampled by an analog-to-digital converter so that the signal can be accurately reproduced?

Take the case where the sampling frequency is less than that of the analog sine wave as shown in **Figure 6.30A**. The sampled output in Figure 6.30B no longer matches the input waveform. Notice that the sampled signal retains the general shape of a sine wave, but at a frequency lower than that of the input. This is an *alias* of the sampled signal and from the perspective of the sampled data, the alias is just as real as the input signal and cannot be filtered out.

To avoid creating aliases by *undersampling*, the sampling frequency, f_s, must be at least twice the highest frequency component of the signal. [**E7F05**] This requirement is known as the *Nyquist sampling theorem*.

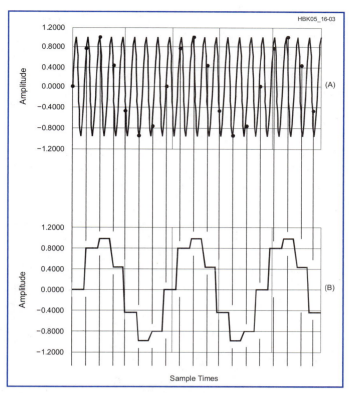

Figure 6.30 — Sine wave of frequency greater than the sampling frequency (A). Alias sine wave created by under-sampling (B).

To avoid creating aliased signals we must limit the bandwidth of the input signal with an *anti-aliasing* low-pass filter that removes any signal components with frequencies higher than one-half the sampling frequency, f_s.

DATA CONVERTERS

E7F06 — What is the minimum number of bits required for an analog-to-digital converter to sample a signal with a range of 1 volt at a resolution of 1 millivolt?

E7F11 — What sets the minimum detectable signal level for a direct-sampling SDR receiver in the absence of atmospheric or thermal noise?

E8A04 — What is "dither" with respect to analog-to-digital converters?

E8A09 — How many different input levels can an analog-to-digital converter with 8-bit resolution?

E8A10 — What is the purpose of a low-pass filter used in conjunction with a digital-to-analog converter?

E8A11 — Which of the following is a measure of the quality of an analog-to-digital converter?

The device used to perform sampling is called an *analog-to-digital converter* (ADC). For each sample, an ADC produces a binary number that is directly proportional to the input voltage. The number of bits in the binary numbers limits the number of discrete voltage levels that can be represented. An 8-bit ADC, for example, can produce one of $2^8 = 256$ values. [**E8A09**] This means the amplitude reported is not the exact amplitude of the input, but only the closest value from those available. The ADC's measurement can be improved over time by "dither," adding small amounts of noise to the input signal. This causes the ADC's average output value to be more precise over time. [E8A04]

The resolution of the ADC is determined by the reference voltage corresponding to the maximum digitized value and the number of bits representing each sample. As another example, if a 10-bit ADC has a reference voltage of 1 V, the resolution is $1 \text{ V} / 2^{10} = 1 / 1024 = 0.976$ mV ≈ 1 mV. [E7F06] Assuming atmospheric or thermal noise are not higher, the resolution of the ADC determines the minimum detectable signal level for an SDR. [E7F11]

The process of converting the input signal to digital form results in data that may be very close to the original analog signal but is not an exact copy. Other effects on the digitization process include the bandwidth and slew rate of the converter input circuits. The result is a small amount of distortion of the signal that is measured as *total harmonic distortion* or *THD* which is a measure of converter quality. [E8A11]

A *digital-to-analog converter* (*DAC*) performs the conversion of binary numbers back into analog voltages — the reverse operation of an ADC. The structure of a DAC allows it to operate at very high speed but it digitizes the signal output levels, just as an ADC can only output a fixed number of digital values. If a sine wave is created by a DAC, it will have a "stairstep" appearance as each successive voltage level is created. The steps between values create unwanted harmonics that are removed with a low-pass filter also called a *reconstruction filter*. [E8A10]

FOURIER TRANSFORMS

E7F07 — What function is performed by a Fast Fourier Transform?

The Fourier transform is the software equivalent of a hardware spectrum analyzer which is described in the chapter on Radio Signals and Measurements. It takes in a signal in the time domain of amplitude versus time and outputs the signal translated to the frequency domain of amplitude versus frequency. This shows the spectral content of the input signal. The Fourier transform is performed on a digitized signal by a DSP algorithm, producing output data that represents the spectrum of the input signal. (An *inverse Fourier transform* reverses the process, turning a frequency domain signal back into a time domain signal.)

Calculating the Fourier transform requires a large number of calculations. The *Fast Fourier Transform* (*FFT*) is a special algorithm that reduces the number of calculations required for a 1024-sample data set by a factor of more than 100 compared to the original methods. The FFT is key to translating signals from the time domain to the frequency domain. [E7F07]

DECIMATION AND INTERPOLATION

E7F08 — What is the function of decimation?
E7F09 — Why is an anti-aliasing digital filter required in a digital decimator?

DSP processing can also perform functions on a signal that are impossible to do effectively in analog systems. For example, changing the effective sample rate shifts the frequency of the digital signals. The process of *decimation*, removes every nth sample, reducing the effective sample rate by the same factor. [E7F08] In order to prevent generating aliases due to the new, lower sample rate, a digital low-pass anti-aliasing filter must be applied before decimation. [E7F09]

A similar process called *interpolation* inserts new samples between existing samples to increase the effective sample rate. No anti-aliasing filter is required in this case because the effective sampling rate is increased, not decreased.

Software-Defined Radio (SDR) Systems

What is a software-defined radio? To be as comprehensive as possible, we can state that a

Chapter 6

software radio is a radio:

1. Whose hardware is able to handle almost any modulation format, signal bandwidth, and frequency desired.

2. Whose functionality may be altered at will by downloading new software.

3. That performs radio signal processing and functions with DSP.

What is important to understand about DSP and SDR systems is that they can perform any mathematically defined signal processing function if hardware is available to adequately sample the signal and perform the required math operations quickly enough. That includes modulation, demodulation, filtering, speech processing and so on.

Some SDRs use a PC to do the computational heavy lifting and external hardware to convert the transmitted and received RF signals to lower-frequency signals that the computer's sound card can digitize. Some SDRs avoid the use of the sound card by including their own audio codec and transferring the data to the PC via a USB port. The can also run logging programs and other software while simultaneously doing the signal processing required by the SDR.

Self-contained SDRs look more like conventional radios with everything contained in one enclosure, which makes for a neater, more compact installation. The signal processing is done with one or more embedded DSP ICs. These compact radios have dedicated controls for specific functions, just like their traditional counterparts, and don't require a separate PC to display a virtual front panel or and control the radio's functions

Either method offers all the most-important advantages of applying DSP techniques to signal processing. Once the signal is in the digital domain sophisticated DSP algorithms can be applied such as automatic notch filters, adaptive channel equalization, noise reduction, noise blanking, and feed-forward automatic gain control. Correcting bugs, improving performance or adding new features is as simple as downloading new software.

SDR HARDWARE

E7F01 — What is meant by direct digital conversion as applied to software defined radios?

E7F10 — What aspect of receiver analog-to-digital conversion determines the maximum receive bandwidth of a Direct Digital Conversion SDR?

E8A02 — Which of the following is a type of analog-to-digital conversion?

E8A08 — Why would a direct or flash conversion analog-to-digital converter be useful for a software defined radio?

The transition between analog and digital signals can occur at any of several places in the signal chain between the antenna and the human interface. With the wide variety of affordable equipment available today, the choice is largely one of convenience.

For example, many popular digital communication software packages use the sound card of a PC connected to the audio input and output of a transceiver. The digital signals are received and transmitted over the air using the same process as regular microphone audio. All RF signals are processed within the transceiver, whether it is an analog superheterodyne or a state-of-the-art SDR.

In some software defined radios, the analog-to-digital transition is made at an IF stage where an ADC samples receive signals and a DAC creates the transmit signal. These digital signals can be processed by DSP in the transceiver or sent to a PC where software performs modulation and demodulation.

The current state-of-the-art SDR architecture is to make the transition between the analog and digital domains right at the frequency to be transmitted or received with no mixer converting the frequency of received or transmitted signals. This is called *direct digital conversion* (*DDC*) and the block diagram of a typical DDC transceiver is shown in **Figure 6.31**. **[E7F01]** DDC requires the ADC and DAC to operate at very high sample rates — at least twice the bandwidth of the transceiver using a *direct conversion* or *flash conversion ADC*.

Radio Circuits and Systems 6-29

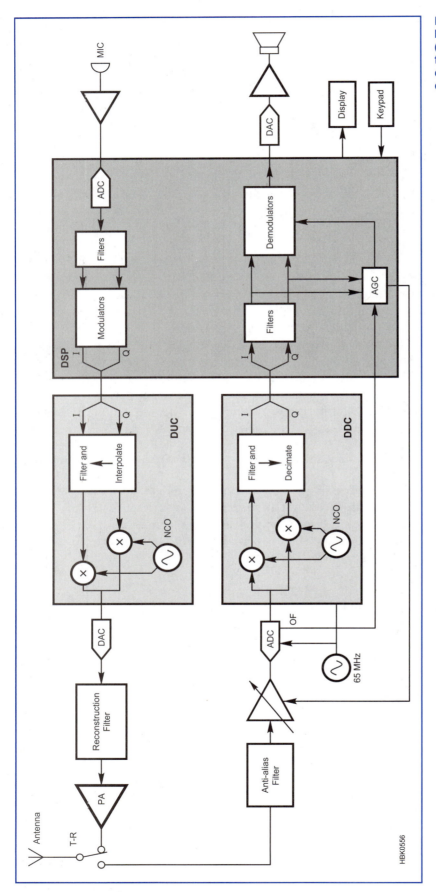

Figure 6.31 — An SDR transceiver based on direct digital conversion (DDC), sampling directly at the RF frequency. DUC stands for digital up-conversion. The block labeled DDC is a digital down-converter.

[**E8A08**] Other types of converters, such as pipeline architectures, are also popular. For digitizing baseband audio or data signals, a much slower converter can be used such as a *successive approximation* or *sigma-delta* ADC. [**E8A02**] Technology is moving very rapidly in this area.

In the DDS transceiver, the only remaining analog components in the signal chain are wide-band anti-aliasing and reconstruction filters, an amplifier to improve the noise figure of the ADC, and power amplifiers for the transmitted signal. The local oscillator, mixer, IF filters, AGC, modulators, demodulators, and other circuitry except audio output and power supply components are all replaced by digital hardware and software.

Regardless of whether the digitization is performed directly on the RF signal or on an IF signal, the receive and transmit bandwidths are limited by the sample rate of the ADC and DAC, respectively. [**E7F10**]

DSP Modulation

A sinusoidal wave of any arbitrary amplitude and phase may be represented by the weighted sum of a sine and cosine wave:

x(t) = I cos(wt) + Q sin(wt)

For mathematical convenience, the I and Q values are combined in a single complex number, x = I + jQ. I and Q are the baseband message information streams that are applied to the two RF carrier signals of the same frequency (ωt) but 90° out of phase. This creates the modulated signal:

$$x(t) = (I + jQ)(\cos(\omega t) - j\sin(\omega t))$$

In this equation, the cos(wt) – *j*sin(wt) portion represents two RF carriers and the I + *j*Q part represents the baseband message. I and Q can be digital data bits (1 or 0) that turn the carriers on and off or they can be analog signals. The process of recovering I and Q from the modulated signal is the same for both.

I/Q MODULATION AND DEMODULATION

E7F03 — What type of digital signal processing filter is used to generate an SSB signal?

E7F04 — What is a common method of generating an SSB signal using digital signal processing?

An I/Q modulator controls the amplitude and phase of an RF signal directly from the I and Q components. See **Figure 6.32** for block diagrams of an I/Q modulator and demodulator. The I and Q signals are two separate baseband streams of data. The I/Q demodulator in Figure 6.32B is basically the same circuit as the modulator, but in reverse. It recovers the I and Q signals that represent the in-phase and quadrature components of the incoming RF signal. Assuming the demodulator's local oscillator is on the same frequency and is in phase with the carrier of the signal being received, the I/Q output of the receiver's demodulator is theoretically identical to the I/Q input at the transmitter end.

In a digital I/Q modulator, the I/Q inputs to the modulator are streams of digital bits from ADCs at the top of the DSP block in Figure 6.31. The mixers, oscillator, phase-shift network and summer are all digital functions in the digital upconverter (DUC) block and converted to RF by the DAC. The resulting digital output signal has perfect unwanted sideband rejection, no carrier feedthrough and no distortion within the dynamic range afforded by the number of bits in the data words.

A similar process applies to a digital demodulator. If the incoming RF signal is first digitized with an ADC, then the demodulation can be done digitally without any artifacts caused by imperfections in analog circuitry. The first step is to recover the I and Q signals in the block labeled DDC in Figure 6.31 as time-domain signals. After filtering, the next step in the DSP block is to apply an FFT to the digitized I and Q signals, changing them to the

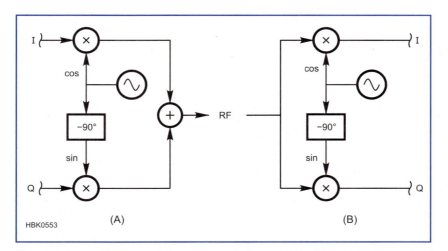

Figure 6.32 — An I/Q modulator (A) and demodulator (B). I and Q can be smoothly varying signals, such as speech, or digital data composed of 1 and 0 bit values.

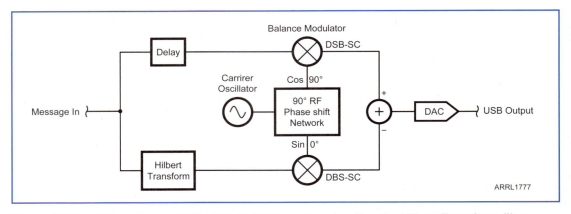

Figure 6.33 — Although very difficult to build in analog circuitry, the Hilbert Transform filter can be implemented by DSP software to generate SSB signals. A delay is required in the non-filtered path so that the message signals remain in phase at the double-balanced modulators.

frequency domain. Demodulation functions can then be performed on the I and Q spectra to recover the baseband signals.

The level of precision required for acceptable 90° audio phase shifting is difficult to obtain with analog circuits. With DSP, however, producing a 90° phase shift over a wide frequency range is easily accomplished using a special combination of filters called the *Hilbert transform*. [**E7F03**] A DSP system using the Hilbert transform as in **Figure 6.33** creates a special combination of filters that produces the necessary signals — phase-shifted message and RF carriers — then performs the balanced modulator function by multiplying the sampled signals together as numbers. The SSB signal is generated by adding the two multiplied sets of data together and using a DAC to turn the numbers back into an analog waveform. This technique makes the quadrature technique a popular method of SSB generation in DSP systems. [**E7F04**]

6.4 Filters and Impedance Matching

FILTER FAMILIES AND RESPONSE TYPES

Filters are used to block, pass, or otherwise modify signals within some defined range of frequencies. While the resonant circuits discussed previously also do this and can be considered a simple filter, the term "filter" generally refers to circuits that act over broader ranges of frequencies with well-defined characteristics.

PASSIVE AND ACTIVE FILTERS

E7C10 — Which of the following filters would be the best choice for use in a 2 meter band repeater duplexer?

In this section you will learn about *passive* filters and *active* filters. Passive filters are made with unpowered components (R, C, or L) and always result in some loss of signal strength. This is called *insertion loss*. Active filters include a powered amplifying device to overcome the filter insertion loss and sometimes even provide signal gain. Some types of filters can only be built using active components.

Passive filters constructed using inductors and capacitors are *LC filters*. There are other types of passive filters, however. For example, *mechanical filters* using internal elements such as disks and rods that vibrate at the frequencies of interest are used as receiver IF filters. *Cavity filters* use the resonant characteristics of a conducting tube or box to act as a filter and

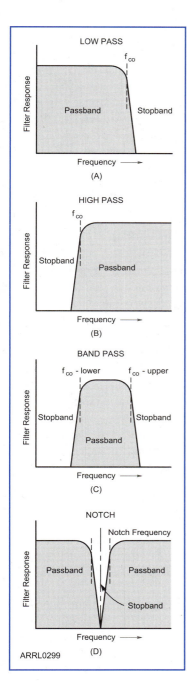

Figure 6.34 — Ideal filter-response curves for low-pass, high-pass, band-pass, and notch filters.

are used in repeater duplexers because of their extremely low loss and sharp tuning characteristics. [**E7C10**]

Filter Classification

Filters are classified into the general groups shown in **Figure 6.34**. A *low-pass filter* is one in which all frequencies below the *cutoff frequency*, f_{co}, (at which the output signal power is one-half that of the input) are passed with little or no attenuation. Above the cutoff frequency, the attenuation generally increases with frequency. A *high-pass filter* is just the opposite; signals are passed above the cutoff frequency, and attenuated below. The range of frequencies that is passed is the *passband* and the range that is attenuated, the *stopband*.

A *band-pass filter* has both an upper and a lower cutoff frequency. Signals between the cutoff frequencies are passed, while those outside the passband are attenuated. The opposite of a band-pass filter is a *band-stop filter*. It attenuates signals at frequencies between the cutoff frequencies. If the stopband is very narrow, that is a *notch filter*.

FILTER DESIGN

E7C05 — Which filter type is described as having ripple in the passband and a sharp cutoff?
E7C06 — What are the distinguishing features of an elliptical filter?
E7C11 — Which of the following describes a receiving filter's ability to reject signals occupying an adjacent channel?

Filter designs use techniques based on certain types of mathematical equations that describe the filter characteristics. You may have heard of filters referred to as *Butterworth*, or *Chebyshev*, or *elliptical* and these names refer to the family of equations used to design that type of filter. Each type of equation results in filters with different characteristics as described below. These three types of filters are the most common that are used in amateur equipment, but there are many others.

Using these equations, it is possible to build an entire catalog of filters with different characteristics. Tables summarizing these computations can be found in *The ARRL Handbook* and other reference books. From the tables, component values can be determined and the filter constructed with confidence that it will perform as expected. A version of Tonnesoft's filter design program *ELSIE* is provided on the *ARRL Handbook's* web page, **www.arrl.org/arrl-handbook-reference**.

Before discussing the different types of filters, we should define the terms we use to describe their behavior. **Figure 6.35** shows *response curves* showing the filter's effect on signal amplitude with frequency. The vertical axis has units of dB representing the ratio of output to input signal, so smaller response values correspond to more attenuation of the signal. Frequency increases from left to right, so all of the filter responses in the figure are low-pass filters, attenuating frequencies above the cutoff frequency where the filter response is −3 dB.

Two additional characteristics describe the response curve: *cutoff transition* and *ripple*. *Cutoff* refers to the steepness with which the response curve moves from the passband to the stopband through the *transition region*. A filter with a steep response curve in the transition region is referred to as "sharp." Ripple refers to variations in the response within the passband and the stopband. A "flat" filter response has small amounts of ripple.

Radio Circuits and Systems 6-33

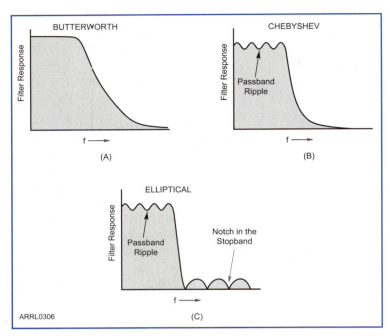

Figure 6.35 — Typical response curves for low-pass Butterworth (A), Chebyshev (B), and elliptical (C) filters.

A filter's *phase response* describes variations in signal phase from input to output at different frequencies. Typically, as signal attenuation increases, so does the amount by which the signal's phase is delayed as it passes through the filter. A *linear phase response* indicates that the change in phase is smooth and does not exhibit ripple at different frequencies. Ripple in either the passband or stopband means that the phase response is *nonlinear*. This can cause distortion of signals passing through the filter.

The three types of filters have characteristics that complement each other.

- *Butterworth*: The passband and stopband are both as flat as possible (*maximally flat*) with no ripple at all. The cutoff transition is smooth, but not steep. Butterworth filters are used when smoothly varying phase response is important to minimize signal distortion.
- *Chebyshev*: The passband has variable amounts of ripple, trading flatness of the passband for a sharper cutoff transition. Chebyshev filters are used when a sharp filter with consistent attenuation in the stopband is more important than maintaining signal phase in the passband. [**E7C05**]
- *Elliptical*: Cutoff is the steepest of all three filter types at the expense of ripple in both the passband and stopband. Elliptical filters are used when the most important characteristic is the sharpness of the filter cutoff. Notches in the stopband are positioned at specific frequencies to make cutoff as sharp as possible. [**E7C06**]

Band-pass, band-stop and notch filters are also characterized by their *bandwidth*, the frequency difference between the filter's cutoff frequencies or other frequencies with specific amounts of attenuation. When selecting a filter for a specific signal type it is important to match the filter and signal bandwidths.

The *notch filter*, used to remove a single frequency (more accurately, a narrow range of frequencies) and band-stop filters are specified to have a response *depth* in dB, with higher numbers indicating more rejection at the notch frequency.

Filters also have a *shape factor* that compares the frequency bandwidth at two levels of attenuation. In amateur equipment, the shape factor is the ratio of the filter's response at the –6 dB and –60 dB points. For example, a filter that has a –6 dB bandwidth of 1.8 kHz and a –60 dB bandwidth of 5.4 kHz has a –6 to –60 dB shape factor of 1.8/5.4 = 3.0 to 1. The portions of a band-pass filter's response curve outside the passband are called the filter's *skirts*. The closer a filter's shape factor is to 1.0, the sharper its cutoff and the steeper its skirts. Smaller shape factor values mean the filter has higher rejection of unwanted signals outside its passband, such as on an adjacent channel. [**E7C11**]

CRYSTAL FILTERS

E7C08 — Which of the following factors has the greatest effect on the bandwidth and response shape of a crystal ladder filter?

E7C09 — What is a crystal lattice filter?

The IF section of an analog superheterodyne receiver requires very good band-pass filters

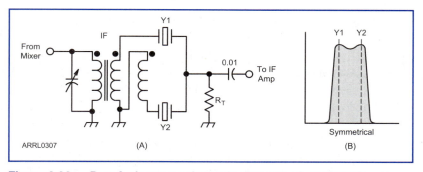

Figure 6.36 — Part A shows a schematic diagram of a half-lattice crystal filter. (B) shows a typical response curve for this type of filter. Note the steep skirts on the response between y$_1$ and y$_2$, indicating good rejection of signals outside the passband.

to provide the narrow bandwidth needed to separate one signal from the many on the band. These filters cannot be built using individual inductors and capacitors. Filters using piezoelectric quartz crystals (discussed previously in the section on oscillators) can provide the high-Q, narrow-bandwidth characteristics required. A hybrid analog-DSP receiver may use band-pass *roofing* filters to keep strong signals outside the desired band from degrading receiver performance.

Although single crystals can be used as filtering devices, the normal practice is to connect two or more together in various configurations to create the desired response. *Crystal-lattice* or *crystal-ladder filters* can provide narrow-bandwidth filtering at the frequencies above 500 kHz encountered in a transceiver's signal processing circuits. For example, analog SSB transmitters use crystal-ladder filters after the balanced modulator to attenuate only the unwanted sideband from the closely-spaced sideband pair.

Figure 6.36 depicts a filter with crystals in a configuration known as the *half-lattice*. In this arrangement, crystals Y1 and Y2 are on different frequencies. The overall bandwidth of the crystal half-lattice filter is equal to approximately 1 to 1.5 times the frequency separation of the crystals. The closer the crystal frequencies, the narrower the bandwidth of the filter.

In general, a crystal filter has narrow bandwidth and steep response skirts, as shown in Figure 6.36B. [**E7C09**] The bandwidth and response shape of crystal filters depend on the relative frequencies of the crystals. [**E7C08**]

The typical bandwidth for a crystal filter used for single-sideband (SSB) signals is 2.4 kHz at the –6 dB points. For CW use, crystal filters typically have 250- to 500-Hz bandwidths.

ACTIVE FILTERS

E7G02 — What is ringing in a filter?
E7G05 — How can unwanted ringing and audio instability be prevented in an op-amp RC audio filter circuit?

An *active filter* is one that uses an amplifier to create its frequency response. In general, passive LC filters that contain inductive and capacitive elements have a fixed frequency response and exhibit insertion loss. LC filters are usually physically larger and heavier than their active counterparts. Active filters have a number of advantages over LC filters at audio frequencies. They provide gain and good frequency-selection characteristics. They do not require the use of inductors, and they can be accurately tuned to a specific design frequency using a potentiometer.

Op amps are often used to build an active filter because the gain and frequency response of the filter can be controlled by a few resistors and capacitors connected externally to the op amp. There are a few disadvantages to using active filters beyond requiring a source of power. Low-cost op amps limit the useful upper frequency to a few hundred kilohertz. Their output voltage swing must be less than the dc supply voltage. Strong out-of-band input signals may overload the op amp and distort the output signal. The op amp may add some noise to the signals, resulting in a lower signal-to-noise ratio than you would have with an LC filter.

Active Audio Filters

Figure 6.37 shows a simple *RC active filter*. The component values and circuit configuration are for a band-pass filter suitable for CW use. Active filters are only useable with low-power signals in the audio and very low RF range. Their principal use in amateur radio is as receiver audio filters to provide additional selectivity. Not only does a well-designed RC filter help to reduce QRM but it also improves the signal-to-noise ratio by reducing unwanted high- and low-frequency noise.

The filter circuit in Figure 6.37 is a *single-section* band-pass filter. (A section refers to one circuit that performs a specific filtering function.) Individual filter sections can be *cascaded* (connected in series) for greater selectivity. One or two sections may be used as band-pass or low-pass filters for improving the audio-channel passband characteristics during SSB or AM reception. Up to four filter sections are frequently cascaded to obtain selectivity for CW or RTTY reception. The greater the number of filter sections, up to a practical limit, the steeper the filter skirts will be.

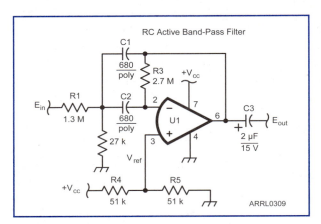

Figure 6.37 — This RC multiple-feedback band-pass filter is typical of active filters. This filter has a center frequency of 900 Hz and would be good for use with CW signals. Increasing C1 and C2 would lower the center frequency.

To select the component values for a specific filter, you must first determine the band-pass characteristics: desired filter Q (the ratio of center frequency to bandwidth), voltage gain (A_v), and center frequency (f_0). The component values for the circuit were calculated based on an f_0 = 900 Hz, an A_v of 1 and a Q of 5. These are typical values of A_v and Q for filters of this type. Both can be increased for a single-section filter, but it is best to restrict the gain to 2 or less and limit the Q to no more than 5 to prevent unwanted filter ringing and audio instability. [**E7G05**] *Ringing*, in which oscillations are sustained beyond the duration of the original signal, can be a problem in high-gain or very narrow bandwidth circuits. [**E7G02**]

DIGITAL SIGNAL PROCESSING (DSP) FILTERS

E7F02 — What kind of digital signal processing audio filter is used to remove unwanted noise from a received SSB signal?

E7F12 — Which of the following is an advantage of a Finite Impulse Response (FIR) filter vs an Infinite Impulse Response (IIR) digital filter?

E7F13 — What is the function of taps in a digital signal processing filter?

E7F14 — Which of the following would allow a digital signal processing filter to create a sharper filter response?

There are numerous advantages to filtering signals digitally. DSP can create a number of filters that are impractical or impossible to build with physical components. For example, "brick wall" filters with extremely steep cutoffs would require expensive precision components and an impractical number of sections to implement with active filters. These are fairly easy to build using DSP techniques, however.

The drawback of DSP filters is that they require the necessary computing hardware to implement them. However, changing the filter characteristics is simply a matter of changing the program. In fact, this leads to some interesting applications because the program can respond differently to different types of signals or conditions. This is called *adaptive processing*.

An *adaptive filter* can be useful for removing unwanted noise from a received SSB signal, for example. [**E7F02**] An adaptive or *automatic notch filter* might automatically identify an interfering tone from a carrier, lock onto that signal and remove it from the received audio. Such a filter can even track the interfering signal as it moves through the receiver passband!

Any type of filter response characteristics that can be created by passive or active components can be implemented in a digital filter. The limits of DSP technique are mainly the sampling rate and resolution with which the input signal can be sampled as discussed in the previous section on DSP and SDR.

Digital filters are categorized according to their response to a narrow pulse. As a pulse becomes narrower and narrower, its frequency spectrum spreads out more and more. This happens because higher and higher frequency components are required to form the sharp edges of the pulse. If the pulse becomes infinitely narrow, the spectrum becomes flat from zero hertz to infinity. An infinitely narrow pulse is called an *impulse* and is a very useful concept because of its flat frequency spectrum. The response of the filter to an impulse is called its *impulse response*. The two primary DSP filter categories are described below.

Finite Impulse Response (FIR) Filters

A *finite impulse response* (*FIR*) filter is a filter with an impulse response that is finite, ending in some fixed time. Note that analog filters have an infinite impulse response — the output theoretically lasts forever. Even a simple RC low-pass filter's output decays exponentially toward zero but theoretically never quite reaches it. In contrast, an FIR filter's impulse response becomes exactly zero at some time after receiving the impulse and stays zero forever (or at least until another impulse comes along).

Figure 6.38 shows the block diagram of an FIR filter. The input signal is stored in a shift register. Each block labeled "Delay" represents a delay of one sample time. At each sample time, the signal is shifted one register to the right. Each output from each incremental delay block is called a *tap*.

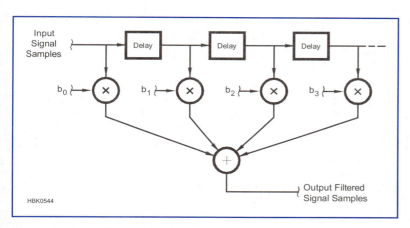

Figure 6.38 — A 4-tap FIR filter. The b_n values are the filter coefficients.

Each register feeds a multiplier. The values by which the signal from each tap is multiplied are called *filter coefficients*. The resulting multiplier outputs are all added together and create the filter output. The combination of the number of taps and the filter coefficient values is the *filter algorithm*. [E7F13]

For an input signal consisting of a single impulse, the shift register will eventually cause all tap outputs to be zero so that the filter output will also be zero, so we have created a finite impulse response filter.

The higher the number of taps, the more precisely the filter output can be calculated. [E7F14] This allows filters with extremely sharp rolloff and narrow bandwidth to be implemented without the ringing or distortion of an equivalent analog filter. The tradeoff is that each tap requires a delay of one sample time, so more taps results in a longer delay as the signal moves through the filter.

Infinite Impulse Response (IIR) Filters

An *infinite impulse response* (*IIR*) filter is a filter with an impulse response that lasts forever. This is because an IIR filter contains feedback and feed-forward loops in its design as shown in **Figure 6.39**. After an impulse is applied to the input, theoretically the output never goes to zero, just like an analog filter. In practice, of course, the signal eventually does decay

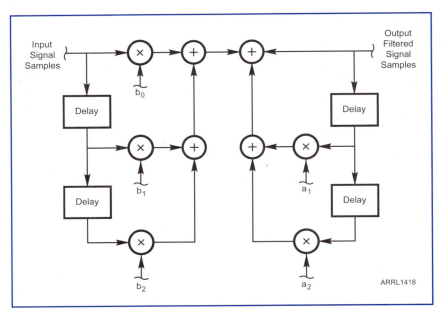

Figure 6.39 — An IIR filter with three feed-forward taps and two feedback taps.

until it becomes smaller than one least significant bit of the filter's resolution.

Unlike a symmetrical FIR filter, frequency components of the input signal can be delayed by different amounts. [**E7F12**] On the other hand, many fewer adders and multipliers are typically required to achieve the same passband and stop band filter response, so IIR filters are often used where computations must be minimized.

IMPEDANCE MATCHING

E7B09 — Which of the following describes how the loading and tuning capacitors are to be adjusted when tuning a vacuum tube RF power amplifier that employs a Pi-network output circuit?

E7C01 — How are the capacitors and inductors of a low-pass filter Pi-network arranged between the network's input and output?

E7C02 — Which of the following is a property of a T-network with series capacitors and a parallel shunt inductor?

E7C03 — What advantage does a series-L Pi-L-network have over a series-L Pi-network for impedance matching between the final amplifier of a vacuum-tube transmitter and an antenna?

E7C04 — How does an impedance-matching circuit transform a complex impedance to a resistive impedance?

E7C07 — Which describes a Pi-L-network used for matching a vacuum tube final amplifier to a 50-ohm unbalanced output?

E7C12 — What is one advantage of a Pi-matching network over an L-matching network consisting of a single inductor and a single capacitor?

When most hams talk about impedance matching circuits or *networks*, they are probably thinking of a piece of equipment — an *impedance matching unit* — used between a transmitter or transceiver and an antenna system. Its basic purpose is to convert or match the input impedance of the antenna feed line to the output of a transceiver or power amplifier so the amplifier has the proper resistive load. These are also known as *antenna couplers*, *transmatches*, *matchboxes*, *antenna tuners*, and *impedance matchers*. Impedance matching

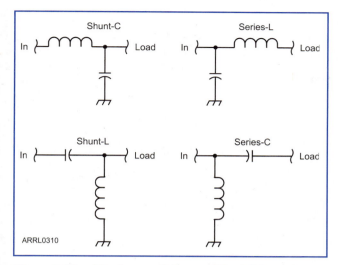

Figure 6.40 — The four variations of the LC impedance-matching L-network. Shunt or series refers to the connection of the component closest to the impedance to be matched.

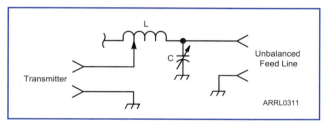

Figure 6.41 — An L-network antenna coupler, useful for an unbalanced feed line, such as coaxial cable. This circuit can transform impedances at the feed line input to the 50-Ω impedances preferred by most transceivers.

networks are also used inside radio equipment to convert impedances from one value to another.

In the case of an antenna coupler, the impedance matching circuit is usually required to transform a complex load impedance with both resistance and reactance to a purely resistive value, usually 50 Ω. To perform this task, the circuit cancels the reactive part of the impedance and then transforms the remaining resistive portion to the desired value. [E7C04]

An impedance matching network performs the transformation by exchanging energy between the inductor and capacitor in such a way that the ratio of voltage and current (the impedance) is changed between the input and output connections. Aside from small resistive losses, no energy is lost in the transformation. Only the ratio of voltage and current are changed. A mechanical analog is the gearbox in which power at one combination of speed and torque is changed to a different combination of speed and torque.

L-Networks

The simplest LC impedance matching network is the *L-network*. **Figure 6.40** shows its four variations that have both an inductor and capacitor. (There are four additional variations that either have two inductors or two capacitors, but they are less common.) The choice of circuit to be used is determined by the ratio of the two impedances to be matched and the practicality of the component values that are required.

The L-network in **Figure 6.41** will transform to 50 Ω any higher impedance presented at the input to the feed line. (At least it will if you have an unlimited choice of values for L and C.) Most antennas and feed lines will present an impedance that can be matched with an L-network.

To adjust this L-network for a proper match, the coil tap is moved one turn at a time, adjusting C for lowest SWR at each step. Eventually a combination should be found that will give an acceptable SWR value. If the impedance at the input to the feed line is lower than 50 Ω, the circuit can be "turned around" to reverse the transformation ratio. Matching networks made entirely of inductors and capacitors work equally well in either direction!

The major limitation of an L-network is that a combination of inductor and capacitor is normally chosen to operate on only one frequency band because a given LC combination has a relatively small impedance-matching range. If the operating frequency varies too greatly, a different set of components will be needed.

Pi and Pi-L Networks

Most tube-type amplifiers use *pi-network* output-coupling circuits as shown in **Figure 6.42**. The most common form of this network consists of one capacitor in parallel with the input and another capacitor in parallel with the output. An inductor is in series between the two capacitors. [E7C01] The circuit is called a pi-network because it resembles the Greek letter pi (π) — if you use your imagination a bit — with the two capacitors drawn vertically at the ends of the horizontal inductor. Using this circuit, very wide values

Radio Circuits and Systems 6-39

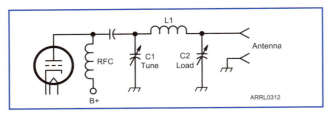

Figure 6.42 — A pi-network output-coupling circuit. C1 adjusts the circuit's tuning to resonance (TUNE) and C2 adjusts the load impedance presented to the tube (LOAD).

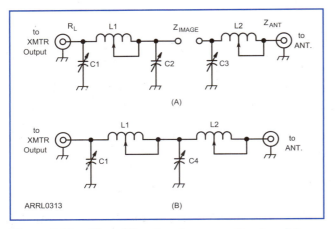

Figure 6.43 — The pi-L-network uses a pi-network to transform the transmitter output impedance (R_L) to an intermediate "image" impedance (Z_{IMAGE}). An L-network then transforms Z_{IMAGE} to the antenna simpedance, Z_{ANT}. C2 and C3 are then combined into a single component, C4.

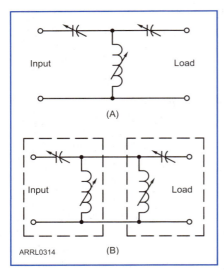

Figure 6.44 — The T-network at (A) can be thought of as two L-networks back-to-back as shown at (B). This network has low losses and is a widely used circuit in amateur antenna couplers.

of load impedance can be matched, providing a greater impedance-transformation range than the L-network. The pi-network can be thought of as two L-networks back-to-back with L1 split into two inductors in series.

To adjust the pi-network in a power amplifier for proper operation, the tuning capacitor (C1) is adjusted for minimum plate current (called "dipping the plate"), and the loading capacitor (C2) is adjusted for maximum permissible plate current. The adjustments interact, so this procedure is usually performed several times to reach the optimum settings. [E7B09]

Because of the series coil and parallel capacitors, this circuit acts as a low-pass filter to reduce harmonics as well as acting as an impedance-matching device. (A pi-network with two coils shunted to ground and a series capacitor would make a high-pass filter and is virtually never used as an amateur output-coupling circuit.) Harmonic suppression with a pi-network depends on the impedance-transformation ratio and the circuit Q. While the L-network's Q is determined by the impedance transformation ratio, the pi-network's Q can be controlled by selecting different combinations of component values. [E7C12] Circuit design information for pi-networks appears in the *ARRL Handbook*.

If you need more attenuation of the harmonics from your transmitter, you can add an L-network in series with a pi-network, to build a pi-L-network. **Figure 6.43** shows a pi-network and an L-network connected in series. [E7C07] It is common to combine the value of C2 and C3 into a single variable capacitor as shown in Figure 6.43B as C4. The pi-L-network thus consists of two series inductors and two shunt capacitors. The pi-L-network provides the greatest harmonic attenuation of the three most-used matching networks — the L, pi, and pi-L-networks. [E7C03]

T-Networks

A T-network as shown in **Figure 6.44** consists of two capacitors in series with the signal lead and a parallel, or shunt-connected inductor between them to ground. The T-network can also be thought of as two L-networks as shown in the figure.

This circuit is commonly used in antenna coupling equipment because the series capacitors and shunt inductor have lower loss than a pi-network. While this type of T-network will transform a wide range of impedances it also acts as a high-pass filter and provides little harmonic rejection. [E7C02]

6.5 Power Supplies

Almost every electronic device requires some type of power supply. The power supply must provide the required voltages when the device is operating and drawing a certain current. The output voltage of most simple power sources, such as batteries or basic rectifier circuits, varies inversely with the load current. If the device starts to draw more current, the supplied voltage will drop. In addition, the operation of most circuits will change as the power-supply voltage changes. For this reason there is a *voltage-regulation* circuit included in the power supply of almost every electronic device. The purpose of this circuit is to stabilize the power-supply output voltage and/or current under changing load conditions.

LINEAR VOLTAGE REGULATORS

E7D01 — How does a linear electronic voltage regulator work?
E7D03 — What device is typically used as a stable voltage reference voltage in a linear voltage regulator?
E7D05 — Which of the following types of linear voltage regulator places a constant load on the unregulated voltage source?
E7D06 — What is the purpose of Q1 in the circuit shown in Figure E7-2?
E7D07 — What is the purpose of C2 in the circuit shown in Figure E7-2?
E7D08 — What type of circuit is shown in Figure E7-2?
E7D11 — What is the function of the pass transistor in a linear voltage regulator circuit?
E7D12 — What is the dropout voltage of an analog voltage regulator?

Linear voltage regulators make up one major category of voltage regulator. In these circuits, regulation is accomplished by varying the conduction of a *control element* in some proportion to the load current. The control element conduction is varied so as to maintain the output voltage at a constant level. [**E7D01**]

Shunt and Series Regulators

In Zener diode regulator circuits the control element is a Zener diode (D1) that varies the current through a fixed resistor (R1) as shown in **Figure 6.45**. Because the Zener diode's reverse-breakdown voltage is relatively constant, varying load currents do not cause the regulated output voltage to change as long as enough current flows through the Zener diode. Because the Zener diode controls the output voltage by drawing current from the power source in parallel with the load, it is called a *shunt regulator*. Shunt regulators are most useful when a constant load on the input voltage source results in constant output voltage. [**E7D05**]

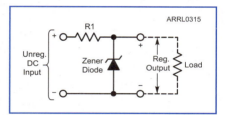

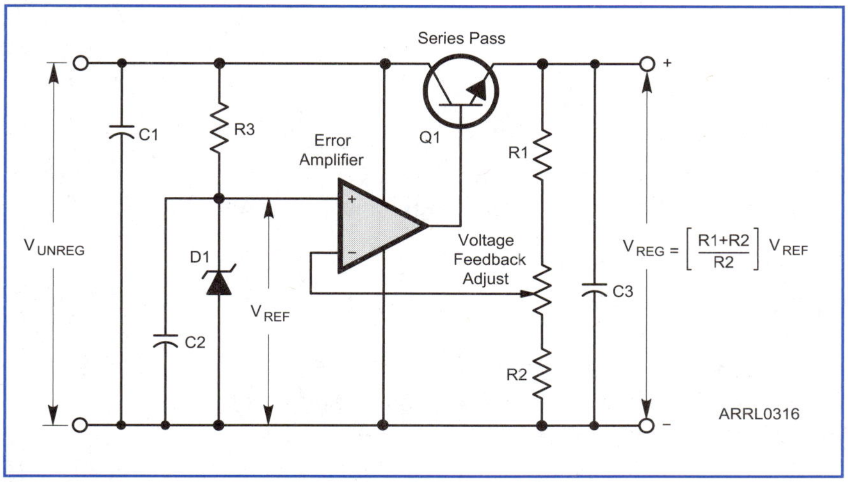

Figure 6.45 — A Zener-diode voltage regulator circuit in which R1 is the control element.

Figure 6.46 — The control element is a pass transistor ("Series Pass") whose base current is varied by the Error Amplifier. The Error Amplifier compares a portion of the output voltage to the voltage reference and adjusts the pass transistor base current until the output voltage has the correct value.

Figure 6.46 shows an example of a *linear series regulator* circuit, also shown on the exam

Radio Circuits and Systems   6-41

in Figure E7-2. [**E7D08**] The control element is a pass transistor (Q1) with base current controlled by the error amplifier. [**E7D06**] The error amplifier compares a fraction of the output voltage to that of the voltage reference and adjusts the pass transistor base current until the output voltage has the correct value. By varying the dc current in the base of the transistor its collector-emitter output current supplied to the load may be varied as necessary to hold the output voltage constant. [**E7D11**]

A series regulator also requires a minimum level of voltage drop across the pass element so that it can respond to changes in the load current and maintain a constant output voltage. If the input voltage to the regulator becomes too low, the regulator will be unable to cause the pass element to supply enough load current and output voltage will begin to fall below the regulated value. The minimum input-to-output voltage is regulator's *drop-out voltage*. [**E7D12**]

The output resistive divider of R1, R2 and the potentiometer both provide a sample of the output voltage and place a small load on the regulator at all times. C1 serves to filter the unregulated input supply voltage. This capacitor is often the rectifier output filter capacitor of an unregulated rectifier supply. The *voltage reference*, usually a Zener diode (D1) as shown in the figure, provides a stable reference for comparing output voltage to the desired value or *set point*. [**E7D03**] R3 supplies current to the Zener diode. C2 across the Zener diode serves to bypass hum, ripple, and noise from the rectifier output around the voltage reference. [**E7D07**] C3 across the output terminals prevents the regulator from oscillating if the load is removed or is very small.

EFFICIENCY AND POWER DISSIPATION

E7D04 — Which of the following types of linear voltage regulator usually make the most efficient use of the primary power source?

E7D13 — What is the equation for calculating power dissipated by a series linear voltage regulator?

As for any system, efficiency for voltage regulators is calculated as:

$$\text{Efficiency (in \%)} = 100\% \times \frac{\text{Power Out}}{\text{Power In}}$$

(Equation 6.5)

In effect, series regulators create "smart resistors" with a value that is varied to create just the right amount of voltage drop, maintaining a constant output voltage. As such, they also dissipate power just like a resistor, calculated as the voltage drop across the regulator multiplied by the load or output current [**E7D13**]:

$$P_{DISS} = (V_{IN} - V_{OUT}) \times I_{OUT}$$

(Equation 6.6)

Both series and shunt regulators dissipate a significant amount of the supply's input power as heat in order to maintain the constant output voltage. Because the series regulators control the load current directly, however, they are more efficient than shunt regulators. [**E7D04**]

BATTERY CHARGING REGULATORS

E7D09 — What is the main reason to use a charge controller with a solar power system?

A special type of voltage regulator called a *charge controller* is used to charge rechargeable batteries. Each different type of battery chemistry requires a different type of controller. The controller applies charging current at the right rate until the desired battery voltage is reached. The controller then supplies just enough charge to maintain the battery at an optimum voltage. This is particularly important if the battery is supplied by an intermittent or variable power source, such as a wind or solar power system. If the power source were connected directly to the battery, it might overcharge and damage the battery. [**E7D09**]

As alternative energy systems become more common, battery charge controllers are becoming an important type of regulator system. Visit **www.batteryuniversity.com** for more information about batteries and charging.

SWITCHING REGULATORS

E7D02 — What is a characteristic of a switching electronic voltage regulator?

E7D10 — What is the primary reason that a high-frequency switching type high-voltage power supply can be both less expensive and lighter in weight than a conventional power supply?

The second major category of voltage regulators is the *switching regulator* in which the control device is switched on and off electronically. The switching regulator works by storing energy in the magnetic field of an inductor or transformer, then releasing it to an output filter circuit. The duty cycle of the control element controls the rate at which energy is stored and released and is automatically adjusted to maintain a constant average output voltage. [**E7D02**]

Switching frequencies of tens of kilohertz or more reduce the size of the transformer or energy storage inductor and of the capacitors needed to filter the output voltage. In an inverter-style dc-to-ac power supply, the savings in weight and component cost can be substantial. [**E7D10**] Switching regulators also have a very high efficiency compared to linear regulators, justifying the higher expense and complexities of their design.

HIGH VOLTAGE TECHNIQUES

E7D14 — What is the purpose of connecting equal-value resistors across power supply filter capacitors connected in series?

E7D15 — What is the purpose of a step-start circuit in a high-voltage power supply?

The construction of high-voltage supplies poses special considerations in addition to the normal design and construction practices used for lower-voltage supplies. In general, remember that physical spacing between leads, connections, parts and the chassis must be sufficient to prevent arcing. Also, the series connection of components such as capacitor and resistor strings must be done with consideration for voltage stresses in the components.

Capacitors

Capacitors are often connected in series strings to form an equivalent capacitor with the capability to withstand the applied voltage. When this is done, equal-value resistors need to be connected across each capacitor in the string in order to distribute the voltage equally across each capacitor. The *equalizing resistors* should have a value low enough to equalize differences in capacitor leakage resistance between the capacitors but high enough not to dissipate excessive power. The equalizing resistors also serve *bleeder resistors* to discharge the filter capacitors when power is removed and place a constant, light load on the supply to prevent excessive voltage with no load connected. [**E7D14**]

Capacitor bodies and cases in high-voltage strings need to be insulated from the chassis and from each other by mounting them on insulating panels to prevent arcing to the chassis or other capacitors in the string.

In order to reduce stress on the power supply high-voltage transformer and rectifier circuits when the supply is turned on, a "step-start" function is often used to charge the filter capacitors gradually. This consists of a resistor in the primary circuit of the power transformer that limits the input current to the supply. After a short period of a second or two, the resistor is switched out with a relay and the supply charges to its full output. [**E7D15**]

Avoid older oil-filled capacitors. They may contain polychlorinated biphenyls (PCBs), a known cancer-causing agent. Newer capacitors have eliminated PCBs and have a notice on the case to that effect. Should you encounter old oil-filled capacitors, contact your local power utility as they often have the means to safely dispose of them.

Radio Circuits and Systems

Chapter 7

Radio Measurements and Performance

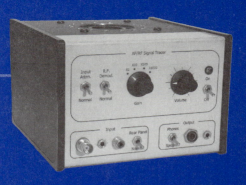

In this chapter, you'll learn about:
- Test equipment used in ham radio
- Oscilloscopes and spectrum analyzers
- Receiver sensitivity and noise
- Dynamic range and intercept point
- Phase noise
- Transmitter intermodulation
- Atmospheric, power line, and vehicle noise
- Noise reduction

This chapter covers the instruments used to measure signals and radio performance. Like other topics on the Extra class exam, you were introduced to many of these concepts as you studied for your Technician and General class licenses. For the Extra class exam, we'll dive deeper into these topics.

The Extra class license exam includes basic questions about test equipment, receiver performance, intermodulation, noise, and interference. These topics are somewhat related and that is why they are grouped together in this chapter. Questions about these topics are located in several parts of the question pool, so be sure to review each one before moving on. If you need a refresher on ac waveforms, read "AC Waveform Supplement" that is provided on this manual's website: **arrl.org/extra-class-license-manual**. It covers basic waveform terms and measurements that are important for understanding the material in this chapter.

7.1 Test Equipment

The following types of test equipment are just a few of the many different instruments used in radio. You have been introduced to the multimeter in studies for your previous license exams and the Extra class license goes a bit further. The other types of test equipment covered in this section may be new to you, but are often encountered in building and testing amateur equipment and antennas.

INSTRUMENTS AND ACCURACY

E4B02 — What is the significance of voltmeter sensitivity expressed in ohms per volt?
E8A05 — What of the following instruments would be the most accurate for measuring the RMS voltage of a complex waveform?

Multimeters

The multimeter is a basic piece of test equipment that makes variety of measurements. The *digital multimeter (DMM)* is microprocessor-controlled and most are *autoranging* so they automatically choose the right range to display voltage, resistance, and current values. While most multimeters are digital, moving-needle analog meters are still common. Regardless of whether the meter is digital or analog, their basic specifications are similar.

The *accuracy* of most meters is specified as a percentage of full scale. If the specification states that the meter accuracy is within 2% of full scale, the possible error anywhere on a scale of 0 to 10 V is 2% of 10 V, or 0.2 V. The *resolution* of almost any multimeter sold today is sufficient for general measurements in radio equipment. Most offer "3½ digit" displays, meaning that the left-most of four digits is 1 or blank. Such a meter has a resolution of

Radio Measurements and Performance   7-1

Accuracy, Precision, and Resolution

The terms *accuracy*, *precision* and *resolution* are often confused and used interchangeably, when they have very different meanings. When dealing with measurements and test instrumentation, it's important to keep them straight.

• **Accuracy** is the ability of an instrument to make a measurement that reflects the actual value of the parameter being measured. An instrument's accuracy is usually specified in percent or decibels referenced to some known standard.

• **Precision** refers to the smallest division of measurement that an instrument can make repeatedly. For example, a metric ruler divided into mm is more precise than one divided into cm.

• **Resolution** is the ability of an instrument to distinguish between two different quantities. If the smallest difference a meter can distinguish between two currents is 0.1 mA, that is the meter's resolution.

It is important to note that the three qualities are not necessarily mutually guaranteed. That is, a precise meter may not be accurate, or the resolution of an accurate meter may not be very high, or the precision of a meter may be greater than its resolution. It is important to understand the difference between the three when selecting and using test instruments.

0.05% at full scale, plenty good for amateur use! See the sidebar "Accuracy, Precision, and Resolution" for more information on these important topics.

Another useful specification is the meter's sensitivity. A sensitive meter draws very little current from the circuit being tested. Sensitivity is often specified in ohms-per-volt (Ω/V), the input impedance in ohms divided by the full-scale reading in volts. The input impedance can also be calculated by multiplying the full-scale meter reading by the sensitivity in Ω/V. [**E4B02**] Digital meters may also specify their input impedance directly.

RMS Measurements

Some meters are specified as measuring "true RMS" values of voltage or current. What does "RMS" mean? When an ac voltage is applied to a resistor, the resistor will dissipate heat, just as if the voltage were dc. The dc voltage that would cause an identical amount of heating as the ac voltage is called the *root-mean-square* (*RMS*) or *effective value* of the ac voltage. (RMS refers to the mathematical method to calculate RMS values.)

The RMS value of any waveform, voltage, or current can be determined by making a large number of point-by-point measurements and then calculating the RMS value. (It can also be determined by measuring the waveform's heating effect on a resistor.) For this reason, "true RMS" calculating meters are the most accurate for determining the RMS value of any waveform, no matter how complex. [**E8A05**]

Meters that don't perform the full calculations usually assume the waveform is a sine wave and convert the measurement to an equivalent RMS value. These meters are not accurate for non-sinusoidal waveforms, pulses, or waveforms with a dc offset.

Fortunately, for common symmetric ac waveforms the conversions between peak, peak-to-peak, average, and RMS are simple. **Table 7.1** shows how to convert between peak, peak-to-peak, average and RMS waveforms of sine and square waves. You will make frequent use of the sine wave conversions.

Table 7.1
AC Measurements for Sine and Square Waves

	Sine Wave	*Square Wave*
Peak-to-Peak	2 × Peak	2 × Peak
Peak	0.5 × Peak-to-Peak	0.5 × Peak-to-Peak
RMS	0.707 × Peak	Peak
Peak	1.414 × RMS	RMS
Average	0 (full cycle)	0 (full cycle)
	0.637 × Peak (half cycle)	0.5 × Peak (half cycle)

7-2   Chapter 7

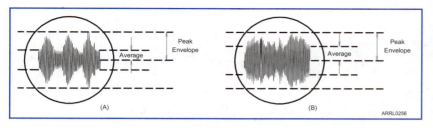

Figure 7.1 — Two modulation envelope patterns that show the difference between average and peak power levels. In each case, the RF amplitude is plotted as a function of time. In B, the average power level has been increased compared to the peak value.

RF Wattmeters

E8A06 — What is the approximate ratio of PEP-to-average power in a typical single-sideband phone signal?

E8A07 — What determines the PEP-to-average power ratio of a single-sideband phone signal?

Nearly all commercial radios have some kind of built-in capability to measure and display RF output power. Standalone wattmeters are installed in the feed line between the transceiver and antenna. Most have the ability to measure both forward and reflected power so that SWR can be displayed. (See the Antennas and Feed Lines chapter for more information.)

Most wattmeters respond to the average of the RF power over many cycles of the modulation envelope. When amateurs refer to *average output power* in this context, they are referring to this long-term average of power.

When using an AM voice mode such as SSB, it is important to know the peak envelope power (PEP) of the transmitted signal because that is how the rules specify power levels. Envelope peaks occur too quickly for meter readings to accurately represent their values, though. *Peak-hold* or *peak-reading wattmeters* have special circuits that measure and display the peak value of a signal.

The PEP of an AM or SSB signal will be several times greater than the average power. The ratio of peak-to-average power in SSB voice signals varies widely with voices of different characteristics. [**E8A07**] **Figure 7.1** shows two typical envelope patterns. In the case shown in Figure 7.1A, the average power (estimated graphically) is such that the peak-to-average ratio is almost 3:1. Depending on the type of voice and manner of speaking the ratio may be more than 10:1. By using a *compressor* circuit to increase the minimum modulated signal levels, the average output power can be increased as in Figure 7.1B. Typical ratios of PEP to average power are about 2.5:1. [**E8A06**] Remember that FM and PM are *constant-power modes* for which the envelope has a constant amplitude during transmission.

Frequency Counters and References

E4A05 — What is the purpose of the prescaler function on a frequency counter?

E4B01 — Which of the following factors most affects the accuracy of a frequency counter?

One of the most accurate means of measuring frequency is the *frequency counter*. This instrument counts cycles of the input signal over a specified time and displays the frequency of the signal on a digital display. For example, if an oscillator operating at 14.230 MHz is connected to the counter input, 14.230 would be displayed. Some counters are usable well into the gigahertz range. Advanced frequency counters can also measure pulse widths and signal periods, count pulses, and measure time intervals. **Figure 7.2** shows a typical frequency counter. Lab-grade counters are widely available as surplus or used equipment and as inexpensive new equipment and kits.

Most counters that are used at such high frequencies use a *prescaler* ahead of a lower-frequency counter. A special type of frequency divider circuit, the prescaler reduces a signal's frequency by a factor of 10, 100, 1000, or some other integer divisor so that a low-frequency

Figure 7.2 — Typical frequency counter used by hams for troubleshooting and construction. Many lab-grade counters are available as surplus or used equipment. Frequency counters are also available in kit form and as new equipment. [Courtesy of B&K Precision]

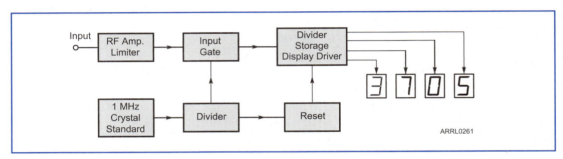

Figure 7.3 — This block diagram shows the basic parts of a frequency counter.

counter can display the input frequency. [E4A05]

The internal circuits of a typical counter are illustrated as the block diagram in **Figure 7.3**. Frequency-counter accuracy depends on an internal crystal-controlled reference oscillator, also called the *time base*, shown in the figure as the 1 MHz crystal standard. Any variation of the time base oscillator frequency affects the counter accuracy and precision. The more accurate the crystal reference, the more accurate the counter readings will be. [E4B01] A crystal frequency of 1 MHz has become more or less standard for use in the reference oscillator. The crystal should have excellent temperature stability so the oscillator frequency won't change appreciably as temperature changes.

A frequency counter will measure the frequency of the strongest signal at its antenna or input connector. Counters can acquire their input signal by an antenna placed close to a transmitter rather than having a direct connection. For low-level signals, however, a probe or other input connection may be used.

Although usually quite accurate, a frequency counter should be checked regularly against WWV, WWVH, or some other *frequency standard*. The accuracy of frequency counters is often expressed in parts per million (ppm). Even after checking the counter against WWV, you must take this possible error into account. The counter error can be as much as:

$$\text{Error in Hz} = \frac{\text{f (in Hz)} \times \text{counter error in ppm}}{1{,}000{,}000} \qquad \text{(Equation 7.1)}$$

This is the maximum displayed error.

THE OSCILLOSCOPE

Direct observation of high-speed signals and waveforms is not possible using any kind of meter or numeric instrument. There is just too much information to be conveyed at too high a rate. Enter the *oscilloscope*, or "*scope*" — the amateur's electronic eyes. A scope is used to display a signal's amplitude versus time so that the shape and other characteristics of the waveform can be seen and measured, even if the signal is changing very quickly.

While digital scopes display the digitized input signal on a computer-type screen, the terminology was developed for analog scopes. For example, the horizontal axis time calibration is often referred to as "sweep speed" because an electron beam was "swept" across the face of a *cathode-ray tube (CRT)*. In a digital scope, nothing is "swept" but the term and others like it remain in use. Because many CRT-based scopes remain in service (and are quite capable, even preferred in some applications) this section will retain the original analog terminology, noting differences with the newer digital instruments as needed.

Oscilloscope Basics

A sawtooth-type *ramp* waveform with a slow rise time and a sudden fall time causes a spot created by the electron beam to move from left to right, creating a narrow line of light (called the *trace*) on the face of the CRT as shown in **Figure 7.4**. On a digital scope, the trace appears as a line of pixels on a digital display.

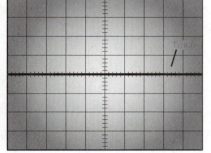

Figure 7.4 — Without any voltage applied to the vertical channel of the oscilloscope, a "flat" trace is displayed. The input voltage moves points on the trace up and down for positive and negative voltage. Grid markings (graticule) allow the operator to make amplitude and time measurements.

The rate at which the beam moves across the CRT is called the *sweep speed* and it is selected by the operator. For digital scopes, a sweep speed or time base value determines the time per horizontal division. The oscillator or clock systems that control the timing of the display are called the scope's *time base*.

The input signal to be analyzed is applied to the scope's *vertical channel* input. Positive voltage moves the trace upward. If a sine wave signal is applied to the vertical input and the appropriate sweep speed is selected, the trace will form a sine wave as it moves up and down simultaneously with its movement across the tube.

An important limitation to the accuracy, frequency response and stability of an oscilloscope is the bandwidth (frequency response) of the scope's vertical channel amplifiers. Scopes are often specified in terms of this bandwidth.

Scopes are also specified by how many vertical channels they have (special circuits can make it appear as if there are separate traces for each channel), so you might see a particular model listed as a "20 MHz dual-channel scope."

Another important performance limitation is the accuracy and linearity of the scope's time base. Unless the time base is stable, frequency and timing measurements made with the scope will not be accurate.

The grid of marks on the face of the tube is called a *graticule* and each line is called a *division*. The graticule's vertical axis is calibrated in *volts/division* or *V/div* and various scales are selectable by the operator. The horizontal axis is calibrated in *time/division* or *seconds/division (s/div)*. Vertical scales are often available from mV/div to tens of V/div. Most scopes offer horizontal axis calibrations of a few s/div to ns/div. This allows the display of signals with frequencies of less than 1 Hz to hundreds of MHz and with amplitudes ranging from mV to tens of V.

By using a positioning control, the amplitude and period of a signal can be compared to the fine divisions on the graticule's central axes. The easiest amplitude measurement to make with an analog scope is an ac signal's peak-to-peak voltage by using the graticule lines as

Radio Measurements and Performance   7-5

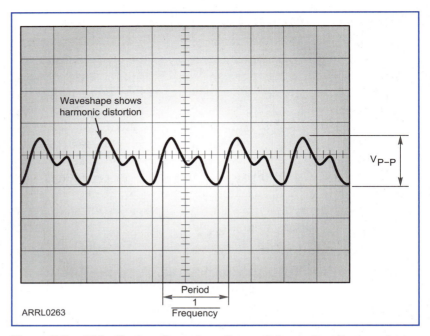

Figure 7.5 — This display shows how the peak-to-peak voltage and period of a complex waveform are measured on an oscilloscope. It can also be seen that the waveform contains enough harmonic energy to cause significant distortion.

shown in **Figure 7.5**. A digital scope can usually make these measurements automatically and display the results on-screen.

There are many uses for an oscilloscope in an amateur station. This instrument is often used to display a transmitter's output waveform. Such a test can help you determine if the amplifier stages in your transmitter are operating properly. An oscilloscope can also be used to display signal waveforms during troubleshooting procedures. For example, consider the waveform display of Figure 7.5. The sine wave can be seen to have some significant distortion due to the presence of harmonics adding to the fundamental waveform.

Oscilloscope Probes

E4A04 — How is the compensation of an oscilloscope probe typically adjusted?
E4A09 — Which of the following is good practice when using an oscilloscope probe?

Oscilloscopes are connected to the circuits being tested using a special *oscilloscope probe* or *scope probe*. Such probes are specially constructed to change the input signal frequency and timing characteristics as little as possible. Each probe has its own ground lead that is connected to the circuit being tested. For the most accurate measurements at high frequencies, it is important to keep the ground connection as short as possible. [**E4A09**]

The probe incorporates a high-impedance voltage divider that loads the circuit as little as possible. The signal's amplitude at the output of the probe (the input to the scope) is typically divided by 10 — this is a *times-10 (×10)* probe. (There are also ×1 and ×100 probes, but they are used much less frequently.) Most scopes assume a ×10 probe is used when displaying the vertical scale in V/div.

Because the frequency response of the probe can affect the frequency response of the signal being displayed on the scope, it is important to ensure that the probe is adjusted properly. This is referred to as the probe's *compensation*. Most scopes provide a square wave *calibration* signal of a few kHz and with an amplitude of 0.1 to 1 V. The probe is connected to this signal and its compensation control adjusted until the square wave's horizontal portions are flat and the corners are sharp. (The scope's user manual will describe this procedure in detail.) [E4A04]

Digital Oscilloscopes

E4A01 — Which of the following limits the highest frequency signal that can be accurately displayed on a digital oscilloscope?

E4A06 — What is the effect of aliasing on a digital oscilloscope caused by setting the time base too slow?

A digital scope samples the input signal, converting it to digital data with an analog-to-digital converter (ADC). All display and calculation functions are then performed using that data. Because digital scopes sample the input signal, they have all of the same concerns regarding bandwidth and frequency response as the DSP systems discussed in the Radio Circuits and Systems chapter. Review the DSP material, if necessary.

Similar to an analog scope, a digital scope has a specified bandwidth for displaying signals accurately. The upper limit on bandwidth is determined by the sampling rate of the analog-to-digital converter. [E4A01]

Aliasing, in particular, can be a significant challenge for digital scopes. If the time base and sample rate are too low for a specific signal or a signal has too high a frequency for the scope's ADC, a false, jittery low-frequency alias of the input signal will appear on the scope's display and be treated just as a real signal. [E4A06] To prevent aliasing, signals with frequencies that are too high must be prevented from reaching the ADC input. Scopes use low-pass, anti-alias filters to eliminate these signals. This limits the practical, alias-free bandwidth of a digital scope to a somewhat lower value than one-half the sample rate.

The digital scope can perform many functions automatically that an analog scope user must perform manually. For example, after a trace has been captured, the time base can be expanded to "zoom in" on a signal feature or contracted to see more of a signal. Other functions performed digitally include automatic amplitude and frequency measurements, labeling traces, storage and recall of traces, and greatly enhanced triggering operation.

The Logic Analyzer

E4A10 — Which of the following displays multiple digital signal states simultaneously?

The *logic analyzer* is a special type of oscilloscope specifically for observing and measuring digital signals, such as from a microprocessor. Instead of a linear vertical amplifier, the logic analyzer senses and displays logic levels. A typical logic analyzer has at least 16 and sometimes many more input channels so that all bits in a data or address bus can be captured and displayed simultaneously along with auxiliary enable/disable and clock signals. [E4A10] The logic analyzer can apply sophisticated logical tests on the signals to perform the trigger function and can capture and record signal states. If you do a lot of microprocessor or digital circuit development, a logic analyzer is a very handy piece of test gear.

THE SPECTRUM ANALYZER

Time and Frequency Domains

An oscilloscope displays signals in the *time domain* with amplitude on a vertical axis and time on a horizontal axis. A spectrum analyzer is very similar to an oscilloscope but displays signal frequency on the horizontal axis instead. Displaying amplitude versus frequency is the *frequency domain*. This type of measurement is useful when testing the frequency content of signals from amplifiers, oscillators, detectors, modulators, mixers, and filters. You can find a

Radio Measurements and Performance

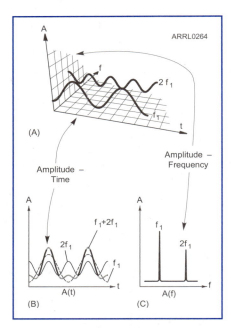

Figure 7.6 — This diagram shows how a complex waveform may be displayed in either the time domain or frequency domain. Part A is a three-dimensional display of amplitude, time, and frequency. At B, this information is shown in the time domain as on an oscilloscope. At C, the signal's frequency domain information is shown as it would be displayed on a spectrum analyzer.

more detailed discussion of spectrum analyzer measurement techniques in *The ARRL Handbook*.

To better understand the concepts of time and frequency domains, refer to **Figure 7.6**. In Figure 7.6A, the three-dimensional coordinates show time (as the line sloping toward the bottom right), frequency (as the line sloping toward the top right), and amplitude (as the vertical axis). The signal frequencies shown are harmonically related (f_1 and $2f_1$). The time domain is represented in Figure 7.6B, in which both signals are shown together. If the two signals were applied to the input of an oscilloscope, we would see the bold line that represents the amplitudes of the signals added together.

The display in Figure 7.6C is typical of a spectrum analyzer presentation of a complex waveform. Here the signals are separated into its individual frequency components, and a measurement made of the amplitude of each signal. A signal's amplitude can be represented on a spectrum analyzer's vertical scale as its voltage or as its power.

The frequency domain contains information not found in the time domain, and vice versa. Hence, the spectrum analyzer offers advantages over the oscilloscope for certain measurements, but for measurements in the time domain, the oscilloscope is an invaluable instrument.

Waveform Spectra

E8A01 — What is the name of the process that shows that a square wave is made up of a sine wave plus all its odd harmonics?

E8A03 — What type of wave does a Fourier analysis show to be made up of sine waves of a given fundamental frequency plus all its harmonics?

Fourier analysis is a mathematical method of analyzing ac waveforms. It breaks or decomposes a waveform into sine and cosine waves of a fundamental frequency and harmonic frequencies. For example, analyzing the distorted sine wave in Figure 7.5 might show it to be made up of a sine wave at some fundamental frequency plus a second harmonic with a somewhat smaller amplitude. That information would help the circuit designer to modify or adjust its operation to reduce the unwanted harmonic.

A spectrum analyzer performs Fourier analysis either with analog electronics as described in the following sections or by digitizing the input signal and doing the analysis on the resulting digital data.

Square Waves

A square wave is one that abruptly changes back and forth between two voltage levels and remains an equal time at each level as in **Figure 7.7**. (If the wave spends an unequal time at each level, it is known as a *rectangular wave*.) Fourier analysis shows the square wave to be made up of a sine wave at the square wave's fundamental frequency and all of the sine wave's odd harmonics as shown in Figure 7.7. [**E8A01**] You can see this by looking at the square wave with a spectrum analyzer.

Sawtooth Waves

A *sawtooth* waveform, as shown in **Figure 7.8**, has a significantly faster *rise time* (the time it takes for the wave to reach a maximum value) compared to its *fall time* (the time it takes for the wave to reach a minimum value). A sawtooth wave is made up of a sine wave

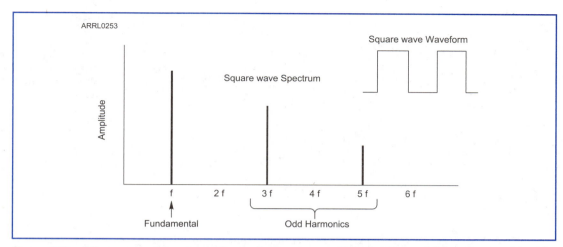

Figure 7.7 — The square wave is made up of sine waves at the fundamental frequency and only the odd harmonics. The amplitude of the harmonics decreases as their frequency increases.

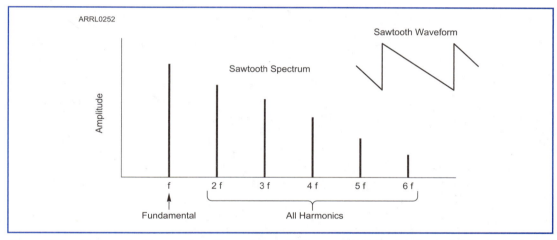

Figure 7.8 — The sawtooth waveform is made up of sine waves at the fundamental frequency and all of its harmonics. The amplitude of the harmonics decreases as their frequency increases.

at its fundamental frequency and all of its harmonics as shown in Figure 7.8. The *ramp* waveform is similar to the sawtooth but with a slow rise time (the ramp) and a fast fall time. [E8A03]

Spectrum Analyzer Basics

E4A02 — Which of the following parameters does a spectrum analyzer display on the vertical and horizontal axes?

E4A03 — Which of the following test instruments is used to display spurious signals and/or intermodulation distortion products generated by an SSB transmitter?

E4B10 — Which of the following methods measures intermodulation distortion in an SSB transmitter?

Just as for oscilloscopes, digital technology is replacing analog designs but many analog spectrum analyzers are still in use by amateurs. Both digital and analog analyzers share

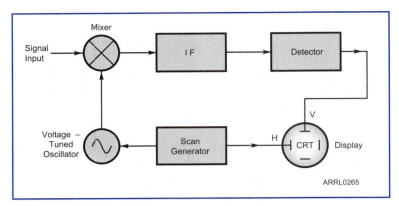

Figure 7.9 — This simplified block diagram illustrates the operation of swept-superheterodyne spectrum analyzer. It is basically an electronically tuned, narrow-bandwidth receiver with a CRT displaying signal amplitude and frequency. A digital spectrum analyzer uses DSP to analyze the signal and replaces the CRT with a computer-type display.

similar concepts and terminology. A simplified block diagram of an analog CRT-based *swept superheterodyne* analyzer is shown in **Figure 7.9**. Digital analyzers are constructed similarly to an SDR as described in the Radio Circuits and Systems chapter. The digitized signal's spectrum is analyzed and displayed on a computer-type screen. As with oscilloscopes, we will use the analog terminology, noting any differences with digital analyzers.

The analyzer is basically a narrow-band receiver with a frequency that is electronically-tuned. Tuning is accomplished by applying a linear ramp voltage to a *voltage-controlled oscillator* (*VCO*) or using a microprocessor to control a *numerically-controlled oscillator* (*NCO*). The same ramp voltage or function simultaneously controls horizontal position on the display. This means the horizontal axis of the spectrum analyzer displays frequency and the vertical axis displays signal amplitude. The resulting spectrum analyzer display shows amplitude versus frequency. [**E4A02**]

Spectrum analyzers are calibrated in both frequency and amplitude for relative and absolute measurements. The *center frequency* control sets the center of the range swept by the receiver. The *scan width* control, is calibrated in hertz, kilohertz, or megahertz per division on the graticule and controls how far the receiver is tuned on either side of the center frequency. For example, if the center frequency were set to 146 MHz and the 10-division horizontal axis scan width were set to 100 kHz per division, the receiver would be tuned across a 1 MHz range from 145.500 MHz to 146.500 MHz,

The vertical axis of the display is commonly calibrated as 1 dB, 2 dB, or 10 dB per division. (Linear scales in V/division are also available, but not used as frequently.) For transmitter testing, 10 dB/div is commonly used because it allows you to view a wide range of signal strengths, such as those of the fundamental signal, harmonics, and spurious signals.

Transmitter Testing with a Spectrum Analyzer

Among other practical uses, the spectrum analyzer is ideally suited for checking the output from a transmitter or amplifier for spectral quality. Within the limits of the receiver, you can test a transmitter over any frequency range. Whether you are testing an HF or a VHF transmitter, the spectrum analyzer displays all frequency components of the transmitted signal. You can easily see any spurious signals from the transmitter on a spectrum analyzer display. [**E4A03**]

Figure 7.10 shows two test setups commonly used for transmitter testing. The setup at B is the more accurate approach for broadband measurements because most transmission line power sampling devices do not have a constant-amplitude output across a broad frequency spectrum. The ARRL Headquarters Laboratory staff uses the setup shown in B.

Another area of concern about transmitter spectral purity has to do with the *intermodulation distortion (IMD)* levels of SSB transmitters and amplifiers. The test setup in **Figure 7.11** shows how a spectrum analyzer is used for SSB transmitter IMD testing. The

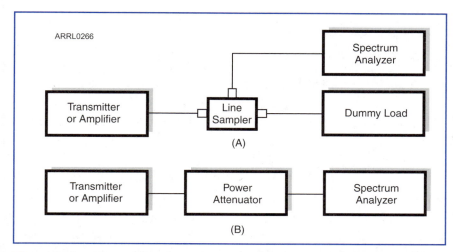

Figure 7.10 — These diagrams show two commonly used test setups to observe the output of a transmitter or amplifier on a spectrum analyzer. The system at A uses a transmission line power sampler to obtain a small amount of the transmitter or amplifier output power. At B, the majority of the transmitter output power is dissipated by the power attenuator, leaving a small amount to be measured by the spectrum analyzer.

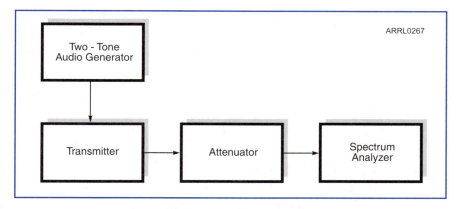

Figure 7.11 — This diagram illustrates the test setup used in the ARRL Laboratory for measuring the IMD performance of SSB transmitters and amplifiers.

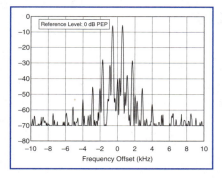

Figure 7.12 — This is the spectrum analyzer display showing the result of a two-tone intermodulation distortion test of an SSB transmitter. Each horizontal division is equal to 2 kHz and each vertical division is 10 dB. The transmitter's PEP output for a single tone is represented by the top line of the display. The sideband's components representing the two audio tones are the large peaks at the center of the display. IMD products can be observed 28, 47 and 52 dB below the PEP output level.

transmitter is first adjusted to produce full PEP output with a single audio tone input. Two equal-amplitude, but non-harmonically related, audio tones are then input to the transmitter and the RF output displayed on a spectrum analyzer. [**E4B10**] (The ARRL Lab uses 700 and 1900 Hz tones.)

Figure 7.12 shows a typical display during a transmitter IMD test. Signals other than from the two individual tones (the large peaks at the center of the display) are distortion products. In this example, IMD products are observed at approximately 28, 47, and 52 dB below the transmitter's single-tone PEP output level, which is at the very top line of the display. Sidebands from the two input tones are 6 dB below the PEP output.

Radio Measurements and Performance 7-11

7.2 Receiver Performance

Effective radio communication depends on quality reception of signals, as free of noise and distortion as possible, able to ignore unwanted signals, and sensitive enough to hear the weakest. That is why it's important to be able to measure and evaluate receiver quality.

When you read product reviews of new radios, you'll notice that a great deal of time is spent evaluating the receiver. Its performance will be measured using the parameters you're about to study. By learning more about each one, you'll be able to compare receivers and make informed decisions about which radio design is better. We'll begin with the receiver's ability to hear and then move on to how well it hears a particular signal.

SENSITIVITY AND NOISE

E4C05 — What does a receiver noise floor of –174 dBm represent?

E4C06 — A CW receiver with the AGC off has an equivalent input noise power density of –174 dBm/Hz. What would be the level of an unmodulated carrier input to this receiver that would yield an audio output SNR of 0 dB in a 400 Hz noise bandwidth?

E4C07 — What does the MDS of a receiver represent?

E4C11 — Why can an attenuator be used to reduce receiver overload on the lower frequency HF bands with little or no impact on signal-to-noise ratio?

One of the fundamental receiver specifications is *sensitivity* or *minimum discernible signal* (*MDS*). The MDS of a receiver represents the strength of the smallest discernible input signal. [**E4C07**]

Another important concept is *signal-to-noise ratio* or *SNR* which is defined as signal power divided by noise power and is expressed in dB. An SNR of greater than 0 dB means the signal is stronger than the noise.

Another type of signal-to-noise ratio is *signal-to-noise-and-distortion* (*SINAD*). This figure includes the ability of the receiver to accurately detect or demodulate the input signal. Any distortion of the signal is added to the noise in the sense that it lowers the ratio of the desired signal's energy to that of the undesired energy.

The MDS is also called the receiver's *noise floor*, because it represents the strength of a signal that produces the same audio output power as the receiver noise. You can measure receiver MDS by measuring its audio output power when the antenna input is connected to a dummy load of the proper impedance. (The AGC system must be disabled to keep it from affecting the receiver's gain.) After making that measurement, feed an unmodified carrier signal from a calibrated signal generator into the receiver antenna input. When the audio output power is twice what it was without an input signal (a 3 dB increase), then the input signal is just strong enough to produce an audio output equal to the receiver's internal noise. Since the signal and noise powers are equal, the signal-to-noise ratio is 0 dB. The strength of that signal is equal to the MDS for that receiver. The lower the MDS, the more sensitive the receiver. (It is also necessary to specify the receiver bandwidth as discussed below.)

MDS and other receiver performance specifications are often given in dBm. This abbreviation means "decibels with respect to one milliwatt." 0 dBm is the same as 1 mW, +10 dBm is 10 mW, –20 dBm is 0.01 mW (or 10 μW), and so forth. Using dBm allows us to discuss an extremely wide range of signal power levels.

MDS may also be given in μV, such as 0.5 μV. This can be converted to power if the receiver input impedance is known — usually 50 Ω. Power, $P = V^2/50$ in this case. The equivalent in dBm = 10 log (P/0.001). For example, an MDS of 0.5 μV equals an MDS of –113 dBm. This is a practical MDS on the HF bands where natural noise is present.

It is useful to know that the theoretical noise power at the input of an ideal receiver, with an input-filter bandwidth of 1 Hz, is –174 dBm at room temperature. [**E4C05**] (*The ARRL*

7-12   Chapter 7

Handbook contains more detailed information about how to calculate this number and why it depends on temperature.) This is considered to be the theoretical best (lowest) noise floor any receiver can have. In other words, for this ideal receiver, the strength of any received signal would have to be at least –174 dBm to be detected. Because the noise power increases linearly with bandwidth, the theoretical MDS with a 1 Hz bandwidth filter is specified as –174 dBm/Hz.

Of course –174 dBm is an incredibly small power level — only four billionths of a billionth of a milliwatt! With an antenna attached, an HF receiver's noise floor is actually determined by atmospheric noise which is far higher than the theoretical noise floor. Atmospheric noise, therefore, is the limiting factor for sensitivity of receivers on the HF bands. This is why turning on an HF receiver's attenuator — to reduce overload, for example — has little or no impact on signal-to-noise ratio. Both signal and noise are attenuated equally. [**E4C11**]

A receiver bandwidth of 1 Hz is impractical, but is used as a reference for comparing wider filters. An actual receiver might have a 500-Hz bandwidth for CW operation, or even wider for SSB or FM voice. As the filter gets wider, more noise will be received by a factor equal to the ratio of the filter used to 1 Hz. For example, a 500-Hz bandwidth increases the received noise power by a factor of 500 over the 1 Hz bandwidth, a ratio of 27 dB = 10 log 500. That increase in filter width also increases the noise floor of this theoretical receiver to –174 dBm + 27 dB = –147 dBm.

For whatever filter width the receiver uses, you can calculate the theoretical MDS for that receiver by calculating the log of the bandwidth and multiplying that value by 10. Add the result to –174 dBm which is the 1-Hz bandwidth value for MDS.

Example 7.1

What is the MDS for a receiver with a –174 dBm/Hz noise floor if a 400 Hz filter bandwidth is used with the AGC turned off? What would be the level of an unmodulated carrier input to this receiver that would yield an audio output SNR of 0 dB in a 400 Hz noise bandwidth? [**E4C06**]

Step 1 — Calculate the bandwidth ratio in dB = 10 log (400 Hz / 1 Hz) = 26 dB
Step 2 — To get MDS, add that figure to –174 dBm = –174 + 26 = –148 dBm

Noise Figure

E4C04 — What is the noise figure of a receiver?

Noise figure is a "figure of merit" for the receiver — expressed in dB, it is the ratio of the noise generated internally by the receiver to the theoretical MDS. [**E4C04**] Noise figure evaluates how much noise the receiver's internal circuits contribute.

The higher a receiver's noise figure, the more noise that is generated in the receiver itself. This also means the receiver will have a higher noise floor. Lower noise figures are more desirable.

The receiver's internal noise raises the noise floor by increasing the power that actual signals must have to be heard. You can calculate the actual noise floor of a receiver by adding the noise figure (expressed in dB) to the theoretical best MDS value in dBm.

Actual Noise Floor = Theoretical MDS + noise figure   (Equation 7.2)

For example, suppose our 500-Hz-bandwidth receiver has a noise figure of 8 dB. We can use Equation 7.2 to calculate the actual noise floor of this receiver.

Actual Noise Floor = –147 dBm + 8 dB = –139 dBm

The noise figure of a receiver is related to the signal-to-noise ratio (SNR) of the input and output signals. Lowering a receiver's noise figure lowers its actual noise floor and improves

Radio Measurements and Performance   7-13

weak signal sensitivity. By lowering noise without changing the input signal level, the SNR ratio will be increased, meaning that the signal is easier to copy.

SELECTIVITY

E4C02 — Which of the following receiver circuits can be effective in eliminating interference from strong out-of-band signals

E4D09 — What is the purpose of the preselector in a communications receiver?

The perfect receiver can tune any frequency and reject every signal except the one you want to receive. That is one definition of *selectivity* — the ability to select a specific signal. That broad definition of selectivity has more specific meanings in the different parts of the receiver. For example, selectivity in a receiver's front end may apply to rejection of strong out-of-band signals, such as shortwave broadcasts or nearby public safety dispatch transmitters. Selectivity is a term that depends on where it is applied.

Band-pass *front-end filters* that pass an entire amateur band (or a significant portion of it) are used at the receiver's antenna input. They provide *front-end selectivity* and their purpose is to keep strong out-of-band signals of nearby transmitters or broadcast stations from overloading the sensitive input circuits. A *preselector* is a tunable input filter adjusted to pass signals at the desired frequency and increase rejection of out-of-band unwanted signals. [**E4D09**] Both improve receiver performance by rejecting signals that can cause image responses in an analog receiver or eliminate overload in an SDR. [**E4C02**]

The degree of selectivity for an analog receiver is determined by the bandwidth of the receiver's entire filter chain, from the front end to the audio output. (Filters are discussed in the Radio Circuits and Systems chapter.) In superheterodyne receivers, there are several filters in the signal path. As the signal passes through the receiver, it encounters progressively narrower filters that remove more and more of the unwanted signals.

In an SDR, filtering is performed mathematically on digitized signals but the definition of filter parameters are the same. Many SDRs and digital mode software packages are capable of receiving, demodulating, and decoding multiple signals at once so the concept of filtering out all but the desired signal is not valid. Selectivity for this type of receiver can also refer to the ability to obtain the information from one signal when others are present.

Analog Receiver IF Filters

E4C09 — Which of the following choices is a good reason for selecting a high frequency for the design of the IF in a superheterodyne HF or VHF communications receiver?

E4C10 — What is an advantage of having a variety of receiver IF bandwidths from which to select?

E4C13 — How does a narrow-band roofing filter affect receiver performance?

E4C14 — What transmit frequency might generate an image response signal in a receiver tuned to 14.300 MHz and that uses a 455 kHz IF frequency?

Farther along the signal's path, analog receivers often use relatively wide filters in each IF amplifier circuit. These may be LC filters, quartz crystal filters, or ceramic resonator filters with characteristics similar to crystal filters. These are used to reject unwanted mixing products and to prevent spurious signals from slipping into the receiver's signal path. These filters pass many signals on or near the desired frequency.

Increasing a superheterodyne receiver's IF improves selectivity. As the IF is increased, the frequency at which image responses occur becomes farther from the desired signal and easier to filter out. [**E4C09**] This filtering can be performed in the receiver's front-end circuitry and at each conversion stage in the receiver. For example, if a low IF such as 455 kHz is used to receive a signal on 14.300 MHz, the BFO could be tuned to 14.3 + 0.455 =

7-14   Chapter 7

14.755 MHz. (The BFO could also be tuned to 14.3 − 0.455 = 13.845 MHz.) The receiver would also receive an image signal on 15.210 MHz because 15.210 − 14.755 = 0.455 MHz. **[E4C14]** If the IF were raised to 9 MHz, the BFO would be set to 14.3 + 9 = 23.3 MHz and the image frequency would be 23.3 + 9 = 32.3 MHz, which is much farther away and easier to filter out.

At the input to each IF stage where the most strenuous filtering is performed, a *roofing filter* is often used. (Most superheterodyne receivers have two or more IF stages.) Roofing filters are high-performance filters (generally crystal filters) that have a bandwidth wider than that of the widest signal that will be received. Their purpose is to increase receiver dynamic range by rejecting as many as possible of the strong signals on adjacent frequencies without affecting the desired signal. **[E4C13]** Such signals can cause amplifiers to overload or affect the AGC system so as to distort the desired signals. **Figure 7.13** illustrates why the filters are called roofing filters — their broad response acts as a "roof" over the narrower filters intended to pass just a single signal.

In the final IF stage of an analog receiver, narrow filters are used to select only one signal from the many that may be present. Crystal or mechanical resonator filters are used for this purpose. The bandwidth of these filters is selectable for the type of desired signal.

It is important to match the filter bandwidth with that of the desired signal. **Table 7.2** shows typical filter bandwidths for receiving a single amateur signal of various types. If the filter is too wide, unwanted signals will be received. If the filter is too narrow, the desired signal will be distorted. For example, using a 500-Hz CW filter to receive an SSB signal would render it nearly unintelligible. In general, selecting an IF filter bandwidth that is slightly greater than the bandwidth of the modulated signal you want to receive maximizes signal-to-noise ratio while minimizing interference. **[E4C10]**

RECEIVER DYNAMIC RANGE

Dynamic range is an important receiver parameter that refers to the ability of the receiver to operate properly in the presence of strong signals. The general definition of receiver dynamic range is the span in dB between the MDS and the largest input signal that does not cause audible distortion products. There are several types of dynamic range measurements used to describe receiver performance, based on input signal levels in dBm (dB with respect to 1 mW).

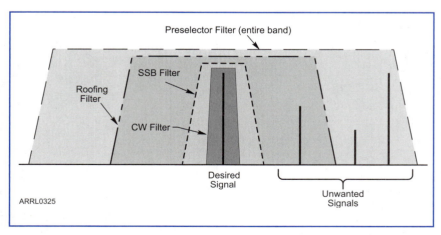

Table 7.2
Typical IF Filter Bandwidths for Single Signals

RTTY	300 Hz
CW	200 to 500 Hz
SSB	1.5 to 2.7 kHz
AM	6 kHz
VHF FM	15 kHz

Figure 7.13 — An analog multiple-conversion superheterodyne receiver has several stages of filtering. Preselector filters reject out-of-band signals. Roofing filters at the input to each IF further restrict receiver bandwidth, attenuating strong in-band signals that might overload the IF amplifiers. In the final IF stage, single-signal filters are used to select just the desired signal.

SDR Dynamic Range

E4C08 — An SDR receiver is overloaded when input signals exceed what level?

E4C12 — Which of the following has the largest effect on an SDR receiver's dynamic range?

The performance of an SDR receiver in the presence of strong signals, which are common on the busy amateur bands, is largely determined by its *sample width*. That is, the number of bits in each sample of the input signal to the analog-to-digital converter. The larger the number of bits, the larger the range of signals the SDR can process linearly, meaning that the strong signals do not overload the receiver. [**E4C12**]

Once a signal has been digitized in an SDR, there are no analog circuits in its path to distort or otherwise compromise the signal. While that provides superior filtering and rejection of unwanted signals, the analog-to-digital converter's maximum input signal level which is set by the reference voltage must not be exceeded or the receiver is overloaded. [**E4C08**] The receiver input includes all of the signals present at the receiver's antenna terminals except for those removed by a front-end filter or preselector. If all of those signals add together to exceed the maximum input voltage, distortion is produced and the receiver's performance begins to degrade. The SDR's dynamic range is approximately the range between its internal noise floor or minimum encoding voltage, whichever is lower, and the analog-to-digital converter's maximum input signal.

Blocking Dynamic Range

E4D01 — What is meant by the blocking dynamic range of a receiver?

E4D07 — Which of the following reduces the likelihood of receiver desensitization?

E4D12 — What is the term for the reduction in receiver sensitivity caused by a strong signal near the received frequency?

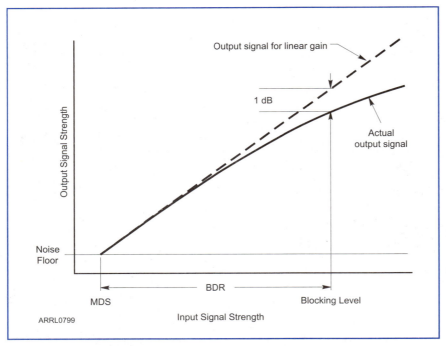

Figure 7.14 — Gain compression occurs in an analog receiver when the input signal is too strong for the receiver to develop full gain. Blocking Dynamic Range (BDR) is measured in dB from the MDS to a level at which the input signal strength causes a 1 dB drop in receiver gain, called the blocking level.

An input signal can be strong enough that an analog receiver no longer responds linearly and its gain begins to drop. (SDR receivers behave differently as described above.) This reduction in gain due to the strong signal causes weaker signals to appear to fade. This reduction in gain is called *gain compression* or *blocking*. Blocking may be observed as *desensitization* or *desense* — the reduction in apparent strength of a desired signal caused by a nearby strong interfering signal. [**E4D12**]

A receiver's *blocking level* is the power of an input signal that causes 1 dB of gain compression. *Blocking dynamic range* (BDR) as illustrated in **Figure 7.14** is the difference between the level of the receiver's MDS and the blocking level. When the blocking dynamic range is

7-16 Chapter 7

exceeded, the receiver begins to lose the ability to amplify weak signals. [**E4D01**] If the interfering signal is far enough from the desired signal, it may be possible to reduce desensitization by reducing the receiver's RF bandwidth to reject the strong signals. [**E4D07**]

INTERMODULATION (IMD)

E4D05 — What transmitter frequencies would cause an intermodulation-product signal in a receiver tuned to 146.70 MHz when a nearby station transmits on 146.52 MHz?

E4D11 — Why are odd-order intermodulation products, created within a receiver, of particular interest compared to other products?

The following discussion applies only to analog receivers. SDR receivers respond differently to strong signals. (This discussion touches on several key points regarding IMD products and receiver linearity but is not meant to be complete. For additional detailed information about receiver performance, see *The ARRL Handbook*.)

A perfectly linear receiver will produce an output signal with a strength that changes exactly as the input signal. If the input signal changes by 1 dB, the output signal will also change by 1 dB. This is called the *first order* response and is shown as the dashed line in Figure 7.14. No receiver is perfectly linear, however. As the input signal strength increases, the receiver's response becomes nonlinear and IMD products or "intermod" are created.

The frequency and amplitude of the IMD products depends upon the *order* of the IMD response. IMD products are created at frequencies which are the sum and difference of the input signals and their harmonics.

$$f_{IMD} = n f_1 \pm m f_2 \qquad\qquad \text{(Equation 7.3)}$$

where:

f_1 and f_2 are the input signal frequencies
n and m are positive integers; 1, 2, 3, etc

Even-order IMD products result if the sum of n and m is even, and *odd-order* IMD products result if the sum is odd. *Second-order* IMD products are created for n + m = 2 (both n and m equal to 1). *Third-order* IMD products are created if n + m = 3.

The frequencies of even-order IMD products caused by signals that are close together are far from the frequency of either input signal and so are generally not a problem if caused by signals within an amateur band. Second-order IMD products are the strongest of the even-order family and can be created in an amateur band by strong out-of-band signals such as from shortwave broadcast stations. Preselectors and front-end band-pass filters can reduce or eliminate second-order IMD products caused by those signals.

There are four third-order IMD product frequencies. Two are additive (f_{IMD1} and f_{IMD3}) and two are subtractive (f_{IMD2} and f_{IMD4}):

$$f_{IMD1} = 2f_1 + f_2 \qquad\qquad \text{(Equation 7.4)}$$

$$f_{IMD2} = 2f_1 - f_2 \qquad\qquad \text{(Equation 7.5)}$$

$$f_{IMD3} = 2f_2 + f_1 \qquad\qquad \text{(Equation 7.6)}$$

$$f_{IMD4} = 2f_2 - f_1 \qquad\qquad \text{(Equation 7.7)}$$

where:

f_{IMD} is the frequency of the IMD product
f_1 and f_2 are the input signals

Radio Measurements and Performance 7-17

If the frequencies of the signals causing the IMD products are close together, such as in the same amateur band as the desired signal, the subtractive IMD products (f_{IMD2} and f_{IMD4}) could possibly be very close to the desired signal frequency. This is true of all odd-order IMD products although the third-order products are the strongest. [**E4D11**] Therefore, the third-order IMD performance of a receiver is an important receiver specification.

Here's an example of third-order IMD performance being important. Let's say your receiver is tuned to 146.70 MHz. Whenever a nearby station is transmitting on 146.52 MHz, you receive intermittent bursts of garbled speech. This is likely to be a third-order intermodulation product generated in your receiver which is very sensitive but becomes nonlinear for very strong input signals.

What are the likely frequencies for a second strong signal that could combine with the one on 146.52 MHz to produce the IMD product you hear on 146.70 MHz? [**E4D05**] You know that the subtractive products are the likely source of the interfering signal because one of the signals causing the interference is close to the desired frequency. If the frequency of the IMD product is 146.70 MHz and you know one of the strong signal frequencies, f_1 = 146.52 MHz, you can solve for f_2 using Equation 7.5:

$$f_{IMD2} = 2f_1 - f_2$$

$$f_2 = 2f_1 - f_{IMD2} = 2 \times 146.52 \text{ MHz} - 146.70 \text{ MHz} = 146.34 \text{ MHz}$$

This is a common repeater input frequency! Solving Equation 7.7 for f_2 using f_{IMD4} and strong signal frequency for f_1, you'll find the other possible frequency to be (146.70 + 146.52) / 2 = 146.61 MHz. It would not be practical to filter out these strong input signals because they are *in-band signals*, close to your operating frequency. It would be better to use a receiver with a high enough dynamic range to accommodate these signals linearly and not produce the IMD products. (If the input signals are simply too strong, an attenuator at the receiver input may reduce the signal levels to a level at which they do not create IMD products.)

Another example from the HF bands will help illustrate the problem. If the interfering IMD product occurs at 14.020 MHz whenever a strong station is transmitting at 14.035 MHz, you can expect to find the other strong signal at:

$$f_2 = 2 \times 14.035 - 14.020 = 14.050 \text{ MHz}$$

or

$$f_2 = (14.035 + 14.020) / 2 = 14.0275 \text{ MHz}$$

With many strong signals closely spaced on a typical amateur band, IMD products can be a real problem! Reducing intermodulation is another reason to use roofing filters. A 6-kHz-wide roofing filter would significantly reduce the level of any signal at all three of these frequencies at which the IMD product could be generated. Other remedies include adding attenuation as mentioned previously or reducing RF gain. By eliminating (or at least reducing) strong in-band signals near the desired signal, the receiver's dynamic range is improved and IMD is reduced.

Intercept Points

E4D02 — Which of the following describes problems caused by poor dynamic range in a receiver?

E4D10 — What does a third-order intercept level of 40 dBm mean with respect to receiver performance?

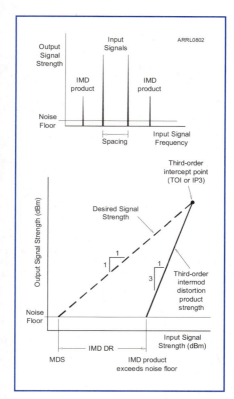

Figure 7.15 — Analog receiver output power for a desired signal and for third-order distortion products varies with changes of input signal power. The input signal consists of two equal-power sine-wave signals. Higher intercept points represent better receiver IMD performance.

Second-order IMD products occur at the sum and the difference of the input signal frequencies and their amplitude changes 2 dB for every 1 dB of input signal change. Third-order IMD product amplitudes change 3 dB for every 1 dB of input-signal change. (This assumes that the input signals have equal amplitudes.) **Figure 7.15** shows the output power of the desired signal versus the output power of the *third-order* distortion products at different input signal power levels. (The graph of second-order product strength would be similar but have a slope of 2 instead of 3.)

The input signal power at which the level of the distortion products equals the output level for the desired signal (where the lines cross) is the receiver's *intercept point*. There is a separate intercept point for each order of IMD product — second-order intercept point (SOI or IP2), third-order intercept point (TOI or IP3), and so on. By measuring signal and IMD product amplitudes, the intercept points can be calculated or estimated graphically. Specifically, the output power level at which the curves intersect is the *output intercept*. Similarly, the input power level corresponding to the point of intersection is called the *input intercept*.

For example, a 40 dBm third-order intercept point means that a pair of 40 dBm signals would theoretically produce a third-order IMD product of the same 40 dBm level. [**E4D10**] A level of 40 dBm is equal to 10 W so calculating the intercept point level is only a method of evaluating receiver performance and not a specification of how much signal the receiver can actually accept.

Although signals on the air are not strong enough to reach the intercept point levels, the intercept point values are useful for assessing receiver linearity. The higher the intercept point, the lower the amplitude of the IMD products generated by the receiver due to nonlinearities at actual received signal levels. Third-order intercept performance of a receiver usually gets worse as the frequencies of the strong signals (f_1 and f_2 in the equations above) get closer together. That is why the IMD performance of a receiver is typically given for several spacings of the input signals.

Intermodulation distortion dynamic range measures the ability of the receiver to avoid generating IMD products. When input signal levels exceed the IMD dynamic range, IMD products will begin to appear along with the desired signal. We can also calculate the third-order IMD dynamic range using the third-order intercept point and the receiver noise floor or MDS value.

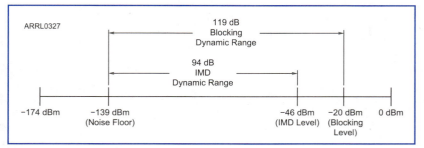

Figure 7.16 — The dynamic-range performance of a typical high-quality analog amateur receiver. The noise floor is –139 dBm, blocking level is –20 dBm and the IMD level is –46 dBm. This corresponds to a receiver blocking dynamic range of 119 dB and an IMD dynamic range of 94 dB.

$$\text{IMD DR}_3 = (2/3) (\text{IP3} - \text{MDS}) \qquad \text{(Equation 7.8)}$$

where:
 IMD DR_3 is the third-order intermodulation distortion dynamic range in dB.
 IP3 is the third-order input intercept point in dBm.
 MDS is the noise floor or MDS of the receiver in dBm.

Figure 7.16 illustrates the relationship between the input signal levels, noise floor, blocking dynamic range, and IMD dynamic range. Top-quality receivers have blocking dynamic ranges of more than 100 dB, so the receiver we have used in this example could be any typical modern receiver. If a receiver has poor dynamic range, cross-modulation or IMD products will be generated and desensitization (blocking) from strong adjacent signals will occur. [E4D02]

PHASE NOISE

E4C01 — What is an effect of excessive phase noise in a receiver's local oscillator?
E4C15 — What is reciprocal mixing?

Phase noise is a problem that has become more apparent as receiver improvements have reduced the noise floor and increased dynamic range. Today's commercial transceivers largely use direct digital synthesis (DDS) frequency synthesizers to create VFOs and other signal sources in a transceiver. DDS synthesizers exhibit small variations in frequency based on the clock signal and artifacts from creating the output waveform as small discrete steps. These result in the phase of the output signal continuously and randomly shifting back and forth a slight amount, creating *phase noise*. (DDS and PLL synthesizers are discussed in the Radio Circuits and Systems chapter.)

Phase noise creates a random collection of low-level sidebands that are increasingly stronger close to the desired signal frequency. On a transmitted signal, phase noise sounds like a strong hiss that can often be heard across an entire band to a nearby receiver, even on other bands. This can be a serious problem for other stations on the band if the transmitted signal is strong.

On receive, phase noise from the receiver's local oscillator mixes with the received signals just as the desired signal does, but the result is random noise, not the desired mixing product. Thus, as you tune toward a strong signal, the receiver noise floor appears to increase. In other words, you hear an increasing amount of noise in an otherwise quiet receiver as you tune toward the strong signal. This is called *reciprocal mixing*. [E4C15]

Excessive phase noise in a receiver local oscillator allows strong signals on nearby frequencies to interfere with the reception of a weak desired signal. [E4C01] This increased receiver noise can cover a weak desired signal, or at least make copying it more difficult.

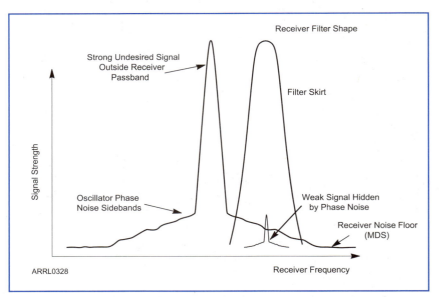

Figure 7.17 — In a receiver with excessive phase noise, a strong signal near the receiver passband can raise the apparent receiver noise floor in the passband. The increased noise can cover a weak signal that you are trying to receive. This is called reciprocal mixing.

Figure 7.17 illustrates how phase noise can cover a weak signal.

A transmitter with excessive phase noise on its output signal also causes noise to be received up and down the band for some range around the desired transmit frequency. This additional noise can fall within the passband of a receiver tuned to some weak signal. So even if you have a receiver with very low phase noise, you can be bothered by this type of interference!

Phase noise can be a serious problem when trying to operate two stations close together, such as during Field Day. The phase noise from one transmitter can cause severe interference to nearby receivers, even if they are operating on a different band! In this case, a band-pass filter at the transmitter is required to eliminate the interference to receivers on other bands.

CAPTURE EFFECT

E4C03 — What is the term for the suppression in an FM receiver of one signal by another stronger signal on the same frequency?

One of the most notable differences between an amplitude-modulated (AM) receiver and a frequency-modulated (FM) receiver is how noise and interference affect an incoming signal. The limiter and discriminator stages in an FM receiver can eliminate most of the atmospheric or impulse-type noise, unless the noise has frequency-modulation characteristics.

FM receivers perform quite differently from AM, SSB, and CW receivers when QRM is present, exhibiting a characteristic known as the *capture effect*. The loudest signal received, even if it is only a few dB stronger than other signals on the same frequency, will be the only signal demodulated, suppressing all weaker signals. [**E4C03**]

It is common that when two stations transmit an FM signal on the same frequency, the receiving station will hear only the stronger signal, with only buzzing or clicks remaining of the weaker signal. This can be a problem on busy repeaters, particularly during nets with multiple stations attempting to check in or pass messages.

7.3 Interference and Noise

Interference and noise are the bane of receivers everywhere. Interference, or QRM, is the term given to unwanted signals that have the characteristics of transmitted signals, whether they are actually a transmitter output or not. Noise, or QRN, is either randomly generated by natural processes or is the unintentional output of non-transmitting equipment. Confronted with a mix of both when trying to receive a weak signal, the distinction can seem a bit strained. Different techniques can be applied to reduce, filter or eliminate each type. This section touches on a few of the different types of interference and noise and how to manage them.

TRANSMITTER INTERMODULATION

E4D03 — How can intermodulation interference between two repeaters occur?

E4D04 — Which of the following may reduce or eliminate intermodulation interference in a repeater caused by another transmitter operating in close proximity?

E4D06 — What is the term for spurious signals generated by the combination of two or more signals in a non-linear device or circuit?

E4D08 — What causes intermodulation in an electronic circuit?

E4E11 — What could cause local AM broadcast band signals to combine to generate spurious signals in the MF or HF bands?

Intermodulation (IMD) is discussed at several points in this manual. IMD is a serious problem because it generates interfering signals both internally and externally to equipment. To a receiver, these signals are no different than signals received over the air and so cannot be filtered out. It is important that IMD be reduced or eliminated as much as possible.

Nonlinear circuits or devices can cause intermodulation distortion in just about any electronic circuit. IMD (also called *cross-modulation*) often occurs when signals from several

Radio Measurements and Performance **7-21**

transmitters, each operating on a different frequency, are mixed in a nonlinear manner, either by an active electronics device or a passive conductor that happens to have nonlinear characteristics. [**E4D08**] The mixing, just like in a mixer circuit, produces mixing products that may cause severe interference in a nearby receiver. Harmonics can also be generated and those frequencies will add to the possible mixing combinations. The *intermod*, as it is called, is radiated and received just like the transmitted signal. [**E4D06**]

For example, suppose an amateur repeater receives on 144.85 MHz. Nearby, are relatively powerful non-amateur transmitters operating on 181.25 MHz and on 36.4 MHz (see **Figure 7.18**). Neither of these frequencies is harmonically related to 144.85. The difference between the frequencies of the two non-amateur transmitters, however, is 144.85 MHz. If the signals from these transmitters are somehow mixed, an intermod mixing product at the difference frequency could be received by the amateur repeater, demodulated as if it were a desired signal, and retransmitted on the repeater's output frequency. A listening station would hear a signal with the modulation of both non-amateur transmitters.

Intermodulation interference can be produced when two transmitted signals mix in the final amplifiers of one or both transmitters and unwanted signals at the sum and difference frequencies of the original signals are generated. [**E4D03**] In this example, the two signals could be perfectly clean, but if they are mixing in one of the transmitters, the intermod signal may actually be transmitted along with the desired signal from that transmitter.

Two devices that are highly effective in eliminating this type of intermod are *isolators* and *circulators*. Circulators and isolators are ferrite components that function like a one-way valve to radio signals. Very little transmitter power is lost as RF travels to the antenna, but a considerable loss is imposed on any power coming back down the feed line to the transmitter. Circulators can also be used to allow two or more transmitters to use a single antenna. Thus, circulators and isolators at a transmitter's output effectively reduce intermod problems. [**E4D04**] Another advantage of a circulator is that it provides a matched load to the transmitter output, regardless of what the antenna-system SWR might be, by routing reflected power to a dummy load.

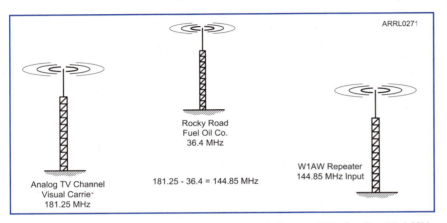

Figure 7.18 — A potential intermod situation. It is possible for the 36.4 MHz and 181.25 MHz signals to mix together in the output stages of either transmitter. While the sum of the two frequencies (36.4 + 181.25 = 217.65 MHz) is not in an amateur band, the difference frequency is (181.25 − 36.4 = 144.85 MHz) and such a signal is on the repeater's input frequency. The repeater would treat this signal as a desired signal, attempt to demodulate it, and retransmit the audio on its output channel. An isolator or circulator should be installed at the outputs of the transmitters to prevent the creation of intermodulation products.

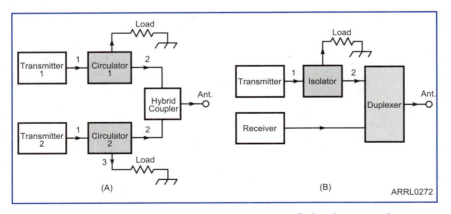

Figure 7.19 — This block diagram shows the use of circulators and an isolator. Circulators may be used to share one antenna with two transmitters, as in A, particularly at multi-transmitter sites. Duplexers are also used along with the circulators. In B, an isolator is placed between the transmitter and duplexer to reduce intermodulation.

Figure 7.19 illustrates how circulators or isolators may be included in a repeater system.

Low-pass and band-pass filters usually are ineffective in reducing intermod problems, because at VHF and UHF they are seldom sharp enough to suppress the offending signal without also weakening the wanted one.

Intermod, of course, is not limited to repeaters. An intermod problem can develop anywhere two relatively powerful and close-by transmitter fundamental-frequency outputs or their harmonics can combine to create a sum or a difference signal at the frequency on which any other transmitter or receiver is operating. Any nonlinear device or conductor in which energy from the two transmitters can combine will generate both harmonics and IMD products from the two signals. For example, corroded metal joints are very nonlinear. If two strong AM broadcast stations are nearby, the signals can mix in the joint and generate intermodulation products across a wide range of frequencies, some in the ham bands. [**E4E11**] Tracking down such a problem can be quite time-consuming, but fortunately, the harmonics and other products are usually fairly weak and not received over a wide area.

Another IMD topic mentioned earlier has to do with transmitter spectral-output purity. When several audio signals are mixed with the carrier signal to generate the modulated signal, spurious signals will also be produced. These are normally reduced by filtering after the mixer, but their strength will depend on the level of the signals being mixed, among other things, and they will be present in the transmitter output to some extent. You can test your transmitter's output signal purity by performing a transmitter two-tone test as described earlier in this section. This is important because excessive intermodulation distortion of an SSB transmitter output signal results in *splatter* being transmitted over a wide bandwidth. The transmitted signal will be distorted, with spurious (unwanted) signals on adjacent frequencies. This is not a good way to make friends on the air!

POWER LINE NOISE

E4E05 — How can radio frequency interference from an AC motor be suppressed?
E4E10 — What might be the cause of a loud roaring or buzzing AC line interference that comes and goes at intervals?

Electrical line noise can be particularly troublesome to operators working from a fixed location. The loud buzz or crackling sound of power line noise can cover all but the strongest signals. Most of this man-made interference is produced by some type of electrical arc. An electrical arc generates varying amounts of RF energy across the radio spectrum. (In the early days of radio, amateurs used spark gap transmitters to generate their radio signals.) Another type of noise present on the ac line is *transients* — short impulses caused by a spark or some amount of energy being dumped or coupling to the ac line from lightning, motors being turned on and off, or even loose connections.

When an electric current jumps a gap between two conductors as in **Figure 7.20**, an arc is produced as the current travels through the air. To produce such an arc, the voltage must be large enough to ionize the air between the conductors. Once an ionized path is established, there is a current through the gap. The electron flow through this gap is highly irregular compared to the smooth flow through a conductor. The resistance of the ionized air varies constantly, so the instantaneous current is also changing. This causes radio-frequency energy to be radiated and the noise can be conducted along the power wires that act as an excellent antenna. The longer the width of the gap and the higher the voltage, the greater the interference the arc causes. Because of the high voltages and power available, poor connections or defective insulators in the power distribution system are frequent sources of potentially severe line noise.

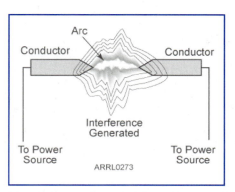

Figure 7.20 — An electrical arc can form through the air gap between conductors. Arcs radiate noise across a wide RF spectrum.

Radio Measurements and Performance 7-23

Small arcs are created in a variety of electrical appliances, especially those using brush-type motors. Electric shavers, sewing machines, and vacuum cleaners are just a few examples. In addition, devices that control voltage or current by opening or closing a circuit can produce momentary arcs that cause interference as pops or clicks. Light dimmers, heater elements, and blinking advertising displays are a few examples of devices like this. Defective or broken appliances and wiring inside the home can also generate line noise. Doorbells or doorbell

Locating Noise and Interference Sources

Perhaps one of the most frustrating facts about power line noise is its intermittent nature. The noise will come and go without warning, usually with no apparent pattern or timing (except that it occurs when you are trying to operate). This can make it extremely difficult to locate the source of the interference. There are two ways that noise interference can find its way into a receiver. If the interference source is located in the same building as the receiver, it's likely that the noise will flow along the house wiring. For example, if a furnace's blower motor is at fault, the interference may be carried by the ac wiring from the furnace to wherever the receiver is located. Interference from outside the home is usually picked up through an antenna and feed line.

The first step in tracking down the interference is to determine if it is being generated in your own house. Check this by opening the mains circuit breaker, powering down your entire house. You will need a battery-powered receiver for this test, but a portable AM receiver should work fine. Check to be sure you can hear the noise on your portable receiver. Tune to a clear frequency and listen for the interference. An FM receiver, with or without a directional antenna, will not be helpful in tracking down the noise source because FM receivers are not affected by line noise.

If the noise goes away when you turn off all power to your house, check to see if it comes back when you turn the power on again. It's possible that the offending device only produces the noise after it has been operating for a while, so the interference may not come back immediately when you restore power. If it doesn't return, continue your investigation later.

After you are reasonably confident that the noise is being produced in your house, proceed with your investigation by removing power from one circuit at a time. When you narrow it down to a particular circuit, unplug the appliances or other electrical devices one by one. Be persistent. You may have to continue your investigation over a long time until you discover the culprit device.

If the interference is not being produced in your house, the search will be a bit more difficult. The problem may be an arc in the utility company's power distribution system, a neighbor's appliance or any of a number of other items. Your portable AM receiver can be used to "sniff" along the power lines, looking for stronger interference. Driving around with your car's AM radio tuned between stations is sometimes effective. A directional antenna may help you locate the noise source. When you get close to the noise source you may find that the null (direction of weakest signal in the antenna pattern) is more helpful in pinpointing the direction to the source.

When you think you've located the source, contact the power company and explain the problem to them. Be as specific as possible about where you believe the interference is originating. Note the identification numbers on the utility pole, for example. The power company may send a technician with even more sophisticated equipment to help pinpoint the location and source of the interference. The technician may have a handheld "RF sniffer" that will pick up the radio frequency noise. They may also use an ultrasonic transducer that uses a parabolic reflector antenna and an amplifier to listen for the sound of an arc at frequencies just above the audio spectrum.

For more information about electrical power line noise and other types of interference, see *The ARRL RFI Book*. That book contains detailed information about how noise and interference are generated, how to locate the source, and how to cure it.

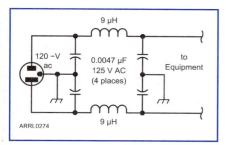

Figure 7.21 — A "brute force" ac line filter may reduce or eliminate line noise when installed in the power leads to an ac brush-type motor.

transformers and other types of devices that are powered continuously often generate low-level arcs (and fire hazards!) when they begin to fail. The typical interference is a loud roaring or buzzing ac line noise that comes and goes at intervals. [**E4E10**]

One effective way to reduce electrical noise produced by an electric motor is to use a "brute force" ac line filter in series with its power leads. This filter will block the noise from being conducted along the power wiring away from the motor. [**E4E05**] **Figure 7.21** shows the schematic diagram for such a filter. All components must be ac line rated, and able to carry the current required by the motor or appliance connected to the filter. UL-listed commercial filters are recommended for this application. Note that this type of filter will not prevent noise from being radiated by the wiring between the filter and the motor.

INTERFERENCE FROM STRONG SIGNALS

E4E07 — Which of the following can cause shielded cables to radiate or receive interference?

E4E08 — What current flows equally on all conductors of an unshielded multi-conductor cable?

A transmitted signal may also cause interference to devices such as a TV, radio, audio system, or telephone. That signal may be from your station or it may be from a nearby non-amateur transmitter. Such interference is often caused by *common-mode* signals picked up on the shields of cables and on unshielded cables such as the ac and telephone wiring in your house. The wiring may pick up your signals and conduct them to the device or re-radiate the signal and create interference that way. [**E4E07**] Common-mode means the that signal flows in the same direction on all of wires in a multiconductor cable such as the power or phone line rather than in opposite directions along the wires as it would on a transmission line. [**E4E08**] You'll need a common-mode choke to cure this type of interference. Wind several turns of the power cord or phone line around a ferrite toroid core. Ferrites made of type 31 and 73 material are a good choice for most HF problems. Type 43 material works best on the upper HF bands and at VHF and UHF.

COMPUTER INTERFERENCE

E4E06 — What is one type of electrical interference that might be caused by a nearby personal computer?

Interference may also be generated by computer and network equipment and switching power supplies. The characteristics of noise from this type of device are unstable modulated or unmodulated signals at specific frequencies. [**E4E06**] The signal may change as the device performs different functions.

VEHICLE NOISE

E4E04 — How can conducted and radiated noise caused by an automobile alternator be suppressed?

One of the most significant deterrents to effective signal reception during mobile or portable operation is electrical *impulse noise* from an engine's ignition system, including gasoline-powered portable generators. This form of interference can completely mask a weak signal. Other sources of noise include conducted interference from the vehicle battery-charging system, instrument-caused interference, static and corona discharge from the mobile antenna.

The first rule when installing a mobile transceiver in a modern vehicle is to follow the manufacturer's recommended procedures. Some manufacturers provide detailed installation guidelines for installing transmitting equipment. Others may recommend against installing any transmitters, or provide little instruction. If possible, you should check with the manufacturer before buying a vehicle. There is a lot of information about mobile installations from the ARRL's Technical Information Service at **www.arrl.org**.

Ferrite beads and cores are a possible means for RFI reduction in modern vehicles. Both primary and secondary ignition leads are candidates for beads. Install them liberally then test the engine under load to ensure adequate ignition system performance.

Conducted noise and radiated noise from the vehicle's battery charging system can be minimized by connecting the radio power leads directly to the battery, as this point has the best voltage quality and regulation in the system. Connect both the positive and negative leads directly to the battery, with a fuse rated to carry the transmit current installed in each lead. Coaxial capacitors in series with the alternator leads may also help. [**E4E04**]

NOISE REDUCTION

E4E01 — **What problem can occur when using an automatic notch filter (ANF) to remove interfering carriers while receiving CW signals?**

E4E02 — **Which of the following types of noise can often be reduced with a digital signal processing noise filter?**

E4E03 — **Which of the following signals might a receiver noise blanker be able to remove from desired signals?**

E4E09 — **What undesirable effect can occur when using an IF noise blanker?**

Once inside the receiver, noise is very difficult to eliminate. Two basic types of noise reduction techniques are widely used. The first, *noise blanking*, works by detecting noise pulses and muting the receiver when they are present. The second, *noise reduction*, uses special DSP techniques to separate noise from the desired signal.

Noise Blankers

Special IF circuits detect the presence of a noise impulse and open or mute the receive signal path just long enough to prevent the impulse from getting through to the audio output stages where it is heard as a "pop" or "tick." This technique, called *gating*, is particularly effective on power line and mobile ignition noise.

A diode or transistor is used as a switch to control the signal path. An important requirement is that the IF signal must be delayed slightly, ahead of the switch, so that the switch is activated precisely when the noise arrives at the switch. The circuitry that detects the impulse and operates the switch has a certain time delay, so the signal in the mainline IF path must be delayed also.

To detect the sharp noise pulses, the noise blanker must detect signals that appear across a wide bandwidth. [**E4E03**] This usually means that the noise blanker cannot be protected by the narrow receive filters. As a consequence, the noise blanker can be fooled by strong signals into shutting down the receiver as if they were noise pulses. This can cause severe distortion of desired signals, even if no noise is present. It might sound as if the strong signal is very "wide" with lots of spurious signals. Before getting upset at the station with the strong signal, make sure your noise blanker is turned off. [**E4E09**] A preamp can make your receiver easier to overload and cause similar problems.

DSP Noise Reduction

DSP noise reduction filters operate by using *adaptive filter* techniques in which software programs search for signals with the desired characteristics of speech or CW or data and remove everything else, such as impulse noise and static. These techniques work particularly well at removing broadband audio "white" noise. DSP noise reduction also works on impulse noise such as ignition and power line noise. [**E4E02**]

Automatic notch filters are a particularly useful feature of these systems. Modern DSP auto-notch features can track and remove several interfering tones from an audio channel! This is very useful under crowded band conditions or on shared allocations where carriers from shortwave broadcast stations can be quite strong. One drawback of these systems is that they sometimes confuse CW or low-rate digital signals with an interfering tone and attempt to remove them, as well! [**E4E01**]

Chapter 8

Modulation, Protocols, and Modes

In this chapter, you'll learn about:
- **Emission designators and types**
- **FM and types of digital modulation**
- **Digital codes, protocols, and modes**
- **Spread spectrum techniques**
- **Fast-scan amateur television**
- **Slow-scan amateur television**

Building on your growing knowledge of radio terminology, circuits, and signals, this chapter deals with FM and digital modulation techniques. Those are the foundation for digital protocols and modes, the fastest-growing part of amateur radio. We'll then review the two most common image modes, fast-scan and slow-scan amateur TV.

While the amateur radio Extra class license exam has groups of questions on each of these topics — modulation, digital communications, and image transmission — there is also a sprinkling of questions from other parts of the question pool that are easier to discuss in the context of these topics. As in all of the sections, it's a good idea to review all of the listed questions before moving on. If you have trouble with any group of questions, review material from earlier chapters and make use of the online references at **www.arrl.org/extra-class-license-manual**.

8.1 Modulation Systems

The process of adding information to and recovering information from signals is what radio is all about! You've already studied AM techniques for the General class exam, and modulators and demodulators for AM signals were covered in the Radio Circuits and Systems chapter. In this section, we'll cover important signal definitions for FM, the most popular analog mode. We'll also take a look at multiplexing (a method of combining more than one flow of information in a single signal).

FCC EMISSION DESIGNATIONS AND TYPES

Although the question pool does not include any direct questions on the system of emission identifiers used by the FCC, it is a good idea to be familiar with them, since they are frequently used in spectrum allocations and regulations.

The International Telecommunication Union (ITU) has developed a special system of identifiers to specify the types of signals (emissions) permitted to amateurs and other users of the radio spectrum. This system designates emissions according to their necessary bandwidth and their classification. While a complete *emission designator* might include up to five characters, generally only three are used.

The designators begin with a letter that tells what type of modulation is being used. The second character is a number that describes the signal used to modulate the carrier. The third character specifies the type of information being transmitted. **Table 8.1** lists emission designators for the most common modes used in amateur radio.

Table 8.1
Emissions Designators for Common Amateur Modes

(1) First Symbol — Modulation Type
(2) Second Symbol — Nature of Modulating Signals
(3) Third Symbol — Type of Transmitted Information

N0N — Unmodulated carrier
A1A — Morse code telegraphy using amplitude modulation
A3E — Double-sideband, full-carrier, amplitude-modulated telephony
J3E — Amplitude-modulated, single-sideband, suppressed-carrier telephony
J3F — Amplitude-modulated, single-sideband, suppressed-carrier television
F3E — Frequency-modulated telephony
G3E — Phase-modulated telephony
F1B — Telegraphy using frequency-shift keying without a modulating audio tone (FSK RTTY).
F1B is designed for automatic reception.
F2B — Telegraphy produced by modulating an FM transmitter with audio tones (AFSK RTTY).
F2B is also designed for automatic reception.
F1D — FM data transmission, such as packet radio

The amateur radio regulations in Part 97 refer to *emission types* rather than emission designators. The emission types are CW, phone, RTTY, data, image, MCW (modulated continuous wave), SS (spread spectrum), pulse, and test. Any signal may be described by either an emission designator or an emission type. Emission types are what amateurs refer to as "modes."

While emission types are fewer in number and easier to remember, they are a somewhat less descriptive means of identifying a signal. There is still a need for emission designators in amateur radio and they are used in Part 97.

FM/PM MODULATION AND MODULATORS

E1C09 — What is the highest modulation index permitted at the highest modulation frequency for angle modulation below 29.0 MHz?

E8B01 — What is the modulation index of an FM signal?

E8B02 — How does the modulation index of a phase-modulated emission vary with RF carrier frequency?

E8B03 — What is the modulation index of an FM-phone signal having a maximum frequency deviation of 3000 Hz either side of the carrier frequency when the modulating frequency is 1000 Hz?

E8B04 — What is the modulation index of an FM-phone signal having a maximum carrier deviation of plus or minus 6 kHz when modulated with a 2 kHz modulating frequency?

E8B05 — What is the deviation ratio of an FM-phone signal having a maximum frequency swing of plus-or-minus 5 kHz when the maximum modulation frequency is 3 kHz?

E8B06 — What is the deviation ratio of an FM-phone signal having a maximum frequency swing of plus or minus 7.5 kHz when the maximum modulation frequency is 3.5 kHz?

E8B09 — What is deviation ratio?

You learned about direct and indirect FM modulation in the Radio Circuits and Systems chapter. With FM, the signal frequency is varied above and below the carrier frequency at a rate equal to the modulating signal frequency. (Carrier frequency refers to the frequency of the FM signal with no modulation applied.) For example, if a 1000-Hz tone is used to modulate a transmitter, the modulated signal's frequency will vary above and below the carrier frequency 1000 times per second.

The amount of frequency change, however, is proportional to the modulating signal amplitude. This frequency change is called *deviation*. Let's say that a certain modulating signal produces a 5-kHz deviation. If another signal, with only half the amplitude of the first, were used to modulate the same transmitter, it would produce a 2.5-kHz deviation.

To more completely describe an FM signal, you will need to understand two terms that refer to FM systems and operation: *deviation ratio* and *modulation index*. They may seem to be almost the same — indeed, they are closely related.

Deviation Ratio

In an FM system, the ratio of the maximum frequency deviation to the highest audio modulating frequency is called the *deviation ratio*. [**E8B09**] It is a constant value for a given modulator and transmitter, and is calculated as:

$$\text{Deviation Ratio} = \frac{D_{MAX}}{M} \qquad\qquad \text{(Equation 8.1)}$$

where:

D_{MAX} = peak deviation in hertz
M = maximum modulating frequency in hertz.

Peak deviation is defined as half the difference between the maximum and minimum signal frequencies. That is, a sine-wave modulating signal will cause the signal frequency to move symmetrically higher and lower about the carrier frequency. If maximum deviation is specified as ±5 kHz, a total difference of 10 kHz between maximum and minimum frequency, the peak deviation is one-half that value, or 5 kHz.

Peak deviation is usually controlled by setting an audio gain control in the FM modulator's circuit. Because it is fixed for that transmitter, there is no microphone gain control on an FM transmitter's front panel.

Example 8.1

In the case of narrow-band FM (the type used in amateur analog FM voice communications), peak deviation at 100% modulation is typically 5 kHz. What is the deviation ratio if the maximum modulating frequency is 3 kHz? [**E8B05**]

$$\text{Deviation Ratio} = \frac{D_{MAX}}{M} = \frac{5 \text{ kHz}}{3 \text{ kHz}} = 1.67$$

Example 8.2

If the maximum deviation of an FM transmitter is 7.5 kHz and the maximum modulating frequency is 3.5 kHz, what is the deviation ratio? [**E8B06**]

$$\text{Deviation Ratio} = \frac{D_{MAX}}{M} = \frac{7.5 \text{ kHz}}{3.5 \text{ kHz}} = 2.14$$

Notice that since the frequencies of D_{MAX} and M were given in kilohertz we did not have to change them to hertz before doing the calculation. The important thing is that they both be in the same units.

Modulation Index

The ratio of the maximum signal frequency deviation to the instantaneous modulating frequency is called the *modulation index*. [**E8B01**] Modulation index is a measure of the relationship between deviation and the modulating signal's frequency. That is:

$$\text{Modulation Index} = \frac{D_{MAX}}{m} \qquad\qquad \text{(Equation 8.2)}$$

where:

D_{MAX} = peak deviation in hertz.
m = modulating frequency in hertz at the same time.

Modulations, Protocols, and Modes 8-3

Example 8.3

If the peak deviation of an FM transmitter is 3000 Hz, what is the modulation index when the carrier is modulated by a 1000-Hz sine wave? [**E8B03**]

$$\text{Modulation Index} = \frac{D_{MAX}}{m} = \frac{3000 \text{ Hz}}{1000 \text{ Hz}} = 3$$

When the same transmitter is modulated with a 3000-Hz sine wave that results in the same peak deviation (3000 Hz), the index would be 1; with a 100-Hz modulating wave and the same 3000-Hz peak deviation, the index would be 30, and so on.

Example 8.4

If the peak deviation of an FM transmitter is 6 kHz, what is the modulation index when the carrier is modulated by a 2 kHz sine wave? [**E8B04**]

$$\text{Modulation Index} = \frac{D_{MAX}}{m} = \frac{6 \text{ kHz}}{2 \text{ kHz}} = 3$$

As Equation 8.1 shows, if the peak deviation is kept the same, modulation index varies inversely with the modulating frequency. A higher modulating frequency results in a lower modulation index. In a frequency modulator, the actual deviation depends only on the amplitude of the modulating signal and is independent of frequency. Thus, a 2-kHz tone will produce the same deviation as a 1-kHz tone if the amplitudes of the tones are equal. The modulation index in the case of the 2-kHz tone is half that for the 1-kHz tone.

By contrast, in a phase modulator, deviation increases with the modulating frequency. If the modulating signal amplitude stays constant, the modulation index in this modulator will also remain constant. In other words, a 2-kHz tone will produce twice as much deviation as a 1-kHz tone if the amplitudes of the tones are equal.

With either an FM or a PM system, the deviation ratio and modulation index are independent of the frequency of the modulated RF carrier. [**E8B02**] It doesn't matter if the transmitter is a 10 meter or 2 meter FM radio.

What are deviation ratio and modulation index used for? Since deviation ratio is fixed, it is used to describe and specify an FM or PM modulator. The deviation ratio of the transmitter is set during the manufacturing process and not adjusted during operation.

Modulation index, on the other hand, varies with the input signal's amplitude (because it changes D_{MAX}) and frequency. The actual spectrum of an FM signal is quite complex and modulation index provides a way to describe how the energy is distributed within that signal.

Just as with an AM signal, modulation index is a way to describe a modulated signal's bandwidth. Increasing the modulation index results in signal components farther and farther from the carrier frequency where they can cause interference to signals on adjacent channels. The solution is reducing the modulating signal's amplitude, particularly at low frequencies. Controlling modulation index is another reason why pre-emphasis and de-emphasis are used in FM modulators and demodulators. To limit the bandwidth of FM signals on the narrow HF amateur bands, the maximum modulation index allowed below 29.5 MHz by FCC rules is 1.0. [**E1C09**]

Multiplexing

E8B10 — What is frequency division multiplexing?
E8B11 — What is digital time division multiplexing?

Multiplexing means to combine more than one stream of information into one modulated signal. This allows one RF transmitter and transmitted signal to carry more than one information stream. There are two common methods of multiplexing, *frequency division multiplexing* (*FDM*) and *time division multiplexing* (*TDM*).

FDM uses more than one *subcarrier*, each modulated by a separate message signal. The subcarriers are combined into a single *baseband* signal that then modulates the RF carrier. [**E8B10**] Amateurs use a form of FDM with multi-carrier digital modes that will be described later in this chapter.

TDM is the transmission of two or more signals over a common channel by interleaving so that the signals occur in different, discrete *time slots* of a digital transmission. [**E8B11**] In amateur radio, the most popular TDM mode is DMR (Digital Mobile Radio). DMR repeater systems use a computer-controlled system that divides a channel into two alternating time slots. TDM is also used for telemetry, such as from amateur satellites and remote repeaters.

8.2 Digital Protocols and Modes

Before plunging into details (and there are many) of the digital transmissions becoming so popular with amateurs, we'll clearly define some digital terminology. It's very easy to misapply a term or to confuse the meaning of one term with another, leading to certain confusion.

SYMBOL RATE, DATA RATE, AND BANDWIDTH

E8C02 — What is the definition of symbol rate in a digital transmission?
E8C10 — How may data rate be increased without increasing bandwidth?
E8C11 — What is the relationship between symbol rate and baud?

Two of the most important and useful characteristics of a digital communications system concern the speed with which data is transferred. "Speed" actually means the rate of information transfer, so it is measured in units of data per second as described below. It also depends on where the rate is measured in the total communications system — from the generation of data to the delivery of data.

There are two ways of defining digital signal speed, depending on whether you are referring to the *air link* — meaning the actual transmitted signal — or the *data stream* that occurs within the computer equipment that handles the digital data. For example, when listening to an RTTY signal's characteristic two-tone warble on the air by ear, you are listening to the RTTY communication system's air link. If you are reading the received characters from an RTTY decoder box or a software program, that is the data stream. The speed of the data stream is often called *data throughput*, meaning the overall speed with which the entire communications system transfers data.

When discussing the speed of the air link, the unit of speed is *baud* (*Bd*) or *bauds*. Baud, like hertz, refers to a quantity of events per second. Just as frequency counts the cycles of an ac waveform, baud counts *signaling events* that refer to changes in the transmitted signal representing information. Each signaling event transfers one *symbol* across the air link to the receiving station. Thus, *symbol rate* refers to the rate at which the transmitted waveform changes in order to convey information. [**E8C02**] Baud and symbol rate are the same, so a rate of one baud means that one symbol is transmitted every second. [**E8C11**] (Baud is also referred to as *baud rate* but that is redundant because baud is already a rate! Just say "baud" or "bauds.")

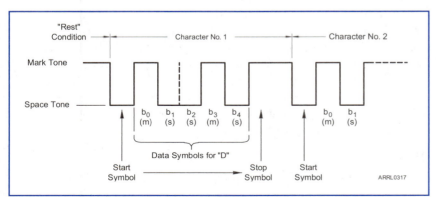

Figure 8.1 — The pattern of symbols transmitted for the letter "D" in Baudot code. Each "bit" of the code is transmitted as one symbol.

In an RTTY signal's air link, for example, (see **Figure 8.1**) the symbols *mark* and *space* are pulses of transmitted tones with separate frequencies. The tone frequency during the pulse is the signaling event. Sometimes the tone changes from one symbol to the next (such as from b_0 to b_1 in the figure) and sometimes it doesn't (such as from b_1 to b_2). The signal's baud is the number of those events — the transmitted symbols — per second.

Data Rate vs Symbol Rate

Figure 8.1 serves two purposes. It describes how the air link would sound if you listened to an RTTY signal and it also shows a possible data stream waveform. The figure shows how the five bits (b_0 through b_4) are encoded as mark or space symbols, one after the other. Each symbol on the air link corresponds one-to-one with a bit in the data stream. In many simple transmission systems — such as RTTY and 1200-baud packet — the system *data rate*, which is measured in *bits per second* or *bps*, is exactly the same as the symbol rate in baud. This is not always the case.

In some transmission systems — such as 9600-baud packet or D-STAR's digital voice and data systems — each data symbol can encode more than one bit at a time. This can be done by simultaneously transmitting more than one signal or by varying more than one attribute of the signal over the air link. When each transmitted symbol encodes more than one bit of information, the data rate is higher than the symbol rate and *coding system efficiency* is increased. This means more data can be sent in the same bandwidth than when a less efficient code is used. [**E8C10**]

PROTOCOLS AND CODES

Before studying the characteristics of digital mode communications, some definitions are in order. A *protocol* is the set of rules that controls the encoding, packaging, exchanging and decoding of digital data. For example, packet radio uses the AX.25 protocol. The AX.25 standard specifies how each packet is constructed, how the packets are exchanged, what characters are allowed, and so forth.

The protocol standard may allow different types of modulation, such as SSB packet (on HF) or FM packet (at VHF and above). The actual transmitted signal characteristics are determined by conventional operating practice and regulation.

Control Operator Responsibilities

E2E12 — How do ALE stations establish contact?

Practical digital protocols require that the receiving and transmitting station operate under the control of microprocessors to perform all of the encoding, decoding, error correction, and so forth. If other signals are present on the channel, the system executing the protocol usually assumes other communications to be interference and just keeps trying until it gets through or fails. For example, stations using the Automatic Link Establishment (ALE) protocol constantly scan a list of frequencies looking for their call sign and activate or respond automatically when they receive it whether other signals are present or not. [**E2E12**] Because of this potential for interference, specific segments of each amateur band are designated for stations that operate under fully automatic control. As in any part of the amateur bands, it is quite important for the control operator who initiates the contact to be sure the transmissions won't cause interference to other stations. Listen first — don't assume that because a modem's BUSY light is off there are no other stations using the channel!

Codes

A *code* is the method by which information is converted to and from digital data. The individual symbols that make up a specific code are its *elements*. The elements may be numbers, bits, tones or even images (think of the code "one if by land and two if by sea" used by Paul Revere).

A digital code doesn't specify how the data is transmitted, the rules for its transmission or the method of modulation. The code doesn't control those things — it's only a set of rules for changing information from one form to another. Certain types of codes are more suitable for different applications, a matter of preference by the communications system designer. Amateur radio uses three common types of codes: varicodes (Morse and PSK31's Varicode), Baudot, and ASCII.

Morse and Varicode

E2E09 — Which of the following HF digital modes uses variable-length coding for bandwidth efficiency?

While most types of digital codes use a fixed number of identical length bits to make up each character, *variable-length codes* or *varicodes* can vary both the length of the bits and the number of bits in each character or symbol. The length of the codes for a character varies with its frequency of use to save transmission time. Varicode is a more efficient code than fixed-length codes and maximizes *bandwidth efficiency* or *spectral efficiency.*

Morse is a varicode because the elements and characters are of different lengths. Morse code is constructed from the dot and its absence, the inter-element space. The dash and longer spaces are made up of multiple dots and inter-element spaces. This creates the elements of Morse: dot, dash, and three lengths of spaces between elements, characters and words.

PSK31 uses a type of varicode that is also named Varicode, invented by Peter Martinez, G3PLX. The elements of the PSK31 Varicode are the same length but its characters have different lengths with the most-common text character "E" being the shortest, similar to Morse code. [E2E09]

Baudot

The Baudot code is used by RTTY systems and has two elements — mark and space — each the same length. The code is made up of different combinations of five mark and space elements as illustrated in Figure 8.1. (The mark tone can also be transmitted continuously when the system is idle, but no information is being sent during that period.) Each combination of elements always has the same length and each element represents one data bit.

Figure 8.1 also shows additional elements called *start* and *stop* bits at the beginning and end of the group of five that represents the character. These are called *framing* bits and allow the receiving system to synchronize itself with the transmitted codes. A complete received character, including the framing bits, is called a *frame.*

ASCII

E8D06 — What is the advantage of including parity bits in ASCII characters?
E8D10 — What are some of the differences between the Baudot digital code and ASCII?
E8D11 — What is one advantage of using the ASCII code for data communications?

ASCII stands for American National Standard Code for Information Interchange and is the most commonly used code in computer systems. Basic ASCII code uses seven information bits, so 128 characters are possible (2^7). This makes it possible for the ASCII character set to include upper- and lower-case letters, numbers, punctuation and special *control* characters. [E8D11] The ASCII code does not need a shift character, like Baudot, to change between the letters and figures characters. Some systems use an eighth bit for another data bit, providing 256 possible characters with each representing a full byte of data. [E8D10]

The eighth bit can also be a *parity bit* that is used to detect transmission errors. The parity bit is set to 1 or 0 to maintain either an even number of 1s (even parity) in the character, or an odd number of 1s (odd parity). Parity bits allow the receiving system to detect *single-bit*

Modulations, Protocols, and Modes   8-7

errors. The receiving system checks the number of 1s in the received character. If the total, including the parity bit, does not match the system convention of odd or even, the receiving system knows an error has occurred during transmission and it can reject the character. **[E8D06]**

ASCII codes also have framing bits — one start bit at the beginning of the code and one or two stop bits at the end. A full ASCII character is thus 10 or 11 bits long: 7 data bits, one parity bit (or extra data bit), one start bit, and one or two stop bits. It is necessary to preset both the transmitting and receiving systems so they agree on these conventions or the recovered data will be garbled.

When transferring ASCII data, remember to account for the difference between bits per second and bytes per second. If each ASCII character includes two framing bits, the *byte rate* of data transfer will be about one-tenth the *bit rate*.

Gray Code

E8C09 — Which digital code allows only one bit to change between sequential code values?

Table 8.2
3-bit Gray Code

Decimal	Binary	Gray
0	000	000
1	001	001
2	010	011
3	011	010
4	100	110
5	101	111
6	110	101
7	111	100

In a table of binary patterns for ASCII or Baudot characters, you'll notice that if you count from all zeroes through the maximum value of all ones, there are some consecutive values where many bits change at once. If your communications system is trying to send those two values back-to-back, the large number of changing bits is harder to decode properly. That makes it more likely an error will occur. Gray codes are arranged so that only a single bit changes between any consecutive value as shown in **Table 8.2**. **[E8C09]** For example, between decimal values 3 and 4, in ASCII three binary bits change from 0 to 1 or from 1 to 0, but only one bit changes in the Gray code.

DIGITAL MODES

A digital *mode* consists of both a protocol and a method of modulation. Since digital protocols can be used to convey speech, video or data files, each specific use forms a different *emission* or mode, as referred to in amateur radio. The FCC includes the type of data being transmitted — voice, text, data or image — in the definition of each emission, assigning a different emission designator to each. FCC emission designators were discussed earlier in this chapter.

Digital modes have become very popular in all types of communications. The following sections present the most common amateur digital modes and compare some of their basic characteristics. Digital protocol development is a hotbed of innovation in the amateur community with new modes and variations on existing modes appearing all the time. If you are interested in digital mode technology in amateur radio, the TAPR group (**www.tapr.org**) should be your first stop on the internet.

Digital Signal Bandwidth

It is important to know the bandwidth of a digital signal when transmitted over the air link. *Shannon's Information Theorems*, fundamental laws of information transmission, link the symbol rate of a signal to its bandwidth. The higher the signal's baud, the wider the air link signal will be. This relationship determines the signal's *necessary bandwidth* which is the minimum bandwidth required to send a certain number of symbols per second using a particular type of modulation.

A calculation of bandwidth can be quite complex so the ITU has done the job for us and published a set of tables and formulas relating bandwidth to baud, including other factors such as fading or keying shift. (ITU-R SM.1138-2 is the document with the information.) The bandwidth formulas in the following sections are taken from that reference.

CW

E8C05 — What is the approximate bandwidth of a 13-WPM International Morse Code transmission?

E8C12 — What factors affect the bandwidth of a transmitted CW signal?

E8D04 — What is the primary effect of extremely short rise or fall time on a CW signal?

E8D05 — What is the most common method of reducing key clicks?

A CW signal produced by turning an AM transmitter on and off is described by the emission designator A1A. The bandwidth of a CW signal is determined by two factors: the speed of the CW being sent and the shape of the keying envelope. **[E8C12]** The ITU equation for the necessary bandwidth of CW is:

$$BW = B \times K \tag{Equation 8.3}$$

where:
  BW is the necessary bandwidth of the signal
  B is the speed of the transmission in baud
  K is a factor relating to the shape of the keying envelope.

The ITU bandwidth tables use a value of 0.8 for the conversion between baud and WPM. The second variable, K, reflects the abruptness of the keying waveform with typical values of 3 to 5 for amateur signals. As CW rise and fall times get shorter (more abrupt, harder keying), K gets larger. This is because signals with short rise and fall times contain more harmonics than longer, softer envelopes. (Remember that a square wave contains an infinite number of odd harmonics.) The more harmonics required to construct the keying envelope, the greater the bandwidth of the resulting CW signal must be. The ITU standard suggests a typical value for K of 5 on an HF channel where the signal is subjected to fading. Thus, for CW signals:

$$BW = (WPM \times 0.8) \times 5 \tag{Equation 8.4}$$

Suppose you are sending Morse code at a speed of 13 WPM. According to Equation 8.4, the necessary bandwidth of the transmitted signal is:

$$BW = WPM \times 0.8 \times 5 = 10.4 \times 5 = 52 \text{ Hz } \textbf{[E8C05]}$$

This is why "hard" keying waveforms with extremely short rise and fall times (1 millisecond or less) cause key clicks. **[E8D04]** The burst of harmonics modulating the signal (CW is, after all, an AM signal) appear as signals on nearby frequencies. These can be extremely disruptive to adjacent stations. Disruptions of a smooth rising or falling edge, such as overshoot or glitches, can also be heard as key clicks. To make sure your signal is not generating key clicks, increase the rise and fall time, usually adjustable as a menu selection in today's transceivers, and be sure the waveform is clean throughout the keying cycle. **[E8D05]**

FSK/AFSK

E2E01 — Which of the following types of modulation is common for data emissions below 30 MHz?

E2E04 — What is indicated when one of the ellipses in an FSK crossed-ellipse display suddenly disappears?

E2E11 — What is the difference between direct FSK and audio FSK?

E8C06 — What is the bandwidth of a 170-hertz shift, 300-baud ASCII transmission?

E8C07 — What is the bandwidth of a 4800-Hz frequency shift, 9600-baud ASCII FM transmission?

Modulations, Protocols, and Modes   **8-9**

Most amateur data transmissions on HF use *frequency shift keying* (FSK). [**E2E01**] In FSK systems, the transmitter uses different frequencies to represent 0 and 1. For RTTY, two frequencies are used so this is a *binary FSK* or *2-FSK* system. By shifting between the two frequencies (called the mark and space frequencies), the transmitter creates data symbols. The difference between the mark frequency and the space frequency is called the *shift*.

Binary FSK signals can be generated in two ways. *Direct FSK* is created by shifting a transmitter oscillator's frequency with a digital signal. *Audio FSK* or *AFSK* is created by injecting two audio tones, separated by the correct shift, into the microphone input of a single-sideband transmitter. [**E2E11**] The necessary bandwidth of that signal is determined by the frequency shift used and the speed at which data is transmitted. The bandwidth is not affected by the type of data being transmitted or its code. The equation relating necessary bandwidth to shift and data rate is:

BW = (K × Shift) + B   (Equation 8.5)

where:
   BW is the necessary bandwidth in hertz.
   K is a constant that depends on the allowable signal distortion and transmission path.
      For most practical amateur FSK communications, K = 1.2.
   Shift is the frequency shift in hertz.
   B is the symbol rate in baud.

Example 8.5

What is the bandwidth of a 170-Hz shift, 300-baud ASCII signal transmitted as a J2D emission? [**E8C06**]

BW = (1.2 × 170 Hz) + 300 = 504 Hz

This is a necessary bandwidth of about 0.5 kHz.

Example 8.6

What is the bandwidth of a 4800-Hz shift, 9600-baud ASCII signal transmitted as an F1D emission? [**E8C07**]

BW = (1.2 × 4800 Hz) + 9600 = 15360 Hz = 15.36 kHz

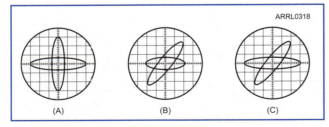

Figure 8.2 — The two tones of an FSK or AFSK signal are represented as a pair of ellipses on a crossed-ellipse display. For the best copy, the signal should be tuned so that the ellipses are of equal size and at right angles as in (A). Displays such as at (B) and (C) indicate a mistuned signal.

RTTY and other FSK/AFSK modes require careful tuning of the SSB transceiver so that the tones of the signals are as close to exactly right as possible. Errors in tuning will result in poor copy and garbled characters, particularly when using RTTY. To assist in tuning, several "cross-style" indicators have been developed. One of the most popular is the crossed-ellipse indicator seen in **Figure 8.2**. The signal should be tuned in so that the ellipses (representing the two FSK tones) are of equal size and at right angles. This display shows *selective fading* very clearly, in which one or both of the tones is severely attenuated for a short period as the ellipse shrinks dramatically. [**E2E04**]

PSK

E2E10 — Which of these digital modes has the narrowest bandwidth?

E8C03 — Why should phase-shifting of a PSK signal be done at the zero crossing of the RF signal?

E8C04 — What technique minimizes the bandwidth of a PSK31 signal?

Peter Martinez, G3PLX, developed PSK31 for real-time keyboard-to-keyboard QSOs. The name derives from the modulation type — *phase-shift keying* (*PSK*) — and data rate, which is actually 31.25 bauds. PSK31 uses the 128-character ASCII and the full 256 ANSI character sets. (ANSI stands for the American National Standards Institute.) Variations of PSK31, such as the faster PSK63, are also used.

PSK31 uses Varicode to encode the characters and as with the Baudot and ASCII codes, there is a need to indicate the gaps between characters. Varicode does this by using "00" to represent a gap.

With Varicode, a typing speed of about 50 words per minute requires a 32 bit/s transmission rate. Martinez chose a rate of 31.25 bps because it is easily derived from the 8-kHz sample rate used in many DSP systems at the time PSK31 was developed.

The bandwidth of PSK31 signals is minimized by the special sinusoidal shaping of the transmitted data symbols and by making shifting the modulation's phase only when the RF carrier signal crosses zero voltage. [**E8C03, E8C04**] This reduces the harmonic content of each symbol. Using Equation 8.1 and K = 1.2, the bandwidth of a PSK31 signal is approximately BW = 31.25 × 1.2 = 37.5 Hz, narrowest of all two-way HF digital modes used by amateurs, including CW. [**E2E10**] The mode's narrow bandwidth and phase-shift keying require stable tuning and careful transmitter adjustment.

Packet Radio and APRS

E2D04 — What technology is used to track, in real time, balloons carrying amateur radio transmitters?

E2D07 — What digital protocol is used by APRS?

E2D08 — What type of packet frame is used to transmit APRS beacon data?

E2D10 — How can an APRS station be used to help support a public service communications activity?

E2D11 — Which of the following data are used by the APRS network to communicate station location?

Packet radio on VHF and UHF uses AFSK modulation to send data at 1200 baud over an FM link. The data is exchanged in packets called *frames* using the methods defined by the AX.25 protocol standard. Complete information on packet radio and the AX.25 protocol are available via the Tucson Amateur Packet Radio (TAPR) website, **www.tapr.org**.

The most common use for packet radio is making connections to *store-and-forward* systems. This refers to a computer system that receives messages, stores them, and makes them available to other users or forwards them to other computer systems. The most common method of creating these systems on VHF and UHF is to use packet radio.

The Automatic Packet Reporting System (APRS) is a messaging system that makes use of packet radio functions by using the AX.25 Amateur Packet Radio Protocol. [**E2D07**]. Position and other data from a station is broadcast to other stations in the region and relayed to internet servers. The network of APRS stations is created by individual stations using packet radio to relay APRS data between other stations in the network.

An APRS station typically uses a 2 meter FM radio operating on the national APRS frequency of 144.39 MHz. The station consists of a packet radio terminal node controller (TNC) and computer system running APRS software. A Global Positioning System (GPS) receiver typically supplies position information.

Modulations, Protocols, and Modes

APRS stations transmit a beacon packet containing the station's location, weather conditions, and short text messages. The data is transmitted in an unnumbered information (UI) frame. [**E2D08**] Other stations act as digipeaters to relay the packets to other stations and to gateways that relay the station's information to internet APRS servers, such as **findu.com**. APRS software running on a personal computer can be used to display the station call sign on a map with a user-selected icon and the transmitted data.

APRS packets are not directed to a specific station and receiving stations do not acknowledge correct receipt of the packets. This simplifies APRS operation because the stations do not all have to remain "connected" for the network to function and transfer data.

Position data is sent to the APRS network as latitude and longitude. [**E2D11**] You can obtain this information:

• From an accurate map in degrees, minutes and seconds, entering it manually into a host computer running APRS software, or

• As an NMEA-0183-formatted text string from a Global Positioning System (GPS) satellite receiver or other suitable navigation system.

NMEA-0183 refers to a data formatting standard from the National Marine Electronics Association. A number of commercially available TNCs support direct connections to a GPS receiver via an RS-232 serial data interface.

Miniature low-power APRS stations called "trackers" incorporate all of the necessary functions (TNC, transceiver, and GPS receiver) into a single package. Trackers make it easy to add location monitoring to portable and mobile platforms when bicycling, hiking, or driving. By adding the altitude data from the GPS receiver, a three-dimensional position can be obtained. This type of APRS system is frequently used to track the position of a high-altitude balloon or rocket in near real-time, aiding in its recovery and linking position with any sensor data being measured. [**E2D04**]

APRS networks can also be used to support public service or emergency communications by providing event managers and organizers continuously-updated location and other information from an APRS-equipped station. [**E2D10**]

HF Packet

E2E06 — What is the most common data rate used for HF packet?
E2E13 — Which of these digital modes has the fastest data throughput under clear
communication conditions?

Packet radio on HF uses the same AX.25 protocol as on VHF, but is limited to 300 baud by regulation to control the signal's bandwidth. Most HF packet transmissions use FSK at 300 baud compared to the 1200-baud AFSK more common on VHF FM packet systems. [**E2E06**] The length of the AX.25 packets (typically 40 bytes) and the distortion and fading of HF propagation combine to make HF packet a niche mode, although under clear communication conditions it is still faster than 45-baud RTTY, PSK31, or AMTOR. [**E2E13**]

PACTOR

E2E05 — Which of these digital modes does not support keyboard-to-keyboard operation?
E2E08 — Which of the following HF digital modes can be used to transfer binary files?

The original PACTOR mode (referred to as PACTOR-I) is an HF digital mode developed by German amateurs Hans-Peter Helfert, DL6MAA, and Ulrich Strate, DF4KV. It was designed to overcome the shortcomings of AMTOR and HF packet. It performs well under both weak-signal and high-noise conditions. The protocol also supports the transfer of binary files, making it quite useful in today's data-intensive world. [**E2E08**] The exchange of data is by formatted packets that include error-detection and correction data. This makes it very reliable but it cannot support keyboard-to-keyboard "chat" operation. [**E2E05**]

The most popular PACTOR modes in use today are PACTOR-III and the related WINMOR, particularly for using e-mail over HF via the Winlink system. (As of early 2020, PACTOR-IV is not legal for US amateurs although the FCC is considering rule changes for digital emissions.)

PACTOR systems automatically evaluate the conditions between receive and transmit stations, *training* to the highest speed supported by the path. As a result, PACTOR-III systems running at better than 5 kbps offer the highest data rate of any amateur HF digital mode. (Note that Winlink is not a mode, it is a system of modes and protocols and internet services that allow e-mail to be exchanged using amateur radio.)

WSJT-X Modes

E2D09 — What type of modulation is used for JT65 contacts?
E2E03 — How is the timing of FT4 contacts organized?

Developed by a team led by Joe Taylor, K1JT, the *WSJT-X* software suite supports several modes designed specifically for weak-signal communication such as Earth-Moon-Earth (JT4 and JT65 for EME or moonbounce), meteor scatter (MSK144), and low signal-to-noise HF contacts (FT8 and FT4).

The *WSJT-X* protocols use advanced multi-tone AFSK modulation and sophisticated codes to recover signals with very low signal-to-noise ratios. [**E2D09**] These protocols use precisely timed sequences of transmitting and receiving to synchronize the stations. For example, FT4 sequences are 7.5 seconds long and JT65 sequences are 1 minute long. [**E2E03**] For more information about the *WSJT-X* protocols, read K1JT's website at **physics.princeton.edu/pulsar/K1JT**.

OFDM Modulation

E8B07 — Orthogonal Frequency Division Multiplexing is a technique used for which type of amateur communication?
E8B08 — What describes Orthogonal Frequency Division Multiplexing?

A special type of modulation used with multi-tone modes creates specially-shaped and spaced signals that interfere with each other as little as possible. **Figure 8.3A** shows the spectrum of one such signal being modulated by a single bit. Note how the spectrum has zero amplitude at regularly spaced intervals. If several of these signals are spaced so their spectra overlap precisely at the zero points as in Figure 8.3B, it becomes much easier to demodulate each signal without interference from the other subcarriers, called *intersymbol*

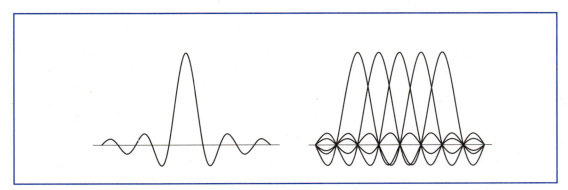

Figure 8.3 — A carrier modulated by a single bit (A) has a spectrum with nulls spaced at regular intervals on either side of the carrier frequency. Multiple signals placed so their nulls coincide (B) exhibit minimal interference with each other and are called orthogonal. Orthogonal frequency division multiplexing (OFDM) divides the data to be transmitted between many such carriers in wide-band digital modes such as the most recent versions of PACTOR.

interference. Signals having this spacing and modulation rate are called *orthogonal* to each other. The combination of multiple tones and the orthogonal signals results in *orthogonal frequency division multiplexing* or *OFDM*. [**E8B08**] OFDM requires precise modulation techniques and clean transmitted signals. It is used by later versions of PACTOR and other advanced wide-band digital modes. [**E8B07**]

Transmitting Digital Mode Signals

E2E07 — Which of the following is a possible reason that attempts to initiate contact with a digital station on a clear frequency are unsuccessful?

E8D07 — What is a common cause of overmodulation of AFSK signals?

E8D08 — What parameter evaluates distortion of an AFSK signal caused by excessive input audio levels?

E8D09 — What is considered an acceptable maximum IMD level for an idling PSK signal?

For digital modes to perform well it is important to pay attention to transmitted signal quality. Especially when using protocols that require very precise phase and amplitude control, it is important to avoid introducing distortion anywhere in the transmit path.

Transmitter ALC systems are designed for voice and will cause distortion of the modulated signal by changing the power level. This creates intermodulation distortion (IMD) and other spurious signals in the transmitter output. Not only do these signals cause interference but they also make it harder for the receiver to recover the data. If the transceiver's ALC meter is active during digital transmission, the signal is overmodulating the transmitter and the audio level should be reduced. [**E8D07**]

To make sure your digital signal is clean, follow the manufacturer's instructions on setting the transmitter controls, especially ALC and microphone gain. Software that receives AFSK digital signals often measures the distortion level of signals, reporting it as an "IMD Level" parameter. [**E8D08**] For example, for a clean PSK signal that is idling (not sending characters), the received IMD level should be –30 dB with respect to the main signal. [**E8D09**] Overmodulation is visible on a waterfall-style display as bars or "ghost signals" alongside the main transmitted signal extending to frequencies above and below it.

Once you have the transmitter and your computer or TNC operating properly, do an "on-air" check yourself or with a nearby station. Measuring your own digital signal can be done in your own station by transmitting into a dummy load, receiving the signal on a second receiver, and feeding the received audio into the sound card of a computer running a demodulation program for that mode. You'll quickly see whether you have the audio and RF levels set properly on the transmitting system.

Along with not being on the right frequency or using the wrong version of a protocol, not being able to establish a digital contact may have nothing to do with your transmitter. The reason is called the "hidden transmitter" problem. On the HF bands, it is common for a station to be inaudible at one location but quite strong at another because of propagation conditions. When this happens, a station receiving signals from two stations that can't hear each other will often be able to decode neither signal well. This situation can prevent you from making contact with another station even on a "clear" channel. When this happens, you may have to just try again later when conditions change. [**E2E07**]

SPREAD SPECTRUM TECHNIQUES

E8D01 — Why are received spread spectrum signals resistant to interference?

E8D02 — What spread spectrum communications technique uses a high-speed binary bit stream to shift the phase of an RF carrier?

E8D03 — How does the spread spectrum technique of frequency hopping work?

The usual measure of efficiency for a modulation scheme is to examine how well it concentrates the signal for a given rate of information — less bandwidth for equivalent data rate is good. While compactness of the signal appeals to the conventional wisdom, spread-spectrum modulation techniques take the exact opposite approach. They spread the signal over a very wide bandwidth by rapidly varying the carrier frequency of the signal in a pre-defined sequence.

Communications signal bandwidth is increased (called *spreading*) by factors of 10 to 10,000 by using a sequence of bits (the *spreading code*) to vary the signal's frequency. The exact techniques are discussed below.

Spreading has two beneficial effects. The first effect is the dilution of the signal energy on a given frequency, so that while occupying a very large bandwidth, the power density at any point within the spread signal is very low (see **Figure 8.4**).

Dilution of the signal across many frequencies causes spread-spectrum signals to appear as wideband noise to a conventional receiver. Spectrum spreading may result in the digital signal being below the noise floor of a conventional receiver, and thus invisible to it, while the signal can still be received with a spread-spectrum receiver.

The second beneficial effect of spectrum spreading is that a spread-spectrum receiver can reject strong undesired signals — even those much stronger than the desired spread-spectrum signal's power density. This is because the receiver uses the spreading code to "de-spread" the signal, a process like following the signal as it changes frequency. Non-spread signals are then suppressed in the processing because they are not consistent with the spreading code. [**E8D01**] The effectiveness of this interference-rejection property has made spread-spectrum a popular communications security technique.

Conventional signals such as narrowband FM, SSB, and CW are rejected by a spread-spectrum receiver, as are other spread-spectrum signals not following the required coding sequence. The result is a type of private channel, one where only spread-spectrum signals using the correct spreading code are received. A two-party conversation can take place, or if the spreading code is known to a number of people, network-type operations are possible.

The use of different spreading codes allows several spread-spectrum systems to operate independently while using the same frequencies. This is a form of frequency sharing called *code-division multiple access* or *CDMA*. If the spreading parameters are chosen judiciously for the propagation conditions that exist on the selected frequencies, conventional users in the same amateur band will experience very little interference from spread-spectrum users. This allows more signals to be packed into a band, but each additional signal (conventional or spread spectrum) will add some interference for all users by raising the received noise level.

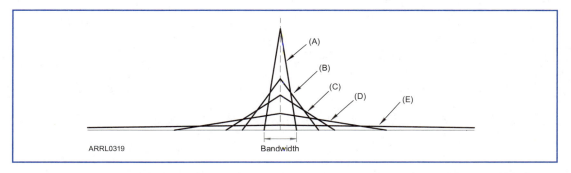

Figure 8.4 — A graphic representation of the distribution of power in a spread spectrum signal as the bandwidth is increased. The unspread signal (A) concentrates most of its energy near a center frequency. As the bandwidth increases (B), the power near the center frequency falls. At C and D, the energy is distributed wider and wider across the spread signal's wider bandwidth. At E, the signal energy is spread over a very wide bandwidth and there is little power at any one frequency.

Modulations, Protocols, and Modes   **8-15**

Types of Spread Spectrum

There are many ways to cause a signal to spread, but all spread-spectrum systems can be viewed as a combination of two modulation processes. First, a conventional form of modulation, either analog or digital, is used to add the information to the carrier. Next, the modulated carrier is again modulated by the spreading code, causing it to spread out over a large bandwidth. Four spreading techniques are commonly used in military and space communications, but amateurs are currently only authorized to use *frequency hopping* and *direct sequence*.

Frequency Hopping

Frequency hopping (FH) is a form of spreading in which the center frequency of a conventional carrier is altered many times per second in accordance with a *pseudorandom* list of channels. [**E8D03**] (Pseudorandom means that the list is not truly random, but is a very long list of numbers that appears random before it repeats.) The same channel list must also be used by the receiving station. The amount of time the signal is present on any single channel is called the *dwell time*. To avoid interference both to and from conventional frequency users, the dwell time must be very short, typically less than 10 milliseconds.

Direct Sequence

In *direct sequence* (*DS*) spread spectrum, a very fast binary bit stream is used to shift the phase of the modulated carrier. [**E8D02**] DS spread spectrum is typically used to transmit digital information.

Like the pseudo-random frequency list of FH systems, the sequence of the bits created by a digital circuit is designed to appear random. This binary sequence can be duplicated and synchronized at the transmitter and receiver. Such sequences are called *pseudo-noise* or *PN*.

Each bit of the PN code is called a *chip* and the rate at which the chips shift carrier phase is called the *chip rate*. If the RF carrier's phase is shifted 0 or 180 degrees, it is called *binary phase-shift keying* (*BPSK*). Other types of phase-shift keying are also used. For example, *quadrature phase-shift keying* (*QPSK*) shifts between four different phases (0, 90, 180 and 270 degrees).

Amateur Spread Spectrum Applications

E2C04 — Which of the following frequencies are sometimes used for amateur radio mesh networks?

E2C09 — What type of equipment is commonly used to implement an amateur radio mesh network?

E2C12 — What technique do individual nodes use to form a mesh network?

Amateurs use spread spectrum techniques to establish mesh networks, such as Broadband-Hamnet (BBHN), Amateur Radio Emergency Data Network (AREDN), and HamWAN. Individual network nodes are constructed from commercial Wi-Fi routers reprogrammed with custom software to meet amateur radio regulations. [**E2C09**] The nodes operate on frequencies shared with various unlicensed wireless data services, such as Wi-Fi. [**E2C04**] When a node is activated, it uses discovery and link establishment protocols, similar to those used by commercial wireless data systems. [**E2C12**] These amateur networks are operational in the 2.4 GHz and 5.6 GHz bands and are a rapidly growing use of amateur microwave frequencies for personal and public service data communications.

ERROR DETECTION AND CORRECTION

E2E02 — What do the letters FEC mean as they relate to digital operation?
E8C01 — How is Forward Error Correction implemented?
E8C08 — How does ARQ accomplish error correction?

Even the best transmitter and receiver systems cannot guarantee 100% accuracy of data transmitted across an air link. There are just too many ways for Mother Nature to disrupt the signal; noise, multipath and fading are just a few causes of errors. To get an idea of what can happen to a data signal, try to copy RTTY signals that are weak or fluttery from ionospheric variations!

In recognition of the realities of radio propagation, data communications engineers have devised a number of strategies. The first challenge is to find out when an error has occurred! This is called *error detection*. Without some clue about what the data should have been, however, there is no way to detect errors. To be able to discern transmission errors, information describing the data is sent along with the original data.

Error detection data can be as simple as the parity bit of ASCII data discussed earlier. Another popular technique of error detection, used by packet radio's AX.25 protocol, TCP/IP networking systems, and Ethernet (among many others) is the *cyclical redundancy check* or (*CRC*). If the CRC of a received packet matches the CRC sent with the packet, the entire packet is judged to be error-free. Using a CRC detects most errors.

Once the system has detected an error, what it decides to do about it is another matter. This moves the process from error detection to *error correction*. The simplest form of error correction is ARQ (Automatic Repeat Request). If the receiving system detects an error, it requests a retransmission of the corrupted packet or message by sending a NAK (Not Acknowledge) message to transmitting station. [**E8C08**] The information is retransmitted until the receiver responds with an ACK (Acknowledge) message. If the errors persist long enough, the system gives up and drops the connection.

Another popular error correction technique is to send some extra data about the information in the packet or message so that the receiving station can actually correct some types of errors. This technique is called *forward error correction* or (*FEC*). [**E2E02**] The term "forward" stems from sending extra error correction data "ahead" with the original information. The combination of the FEC data and the algorithm by which errors are detected and corrected is called an *FEC code*. [**E8C01**]

There are many types of FEC codes: Reed-Solomon, Hamming, BCH, and Golay codes are all used in consumer electronics. FEC data is sometimes spread out over several data packets to account for fading. FEC is used with digitized voice to help preserve the quality of the received speech. This is why digital voice systems (such as mobile phones) tend to have good quality up to a certain error threshold and then become completely garbled — their FEC code fails at that point.

8.3 Amateur Television

Many new hams are surprised to find that amateurs can (and do!) communicate using video television signals "just like broadcast stations." Slow-scan television also has an enduring presence on the amateur bands, taking advantage of computer sound card and signal processing technology. New hams could be equally surprised to find that they may have some of the necessary equipment already available to them.

FAST-SCAN TELEVISION

E2B08 — What technique allows commercial analog TV receivers to be used for fast-scan TV operations on the 70 cm band?

Fast-scan TV (FSTV) can be used by any amateur holding a Technician or higher-class license. FSTV or *amateur TV (ATV)* closely resembles broadcast-quality television, because

Modulations, Protocols, and Modes **8-17**

it normally uses the same technical standards. It is called "fast-scan" because the images are transmitted quickly enough to support full-motion video. Amateurs typically use commercial transmission standards for TV signals, but are not limited to commercial standards. Nevertheless, due to the wide availability of equipment that conforms to those standards, most amateur television is compatible with commercial broadcast equipment for analog signals.

Digital Amateur TV (DATV)

While this section of the chapter focuses on analog television signals, the migration of amateur TV to use the now-standard digital TV mode is underway. Affordable equipment is available as surplus and several groups are experimenting with digital TV around the country. The *ARRL Handbook* chapter on Image Communications covers both analog and digital ATV, as well. There is a lot of analog equipment in active use but expect DATV to become equally popular.

Fast-Scan System Components

A basic ATV station is constructed as shown in **Figure 8.5**. This is where you may recognize some common home entertainment gear. Any camera or camcorder that produces a standard video signal can be used, color or black-and-white. Newer ATV modulators may work directly with HDMI signals from cameras. Converters are available to change analog video signals to HDMI.

Image monitors for "off-air" use can be any TV or computer display that accepts a composite video signal, such as from a camera or video player. Some HDMI-compatible equipment

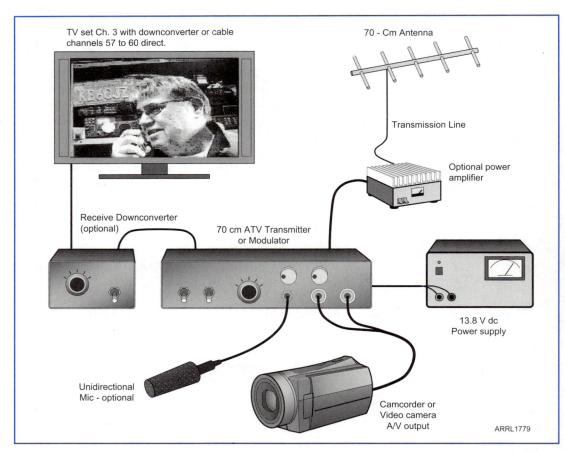

Figure 8.5 — A basic ATV station consists of a camera or camcorder to produce video, an amateur television transmitter or modulator, an optional amplifier, a directional antenna with gain, and a commercial TV receiver. A separate receive converter can also be used for better performance.

may be used, as well. Regular cable-ready TV receivers can be used with a *receive converter* or *down converter* to receive UHF ATV signals that use broadcast transmission standards. The converter shifts the ATV signal to an unused channel shared with cable TV systems for display. [**E2B08**]

An ATV transmitter is one piece of equipment not found in most ham stations. Nevertheless, low-power ATV transmitters, modulators and linear ATV amplifiers with a few watts of output power are available. They take a standard composite video signal as input and produce the UHF ATV signal ready for amplification or connection directly to the antenna.

Antennas for ATV need to have fairly high gain (to boost the strength of the weak, wide bandwidth signal) and a consistent radiation pattern over the wide bandwidth consumed by an ATV signal. Yagis and corner reflectors are popular for ATV stations. At microwave frequencies, both Yagi and dish antennas are used.

Video Signal Definitions

E2B01 — How many times per second is a new frame transmitted in a fast-scan (NTSC) television system?

E2B02 — How many horizontal lines make up a fast-scan (NTSC) television frame?

E2B03 — How is an interlaced scanning pattern generated in a fast-scan (NTSC) television system?

E2B07 — What is the name of the signal component that carries color information in NTSC video?

In analog television, a process called *scanning* translates a picture into a series of horizontal lines, called *scan lines*. The lines are transmitted to a receiver which reproduces the image line by line on a CRT or solid-state display. The combination of video and control signals required to transmit and display the image is called a *raster*, which is defined by a video standard. For digital television, the image is captured as an array of pixels and is transferred as a digital signal with HDMI being the most common interface method.

When transmitting an analog image over the air, US amateurs use the *NTSC* (National Television Standard Committee) standard. For digital TV, either the DVB-T or Cable-QAM standards are used. The exam questions are restricted to analog TV and video signals.

Table 8.3 lists the primary elements of the NTSC standard used for ATV. In the NTSC standard, a total of 525 horizontal scan lines comprise a *frame* to form one complete image. [**E2B02**] Thirty frames are generated each second. [**E2B01**] Each frame consists of two *fields*, each field containing 262½ lines, so 60 fields are generated each

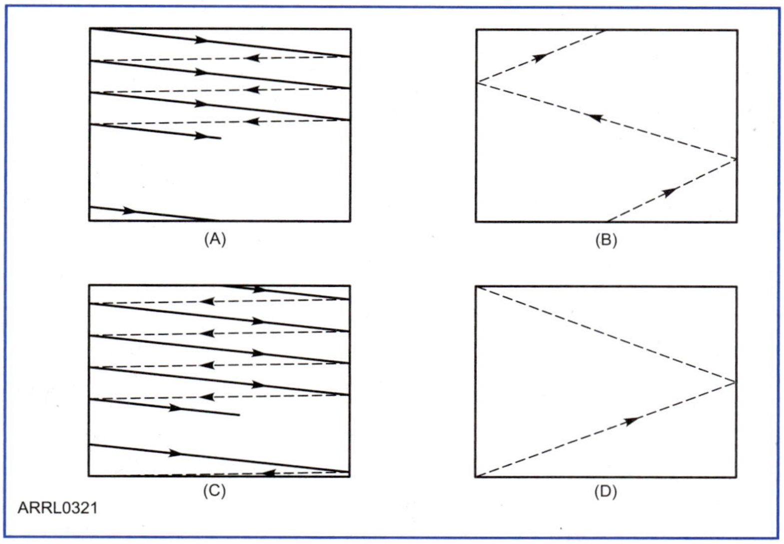

ARRL0321

Figure 8.6 — This diagram shows the interlaced scanning used in TV. In field one, 262½ lines are scanned (A). At the end of field one, the electron scanning beam is returned to the top of the picture area (B). Scanning lines in field two (C) fall between the lines of field one. At the end of field two, the scanning beam is again returned to the top, where scanning continues with field one (D).

**Table 8.3
NTSC Standards for Analog ATV**

Line rate	15,750 Hz
Field rate	60 Hz
Frame rate	30 Hz
Horizontal Lines	262½ per field
	525 per frame
Sound subcarrier	4.5 MHz
Channel bandwidth (VSB-C3F)	6 MHz

Modulations, Protocols, and Modes   **8-19**

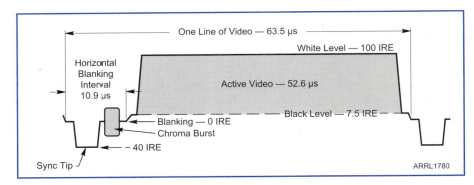

Figure 8.7 — RS-170 waveform for one video line. A full NTSC video frame consists of 525 video lines organized as two interlaced fields. Each field takes 1/60th of a second to transmit, so that 30 frames are transmitted every second. IRE units are used to measure the relative amplitude of the different parts of the video signal.

second. **Figure 8.6** illustrates the concept of scan lines and interlacing. CCD cameras and solid-state displays do not scan or reproduce the image in this way, but the composite video signal is still constructed in the same way. HDMI and other digital video interfaces use a different set of signals and are not discussed here.

Scan lines from one field fall between lines from the next field. This is called *interlacing* and is done to reduce flicker, improve the smoothness of motion from frame to frame, and reduce bandwidth while maintaining adequate image quality. If all 525 scan lines are numbered from top to bottom, then one field contains the even-numbered lines and the alternate field contains the odd-numbered lines. [**E2B03**] This process of slicing up an image is reversed to create the line-by-line image displayed by a TV receiver or computer monitor.

The electronic signal that carries all of the image and display coordination information is called *baseband video* or *composite video* and is described in the standard ANSI RS-170. **Figure 8.7** shows the basic structure of one frame of an RS-170 video signal. Sync signals have a negative voltage and the video portion of the signal a positive voltage. The standard voltage level between white video and the *sync tip* is 1-V peak to peak. Television engineers measure video levels in IRE or IEEE Units.

Vertical sync pulses tell the display electronics when a new field is about to begin. There are two vertical sync pulses per frame, one for each field. *Horizontal sync* pulses occur between each horizontal scan line. Sync pulses are markers to keep the image's information aligned properly.

Blanking refers to the time interval between video lines and frames. Blanking is no longer required of modern displays but the signals are used for timing and synchronization. These time periods are called the *horizontal* and *vertical blanking intervals*.

Between horizontal blanking pulses, the voltage of the signal represents the brightness or *luminance* of the image. In monochrome black-and-white video, higher video voltages create whiter areas on the image. Lower voltages create blacker areas. Note that the sync pulses have lower voltages than the blackest video voltage. Sync pulses are thus termed "blacker than black."

Composite and RGB Video

In a composite color video signal, all of the information about the image is contained in a single waveform. The color (called *chroma*) information is combined with the luminance information through the use of a separate *chrominance subcarrier* signal. [**E2B07**]

In each scan line, between the end of the horizontal sync pulse and the start of the video is a short interval of a 3.5789 MHz signal (see Figure 8.7). This is the chroma subcarrier frequency. The short *chroma burst* helps synchronize the image and chrominance subcarrier timing. If the two are not precisely locked, the hues of the reproduced colors will be wrong.

RF ATV Signal Characteristics

E2B05 — Which of the following describes the use of vestigial sideband in analog fast-scan TV transmissions?

E2B06 — What is vestigial sideband modulation?

A fast-scan color ATV video signal has a bandwidth of about 4 MHz as shown in **Figure 8.8**. (Satisfactory black-and-white signals require somewhat less bandwidth.) The wide bandwidth is necessary to send the information required for full-motion, real-time images. Consequently, ATV is permitted only in the 420 to 450 MHz band and at higher frequencies.

Most ATV activity occurs in the 420 to 450 MHz band. The exact frequency used depends on local custom and band plans. Some populated areas have ATV repeaters. ATV users should avoid interfering with the weak-signal operation (moonbounce, for example) near 432 MHz and with repeater operation above 442 MHz.

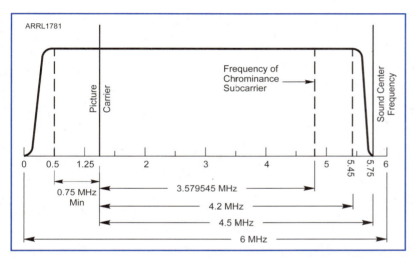

Figure 8.8 —The spectrum of an analog broadcast TV signal. ATV may or may not use the FM subcarrier for sound.

Most amateurs use *vestigial sideband* (VSB) for transmission. VSB is like SSB plus full carrier, except a portion (vestige) of the unwanted sideband is retained. [**E2B06**] In the case of VSB TV, approximately 1 MHz of the lower sideband and all of the upper sideband plus full carrier comprise the transmitted image signal. Figure 8.8 shows the spectrum of a color TV signal. VSB uses less bandwidth than a full DSB AM signal but can still be demodulated satisfactorily by simple video detector circuits. [**E2B05**]

There are at least three ways to transmit voice information with a TV signal. The most popular method is by talking on another band, often 2 meter FM. This has the advantage of letting other local hams listen in on what you are doing. Rather than tie up a repeater for this, it is best to use a simplex frequency.

Commercial TV includes an FM voice sub-carrier 4.5 MHz above the TV image carrier (see Figure 8.8). If your ATV transmitter provides FM audio via a subcarrier 4.5 MHz above the video carrier, the audio can be received easily in the usual way on an analog TV set.

SLOW-SCAN TELEVISION

E1A12 — What special operating frequency restrictions are imposed on slow scan TV transmissions?

Since fast-scan TV systems take several megahertz of bandwidth that must mean there are no images sent on the HF bands, right? Images can't be sent via HF as fast-scan video, but there are two image modes that do work on HF: *facsimile* and *slow-scan television*. You are probably familiar with facsimile, or *fax*, and, yes, amateurs can use the same fax protocols over the airwaves. Amateur facsimile is rarely heard these days because slow-scan television (*SSTV*) is used instead, even though it offers somewhat lower resolution than fax.

As an image mode, SSTV is restricted to the phone segments of all bands (which excludes 30 meters, since phone operation is not allowed there). SSTV signal bandwidth must be no greater than that of a phone signal using the same modulation. [**E1A12**] SSTV is usually

Modulations, Protocols, and Modes 8-21

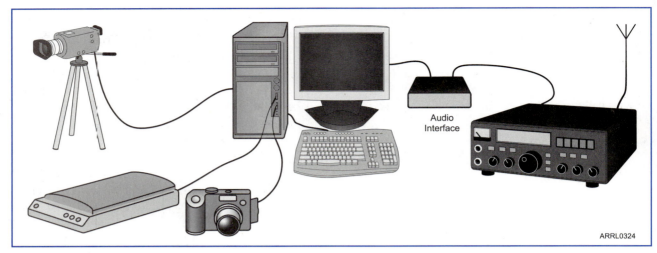

Figure 8.9 — The most common SSTV station uses a video or digital camera, or a document scanner, to generate still images. Software then processes the image and converts it to audio tones that are input to an SSB transmitter.

transmitted on HF because fast-scan is available on UHF and higher bands. Standard HF calling frequencies for SSTV are 3.845, 7.171, 14.230, 21.340 and 28.680 MHz. The most popular bands for SSTV are 20 and 75 meters.

A typical SSTV system shown in **Figure 8.9** uses a computer and software to generate the required audio tones to represent an image for transmission and to decode the received audio tones and display the image on the computer monitor. Computer graphics open up many possibilities for creating your own SSTV images. Software can be used to capture an image from a video or still camera or document scanner for transmission via slow-scan. You can store received images on disk for later retransmission, as well. **Figure 8.10** shows a screen shot of a program used to handle SSTV images.

An important transceiver consideration is that SSTV is a 100%-duty-cycle transmission mode. This means your transmitter will be producing full power for the entire image-transmission time. Most transmitters and amplifiers will have to be run at reduced power output to avoid overheating.

Figure 8.10 — This screen shot shows an SSTV program that uses the computer sound card to send and receive images.

Analog SSTV Signal Basics

E2B04 — How is color information sent in analog SSTV?

E2B10 — What aspect of an analog slow-scan television signal encodes the brightness of the picture?

E2B11 — What is the function of the Vertical Interval Signaling (VIS) code sent as part of an SSTV transmission?

E2B12 — What signals SSTV receiving software to begin a new picture line?

Like fast-scan television, SSTV divides the image into scan lines and frames. (Basic television image terminology is defined in the preceding section on fast-scan TV.) Between each

Table 8.4
Black-and-White Analog SSTV Standards

Frame time	8 seconds
Lines per frame	120
Time to send one line	67 ms
Duration of horizontal sync pulse	5 ms
Duration of vertical sync pulse	30 ms
Horizontal and vertical sync frequency	1200 Hz
Black frequency	1500 Hz
White frequency	2300 Hz

scan line is a horizontal sync pulse to establish the beginning of each line. The beginning of a frame is established by the vertical sync pulse.

Instead of using voltage levels as video and sync signals, SSTV uses frequencies with specific frequencies encoding specific functions. This allows SSTV signals to be transmitted using SSB modulation. For example, sync pulses signifying new lines are sent as bursts of 1200-Hz tones. [**E2B12**] The image's luminance or brightness is transmitted as a tone of varying frequency. [**E2B10**] For monochrome black-and-white SSTV signals, a 1500-Hz signal produces black and a 2300-Hz signal produces white. Frequencies between these limits represent shades of gray. The 1200 Hz of the sync pulses are thus "blacker than black" like their fast-scan counterparts and do not show up on the display.

Table 8.4 summarizes the standard parameters for a black-and-white analog SSTV signal. The horizontal sync pulse is included in the time required to send one line, but the vertical sync pulse adds 30 msec of "overhead." So, it takes a bit more than 8 seconds to actually transmit a black-and-white image frame.

A basic monochrome black-and-white SSTV image takes 8 seconds for one frame and has only 120 scan lines. This works out to 15 scan lines per second. The bandwidth of a monochrome SSTV signal is about 2 kHz. Unlike fast-scan images, the scan lines of SSTV are not interlaced and so each image frame is composed of a single field.

Most SSTV operators today send color images. There are several modes of formatting color images, transmitting 120, 128, 240, or 256 scan lines. Colors are transmitted as sequential lines showing the same image in red, green, and blue. [**E2B04**] The receiving system recombines the three lines into a full-color line.

For receiving equipment and software to discern the mode of the SSTV image, a code is transmitted with each image frame. The code is sent during the vertical sync pulse and is called *vertical interval signaling* or *VIS*. Receiving software reads the code and adjusts its decoding settings to properly capture and display the image. Similarly, the operator can select the mode for transmission and the appropriate code will be incorporated into the image. [**E2B11**]

The various SSTV modes allow the operator to select different resolutions and control image transmission time. Low to moderate resolution images can be transmitted when crowded band conditions favor short transmission times. Higher resolution images can be transmitted when longer transmissions are acceptable.

Digital SSTV

E1B02 — Which of the following is an acceptable bandwidth for Digital Radio Mondiale (DRM) based voice or SSTV digital transmissions made on the HF amateur bands?

E2B09 — What hardware, other than a receiver with SSB capability and a suitable computer, is needed to decode SSTV using Digital Radio Mondiale (DRM)?

To send high-quality digital-encoded program material, shortwave broadcasters developed and released the *Digital Radio Mondiale* (*DRM*) protocol, meaning "Digital World Radio." Amateurs adapted DRM's file transfer capabilities to send digitized SSTV images instead. Since DRM signals can be generated and decoded with software on a PC, no additional special equipment beyond a receiver is required to use DRM for SSTV communications. [**E2B09**] While broadcast DRM signals have bandwidths of 4 kHz or more, amateur DRM signals on HF are restricted to a normal SSB signal's bandwidth of 3 kHz. [**E1B02**]

Modulations, Protocols, and Modes

Chapter 9

Antennas and Feed Lines

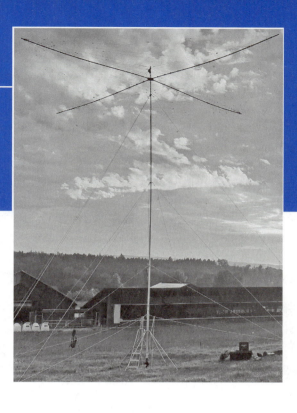

In this chapter, you'll learn about:
- Antenna basics — radiation patterns, gain and beamwidth, polarization, and bandwidth
- The effects of ground on antennas
- Variations of the dipole
- Shortened, multiband and satellite antennas
- Traveling wave antennas and phased arrays
- Receiving and direction finding antennas
- Impedance matching techniques
- Transmission line mechanics and the Smith chart
- Antenna measurements and analyzers
- Antenna modeling and design

Antennas and antenna systems are of primary importance to amateurs. Equipment available to amateurs is of the best quality in the history of radio but even the best radios require an antenna to communicate effectively. It is up to the station owner to select and install antennas and antenna system components that get the most out of the radio equipment.

The topics covered by the license exam touch on many important subjects for amateur radio antenna systems. Basic antenna concepts are explored in more detail than for other license classes. Practical issues affecting antennas, such as ground systems, shortened antennas, and popular multiband designs are covered. We'll also look into the functions of transmission lines as they affect antenna systems. The section concludes with some discussion of using computers to model antennas. The set of exam questions for each subject is listed at the end of each section so that you can review your understanding and build a solid background in antennas. All of these topics are covered in detail in *The ARRL Antenna Book* if you would like more information.

9.1 Basics of Antennas

E9A01 — What is an isotropic antenna?
E9A12 — How much gain does an antenna have compared to a ½-wavelength dipole when it has 6 dB gain over an isotropic antenna?
E9B07 — How does the total amount of radiation emitted by a directional gain antenna compare with the total amount of radiation emitted from a theoretical isotropic antenna, assuming each is driven by the same amount of power?
E9B08 — What is the far field of an antenna?

We'll explore another level of detail beyond your General class license studies, including some background on antenna radiation patterns.

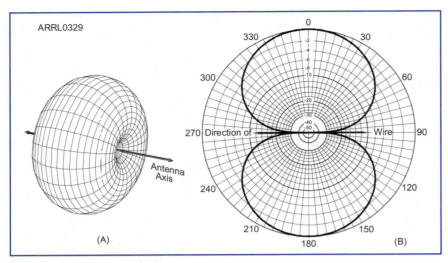

Figure 9.1 — The three-dimensional radiation pattern of a dipole appears at A. B shows a plane radiation pattern of the same antenna — a slice through the pattern at A. The solid line in B represents the relative amount of power radiated at the angles on the outer ring.

ANTENNA RADIATION PATTERNS

An antenna radiation pattern contains a wealth of information about the antenna and its expected performance. By learning to recognize certain types of patterns and what they represent, you will be able to identify many important antenna characteristics and to compare design variations. **Figure 9.1** is an example of a dipole's radiation pattern, including a three-dimensional view (Figure 9.1A).

Figure 9.1B shows one "slice" through the three-dimensional pattern of Figure 9.1A. The orientation of the antenna is shown and the directions around the antenna are shown on the outer circle. The strength scale from the center to the edge of the pattern is usually shown in decibels, but the step size of each ring can be adjusted to show desired pattern details. Make sure the strength scales are the same when comparing the patterns of different antennas.

The radiation pattern is usually drawn so it just touches the outer circle at the point of its maximum strength. The scale then shows the relative strength of signals radiated in any direction with the maximum point representing 0 dB with respect to all other directions. A legend on the chart often gives the value of gain at the outer circle. To compare antennas, use patterns with the same value of gain at the outer circle. Or, if you're modeling the antennas, plot the radiation patterns for all antennas on a common chart.

Antenna radiation patterns describe the antenna's radiated signal in the antenna's *far field* which begins several wavelengths from the antenna and extends to infinity. Antennas also have a *near field* region which is too close to the antenna for the final pattern to emerge. In the far field, the pattern shape is independent of distance. [**E9B08**]

ANTENNA GAIN

In many applications, the antenna's most important property is its ability to concentrate its radiated power in useful directions. This property, however, only has meaning with respect to other antennas, so a reference must be established.

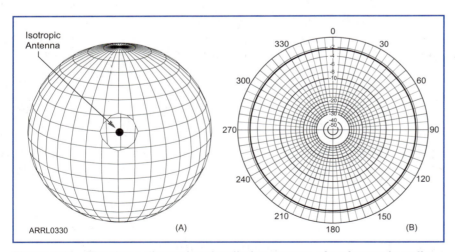

Figure 9.2 — The three-dimensional radiation pattern of an isotropic radiator is shown at A. B shows the radiation pattern of the isotropic antenna in any plane. This plot has been reduced by 2.15 dB to compare the isotropic antenna's radiated signal strength with that in the dipole's peak direction as shown in Figure 9.1.

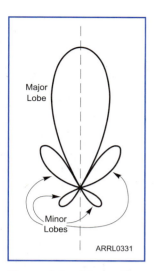

Figure 9.3 — The radiation pattern typical of a VHF beam antenna, illustrating major and minor radiation lobes.

The Isotropic Radiator

An *isotropic radiator* is a theoretical, point-sized antenna that is assumed to radiate equally in all directions. The three-dimensional radiation pattern of an isotropic radiator (see **Figure 9.2**) is a sphere, since the same amount of power is radiated in all directions. This hypothetical antenna has a radiation pattern that is omnidirectional, because the signal is equal in all directions. No such antenna actually exists, but it serves as a useful theoretical gain reference for comparison with real antennas. The isotropic antenna also provides a useful reference for comparing the differences in gain among real antennas. [**E9A01**]

Directional Antennas

Directional antennas are designed specifically to concentrate their radiated power in one (or more) directions. The direction in which most of the power is focused is the *major lobe* or *main lobe* of radiation and is designated to be the *forward* direction. Most directional antennas also have *minor lobes* in the *back* and *side* directions. The directions of minimum radiation between the lobes are the pattern's *nulls*.

Figure 9.3 is an example of a radiation pattern for a typical VHF beam antenna, illustrating major and minor lobes. By reducing radiation in the side and back directions and concentrating it instead in the forward direction, a beam antenna can transmit or receive a stronger signal in that direction.

An antenna's *gain* is the ratio (expressed in decibels) between the signal radiated from an antenna in the direction of its main lobe and the signal radiated from a reference antenna in the same direction and with the same power. A typical beam might have 6 dB of gain compared to a dipole which means that it makes your signal sound four times (6 dB) louder than if you were using a dipole with the same transmitter. The gain of directional antennas is the result of concentrating the radio wave in one direction at the expense of radiation in other directions. There is no difference in the total amount of power radiated. [**E9B07**]

There are two reference antennas used to compare the radiation patterns of other antennas: the half-wavelength dipole and the isotropic antenna. Gain referenced to the isotropic antenna is given in dBi, and gain referenced to the dipole in its direction of maximum radiation is given as dBd. Fortunately, it's simple to convert between gain with respect to an isotropic antenna and gain with respect to a dipole because the dipole has 2.15 dB of gain over an isotropic radiator.

Gain in dBi and dBd are related as follows:

Gain in dBd = Gain in dBi − 2.15 dB (Equation 9.1)

and

Gain in dBi = Gain in dBd + 2.5 dB (Equation 9.2)

where:
  dBd is antenna gain compared to a dipole in its direction of maximum radiation.
  dBi is antenna gain compared to an isotropic radiator.

Example 9.1

If an antenna has 6 dB more gain than an isotropic radiator, how much gain does it have compared to a dipole? [**E9A12**]

Gain in dBd = Gain in dBi − 2.15 dB = 6 dBi − 2.15 dB = 3.85 dBd

Example 9.2

If an antenna has 12 dB more gain than a dipole, how much gain does it have compared to an isotropic antenna?

Gain in dBi = Gain in dBd + 2.15 dB = 12 dBi + 2.15 dB = 14.15 dBi

dBd and dBi gains are *free-space* gains, meaning that there are no reflecting surfaces near the antenna, such as the ground. Nearby reflecting surfaces can dramatically increase or decrease an antenna's gain. When you compare specifications for several antennas, be sure that they all use the same reference antenna for comparison or convert the gains from one reference to another. Specifications should give free-space gain or state the antenna's height.

BEAMWIDTH AND PATTERN RATIOS

E9B01 — In the antenna radiation pattern shown in Figure E9-1, what is the beamwidth?

E9B02 — In the antenna radiation pattern shown in Figure E9-1, what is the front-to-back ratio?

E9B03 — In the antenna radiation pattern shown in Figure E9-1, what is the front-to-side ratio?

In comparing antennas, it is useful to know their *beamwidth*. Beamwidth is the angular distance between the points on either side of the major lobe at which the gain is 3 dB below the maximum. This is also sometimes called the *3 dB beamwidth*. **Figure 9.4** illustrates the idea. The antenna in the figure has a beamwidth of 30°. This means if you turn the antenna plus or minus 15° from the optimum heading, the signal received (and the signal received from your transmitter) will drop by 3 dB.

Figure 9.5 illustrates how to determine beamwidth from a radiation pattern. The major lobe of radiation from the antenna in the figure points to the right, and is centered along the 0° axis. You can make a pretty good estimate of the beamwidth of this antenna by carefully reading the graph. Notice that angles are marked off every 15° and the –3 dB circle is

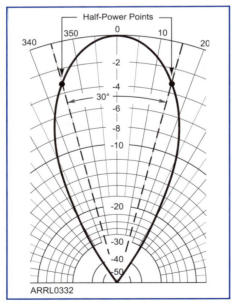

Figure 9.4 — The beamwidth of an antenna is the angular distance between the directions at which the antenna gain is one-half (–3 dB) its maximum value.

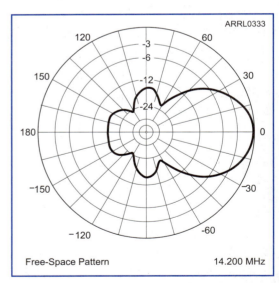

Figure 9.5 — The radiation pattern of a typical HF beam antenna. The text shows how to read beamwidth and other pattern ratios from the graph.

the first one inside the outer circle. The pattern crosses the –3 dB circle at points about 25° either side of 0°. So we can estimate the beamwidth of this antenna as 50°. [**E9B01**]

There are other useful measurements to be taken from an antenna's radiation pattern. Gain is not everything! The ability to reject received signals from unwanted directions is also important. This ability, called a *pattern ratio*, is measured with respect to the directivity of the antenna in the forward or "front" direction. There are three primary pattern ratios:

• *Front-to-back* (*F/B*): the difference in gain in the direction of the major lobe to the gain in the exact opposite (back) direction.

• *Front-to-rear* (*F/R*): the difference in gain in the direction of the major lobe to the average gain over a specified angle centered on the back direction.

• *Front-to-side* (*F/S*): the difference in gain in the direction of the major lobe to the gain at 90° to the front direction. (F/S usually assumes a symmetrical pattern to either side of the major lobe.)

In the example shown in Figure 9.5, find the front-to-back ratio by reading the maximum value of the minor lobe at 180°. This maximum appears to be halfway between the –12 dB and –24 dB circles, so estimate it to be about 18 dB below the major lobe. [**E9B02**]

The pattern in Figure 9.5 has a minor lobe to each side of the antenna whose maximum strength is a bit more than 12 dB below the main-lobe maximum. A front-to-side ratio of 14 dB looks like a pretty good estimate for this pattern. [**E9B03**]

RADIATION AND OHMIC RESISTANCE

E9A03 — What is the radiation resistance of an antenna?
E9A05 — What is included in the total resistance of an antenna system?

The power supplied to an antenna is dissipated in the form of radio waves and in heat losses in the wire and materials nearby that absorb the waves, such as foliage or buildings. The radiated energy is the useful part, of course. The power being radiated away by the antenna can be treated as if the energy were dissipated in a resistance. The antenna can then be modeled as two resistances, one for radiated power and one for power dissipated as heat.

In the case of the heat losses, R_{LOSS} is a *real* or *ohmic* resistance that gets warm. In the case of the radiated power, R_R is an assumed resistance, that if actually present, would dissipate the same power radiated by the antenna. This assumed resistance is called the *radiation resistance*, R_R. [**E9A03**] These resistances are treated as if they are in series so the total power dissipated by current, I, in the antenna is equal to $I^2 (R_R + R_{LOSS})$. The sum of these two resistances forms the total resistance of an antenna system, R_T. [**E9A05**]

Taking as an example an ordinary half-wave dipole antenna for the HF bands, the power lost as heat in the antenna does not exceed a few percent of the total power supplied to the antenna. This is because the RF resistance of copper wire even as small as #14 AWG is very low compared with the radiation resistance if the antenna is reasonably clear of surrounding objects and not too close to the ground. It is reasonable to assume that the ohmic (heat) loss in a reasonably located antenna is negligible and that all of the resistance shown by the antenna is radiation resistance. Such an antenna is a highly efficient radiator of radio waves! If the antenna is very short in terms of wavelength, though, R_R is quite low. For example, for the popular magnetic loops less than 1/10th of a wavelength in circumference, R_R can be much less than 1 Ω! This can result in the applied power being largely dissipated as heat by losses in the materials used to construct the antenna and surrounding materials.

FEED POINT IMPEDANCE

E9A04 — Which of the following factors affect the feed point impedance of an antenna?

An important characteristic of an antenna is the *feed point impedance* presented to the transmission line. Feed point impedance is simply the ratio of RF voltage to current wher-

Antennas and Feed Lines 9-5

ever the transmission line is attached to the antenna. If the voltage and current are in-phase, the feed point impedance is purely resistive and the antenna is resonant, regardless of the value of the resistance. If the voltage and current are not in-phase, the impedance will have some reactance as well and may be inductive or capacitive. Feed point impedance consists of the antenna's radiation resistance, ohmic losses including ground losses, and any reactance caused by the antenna being nonresonant.

Feed point impedance also changes with position on the antenna. Consider a resonant, half-wavelength dipole with maximum current at the mid-point and minimum current at the end. The feed point impedance is lowest in the middle of the antenna where the ratio of voltage to current is lowest. As the feed point is moved toward either end of the dipole, however, the voltage increases and current decreases, causing the feed point impedance to increase. Near the center, feed point impedance is less than 100 ohms but at the ends, feed point impedance can be several thousand ohms. If the same amount of power is transferred to the dipole at both locations for the feed point, the radiated signal will be the same.

The value of an antenna's feed point impedance is also affected by a number of other factors. One is the location of the antenna with respect to other objects, particularly ground. For example, in free space with nothing else near it, the radiation resistance of a resonant ½-wavelength dipole made of thin wire is approximately 73 Ω at its center. As the antenna is lowered closer to the ground, the radiation resistance drops as well, lowering the feed point impedance. Other nearby conducting surfaces such as buildings and other antennas can also affect the antenna's feed point impedance. [E9A04]

Another factor is the *length/diameter ratio* of the conductor that makes up the antenna. As the conductor is made thicker, radiation resistance decreases. For most practical wire sizes, the half-wave dipole's radiation resistance is close to 65 Ω. At VHF and above, the dipole's radiation resistance will be from 55 to 60 Ω for antennas constructed of rod or tubing.

ANTENNA EFFICIENCY

E9A09 — What is antenna efficiency?

Antenna efficiency — the ratio of power radiated as radio waves to the total power input to the antenna — is given by Equation 9.3. [E9A09]

$$\text{Efficiency} = \frac{R_R}{R_T} \times 100\%$$ (Equation 9.3)

where:
  R_R = radiation resistance.
  R_T = total resistance.

Example 9.3

If a half-wave dipole antenna has a radiation resistance, R_R, of 70 Ω and a total resistance, R_T, of 75 Ω, what is its efficiency?

Efficiency = (70 / 75) × 100% = 93.3%

The actual value of the radiation resistance has little effect on the radiation efficiency of a practical antenna. This is because the ohmic resistance is only on the order of 1 Ω with the conductors used for thick antennas. The ohmic resistance does not become important until the radiation resistance drops to very low values — say less than 10 Ω — as may be the case when several antenna elements are very close together or for antennas such as mobile whips and the small loops that are very short in terms of wavelength.

ANTENNA PATTERN TYPES

E9B04 — What is the front-to-back ratio of the radiation pattern shown in Figure E9-2?

E9B05 — What type of antenna pattern is shown in Figure E9-2?

E9B06 — What is the elevation angle of peak response in the antenna radiation pattern shown in Figure E9-2?

E and H Planes

Two types of radiation patterns are often used to picture the overall, three-dimensional radiation pattern — the *E-plane* and the *H-plane* radiation patterns. The E-plane pattern is taken in the plane of the radiated electric field and the H-plane pattern in the plane of the magnetic field. In general, the E-plane pattern is in the plane of the antenna's elements and the H-plane pattern is perpendicular to them.

Azimuthal and Elevation Patterns

For a horizontally polarized antenna, the E-plane pattern is parallel to the surface of the Earth and shows the antenna's radiation pattern in directions around the antenna. This is called an *azimuthal pattern*. The H-plane pattern of the same antenna is called the *elevation pattern* and shows the antenna's radiation pattern at different angles above the Earth. Figure 9.5 in the section on Beamwidth and Pattern Ratios shows a typical azimuthal pattern for a beam antenna.

Figure 9.6 shows a typical elevation pattern with multiple lobes in the forward and back directions. [**E9B05**] The antenna pattern shows four forward lobes (those on the same half of the graph as the largest or main lobe) and three rear lobes. For antennas mounted over ground, an elevation pattern shows only the half of the radiation pattern at positive angles above ground. The angles on the graph represent the vertical angle above ground at which radiated power is measured. The vertical angle at which the antenna's major lobe has its maximum radiation is called the *takeoff angle*. In Figure 9.6, the takeoff angle is 7.5 degrees. [**E9B06**]

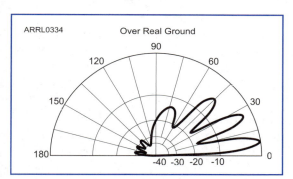

Figure 9.6 — This radiation pattern for a beam antenna mounted above ground is an elevation pattern. The major lobe in the forward direction is about 7.5° above the horizon. There are four radiation lobes in the forward direction and three to the back, all at different vertical angles.

Just as with an azimuthal pattern, an antenna's front-to-back ratio can be determined from the maximum radiation in the forward direction to that in the opposite direction. In an elevation pattern, however, it is necessary to specify the elevation angle at which the pattern is to be measured — usually the takeoff angle. In the case of Figure 9.6, toward the back of the pattern at the takeoff angle antenna gain is about 28 dB below that of the main lobe so front-to-back ratio for this antenna would be 28 dB. [**E9B04**] At other elevation angles, the front-to-back ratio would be quite different.

BANDWIDTH

E9A08 — What is antenna bandwidth?

E9D08 — What happens as the Q of an antenna increases?

Along with the characteristics of the antenna's radiation pattern, feed point impedance, and so forth, it's also important to know how antennas behave over the range of frequencies over which they are expected to operate.

As frequency changes, the electrical size of the antenna and all of its components changes, too. This means the antenna's gain, feed point impedance, radiation pattern, and so forth

Antennas and Feed Lines 9-7

will also change. The antenna is expected to perform to some specified level — these are the *performance requirements* of the antenna. In general, the bandwidth of an antenna is the frequency range over which it satisfies a performance requirement. [**E9A08**]

For example, an antenna may be specified to have an SWR of less than 1.5:1, but without an associated bandwidth, the specification is incomplete. Is the SWR going to be less than 1.5:1 at just a few frequencies or over the whole band? If that antenna's *SWR bandwidth* is specified to be 200 kHz, then you can expect to see an SWR of 1.5:1 or less over a range of 200 kHz. Notice that the frequency of minimum SWR is not known, so the range may shift depending on factors relating to assembly or installation. Other common bandwidth specifications are *gain bandwidth* and *front-to-back bandwidth*.

An antenna's bandwidth is affected by Q, just like the tuned circuits you studied in the Electrical Principles chapter. For an antenna, Q is defined as the energy stored in the fields around the antenna divided by the power it radiates. The antenna's radiation resistance has the same effect as loss resistance in the tuned circuit. The higher an antenna's Q, the narrower its SWR bandwidth will be just as a tuned circuit's bandwidth gets narrower as Q increases. [**E9D08**]

EFFECTS OF GROUND AND GROUND SYSTEMS

E9A10 — Which of the following improves the efficiency of a ground-mounted quarter-wave vertical antenna?

E9A11 — Which of the following factors determines ground losses for a ground-mounted vertical antenna operating in the 3 MHz to 30 MHz range?

E9C11 — How is the far-field elevation pattern of a vertically polarized antenna affected by being mounted over seawater versus soil?

E9C13 — How does the radiation pattern of a horizontally polarized 3-element beam antenna vary with increasing height above ground?

E9C14 — How does the performance of a horizontally polarized antenna mounted on the side of a hill compare with the same antenna mounted on flat ground?

The biggest effect on antenna system efficiency at HF is usually the losses in nearby ground, grounded structures, or the antenna's ground system. The radiation pattern of an antenna over real ground is always affected by the electrical conductivity and dielectric constant of the soil, and most importantly by the height of a horizontally polarized antenna over ground. Signals reflected from the ground combine with the signals radiated directly from the antenna. If the signals are in phase when they combine, the signal strength will be increased but if they are out of phase, the strength will be decreased. These ground reflections affect the radiation pattern for many wavelengths from the antenna.

This is especially true of the *far-field pattern* measured many wavelengths from a vertically polarized antenna operating at HF frequencies. In theory, this type of antenna should produce a major lobe at a low vertical angle — good for DX! Losses caused by low conductivity in the soil near the antenna dramatically reduce signal strength at low angles, however. [**E9A11**] For example, the low-angle radiation from a vertically polarized antenna mounted over seawater will be much stronger than for a similar antenna mounted over rocky soil. [**E9C11**] The far-field, low-angle radiation pattern for a horizontally polarized antenna is not as significantly affected, however.

For antennas that have a ground connection or use a ground system as part of the antenna, the *ground resistance* can be a significant contributor to antenna system losses — even if the antenna itself is constructed from low-loss materials.

In the case of a ¼-wavelength vertical antenna, with one conductor of the feed line connected to ground, the ground resistance is usually not negligible. To be efficient, this type of antenna requires a ground system of *radial* wires to reduce losses that would otherwise result from current flowing in the lossy soil. [**E9A10**]

Height Above Ground

The height at which an antenna is mounted above ground can also have a large effect on an antenna's radiation pattern. First, by moving the antenna away from ground, the amount of power dissipated in the Earth (ground losses) will be reduced. Second, reflections of the radiated signal from the ground will reinforce the direct radiation from the antenna at lower vertical angles, resulting in a stronger low-angle signal. In general, raising an antenna lowers the vertical takeoff angle of peak radiation.

Horizontally polarized antennas, such as dipoles and Yagis, have less ground loss than ground-mounted vertical antennas. As they are raised, ground losses drop to negligible levels. As such an antenna is raised, the vertical angle of maximum radiation drops until a height of ½ wavelength is reached. Raising the antenna farther causes additional lobes to appear in the elevation pattern above the main lobe, which continues to become lower. Figure 9.6 shows an example of these additional lobes. In general, mounting horizontally polarized antennas as high as possible gives the strongest signal at low vertical angles. [**E9C13**]

Terrain

The terrain on which an antenna is mounted affects both the azimuthal and elevation pattern of an antenna. Over flat ground and without nearby obstructions the radiation patterns for the antenna will resemble those in the antenna design books. Once buildings and uneven terrain enter the picture however, the results can be much more complicated!

Nearby buildings can serve as "passive" reflectors (or absorbers) of radio waves, most strongly at VHF and higher frequencies. This can be used to advantage, for example, by aiming an antenna to reflect a signal off a building toward a distant station. (Take care, of course, to avoid exposing the inhabitants to excessive levels of RF.) At HF, small buildings are less of a problem but large buildings can have the same effect as on the higher bands.

Hills and slopes have an effect on both the azimuthal and elevation patterns. A hilltop is highly sought after for radio work because the reflections from the ground's surface are either reduced or are more likely to reinforce the signal at low takeoff angles. This is particularly true for horizontally-polarized antennas; the major lobe's takeoff angle will typically be lower in the direction of a downward slope. [**E9C14**]

GROUND CONNECTIONS

E9D11 — Which of the following conductors would be best for minimizing losses in a station's RF ground system?

E9D12 — Which of the following would provide the best RF ground for your station?

It is important to understand the difference between electrical safety grounding, lightning protection grounding, and bonding equipment together. In most cases, there really isn't a point of zero RF voltage — an "RF ground" — to connect to. See the Safety chapter for more about the different types of ground or earth connections. In the station, equipment should be bonded together to minimize RF voltages between them that create "hot spots" or unwanted RF current flowing on cables. See the *ARRL Handbook* or the ARRL's *Grounding and Bonding for the Radio Amateur* for more information about safety and RF grounding.

Conductors used for lightning protection grounds have several basic requirements: they must be electrically short to avoid acting like an antenna themselves; they must be as straight as possible to avoid creating unnecessary inductance; and they must be a type of conductor that has low RF impedance. Wide, flat copper straps are the standard for this type of ground connection. [**E9D11**] Heavy copper wires also work well.

Once the ground connection reaches the Earth, an RF connection is different than a power system's safety ground. A single ground rod at doesn't offer enough surface area to guarantee a low-impedance connection to the Earth. Several interconnected ground rods (three or four is a good compromise) make a much better connection. [**E9D12**]

Antennas and Feed Lines

9.2 Practical Antennas

Armed with a better theoretical understanding of what affects antenna system efficiency, you are now ready to consider practical antenna systems. Rarely does an antenna get installed in a location without compromises. What happens if the number of radials available for an HF vertical antenna can't be as many as you'd like? What would be the effect of a beam antenna being close to the roof? How will that nearby hill or building affect your signal? These are all questions that amateurs have to deal with every time an antenna is erected.

DIPOLE VARIATIONS

E9C05 — Which of the following is a type of OCFD antenna?
E9C07 — What is the approximate feed point impedance at the center of a two-wire folded dipole antenna?
E9C08 — What is a folded dipole antenna?
E9C09 — Which of the following describes a G5RV antenna?
E9C10 — Which of the following describes a Zepp antenna?
E9C12 — Which of the following describes an Extended Double Zepp antenna?

There are many different variations of the dipole antenna used on the HF bands. Let's meet some of the common designs.

Folded Dipole

A *folded dipole* antenna in **Figure 9.7** is a ½-wavelength dipole with an additional closely-spaced parallel wire connecting the two ends at B and C to form a narrow loop.

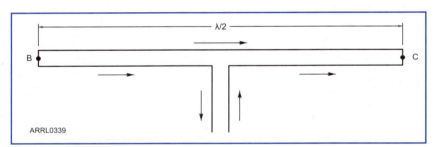

Figure 9.7 — The current distribution on the multiple conductors of a folded dipole. The arrows indicate the direction of current flow during one half-cycle of applied current.

[**E9C08**] The antenna is fed in the middle of one side of the loop. The antenna is ½ λ long from end to end, thus the name "folded dipole." The folded dipole has the same directional characteristics as a regular dipole, but its feed point impedance is four times that of a regular dipole and its SWR bandwidth is wider than for the single-conductor antenna. The higher impedance is useful when it is desirable to feed the antenna with ladder line or twin lead transmission line.

The two wires act as two dipoles connected in parallel. Because of the close spacing the current in each is equal. If the total power at the feed point, P, is the same as for a single dipole, but with only half the current, the folded dipole's feed point impedance, P / I^2, must be four times that of the single-conductor dipole: $4 \times 73 = 292$ Ω, a fairly close match to 300-Ω twin lead or open-wire transmission line. [**E9C07**]

Zepp and Extended Double Zepp Antennas

The original Zepp antenna in **Figure 9.8A** is named for the Zeppelin airship where it was originally used, hanging below the airship. The Zepp is simply a half-wave dipole with an open-wire feed line connected at one end. [**E9C10**] A common term for this antenna today is the *end-fed half-wave* or *EFHW*. The end of a half-wave wire is a point of high-impedance so the length of feed line was chosen to present a lower impedance at the transmitter, generally a quarter-wavelength was used.

The high feed point impedance could be reduced by lengthening the dipole until it was

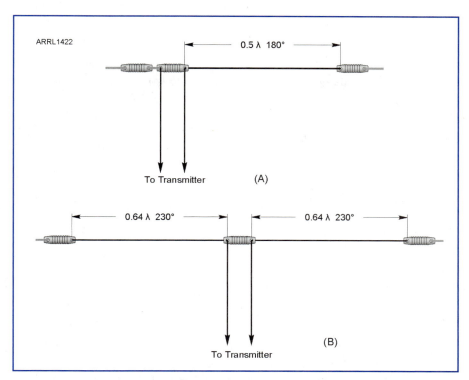

Figure 9.8 — The original Zepp antenna (A) was a half-wave dipole fed at one end with open-wire feed line. The extended double Zepp (B) consists of two ⅝ wavelength extended Zepps. A length of feed line is used as a matching section to create a 50 Ω point at which coax can be attached.

approximately ⅝ wavelengths long creating the *extended Zepp*. Two extended Zepps can also be connected together as in Figure 9.8B. This creates the *extended double Zepp* (EDZ) antenna which is 2 × ⅝ = 1.25 wavelengths long. Fed in the center, a short section of open-wire feed line is used to create a point at which the impedance is approximately 50 Ω. At that point a choke balun is used to allow a coaxial cable to be attached. The EDZ is really a *collinear array* of two half-wave dipoles with their currents in-phase so the pattern is bidirectional, broadside to the antenna. [E9C12]

G5RV Antenna

A variation on the extended Zepp, the design invented by G5RV is shown in **Figure 9.9**. The length of open-wire line is selected to produce a low impedance on at least one band so that a 1:1 choke balun can be used for attaching 50-Ω coax. [E9C09] The antenna may be used from 3.5 through 30 MHz, although the use of an antenna tuner should be expected on any band except 14 MHz. On its fundamental frequency, the G5RV has a four-lobed pattern that is somewhat more omnidirectional than either dipole or doublet. There are a number of variations on this basic antenna configuration that are referred to as "G5RV" antennas. The common features are center-feed with a short length of open-wire line terminated in a choke balun for attachment to coaxial cable.

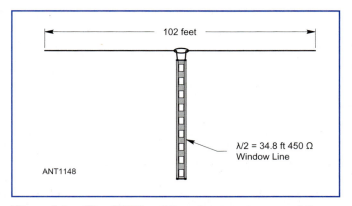

Figure 9.9 — The G5RV multiband antenna covers 3.5 through 30 MHz. There are many similar designs, all with similar names.

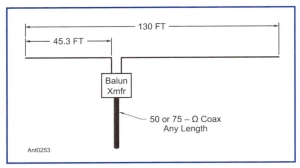

Figure 9.10 — The off-center-fed dipole for 3.5, 7, and 14 MHz. A 4:1 or 6:1 impedance transformer is used at the feed point.

Off-Center Fed Dipole (OCFD)

While usually fed in the center, a ½-wavelength dipole will accept energy from a feed point anywhere along its length. If the feed point is moved away from the center of the dipole as in **Figure 9.10**, the impedance rises because current is dropping while voltage is rising along the antenna. The *off-center-fed dipole* or *OFCD* takes advantage of placing the feed point where the impedance is similar on more than one band, generally in the neighborhood of 150 – 300 Ω. A suitable matching device such as a 4:1 impedance transformer is then used to reduce the feed point impedance to something closer to 50 Ω. [**E9C05**]

LOADED WHIPS

E9D03 — Where should a high-Q loading coil be placed to minimize losses in a shortened vertical antenna?

E9D04 — Why should an HF mobile antenna loading coil have a high ratio of reactance to resistance?

E9D06 — What happens to the SWR bandwidth when one or more loading coils are used to resonate an electrically short antenna?

E9D07 — What is an advantage of using top loading in a shortened HF vertical antenna?

E9D09 — What is the function of a loading coil as used as part of an HF mobile antenna?

E9D10 — What happens to feed-point impedance at the base of a fixed length HF mobile antenna when operated below its resonant frequency?

The most difficult place to achieve effective HF antenna performance is a mobile station. Not only is the antenna system exposed to the mobile environment's vibration, temperature extremes and corrosion, but the antenna's size is quite limited by the constraints on vehicle size and maneuverability. The 10 and 12 meter bands are the only bands on which "full-sized" ¼-wavelength ground-plane antennas can realistically be used for a mobile station.

Practically, mobile antennas for HF are almost all some variation on the *whip* — a flexible, vertical conductor attached to the vehicle with a threaded or magnetic mount. The vehicle acts as a ground-plane for the antenna. Whips are usually 8 feet or less in length. At 21 MHz and lower frequencies, the antenna is "electrically short," meaning less than ¼-wavelength long. As the operating frequency is lowered, the feed point impedance of such an antenna is a decreasing radiation resistance in series with an increasing capacitive reactance as shown by the equivalent circuit in **Figure 9.11**. [**E9D10**] A full-size ¼-wavelength whip's radiation resistance is approximately 36 Ω.

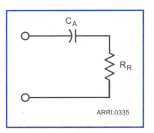

Figure 9.11 — At frequencies below its resonant frequency, the feed point impedance of a whip antenna can be represented as a capacitive reactance, C_A, in series with the antenna's radiation resistance, R_R.

To tune out the capacitive reactance and resonate the antenna, a series inductive reactance, or *loading coil* is used. (Remember that resonance occurs when the feed point impedance is entirely resistive.) [**E9D09**] The amount of inductance required is determined by the desired operating frequency and where the coil is placed along the antenna. **Figure 9.12** shows the loading coil as an inductance in series with the whip.

Base loading (placing the loading coil at the feed point, assumed to be at the base of the antenna) requires the lowest value of inductance for a given antenna length. As the coil is moved along the whip farther from the feed point, the

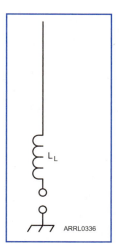

Figure 9.12 — The capacitive reactance of a whip antenna can be cancelled by adding an equivalent amount of inductive reactance as a loading coil in series with the antenna.

required amount of inductance increases. This is because the amount of capacitive reactance increases as the feed point moves closer to the end of the whip. The tradeoff of using loading coils in a shortened antenna is that the SWR bandwidth of the antenna is reduced. [**E9D06**] The reduction in bandwidth occurs because reactance of the tuned system increases more rapidly away from the resonant frequency than does the feed point reactance of a full-size antenna.

One advantage of placing the coil at least part way up the whip, however, is that the current distribution along the antenna is improved, and that increases the radiation resistance. The major disadvantage is that the requirement for a larger loading coil means that the coil losses will be greater, although this is offset somewhat by lower current through the larger coil and is the reason loading coils should have a high Q (ratio of reactance to resistance). [**E9D04**] Assuming a high-Q coil, center loading offers the best compromise for minimizing losses in an electrically-short vertical antenna. [**E9D03**]

Figure 9.13 shows a typical center-loaded whip antenna suitable for operation in the HF range. The antenna could also be mounted directly on the car body (such as a fender or trunk lid). The base spring acts as a shock absorber for the base of the whip, since continual flexing would weaken the antenna. A short, heavy, mast section is mounted between the base spring and loading coil. Some models have a mechanism that allows the antenna to be tipped over for adjustment or for fastening to the roof of the car when not in use. Optional guy lines can be used to stabilize the antenna while in motion.

The popular mobile "Hamstick" antenna consists of a two-section tubular fiberglass base helically wound with the antenna conductor and topped with a short metal whip. The entire base becomes the loading coil. These inexpensive antennas work on a single band, requiring multiple antennas to be carried for operation on

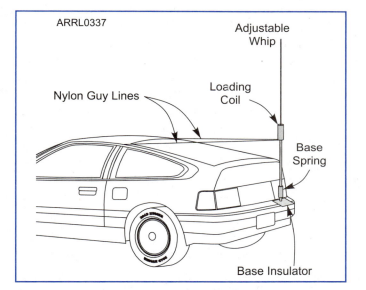

Figure 9.13 — A drawing showing a typical bumper-mounted HF-mobile antenna. Optional guy lines help stabilize the antenna while the vehicle is in motion.

multiple bands. They give good performance for their modest price. At the other end of the price scale are the tunable "screwdriver" antennas similar to that in Figure 9.13 but with a loading coil inductance that can be adjusted from inside the vehicle. (The name derives from the small dc motor used to tune the coil, similar to those found in electric screwdrivers.)

Losses in the loading coil can be reduced if the required loading coil inductance is reduced, allowing a smaller coil. To use a smaller coil, the capacitive reactance that must be tuned out must also be reduced. One method of decreasing capacitive reactance is to increase capacitance. The *top loading* method is one such technique.

Top loading adds a "capacitive hat" above the loading coil, either just above the coil or near the top of the whip. The "hat" usually consists of short wires perpendicular to the whip, often with the ends of the wires connected by a metal ring for additional strength. The added capacitance reduces the resonating value of inductance and the size of the loading coil. Using a smaller loading inductor reduces the loading coil's resistive loss and improves the antenna radiation efficiency. [**E9D07**]

Antennas and Feed Lines 9-13

TRAVELING WAVE ANTENNAS

E9C04 — What happens to the radiation pattern of an unterminated long wire antenna as the wire length is increased?

E9C06 — What is the effect of adding a terminating resistor to a rhombic antenna?

E9H01 — When constructing a Beverage antenna, which of the following factors should be included in the design to achieve good performance at the desired frequency?

E9H02 — Which is generally true for low band (160 meter and 80 meter) receiving antennas?

So far, the antennas you've studied for your license exams have been resonant antennas, based on ½-λ or ¼-λ elements. Another class of antennas lets the power flow along elements up to several wavelengths long, being dissipated as it goes. These are *traveling wave* antennas.

The most common traveling wave antenna is the *long-wire* antenna. Long wires, as the name implies, are just, well, long wires — one wavelength long or longer. They can be fed anywhere along their length, but typically ¼ λ from one end as a dipole with one end extended. The radiation pattern for such an antenna is shown at the left in **Figure 9.14**. The feed point impedance of the long-wire antenna varies dramatically as a consequence of the changing current patterns with frequency.

The long wire has four major lobes (and several minor lobes). The longer the wire, the closer to the direction of the wire the lobes become. [**E9C04**] If two long wires are combined as shown in the figure, and fed out-of-phase with a single transmission line where the wires come together, their major lobes will coincide in two directions, creating the pattern shown at the right of the figure. This is the antenna known as the *Vee beam*. The open wire ends reflect any power that hasn't been dissipated back toward the feed point, creating the "backward" major lobe of the radiation pattern.

Rhombic Antennas

Vee beams can be combined as well, creating the *rhombic antenna*, shown in **Figure 9.15**. The diamond-shaped rhombic antenna can be considered as two Vee beams placed end-to-end; it has four equal-length legs and the opposite angles are equal so the antenna is symmetrical. Each leg is at least one wavelength long. Rhombic antennas are installed horizontally with supports at the four corners.

The radiation pattern of a rhombic is bidirectional as shown by the arrows in Figure 9.15. There are minor lobes in other directions; their number and strength depend on the length of the legs. Notice that the wires at the end opposite the feed point are open. Just as for Vee beams, the open wire ends reflect power that's not dissipated back toward the feed point and create the second major lobe.

In Figure 9.15B, a terminating resistor has been added to form the *terminated rhombic*. The main effect of adding a terminating resistance is to change the pattern to unidirectional by absorbing the power that would

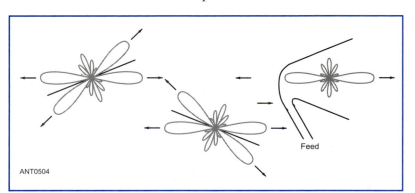

Figure 9.14 — Two long wires and their radiation patterns are shown at the left. If the two are combined to form a V with the major lobes aligned and fed out of phase, the resulting patterns reinforce the major lobes of each wire along the axis of the antenna.

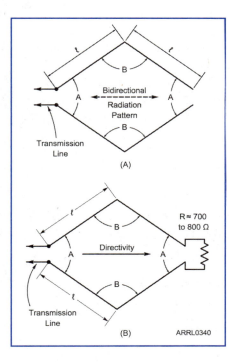

Figure 9.15 — The rhombic at A is a diamond-shaped antenna. All legs are the same length and opposite angles of the diamond are equal. The radiation pattern is bidirectional as shown. Adding a terminating resistor as shown at B makes the pattern unidirectional with no loss of gain in the preferred direction.

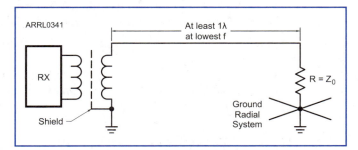

Figure 9.16 — The one-wire Beverage antenna forms a transmission line with the ground and is terminated at the far end. The antenna's preferred direction is to the right in this drawing.

have been reflected to create the unwanted second major lobe. [**E9C06**] The loss of power as heat (about a third of the input power) does not result in lower gain in the desired direction, however.

Beverage Antennas

Nearly every antenna installed by radio amateurs is used for both receiving and transmitting. One common exception is the traveling wave receive antenna invented by H.H. Beverage and shown in **Figure 9.16**. The Beverage antenna acts like a long transmission line, with one lossy conductor (the Earth) and one good conductor (the wire). As for the terminated rhombic antenna, a Beverage antenna has a terminating resistor to ground at the end farthest from the radio. The terminating resistor absorbs the power of signals arriving from the unwanted direction instead of allowing them to be reflected back toward the feed point.

Beverage antennas are effective directional antennas for 160 meters and 80 meters. They are less effective at higher frequencies, however, and are seldom used on 40 meters and shorter wavelength bands. Beverage antennas should be at least one wavelength long at the lowest operating frequency. [**E9H01**] Longer antennas provide increased gain and directivity. Beverage antennas are installed at relatively low heights, normally 8 to 10 feet above ground. They should form a relatively straight line extending from the feed point toward the preferred direction.

The Beverage's chief benefit is in rejecting noise from unwanted directions, not in having high gain. In fact, the Beverage is rather lossy! In general, atmospheric noise is high enough on the lower bands that antenna gain is not important. [**E9H02**] The improvement in signal-to-noise ratio, however, can be dramatic.

PHASED ARRAYS

E9C01 — What is the radiation pattern of two ¼-wavelength vertical antennas spaced ½-wavelength apart and fed 180 degrees out of phase?

E9C02 — What is the radiation pattern of two ¼-wavelength vertical antennas spaced ¼-wavelength apart and fed 90 degrees out of phase?

E9C03 — What is the radiation pattern of two ¼-wavelength vertical antennas spaced ½-wavelength apart and fed in phase?

E9E08 — What is a use for a Wilkinson divider?

E9E11 — What is the primary purpose of phasing lines when used with an antenna having multiple driven elements?

Various pattern shapes can be obtained using an antenna system that consists of two vertical antennas fed with various phase relationships. These are examples of *phased arrays* in which the phase differences of the signals that the antennas receive or transmit create the desired radiation pattern.

Figure 9.17 illustrates the basics of how a phased array pattern is created. In Figure 9.17A, a single antenna (oriented so that the current, I, is flowing perpendicularly to the

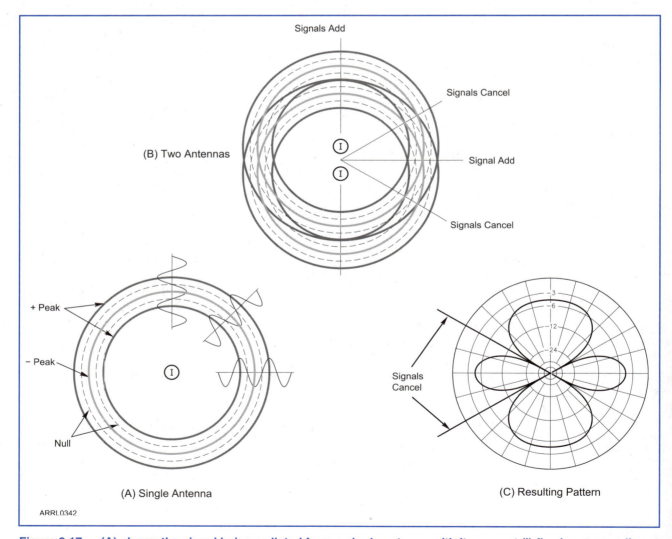

Figure 9.17 — (A) shows the signal being radiated from a single antenna with its current (I) flowing perpendicularly to the page. At (B) a second antenna has been added and the radiated signals reinforce and cancel along different directions, leading to the radiation pattern for the array in (C).

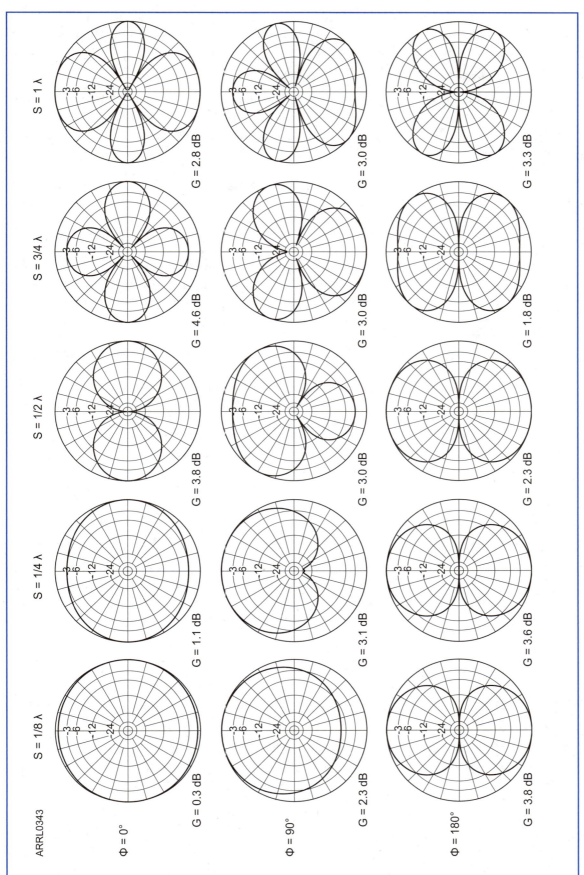

Figure 9.18 — The horizontal radiation patterns of two identical ½ λ dipole or ¼ λ ground plane antennas, physically spaced as indicated by S and with relative phases of their identical feed point currents indicated by φ. The two elements are aligned along a vertical line from top to bottom of the figure and the antenna toward the top of the pattern has the lagging current phase. The gain values, G, are with respect to that of a single antenna.

Antennas and Feed Lines 9-17

page) is radiating a signal as shown by the different circles, representing the positive and negative peaks of the signal. When a second antenna is added at Figure 9.17B, the spacing of the antennas causes the signals to reinforce at some angles to the array and cancel at others. The resulting radiation pattern is shown at Figure 9.17C.

The relative phase between the antennas can be varied by changing their physical spacing or by changing the electrical phase of the currents that feed the antenna. **Figure 9.18** shows patterns for a number of common spacings and current phases. The patterns assume that the antennas are identical ¼-wavelength verticals and are fed with equal magnitudes of current. As in Figure 9.17, the currents in both antennas are perpendicular to the page. The antennas themselves are aligned along a vertical line from top to bottom of the figure. The antenna toward the top of the pattern has the lagging current phase.

By studying the patterns shown in Figure 9.18 you can see that when the two antennas are fed in phase, a pattern that is *broadside* to the elements always results. If the antennas are ¼ wavelength apart and fed in phase, the pattern is elliptical, like a slightly flattened circle. This system is substantially omnidirectional. If the antennas are ½ wavelength apart and fed in phase the pattern is a figure-8 that is broadside to the antennas. [**E9C03**]

At spacings of less than ⅝ wavelength, with the elements fed 180° out of phase, the maximum radiation lobe is in line with the antennas. This is an *end-fire array*. For example, if the antennas are ⅛ wavelength, ¼ wavelength or ½ wavelength apart and fed 180° out of phase the pattern is a figure-8 that is in line with the antennas. [**E9C01**]

With intermediate amounts of phase difference, the results cannot be stated so simply. Patterns evolve that are not symmetrical in all four quadrants. If the antennas are ¼ wavelength apart and fed 90° out of phase, an interesting pattern results. (See the middle pattern of the second row from the left in Figure 9.15.) This is a *unidirectional cardioid* pattern. [**E9C02**] A more complete table of patterns is given in *The ARRL Antenna Book*.

These arrays are constructed by using physically identical antennas that are fed with *phasing lines* that create the necessary phase differences between them. This ensures that each element radiates a signal with the necessary phase to create the desired antenna pattern. [**E9E11**] When the antennas are identical and being fed with identical currents, such as any of the "in-phase" designs with $\phi = 0°$, a *Wilkinson power divider* can be used to split the power from the transmitter into equal portions while preventing changes in the loads from affecting power flow to the other loads. [**E9E08**]

ANTENNAS FOR SPACE COMMUNICATIONS

E9D01 — How much does the gain of an ideal parabolic dish antenna change when the operating frequency is doubled?

E9D02 — How can linearly polarized Yagi antennas be used to produce circular polarization?

Previous chapters discuss amateur radio satellites and their operation. In this section, we consider the gain and polarization requirements for antennas used in space radio communications.

Gain and Antenna Size

At VHF and UHF, Yagi-style antennas are the most common for satellite and EME contacts. For a properly designed Yagi, the longer the boom, the greater the gain. At microwave frequencies, though, a dish antenna may be required because of the higher path loss at these frequencies and the difficulty of constructing Yagis with enough gain.

Parabolic dish antennas are so named because the cross section of the dish is a parabola with the feed point at its focus. While these antennas can be quite large at frequencies of more than a few GHz, at 10 GHz and up, they are quite reasonably sized. For example, at 10 GHz a typical dish antenna need only be a bit more than a foot wide to develop substantial amounts of gain. The gain of a parabolic antenna is directly proportional to the square of the dish diameter and directly proportional to the square of the frequency. That means the gain will increase by 6 dB if either the dish diameter or the operating frequency is doubled. [**E9D01**]

Effects of Polarization

Best results in space radio communication are obtained not by using horizontal or vertical polarization, but by using a combination of the two called *circular polarization*. When two equal waves, one horizontally polarized and one vertically polarized, are combined with a phase difference of 90°, the result is a circularly polarized wave.

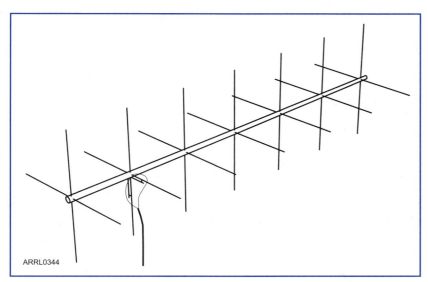

Figure 9.19 — Two Yagi antennas built on the same boom, with elements placed perpendicular to each other create a circularly polarized antenna. The driven elements are located at the same position along the boom and are fed 90° out of phase.

A circularly polarized antenna can be constructed from two dipoles or Yagis mounted at 90° with respect to each other and fed 90° out of phase. [**E9D02**] **Figure 9.19** shows an example of a circularly polarized antenna made from two Yagi antennas. The two driven elements must be at the same position along the boom for this antenna. The driven elements are in the same plane, which is perpendicular to the boom and to the direction of maximum signal.

RECEIVING LOOP ANTENNAS FOR DIRECTION FINDING

E9H03 — What is Receiving Directivity Factor (RDF)?
E9H04 — What is an advantage of placing a grounded electrostatic shield around a small loop direction-finding antenna?
E9H05 — What is the main drawback of a small wire-loop antenna for direction finding?
E9H06 — What is the triangulation method of direction finding?
E9H07 — Why is RF attenuation used when direction-finding?
E9H08 — What is the function of a sense antenna?
E9H09 — What is a Pennant antenna?
E9H10 — How can the output voltage of a multiple-turn receiving loop antenna be increased?
E9H11 — What feature of a cardioid pattern antenna makes it useful for direction finding?

A simple receiving antenna at MF and HF is a small loop antenna consisting of one or more turns of wire wound in the shape of a large open inductor or coil. The loop is usually tuned to resonance with a capacitor. The loop must be small compared to the wavelength — in a single-turn loop, the conductor should be less than 0.08 wavelength long.

Loops used for receiving and *radio direction finding* (*RDF*) are used for their noise rejection and nulls rather than their gain, which is often quite low. The nulls can be used to reject

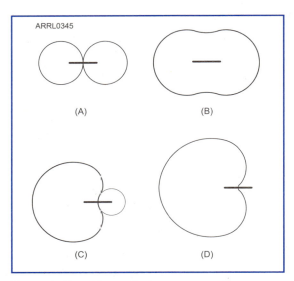

Figure 9.20 — The radiation patterns for a small loop antenna. The heavy line shows the plane of the loop. (A) shows the pattern for an ideal loop and (B) modifies the pattern showing the pattern with appreciable antenna effect present. (C) shows the pattern of a detuned loop with shifted phasing. In (D) the detuning is optimum to produce a single null.

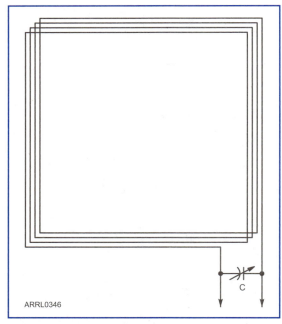

Figure 9.21 — A small loop consisting of several turns somewhat improved sensitivity. The total conductor length must still be less than 0.08 λ to have the same pattern as the single-turn loop.

a local source of noise or interference or they can be used to indicate the direction to a signal source.

An ideal loop antenna has maximum response in the plane of the loop, as **Figure 9.20** shows. The ideal loop has deep nulls at right angles to its plane. Because there are two nulls, the pattern is bidirectional. Such an antenna is a simple one to construct, but the bidirectional pattern is a major drawback for direction finding. You can't tell which of the two directions points to the signal source! Thus, a single null reading with a small loop antenna will not indicate the exact direction toward the transmitter — only the line along which it lies. [**E9H05**]

If the loop is not balanced with respect to ground (meaning that voltages from vertically polarized electric fields will not cancel), it will exhibit two modes of operation. One is the mode of the true loop while the other is that of an omnidirectional, small vertical antenna. This second mode of operation is sometimes called *antenna effect*. An electrostatic balance can be obtained by shielding the loop. This reduces unbalanced capacitive coupling to the loop's surroundings, eliminating the antenna effect so that the response is quite close to the ideal pattern of Figure 9.20A, giving deeper, sharper nulls. [**E9H04**]

For direction finding, it is better for the receiving antenna pattern to have just one null, so there is no question about where the transmitter's true direction lies. A loop may be made to have a single null if a second antenna element, called a *sense antenna*, is added. [**E9H08**] The sense antenna must be omnidirectional, such as a short vertical. If the signals from the loop and the sense antenna are combined with a 90° phase shift between the two, a cardioid radiation pattern results, similar to that of Figure 9.20D. This pattern has a single large lobe in one direction, with a deep, narrow null in the opposite direction. The deep null can help pinpoint the direction of the desired signal.

The loop and sensing-element patterns combine to form the cardioid pattern which has a very sharp single null. [**E9H11**] For the best null in the composite pattern, the signals from the loop and the sensing antenna must be of equal amplitude so that they can cancel completely in the direction of the null. The null of the cardioid will be 90° away from the nulls of the loop, so it is customary to first use the loop alone to obtain a precise bearing line, then switch in the sensing antenna to resolve the ambiguity.

For general receiving on the lower-frequency bands, small, single-turn loops are generally not sensitive enough to be effective. Therefore, multi-turn loops, such as shown in **Figure 9.21**, are often used. This loop may also be shielded and if the total conductor length remains below 0.08 wavelength, the pattern is that of the single-turn loop.

The voltage generated by the loop is proportional to the strength of the magnetic component of the radio wave passing through the coil and to the number of turns in the coil. The

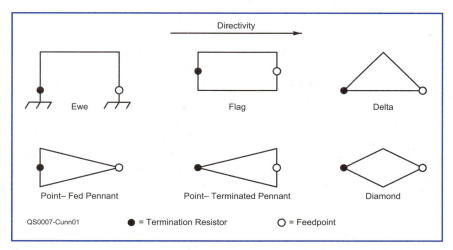

Figure 9.22 — Configurations of single-turn low-frequency receiving antennas showing termination and directivity. Termination resistance is typically around 900 Ω.

action is much the same as in the secondary winding of a transformer. The output voltage of the loop can be increased by increasing the number of turns in the loop or the loop area. [**E9H10**]

Larger single-turn receiving loops known as *pennants* and *flags* have become popular on the lower-frequency bands. **Figure 9.22** shows several configurations of these antennas. All are less than 0.1 wavelength in circumference and are typically installed vertically, sometimes even rotatable, and terminated with around 900 Ω. [**E9H09**] They have a cardioid-type pattern with a modest null to reject noise.

All of these receiving antennas, including Beverage antennas discussed previously, are used because of their ability to reject unwanted noise (or signals) away from a preferred direction. The result is an improved signal-to-noise ratio, even if the amount of signal is lower. This ability is measured by the figure of merit called *receiving directivity factor* (also called relative directivity factor). It is equal to the antenna's gain in the forward direction divided by the gain in all other directions. [**E9H03**] This factor is calculated by antenna modeling software and can be used to compare and optimize antenna designs. (RDF, the factor, should not be confused with RDF, the activity.)

The equipment required for a radio direction finding system is a directive antenna and a device for detecting the radio signal. The signal detector is usually a receiver with a meter to indicate signal strength. Some form of RF attenuation is desirable to allow proper operation of the receiver under high signal conditions, such as when zeroing-in on the transmitter at close range. Otherwise the strong signals may overload the receiver and reduce the pattern nulls. [**E9H07**]

If two or more RDF bearing measurements are made at several locations separated by a significant distance, the bearing lines can be drawn from those positions as represented on a map as in **Figure 9.23**. This technique is

Figure 9.23 — Bearings from three RDF positions are drawn on a map to perform triangulation of a signal source. This technique allows antennas with bidirectional patterns to be used since the lines on multiple bearings will only intersect at the signal source.

Antennas and Feed Lines 9-21

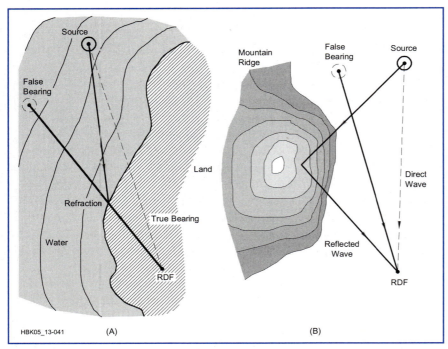

Figure 9.24 — This drawing shows RDF errors caused by refraction (A) and reflection (B).

called *triangulation*. [**E9H06**] It is important that the two DF sites not be on the same straight line with the signal you are trying to find. The point where the lines cross (assuming the bearings are not the same or 180° apart) will indicate a "fix" of the approximate transmitter location.

The effects of refraction and reflection from terrain are shown in **Figure 9.23**. At A, the signal is actually arriving from a direction different from the true direction of the transmitter. This happens because the wave is refracted at the shoreline. Even the most sophisticated equipment will not indicate the true bearing in this instance, as the equipment can only show the direction from which the signal is arriving.

9.3 Antenna Systems

An antenna system is much more than the pieces of metal dangling from a tower or tied off to the local trees. It consists of the antenna, the supports, the connection to the feed line, the feed line itself, plus any metering and impedance-matching devices. Gain and radiated power calculations must include the feed line and antenna gain, as well. System topics like these are the subjects of this section.

EFFECTIVE RADIATED POWER

E9A02 — What is the effective radiated power relative to a dipole of a repeater station with 150 watts transmitter power output, 2 dB feed line loss, 2.2 dB duplexer loss, and 7 dBd antenna gain?

E9A06 — What is the effective radiated power relative to a dipole of a repeater station with 200 watts transmitter power output, 4 dB feed line loss, 3.2 dB duplexer loss, 0.8 dB circulator loss, and 10 dBd antenna gain?

E9A07 — What is the effective isotropic radiated power of a repeater station with 200 watts transmitter power output, 2 dB feed line loss, 2.8 dB duplexer loss, 1.2 dB circulator loss, and 7 dBi antenna gain?

E9A13 — What term describes station output, taking into account all gains and losses?

When evaluating total station performance, accounting for the effects of the entire system is important, including antenna gain. This allows you to evaluate the effects of changes to the station. Transmitting performance is usually computed as *effective radiated power* (*ERP*). ERP is calculated with respect to a reference antenna system — usually a dipole but occasionally an isotropic antenna — and answers the question, "How much power does my station radiate as compared to that if my antenna was a simple dipole?" *Effective isotropic radiated power (EIRP)* results when an isotropic antenna is used as the reference. If no antenna reference is specified, assume a dipole reference antenna.

ERP is especially useful in designing and coordinating repeater systems. The effective power radiated from the antenna helps establish the coverage area of the repeater. In addition, the height of the repeater antenna as compared to buildings and mountains in the surrounding area (*height above average terrain*, or *HAAT*) has a large effect on the repeater coverage. In general, for a given coverage area, with a greater antenna HAAT, less effective radiated power (ERP) is needed. A frequency coordinator may even specify a maximum ERP for a repeater, to help reduce interference between stations using the same frequencies.

ERP calculations begin with the *transmitter power output (TPO)*. (This is assumed to be the output of the final power amplification stage if an external power amplifier is used.) Then the *system gain* of the entire antenna system including the transmission line and all transmission line components is applied to TPO to compute the entire station's output power. **[E9A13]**

There is always some power lost in the feed line and often there are other devices inserted in the line, such as a filter or an impedance-matching network. In the case of a repeater system, there is usually a duplexer so the transmitter and receiver can use the same antenna and perhaps a circulator to reduce the possibility of intermodulation interference. These devices also introduce some loss to the system. The antenna system then usually returns some gain to the system.

ERP = TPO × System Gain (Equation 9.4A)

Since the system gains and losses are usually expressed in decibels, they can simply be added together, with losses written as negative values. System gain must then be converted back to a linear value from dB to calculate ERP. Remember that $\log^{-1}$ is usually called the anti- or inverse-log on calculators.

$$ERP = TPO \times \log^{-1}\left(\frac{\text{System Gain}}{10}\right)$$ (Equation 9.4B)

It is also common to work entirely in dBm and dB until the final result for ERP is obtained and then converted back to watts.

ERP (in dBm) = TPO (in dBm) + System Gain (in dB) (Equation 9.4C)

Suppose we have a repeater station that uses a 50 W transmitter and a feed line with 4 dB of loss. There is a duplexer in the line with 2 dB of loss and a circulator that adds another 1 dB of loss. This repeater uses an antenna that has a gain of 6 dBd. Our total system gain is then:

System gain = –4 dB + –2 dB + –1 dB + 6 dBd = –1 dB

Note that this is a loss of 1 dB total for the system from TPO to radiated power. The effect on the 50 W of TPO results in:

$$ERP = 50 \text{ W} \times \log^{-1}\left(\frac{\text{system gain}}{10}\right) = 50 \times \log^{-1}(-0.1) = 50 \times 0.79 = 39.7 \text{ W}$$

This is consistent with the expectation that with a 1 dB system loss we would have somewhat less ERP than transmitter output power.

As another example, suppose we have a transmitter that feeds a 100 W output signal into a feed line that has 1 dB of loss. The feed line connects to an antenna that has a gain of 6 dBd.

Antennas and Feed Lines **9-23**

What is the effective radiated power from the antenna? To calculate the total system gain (or loss) we add the decibel values given:

System gain = – 1 dB + 6 dBd = 5 dB

and

$$ERP = 100 \text{ W} \times \log^{-1}\left(\frac{\text{system gain}}{10}\right) = 100 \times \log^{-1}(0.5) = 100 \times 3.16 = 316 \text{ W}$$

The total system has positive gain, so we should have expected a larger value for ERP than TPO. Keep in mind that the gain antenna concentrates more of the signal in a desired direction, with less signal in undesired directions. So the antenna doesn't really increase the total available power. If directional antennas are used, ERP will change with direction.

Example 9.4

What is the effective radiated power of a repeater station with 150 W transmitter power output, 2 dB feed line loss, 2.2 dB duplexer loss and 7 dBd antenna gain? [**E9A02**]

System gain = –2 dB – 2.2 dB + 7 dBd = 2.8 dB

$$ERP = 150 \text{ W} \times \log^{-1}\left(\frac{\text{system gain}}{10}\right) = 150 \times \log^{-1}(0.28) = 150 \times 1.9 = 285 \text{ W}$$

Example 9.5

What is the effective radiated power of a repeater station with 200 W transmitter power output, 4 dB feed line loss, 3.2 dB duplexer loss, 0.8 dB circulator loss and 10 dBd antenna gain? [**E9A06**]

System gain = –4 – 3.2 – 0.8 + 10 = 2 dB

$$ERP = 200 \text{ W} \times \log^{-1}\left(\frac{\text{system gain}}{10}\right) = 200 \times \log^{-1}(0.2) = 200 \times 1.58 = 317 \text{ W}$$

Example 9.6

What is the effective isotropic radiated power of a repeater station with 200 W transmitter power output, 2 dB feed line loss, 2.8 dB duplexer loss, 1.2 dB circulator loss and 7 dBi antenna gain? [**E9A07**]

System gain = –2 – 2.8 – 1.2 + 7 = 1 dB

$$ERP = 200 \text{ W} \times \log^{-1}\left(\frac{\text{system gain}}{10}\right) = 200 \times \log^{-1}(0.1) = 200 \times 1.26 = 252 \text{ W}$$

IMPEDANCE MATCHING

E9E01 — What system matches a higher-impedance transmission line to a lower-impedance antenna by connecting the line to the driven element in two places spaced a fraction of a wavelength each side of element center?

E9E02 — What is the name of an antenna matching system that matches an unbalanced feed line to an antenna by feeding the driven element both at the center of the element and at a fraction of a wavelength to one side of center?

E9E03 — What is the name of the matching system that uses a section of transmission line connected in parallel with the feed line at or near the feed point?

E9E04 — What is the purpose of the series capacitor in a gamma-type antenna matching network?

E9E05 — How must an antenna's driven element be tuned to use a hairpin matching system?

E9E09 — Which of the following is used to shunt-feed a grounded tower at its base?

E9G05 — Which of the following is a common use for a Smith chart?

Through the hard work of antenna designers over the years, there are many techniques for matching transmission lines to antennas. This section covers the basic principles of the delta, gamma, hairpin, and stub-matching systems. The *ARRL Antenna Book* contains additional information about these and other impedance-matching techniques.

Matching the antenna feed point and feed line impedances at the antenna eliminates the need for an impedance matching unit at the transmitter. Impedance matching at the antenna is often less expensive than a piece of equipment. In a station with multiple antennas, matching their impedances to that of the feed line eliminates the need to retune an impedance matching unit when changing antennas.

The Delta Match

If you try to match a half-wave dipole to open-wire feed line, you will face a problem. The center impedance of the dipole is too low to be matched directly by any practical type of air-insulated parallel-conductor line. The delta match gives us a way to match a high-impedance transmission line to a lower impedance antenna. The line connects to the driven element in two places, spaced a fraction of a wavelength on each side of the element center. [**E9E01**] The connection points are separated until the impedance between them matches that of the feed line. The antenna is not broken in the center, so there is no center insulator. This principle is illustrated in **Figure 9.25**. The fanned-out section of feed line is triangular, similar to the Greek letter Δ (delta) that gives the technique its name.

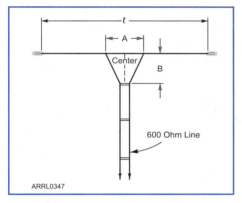

Figure 9.25 — The delta matching system is used to match a high-impedance transmission line to a lower-impedance antenna. The feed line attaches to the driven element in two places, spaced a fraction of a wavelength to each side of the element's center point.

The Gamma Match

A commonly used method for matching a coaxial feed line to the driven element of a parasitic array is the *gamma match*. Shown in **Figure 9.26**, and named for the Greek letter Γ, the gamma match has considerable flexibility in impedance matching ratio. Because this match is inherently unbalanced, a balun is not required to use coaxial cable feed line. The feed line attaches at the center of the driven element and at a fraction of a wavelength to one side of center. [**E9E02**]

Electrically speaking, the gamma conductor and the associated antenna conductor can be considered as a section of transmission line shorted at the end. Since it is shorter than ¼ wavelength the gamma matching section has inductive reactance. This means that if the antenna itself is exactly resonant at the operating frequency, the input impedance of the gamma section will show inductive reactance as well as resistance. The reactance must be tuned out to present a good match to the transmission line. This can be done in two ways. The antenna can be shortened so that its impedance contains capacitive reactance to cancel the inductive reactance of the gamma section, or a capacitance of the proper value can be

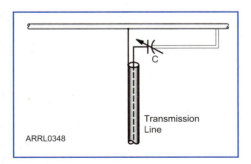

Figure 9.26 — The gamma matching system is used to match an unbalanced feed line to an antenna. The feed line attaches to the center of the driven element and to a point that is a fraction of a wavelength to one side of center.

Antennas and Feed Lines 9-25

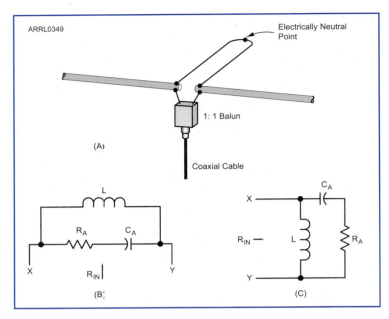

Figure 9.27 — The driven element of a Yagi antenna can be fed with a hairpin matching system, as shown in Part A. B shows the lumped-constant equivalent circuit, where R_A and C_A represent the antenna feed point impedance, and L represents the parallel inductance of the hairpin. Points X and Y represent the feed line connection. When the equivalent circuit is redrawn as shown in (C), L and C_A are seen to form an L-network to match the feed line characteristic impedance to the antenna resistance, R_A.

inserted in series at the input terminals as shown in Figure 9.26. [**E9E04**]

Gamma matches have been widely used for matching coaxial cable to parasitic beams for a number of years. Because this technique is well suited to all-metal construction, the gamma match has become quite popular for amateur antennas. Gamma matches can also be used at the base of a grounded tower to be used as a vertical antenna. [**E9E09**] Called *shunt feeding*, the gamma matched driven element is turned on its side with the missing half supplied by the electrical image created by the ground system.

Because of the system's many variables — driven-element length, gamma rod length, rod diameter, spacing between rod and driven element, and value of series capacitance — a number of combinations will provide the desired match. A more detailed discussion of the gamma match as well as design software can be found in *The ARRL Antenna Book*.

The Hairpin Match

The *hairpin matching system* is a popular method of matching a feed line to a Yagi antenna. **Figure 9.27** illustrates this technique. The hairpin match is also referred to as a *beta match*. To use a hairpin match, the driven element must be split in the middle and insulated from its supporting structure. The driven element is tuned so it has a capacitive reactance at the desired operating frequency. [**E9E05**] (This means the element is a little too short for resonance.) The hairpin adds some inductive reactance, transforming the feed point impedance to that of the feed line.

Figure 9.27B shows the equivalent lumped-constant network for a typical hairpin matching system for a 3-element Yagi. R_A and C_A represent the antenna feed point impedance. When the network is redrawn, as shown in Figure 9.27C, you can see that the circuit is the equivalent of an L network with the hairpin supplying the inductance, L. The center point of the hairpin is electrically neutral and is often attached to an antenna's metal boom for mechanical stability.

The Stub Match

In some cases, it is possible to match a transmission line and antenna by connecting an appropriate reactance in parallel with them at the antenna feed point. Reactances formed from sections of transmission line are called *matching stubs*. Those stubs are designated either as open or closed, depending on whether the free end is an open or short circuit. Using a stub in this way is called a *stub match*. [**E9E03**]

An impedance match can also be obtained by connecting the feed line at an appropriate point along the matching stub, as shown in **Figure 9.28**. The *universal stub* system illustrated here is used mostly at VHF and higher frequencies where the lengths of transmission lines are more manageable.

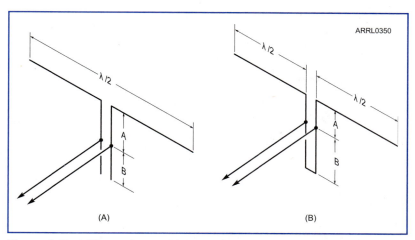

Figure 9.28 — The stub matching system uses a short perpendicular section of transmission line connected to the feed line near the antenna. Dimensions A and B are adjusted to provide a match to the transmitter and feed line.

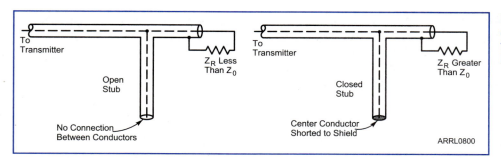

Figure 9.29 — Open and closed stubs can be used for matching to coaxial lines, as well.

A stub match can also be made with coaxial cable, as illustrated in **Figure 9.29**. By varying the length and termination for the stub, it can present a wide variety of impedances at the end connected to the main feed line. By choosing the right position in the main feed line, the impedance of the added stub can transform the impedance to a desired value. The challenge is to find the right stub impedance and the right point in the main feed line to attach it. The Smith chart described in the next section is a tool you can use to determine the length and position of the required impedance matching stub. [E9G05] There are a number of software utilities and online services that help design stub matching systems.

9.4 Transmission Lines

As you've already noticed from operating radio equipment, transmission lines or feed lines seem to make up a significant portion of a radio station! And while they are the "silent partners" in the antenna system, not turning like antennas or pumping out power like a transmitter, every signal you send and receive goes through them. It pays to understand the basics of how they work!

VELOCITY FACTOR AND ELECTRICAL LENGTH

E9F01 — What is the velocity factor of a transmission line?

E9F02 — Which of the following has the biggest effect on the velocity factor of a transmission line?

E9F03 — Why is the physical length of a coaxial cable transmission line shorter than its electrical length?

E9F05 — What is the approximate physical length of a solid polyethylene dielectric coaxial transmission line that is electrically ¼ wavelength long at 14.1 MHz?

E9F06 — What is the approximate physical length of an air-insulated, parallel conductor transmission line that is electrically ½ wavelength long at 14.10 MHz?

E9F09 — What is the approximate physical length of a foam polyethylene dielectric coaxial transmission line that is electrically ¼ wavelength long at 7.2 MHz?

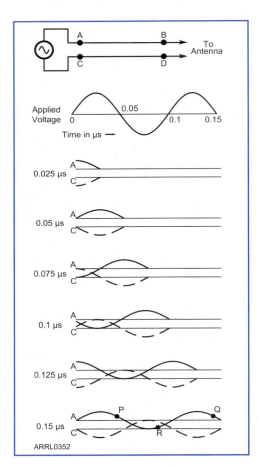

An ac voltage applied to a feed line gives rise to the sort of current shown in **Figure 9.30**. If the frequency of the ac voltage is 10 MHz, each cycle will take 0.1 microsecond. Therefore, a complete current cycle will be present along each 30 meters of line (assuming free-space velocity). This distance is one wavelength.

Current observed at B occurs just one cycle later in time than the current at A. To put it another way, the current initiated at A does not appear at B, one wavelength away, until the applied voltage has had time to go through a complete cycle.

In Figure 9.30, the series of drawings shows how the instantaneous current might appear if we could take snapshots of it at quarter-cycle intervals in time. The current travels out from the input end of the line in waves. At any selected point on the line, the current goes through its complete range of ac values in the time of one cycle just as it does at the input end.

In the previous example, we assumed that energy traveled along the line at the velocity of light. The actual velocity is very close to that of light if the insulation between the conductors of the line is solely air. The presence of dielectric materials other than air reduces the velocity since electromagnetic waves travel more slowly in materials other than in a vacuum. [**E9F02**] Because of this, the length of line that makes one wavelength will depend on the velocity of the wave as it moves along the line. The ratio of the actual velocity at which a signal travels along a line to the speed of light in a vacuum is called the *velocity factor*. [**E9F01**]

$$VF = \frac{\text{speed of wave in line}}{\text{speed of light in vacuum}}$$ (Equation 9.5)

where VF is the velocity factor.

The velocity factor is also related to the dielectric constant, ε, by:

$$VF = \frac{1}{\sqrt{\varepsilon}}$$ (Equation 9.6)

For example, several popular types of coaxial cable have a solid polyethylene dielectric, which has a dielectric constant of 2.3. For those types of coaxial cable, we can use Equation 9.6 to calculate the velocity factor of the line.

$$VF = \frac{1}{\sqrt{\varepsilon}} = \frac{1}{\sqrt{2.3}} = \frac{1}{1.5} = 0.66$$

Because of the slower velocity of propagation, the *electrical length* of a transmission line (or antenna) is not the same as its physical length. The electrical length is measured in wavelengths at a given frequency. Waves move slower in the line than in air, so the physical length of line will always be shorter than its electrical length. [**E9F03**] To calculate the physical length of a transmission line that is electrically one wavelength, use the formulas:

$$\text{Length (meters)} = VF \times \frac{300}{f \text{ (in MHz)}}$$ (Equation 9.7)

or

$$\text{Length (feet)} = VF \times \frac{984}{f \text{ (in MHz)}}$$ (Equation 9.8)

where:
   f = operating frequency (in MHz).
   VF = velocity factor.

Suppose you want a section of RG-8 coaxial cable that is ¼ wavelength long at 14.1 MHz. What is its physical length? The answer depends on the dielectric used in the coaxial cable. RG-8 is manufactured with polyethylene or foamed polyethylene dielectric; velocity factors for the two versions are 0.66 and 0.80, respectively. We'll use the polyethylene line with a velocity factor of 0.66 for our example. The physical length in meters of 1 wavelength of feed line is given by Equation 9.7:

$$\text{Length (meters)} = 0.66 \times \frac{300}{14.1} = 0.66 \times 21.3 = 14.1 \text{ m}$$

To find the physical length for a ¼-wavelength section of line, we must divide this value by 4. A ¼-wavelength section of this coax is 3.52 meters. [**E9F05**]

Example 9.7

What would be the physical length of a typical foam-dielectric coaxial transmission line that is electrically ¼ wavelength long at 7.2 MHz? (Assume a velocity factor of 0.80.)

$$\text{Length (meters)} = 0.80 \times \frac{300}{7.2} = 0.80 \times 41.7 = 34.2 \text{ m}$$

To find the length of the ¼ -wavelength line, divide by 4 = 8.3 meters. [**E9F09**]

Table 9.1 lists velocity factors and other characteristics for some other common feed lines. You can calculate the physical length of a section of any type of feed line, including twin lead and ladder line, at some specific frequency as long as you know the velocity factor.

You may have noted in the table that some of the parallel conductor lines have velocity factors closer to 1 than those of coaxial cables. The air insulation used in the open-wire and ladder lines has a value for ε much closer to vacuum. This means that the electrical and

Antennas and Feed Lines   9-29

Table 9.1

Characteristics of Commonly Used Transmission Lines

Type of Line	Z_0 Ω	VF %	Cap. pF/ft	Diel. Type*	OD inches	Max V (RMS)	Loss (dB/100 ft) at 100 MHz
RG-8	50	82	24.8	FPE	0.405	600	1.5
RG-8	52	66	29.5	PE	0.405	3700	1.9
RG-8X	50	82	24.8	FPE	0.242	600	3.2
RG-58	52	66	28.5	PE	0.195	1400	4.3
RG-58A	53	73	26.5	FPE	0.195	300	4.5
RG-174	50	66	30.8	PE	0.110	1100	8.6
CATV Hardline (Aluminum Jacket)							
½ inch	75	81	16.7	FPE	0.500	2500	0.8
⅞ inch	75	81	16.7	FPE	0.875	4000	0.6
Parallel Lines							
Twin lead	300	80	4.4	PE	0.400	8000	1.1
Ladder line	450	91	2.5	PE	1.000	10000	0.3
Open-wire line	600	95-99	1.7	none	varies	12000	0.2

*Dielectric type: PE = solid polyethylene; FPE = foamed polyethylene
Excerpted from *The ARRL Antenna Book* (23rd edition), Chapter 23, Table 1

physical lengths are more nearly equal in these feed lines.

Example 9.8

What is the physical length of a parallel conductor feed line that is electrically ½ wavelength long at 14.1 MHz? (Assume a velocity factor of 0.95.)

$$\text{Length (meters)} = \text{VF} \times \frac{300}{\text{f (in MHz)}} = 0.95 \times \frac{300}{14.1} = 0.95 \times 21.3 = 20.2 \text{ meters}$$

To find the length of the ½-wavelength line, divide by 2, so the length is 10 meters. [**E9F06**]

FEED LINE LOSS

E9F07 — How does ladder line compare to small-diameter coaxial cable such as RG-58 at 50 MHz?

E9F08 — Which of the following is a significant difference between foam dielectric coaxial cable and solid dielectric cable, assuming all other parameters are the same?

When selecting a feed line, you must consider some conflicting factors and make a few trade-offs. For example, most amateurs want to use a relatively inexpensive feed line. We also want a feed line that does not lose an appreciable amount of signal energy. For many applications, coaxial cable seems to be a good choice but parallel-conductor feed lines generally have lower loss and may provide some other advantages such as being able to operate with high SWR on the line.

Matched line loss (the loss without taking into account any additional loss resulting from SWR) increases as the operating frequency increases, so on the lower-frequency HF bands you may decide to use a less-expensive coaxial cable with a higher loss than you would on 10 meters. Open-wire or ladder-line feed lines generally have lower loss than coaxial cables at any frequency. [**E9F07**] Table 9.1 includes approximate loss values for 100 feet of the various feed lines at 100 MHz. Again, these values vary significantly as the frequency changes. Note that there will be minor variations in specifications for similar cable types from different manufacturers.

While you're looking at Table 9.1 note the difference in *maximum rated voltage* between the coaxial cables with PE (solid polyethylene) and FPE (foamed polyethylene) dielectrics. The addition of air to the dielectric reduces loss and increases velocity factor but the tradeoff is a considerably lower ability to handle high voltages. [**E9F08**]

High SWR results in higher signal losses in the feed line as the power is reflected back and forth between the transmitter and the antenna. Eventually, the power will be transferred to the antenna, but each trip along the feed line results in some additional loss. The higher the matched line loss, the greater the additional loss will be due to impedance mismatches at the antenna. Remember that an impedance matching unit at the transmitter does not reduce the SWR in the feed line to the antenna!

REFLECTION COEFFICIENT AND SWR

E9E07 — What parameter describes the interactions at the load end of a mismatched transmission line?

The *voltage reflection coefficient* is the ratio of the reflected voltage at some point on a feed line to the incident voltage at the same point. It is also equal to the ratio of reflected current to incident current at the same point on the line. The reflection coefficient is determined by the relationship between the feed line characteristic impedance, Z_0, and the actual load impedance, Z_L. The reflection coefficient is a good parameter to describe the interactions at the load end of a mismatched transmission line. [**E9E07**]

The reflection coefficient is a complex quantity, having both a magnitude and phase. It is generally designated by the lower-case Greek letter ρ (rho), although some professional literature uses the capital Greek letter Γ (gamma). The formula for reflection coefficient is:

$$\rho = \frac{Z_L - Z_0}{Z_L + Z_0} \qquad \text{(Equation 9.9)}$$

where:
   Z_0 is the line's characteristic impedance
   Z_L is the impedance of the load.

Evaluate this equation when $Z_L = Z_0$ or $0 \ \Omega$ (shorted) or $\infty \ \Omega$ (open). The only situation where $\rho = 0$, meaning no power is reflected and is all delivered to the load, occurs for $Z_L = Z_0$.

SWR is related to the magnitude of the reflection coefficient by:

$$\text{SWR} = \frac{1 + |\rho|}{1 - |\rho|} \qquad \text{(Equation 9.10)}$$

and conversely the reflection coefficient magnitude may be defined from a measurement of SWR as

$$|\rho| = \frac{\text{SWR} - 1}{\text{SWR} + 1} \qquad \text{(Equation 9.11)}$$

When both the line and load impedances are purely resistive, SWR can be computed directly from the impedances of the line and load:

$$\text{For } Z_L > Z_0, \text{SWR} = \frac{Z_L}{Z_0} \text{ and for } Z_L < Z_0, \text{SWR} = \frac{Z_0}{Z_L} \qquad \text{(Equation 9.12)}$$

Antennas and Feed Lines

POWER MEASUREMENT

E4B06 — How much power is being absorbed by the load when a directional power meter connected between a transmitter and a terminating load reads 100 watts forward power and 25 watts reflected power?

E4B09 — What is indicated if the current reading on an RF ammeter placed in series with the antenna feed line of a transmitter increases as the transmitter is tuned to resonance?

You can use a variety of instruments to tell your relative power output. For example, as you tune a transmitter or antenna tuning unit, the increasing brightness of a neon bulb connected across the feed line or an increased current reading of an RF ammeter tells you that more power is going into the antenna. [**E4B09**]

It is much more convenient, however, to read the forward and reflected power directly, by using a *directional RF wattmeter* or *power meter*. This type of meter can determine total power flowing in either direction in the transmission line. This allows you to adjust an impedance-matching network by observing reflected power.

The reflection coefficient can also be computed from the forward and reflected power:

$$|\rho| = \sqrt{\frac{P_R}{P_F}}$$

(Equation 9.13)

where:

P_R = power in the reflected wave

P_F = power in the forward wave.

Whatever the reflected and forward power may be, the difference between them is the net amount of power being transferred to the load; $P_{LOAD} = P_F - P_R$. Both forward and reflected power can be measured with a directional power meter or directional wattmeter in the transmission line. Remember that the net forward power ($P_F - P_R$) is the power delivered to the load.

Example 9.9

How much power is being absorbed by the load when a directional wattmeter connected between a transmitter and a terminating load reads 100 W forward power and 25 W reflected power? [**E4B06**]

$$P_{LOAD} = P_F - P_R = 100 - 25 = 75 \text{ W}$$

SMITH CHART

E9G01 — Which of the following can be calculated using a Smith chart?

E9G02 — What type of coordinate system is used in a Smith chart?

E9G03 — Which of the following is often determined using a Smith chart?

E9G04 — What are the two families of circles and arcs that make up a Smith chart?

E9G06 — On the Smith chart shown in Figure E9-3, what is the name for the large outer circle on which the reactance arcs terminate?

E9G07 — On the Smith chart shown in Figure E9-3, what is the only straight line shown?

E9G08 — What is the process of normalization with regard to a Smith chart?

E9G09 — What third family of circles is often added to a Smith chart during the process of solving problems?

E9G10 — What do the arcs on a Smith chart represent?

E9G11 — How are the wavelength scales on a Smith chart calibrated?

Before discussing the Smith chart, let's back up a step. All impedances consist of two components: resistance and reactance. Graphically, these components are represented as a pair of axes — the rectangular coordinate system for graphing impedance. The horizontal axis represents resistance — positive to the right of the origin and negative to the left. The vertical axis represents reactance — positive (inductive) above the origin and negative (capacitive) below.

All possible impedances can be plotted as one point (Z) on that graph, corresponding to the values of resistance and reactance. Those two values are the rectangular coordinates of the impedance. When an impedance is connected to a transmission line and a signal of some frequency is applied to the other end of the line, the interaction between the energy in the line and that terminating impedance results in energy being reflected back and forth in the transmission line.

The ratios of voltage and current (that's the definition of impedance) also turn out to change with position along the line. So, if an impedance-measuring meter is inserted at different points along the line, it would observe different values of impedance because of the different values of voltage and current.

Starting at the terminating impedance itself, as it moved farther and farther away along the line, the impedance measured by the meter would change until at ½ wavelength away from the terminating impedance, it would again report the terminating impedance's actual value and the cycle would begin again.

Smith Chart Construction

Plotted on rectangular coordinates, the path of that impedance measurement as the meter's position changed would be pretty messy, described by a fairly involved mathematical equation. What Phillip Smith discovered was that if you distort the rectangular graph in a certain way (called a *mapping*), the path of the impedance point along the transmission line becomes a circle! That concept is shown graphically on a Smith chart.

Imagine the Smith chart as a fun house mirror in reverse. Instead of taking your handsome image and distorting it to look bizarre, it takes the bizarre path of the impedance on the rectangular graph and makes a lovely circle out of it! This is a lot easier to work with. For this reason, the Smith chart is used, among other things, to calculate impedances and SWR anywhere along a transmission line. [**E9G01, E9G03**]

What is this magic mapping? Imagine yourself standing at the origin of the rectangular graph with the positive resistance axis in front of you and the negative behind. The positive reactance axis starts at your feet and goes straight up and the negative straight down. All of the axes extend to infinity.

Now imagine reaching up over your head and bending the positive reactance axis down in front of you in a semicircle whose far end meets up with the far end of the positive resistance axis. Do the same for the negative reactance axis, bending it up instead. The negative resistance axis still extends behind you, as straight as ever. This process is sketched in **Figure 9.31**.

You have created a circle from the two reactance axes, bisected by the only straight line on the chart, the horizontal *resistance axis* through the center. [**E9G07**] The infinity points join together with the infinite point of the positive resistance axis at the right of the chart. All of the points that were once in the right-hand side of the rectangular graph are now somewhere inside or on the boundary of that circle. Points on the left-hand side of the rectangular graph

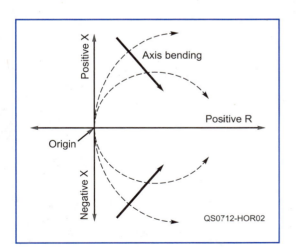

Figure 9.31 — **Distorting or mapping the rectangular graph captures all of its right-hand side impedances inside the circle formed by the bent reactance axes. This is the basis of the Smith chart.**

Antennas and Feed Lines 9-33

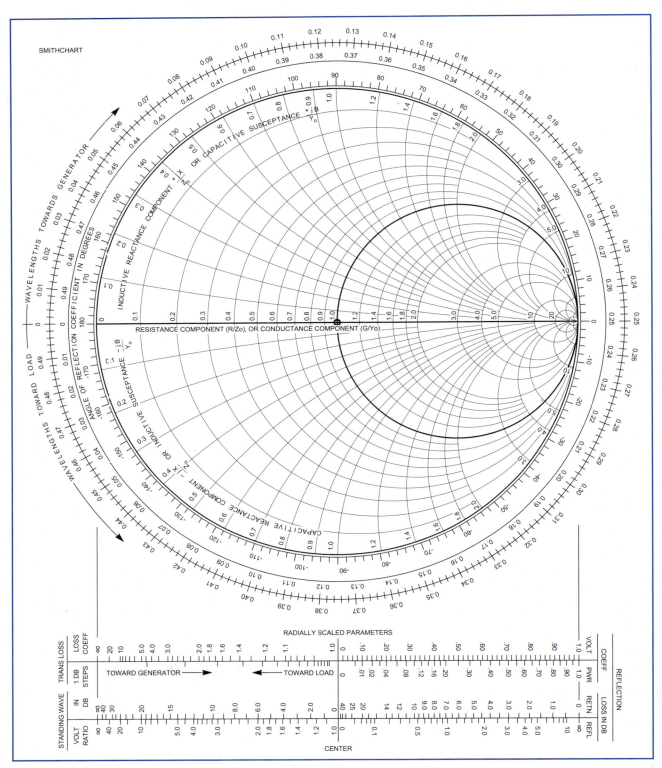

Figure 9.32 — The Smith chart represents all possible impedances in a transmission line. The outer edge of the chart represents pure inductive and capacitive reactance. The horizontal axis through the center of the chart represents pure resistance. Impedances of any value are plotted at the intersection of constant-resistance circles and constant-reactance arcs.

are now spread outside the circle. Nothing has been lost, just squashed or stretched.

The Smith chart shown in **Figure 9.32** only contains the circle and what's inside. It ignores everything outside the circle because of the negative resistance value of those points originally. They were on the left side of the graph, remember? Those impedances cannot be present in a transmission line.

The circles and arcs on the Smith chart show what happens to straight lines on the rectangular graph after remapping. Lines of constant resistance, originally vertical and on which all points had the same value of resistance, are now nested *constant-resistance circles* that come together at the far right of the Smith chart. That should make sense because all of those straight lines originally went where? To infinity — which is now the point at the right side of the Smith chart. Horizontal constant-reactance lines that represented all the points having the same reactance are now bent into *constant-reactance arcs* with one end on the outer circle (the reactance axis) and the other end at infinity! [**E9G02, E9G04, E9G06, E9G10**] This distortion results in the path of the impedance point becoming a circle on the Smith chart as we look at each point along the line.

Normalization

Take a close look at the Smith chart in Figure 9.32. If you look for the impedance point of $50 + j0 \, \Omega$, you will find it squashed way over in the nest of circles at the right-hand side of the chart — not very easy to use. Smith avoided the problem of big numbers by *normalizing* all of the coordinates to the characteristic impedance of the line, Z_0. That impedance point is the *prime center* of the Smith chart.

Normalization reassigns the values of all points according to their ratio to Z_0 at the prime center, in this case dividing them by $50 \, \Omega$. [**E9G08**] So instead of $50 \, \Omega$ being over in the hard-to-read section at the right, it's right in the middle of the chart at 1.0. Much better! From here on, all of the values you plot on the Smith chart will be the value you read on the meter divided by $50 \, \Omega$.

Constant-SWR Circles

If you take all of the normalized impedance points on the Smith chart that create a certain value of SWR in a 50-Ω transmission line, you will find that the points make a circle centered on the point $Z = 1.0 + j0$ that is at the center of the chart. This is called a *constant-SWR circle*. [**E9G09**] Lower SWR makes smaller and smaller circles until at SWR = 1.0, the circle is merely the point at the prime center of the chart, meaning that the terminating impedance is equal to $Z_0 = 50 \, \Omega$.

As SWR increases, the circles increase in size until at SWR = ∞, the circle is the outside edge of the chart. The SWR caused by any impedance can be found by measuring the distance from the center of the chart to the impedance point, then translating that distance onto the linear SWR scale at the bottom of the chart. These scales are called *radially-scaled* because they represent measurements made radially from the center of the chart.

Wavelength Scales

Look carefully at the left side of the Smith chart along the rim and you will see two arrows pointing in opposite directions, labeled "Wavelengths Toward Generator" and "Wavelengths Toward Load." The chart's outer scale, the reactance axis, is marked to show movement in wavelengths along the transmission line.

There are two scales, one starting at 0 and increasing clockwise and the other starting at 0.5 and decreasing clockwise. Both are calibrated in fractions of electrical wavelength inside the transmission line. [**E9G11**] These are used to work out problems that involve the changing impedance along a transmission line as described in the next section.

Antennas and Feed Lines

TRANSMISSION LINE STUBS AND TRANSFORMERS

E9E06 — Which of these feed line impedances would be suitable for constructing a quarter-wave Q-section for matching a 100-ohm loop to 50-ohm feed line?

E9E10 — Which of these choices is an effective way to match an antenna with a 100-ohm feed point impedance to a 50-ohm coaxial cable feed line?

E9F04 — What impedance does a ½-wavelength transmission line present to a generator when the line is shorted at the far end?

E9F10 — What impedance does a ⅛-wavelength transmission line present to a generator when the line is shorted at the far end?

E9F11 — What impedance does a ⅛-wavelength transmission line present to a generator when the line is open at the far end?

E9F12 — What impedance does a ¼-wavelength transmission line present to a generator when the line is open at the far end?

E9F13 — What impedance does a ¼-wavelength transmission line present to a generator when the line is shorted at the far end?

In a transmission line, when a wave of RF voltage and current encounters an impedance different from the characteristic impedance of the transmission line, Z_0, some of the energy in the wave is reflected back toward the wave's source. The phase of the voltage and currents making up the reflected wave will differ from those in the incoming or incident wave depending on the value of the impedance causing the reflection.

The incident and reflected voltage and current waves combine at every point along the line. At each point, the combination results in voltage and current with a phase relationship different from either the incident or reflected waves. It is as if the same energy in the line had been applied to an impedance with values of resistance and reactance that create the same phase relationship. For the Extra class exam we'll examine what impedance an impedance-meter would "see" if it is connected a transmission line with the other end shorted or open.

The first and easiest rule to remember is that if any transmission line is any integer multiple of ½ wavelength long, the impedance at one end will be the same as at the other. It doesn't matter what the terminating impedance is, how many ½ wavelengths of line are involved (neglecting line loss), or even what the characteristic impedance of the line is! Every ½ wavelength along the line, impedance repeats. If the terminating impedance is a short circuit, the impedance meter will see a short every ½ wavelength away and the same situation applies to an open circuit termination. [**E9F04**] You can see this on the Smith chart because the impedance point travels in one complete circle around the chart every ½ wavelength.

The second rule, almost as easy to remember, is that if the transmission line is an odd multiple of ¼ wavelength long, the impedance at one end is inverted from that at the other end. If the terminating impedance is an open, the impedance ¼ wavelength away will be a short, and vice versa. [**E9F12, E9F13**] This behavior repeats at ¾ wavelength, 1¼ wavelengths, 1¾ wavelengths away, and so forth.

The remaining cases for ⅛-wavelength lines are not so easy to remember but with a little study, you'll be able to figure them out. Let's start with an open-circuited transmission line that is very, very short. An impedance meter will view this short piece of transmission line as an open circuit. As the line is lengthened it will exhibit a small capacitive reactance because that's what it is — a small capacitor formed by the inner and outer conductors. When the line reaches ⅛ wavelength, the capacitive reactance will reach the value of $-jZ_0$. In other words, a ⅛-wavelength piece of open 50-Ω transmission line will have an impedance of $-j50 \, \Omega$ at the other end. [**E9F11**]

Working with the opposite case, a shorted line, the impedance of the very, very short short-circuited line looks like — this shouldn't surprise you — a short-circuit. As the line is lengthened, it exhibits inductive reactance because the loop of inner and outer conductor

9-36   Chapter 9

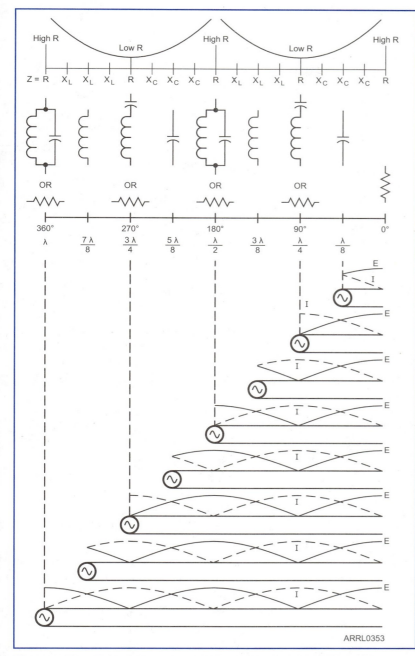

Figure 9.33 — This diagram summarizes the characteristics of open-circuited transmission lines. Voltage standing waves are shown as solid lines above each length of cable, and current standing waves are shown as dashed lines.

form an inductor. When the line reaches 1/8 wavelength long, the impedance meter will read $+jZ_0\ \Omega$ of inductive reactance. [**E9F10**]

Figures 9.33 and **9.34** illustrate the behavior of open and shorted transmission lines up to 1 λ long. You can clearly see the cyclic behavior of the impedances in the line. The impedance "seen" looking into various lengths of feed line is indicated directly above the chart. Curves above the axis marked with R, X_L and X_C indicate the relative value of the impedance presented at the input of the line. Circuit symbols indicate the equivalent circuits for the lines at that particular length. Standing waves of voltage (E) and current (I) are shown above each line. Remember that Z = E / I, by Ohm's Law, so you can use the curves above each piece of line to estimate the input impedance of a given line length.

Synchronous Transformers

There is one special technique of impedance matching using transmission lines that every Extra class ham should learn — the *synchronous transformer* shown in **Figure 9.35**. This method of matching involves setting up a series of reflections of just the right magnitude and phase so that two transmission lines or a transmission line and a load of two different impedances can be connected together without creating any standing waves in the transmission lines!

To match two different impedances such as the transmission line and load shown in the figure, Z_0 and Z_{LOAD}, a ¼-wavelength section of transmission line with a characteristic impedance, Z_1, equal to the *geometric mean* of Z_0 and Z_{LOAD}. In mathematical form:

$$Z_1 = \sqrt{Z_0\ Z_{LOAD}} \qquad \text{(Equation 9.14)}$$

The transformer is called *synchronous* because it must be a certain fraction of a wavelength long (¼ λ) to function. For example, to match a 50-Ω transmission line to a quad antenna with a feed point impedance of 100 Ω, the transmission line used for the synchronous

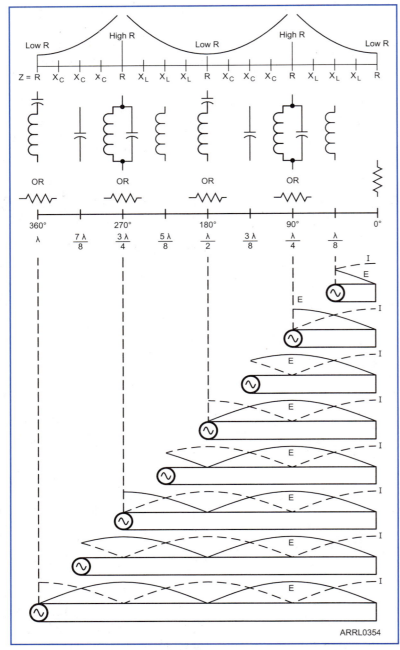

Figure 9.34 — This diagram summarizes the characteristics of short-circuited transmission lines. Voltage standing waves are shown as solid lines above each length of cable, and current standing waves are shown as dashed lines.

transformer should have a characteristic impedance of:

$$Z_1 = \sqrt{50 \times 100} = 70.7 \, \Omega$$

A section of 75-Ω RG-59/U cable will work quite well in this application. [**E9E06**, **E9E10**] The load impedance to be matched can also be the characteristic impedance of another transmission line. Other forms of synchronous transformers with different lengths can match different levels of impedances, too.

SCATTERING (S) PARAMETERS

E4B03 — Which S parameter is equivalent to forward gain?

E4B04 — Which S parameter represents input port return loss or reflection coefficient (equivalent to VSWR)?

E4B07 — What do the subscripts of S parameters represent?

Scattering parameters or S parameters are a way of characterizing a circuit, transmission line, or antenna in terms of voltage waves that are incident, reflected, and transmitted at the connections to the circuit, called *ports*. S parameters assume there are two ports as shown in **Figure 9.36**. The voltage waves are defined with the letters a1, b1, a2, and b2. The a waves are considered to be incident waves on the ports and are the independent variables. The b waves are the result of reflection or "scattering." The subscripts of the parameter show which port received the input wave and which port was the source of the reflected wave. For example, S_{12} is the parameter that shows the wave from port 1 that was caused by the incident wave at port 2. [**E4B07**]

The two most important S parameters used by amateurs are S_{11} — the ratio of the reflected wave to the incident wave at the input, port 1. This is the *input port reflection coefficient*. [**E4B04**] (See the earlier discussion on reflection coefficient.) S_{11} can be converted to return loss (RL — see below) or SWR and vice versa. Most instruments that measure antenna or transmission line impedance can provide values of S_{11}, RL, or SWR.

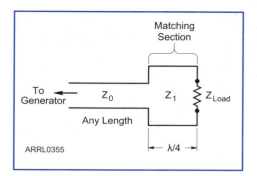

Figure 9.35 — The synchronous transformer creates two sets of reflections to power traveling in both directions in the transmission lines or between the transmission line and a load. These reflections cancel so that the net result is no increase in SWR from the impedance mismatch.

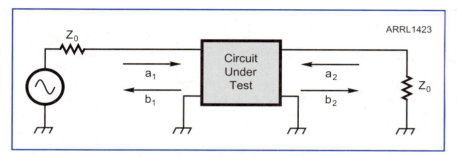

Figure 9.36 — A two-port circuit viewed as being driven by voltage waves. The relationship of the voltage waves, such as gain or SWR, is described by scattering or S parameters.

Similarly, S_{21} is the voltage wave emanating from the output, port 2, as the result of an incident wave at the network input, port 1. In other words, S_{21} represents a forward gain. **[E4B03]**

A closely related measurement provided by transmission line analyzers and used by RF professionals is *return loss (RL)*. Return loss and SWR measure the same thing — how much of the incident power, P_{INC}, in the transmission line is transferred to the load and how much is reflected by it, P_{REFL} — but state the result differently.

$$RL(dB) = -10 \log_{10} \left(\frac{P_{REFL}}{P_{INC}} \right) \qquad \text{(Equation 9.15)}$$

ANTENNA AND NETWORK ANALYZERS

E4A07 — Which of the following is an advantage of using an antenna analyzer compared to an SWR bridge to measure antenna SWR?
E4A08 — Which of the following measures SWR?
E4A11 — How should an antenna analyzer be connected when measuring antenna resonance and feed point impedance?
E4B05 — What three test loads are used to calibrate an RF vector network analyzer?
E4B11 — Which of the following can be measured with a vector network analyzer?

The *antenna analyzer* has made measuring impedance and SWR a simple task since its introduction to amateur radio in the 1990s. The analyzer consists of a tunable RF source, a frequency counter, an impedance bridge, displays and a microprocessor to run them. It is capable of measuring impedance, SWR, reactance, and frequency. **[E4A08]** **Figure 9.37** shows a typical model. Analyzers are battery powered and small enough to be taken into the field or up on an antenna tower.

The analyzer is used by connecting it directly to the impedance to be measured. **[E4A11]**

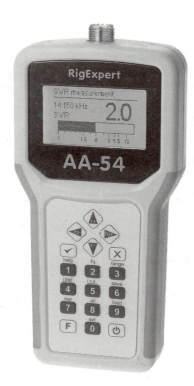

Figure 9.37 — An antenna analyzer, such as this RigExpert AA-54, consists of a tunable signal source, a frequency counter, an impedance bridge, and displays — all microprocessor controlled. The analyzer displays impedance, SWR, and frequency.

The impedance can be a component, a circuit (with power removed), a transmission line or an antenna. Since the analyzer contains its own low-power signal source, it is not necessary to use a transmitter for testing antennas (as you would when an SWR meter is used). [**E4A07**] By using your knowledge of transmission lines (and the device's user manual), an antenna analyzer can be used to measure transmission line length and characteristic impedance. If a defective line is suspected, the analyzer can be used to find the location of a short or open circuit, as well. These handy accessories have become a fixture of ham radio test instrumentation.

A related but more powerful instrument is the *vector network analyzer* or *VNA* The VNA can not only measure complex impedance, S_{11} and return loss, or SWR at a single port, but it can measure the input and output impedance of a circuit or transmission line, as well. [**E4B11**] A VNA can measure all four S parameters by using two ports while an antenna analyzer measures just S_{11}.

For either the antenna or network analyzer, it is a good idea to have calibrated loads, such as a 50 Ω dummy load, on-hand to check their performance. Network analyzers use three known loads of 0 Ω (short circuit), 50 Ω, and an open circuit to calibrate themselves. [**E4B05**]

9.5 Antenna Design

The power of personal computers makes it possible to design and analyze antennas by mathematically modeling the antenna. Computer analysis allows us to study how performance changes as the height of the antenna changes or what effects different ground conditions will have. The antenna's characteristics can be adjusted until the design meets expectations. The accuracy of the programs is excellent in the hands of a reasonably skilled modeler. This saves a lot of time over the cut-and-try method!

ANTENNA MODELING AND DESIGN

E9B09 — What type of computer program technique is commonly used for modeling antennas?
E9B10 — What is the principle of a Method of Moments analysis?
E9B11 — What is a disadvantage of decreasing the number of wire segments in an antenna model below 10 segments per half-wavelength?
E9D05 — What usually occurs if a Yagi antenna is designed solely for maximum forward gain?

There are a number of programs in common use for antenna analysis. Most of them are derived from a program developed at US government laboratories, called *NEC*, short for *Numerical Electromagnetics Code*. This complex program, originally written for mainframe computers, uses a modeling technique called the *method of moments*. [**E9B09**]

In the method of moments, the antenna wires (or tubing elements) are modeled as a series of *segments* and a uniform value of current in each segment is computed. [**E9B10**] The field resulting from the RF current in each segment is evaluated, along with the effects from other

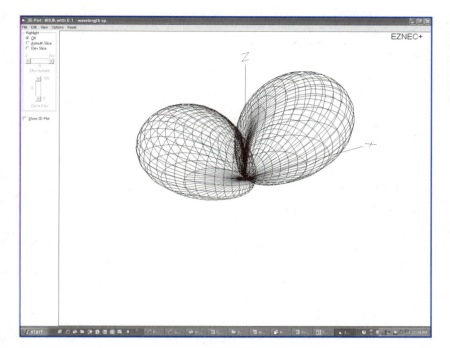

Figure 9.38 — The *EZNEC* program is one of the best known antenna modeling programs used by hams. Written by Roy Lewallen, W7EL, the program models antennas of any size and frequency.

mutually coupled segments. The higher the number of segments, generally the more accurate the modeling results will be. However, most programs have a limit on the number of segments in a particular antenna because of the amount of memory and processing time required to perform the necessary calculations. A lower number of segments will reduce the time required for model calculations but the outputs, such as pattern shape or feed point impedance, will not be as accurate. [E9B11]

The result of an antenna model can take several forms. Of primary interest to most amateurs is the radiation pattern of the antenna. These are given in the standard polar plot format for far-field elevation and azimuthal patterns. Many programs provide a three-dimensional view of the pattern, as well. The programs compute, at a minimum, antenna gain, beamwidth, all the pattern ratios, feed point impedance, and SWR versus frequency "sweep" graphs. Example outputs from the *EZNEC* program by Roy Lewallen, W7EL are shown in **Figure 9.38**.

Along with single-parameter graphs versus frequency, the software can perform advanced evaluations of the antenna pattern. For example, receiving directivity factor that was discussed in the section on receiving antennas would be very difficult to calculate by hand or from sets of radiation patterns. The ability to define and calculate new ways of evaluating performance is an important way that modeling allows amateurs to experiment and innovate in antenna design.

DESIGN TRADEOFFS AND OPTIMIZATION

Any antenna design represents some compromises. You may be able to modify the design of a particular antenna to improve some desired characteristic, if you are aware of the tradeoffs. As mentioned earlier, the method of moments computer modeling techniques that have become popular can be a great help in deciding which design modifications will produce the "best" antenna for your situation.

When you evaluate the gain of an antenna, you (or the computer modeling program) will have to take into account a number of parameters. You will have to include the antenna feed point impedance, any loss resistance in the elements and impedance-matching components, as well as the E-field and H-field radiation patterns.

You should also evaluate the antenna across the entire frequency band for which it is designed. You may discover that gain may change rapidly as you move away from its design frequency. (You may be willing to make that trade-off if all your operating on that band is within a narrow frequency range.) You may also discover that the feed point impedance

Antennas and Feed Lines 9-41

varies widely as you change frequency across the band, making it difficult to design a single impedance-matching system for the antenna. You may also discover that the front-to-back ratio varies excessively across the band, resulting in too much variation in the rearward pattern lobes.

The forward gain of a Yagi antenna can be increased by using a longer boom, spreading the elements farther apart or adding more elements. Of course, there are practical limitations on how long you can make the boom for any antenna! The element lengths will have to be adjusted to retune them as the boom length changes.

You may decide to optimize a Yagi antenna for maximum forward gain, but in that case the front-to-back ratio usually decreases, feed point impedance becomes very low, and the SWR bandwidth will decrease. [**E9D05**] Optimizing performance for one parameter often leads to a reduction in performance in other parameters. In general, the interdependency of gain, SWR bandwidth, and pattern ratios requires compromises by the antenna modeler to achieve realistic goals.

Chapter 10

Topics in Radio Propagation

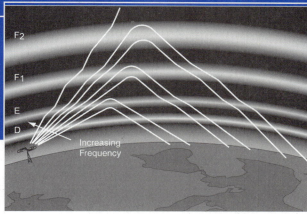

In this chapter, you'll learn about:
- Electromagnetic waves
- Solar effects on propagation
- Ground-wave and sky-wave
- Long-path and gray-line
- The radio horizon
- Tropospheric and transequatorial propagation
- Auroral, meteor scatter, and Earth-Moon-Earth (moonbounce)

What happens after the signal leaves the antenna? Everything between the transmit and receive antennas involves propagation — a great deal of radio's magic! This chapter is a collection of short discussions on various aspects of radio propagation at different frequencies. Following a short introduction to electromagnetic waves, we move on to a review of solar phenomena that affect terrestrial radio propagation. The chapter then covers topics that matter most at HF and those of interest primarily at VHF and higher frequencies. There are a few questions about all three subjects in the Extra class exam. You can check your understanding of propagation by being able to answer all of the questions listed at the end of each topic.

10.1 Electromagnetic Waves

E3A14 — What is meant by circularly polarized electromagnetic waves?

While the exam only has one question on electromagnetic or EM waves, becoming acquainted with their fundamentals will help the material on propagation make a lot more sense. EM waves (what we fondly call radio waves) are created when an electric field or magnetic field changes. These waves move or *propagate* through space carrying both electric and magnetic energy. The electric and magnetic fields in the wave are oriented at right angles to each other as shown by **Figure 10.1A** and vary with time in a sinusoidal pattern.

The direction of the right angle from the positive direction of the electric field to that of the magnetic field is determined by the direction in which the wave travels, as illustrated in the figure. The term "lines of force" in the figure means the direction in which a force would be felt by an electron (from the electric field) or by a magnet (from the magnetic field).

An important point about electromagnetic waves: The electric and magnetic fields making up the wave are not just perpendicular electric and magnetic fields that happen to be in the same place at the same time. The fields are *coupled*; they are both aspects of the same thing — the electromagnetic wave — created by the motion of electrons, such as in a transmitting antenna. The fields cannot be separated although the energy in the wave can be detected as either electric or magnetic force.

In free space, the waves move at the speed of light, approximately 300 million meters

Topics in Radio Propagation   10-1

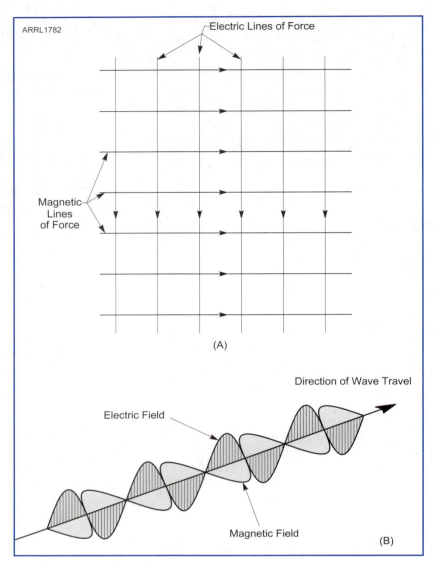

Figure 10.1 —Representation of electric and magnetic fields in an electromagnetic wavefront (A). Arrows indicate the instantaneous directions of the fields for a wavefront in a wave traveling toward you, out of the page. B shows the magnetic and electric field strengths as the wave moves through space as a series of wavefronts. In this drawing, the electric field is oriented vertically and the magnetic field horizontally.

per second (3×10^8 m/s). (Light is an electromagnetic wave of extremely high frequency — many thousands of GHz.) In general, the speed at which electromagnetic waves travel or propagate depends on the characteristics of the medium through which they travel. The speed of light is highest in the vacuum of free space and only slightly lower in air. In materials such as glass or plastic, however, velocity can be quite a bit lower. For example, in polyethylene (commonly used as a center insulator in coaxial cable), the velocity of propagation is about two-thirds (67%) of that in free space.

WAVEFRONTS

To an observer staying in one place, such as a fixed station's receiving antenna, the electric and magnetic fields of the wave appear to oscillate as the wave passes. That is, the fields create forces on electrons in the antenna that increase and decrease in a sine wave pattern.

Some of the energy in the propagating wave is transferred to the electrons as the forces from the changing fields cause them to move. This creates a sine wave current in the antenna with a frequency determined by the rate at which the field strength changes in the passing wave.

If the observer is moving along with the wave at the same speed, however, the strength of the fields will not change. To that observer, the electric and magnetic field strengths are fixed, as in a photograph. This is a *wavefront* of the electromagnetic wave — a flat surface or plane moving through space on which the electric and magnetic fields have a constant value as illustrated in Figure 10.1.

Just as an ac voltage is made up of an infinite sequence of instantaneous voltages, each slightly larger or smaller than the next, an infinite number of wavefronts make up an electromagnetic wave, one behind another like a deck of cards. The direction of the wave is the direction in which the wavefronts move. The fields on each successive wavefront have a slightly different strength, so as they pass a fixed location, the detected field strength changes, too. The result is that the fixed observer "sees" fields with strengths varying as a sine wave.

Figure 10.1B is a drawing of what would happen if we could suddenly freeze all of the wavefronts in Figure 10.1A and measure the electric and magnetic field strengths of each all along the direction the wave is traveling. In this example, the electric field is oriented vertically and the magnetic field horizontally. (Each of the vertical lines in the electric field can be thought of as representing an individual wavefront.) All of the wavefronts are moving in the direction indicated — the whole set of them moves together at the same speed. As the wavefronts move past the receive antenna, the varying field strengths are perceived as a continuously changing wave moving through space.

POLARIZATION

The orientation of the pair of fields in an electromagnetic wave can have any orientation with respect to the surface of the Earth, but the electric and magnetic fields will always be at right angles to each other. The orientation of the wave's electric field determines the polarization of the wave. If the electric field's lines of force are parallel to the surface of the Earth (meaning those of the magnetic field are perpendicular to the Earth), the wave is *horizontally polarized*. Conversely, if the electric field's lines of force are perpendicular to the surface of the Earth, the wave is *vertically polarized*. Knowing the polarization of the wave allows the receiving antenna to be oriented so that the passing wave's E-field results in the most current in the antenna, maximizing received signal strength.

For the most part, the wave's polarization is determined by the type of transmitting antenna and its orientation. For example, a Yagi antenna with its elements parallel to the Earth's surface transmits a horizontally polarized wave. On the other hand, an amateur mobile whip antenna, mounted vertically on an automobile, radiates a vertically polarized wave. If a vertically polarized antenna is used to receive a horizontally polarized radio wave (or vice versa), received signal strength can be reduced by more than 20 dB as compared to using an antenna with the same polarization as the wave. This is called *cross-polarization*. The polarization of radio waves can be altered by being refracted (bent) or reflected so antenna orientation may not have to always match for signals to be received.

It is also possible to generate electromagnetic waves in which the orientation of successive wavefronts — both the electric and magnetic fields — rotates around the direction of travel. This is called *circular polarization*. [**E3A14**] Imagine the wave of Figure 10.1B being twisted so at one point the direction of the electric field is horizontal and a bit further along the wave it is vertical. As the twisted, circularly polarized wave passes the receiving antenna, the polarization of its fields will appear to rotate. The rate at which the polarization changes and the direction or *sense* of the rotation — right-handed (clockwise or CW as the wave travels) or left-handed (counter-clockwise or CCW) — is determined by the construction of the transmitting antenna. Note that the electric and magnetic fields rotate together and the right-angle between them remains fixed. Polarization that does not rotate is called *linear polarization* or

Topics in Radio Propagation **10-3**

plane polarization. Horizontal and vertical polarization are examples of linear polarization.

To best receive a circularly polarized wave, the sense of the receiving antenna should match that of the transmitting antenna. It is particularly helpful to use circular polarization in satellite communication, where polarization tends to shift with the orientation of the satellite and the path of its signal through the atmosphere. Circular polarization is usable with linearly polarized antennas at one end of the signal's path. There will be some small loss in this case, however.

10.2 Solar Effects

The Sun is the biggest source of effects on radio wave propagation here on Earth, even affecting UHF and microwaves at times. HF propagation is dominated by daily and seasonal effects, as well as the 11-year sunspot cycle. With plenty of data about solar activity available on the internet, the radio amateur has better tools to predict and explain propagation than at any time in the past. This section identifies some of the most important measurements and how they are used. A full treatment is available in the *ARRL Handbook* and from numerous online sources.

Most of the measurements are available on websites like the NOAA "Radio Communications Dashboard" website at **swpc.noaa.gov/communities/radio-communications**. There are many such websites, such as **hfradio.org** and **spaceweather.com**. By bookmarking one of these sites for your web browser and visiting it regularly, you'll stay up to date on radio conditions and learn a lot about the Sun, as well.

FLUX AND FLARES

E3C07 — Which of the following descriptors indicates the greatest solar flare intensity?
E3C09 — How does the intensity of an X3 flare compare to that of an X2 flare?
E3C10 — What does the 304A solar parameter measure?

Whether the Sun's outer layers are active or quiet, a great deal of energy is radiated into space at various wavelengths. The range of energies most affecting amateurs is in the extreme ultraviolet (EUV) spectrum at wavelength of 10 – 120 nm (100 – 1200 angstroms). EUV light is completely absorbed in the upper atmosphere, creating the ionosphere.

A number of satellites observe the Sun at many different wavelengths. Photographs of the Sun at different wavelengths show different features that are used to evaluate solar activity. The Solar Dynamics Observatory (SDO) shows a number of these photographs as they are taken at **sdo.gsfc.nasa.gov/data**. They are labeled by the wavelength, such as the AIA 304 image which shows the Sun at a wavelength of 304 angstroms (30.4 nm). [**E3C10**]

Since absorption by the atmosphere makes it impossible to measure EUV levels from the ground, several observatories around the world measure incoming energy at a wavelength of 10.7 cm (2.8 GHz). The level of 2.8 GHz energy tracks the EUV levels well. When you see data for *solar flux* it is an averaged value from those measurements. More solar flux means higher levels of ionization in the ionosphere.

Due to various processes on the surface of the Sun, *solar flares* sometimes occur, releasing large amounts of energy from X rays through EUV and beyond. When the energy reaches Earth's atmosphere a few minutes later, it temporarily increases ionization and can disrupt the geomagnetic field. This also affects HF and low VHF propagation, so amateurs want to know when flares occur. Flares are ranked by classes of intensity: A (small), B, C, M, and X (very large). [**E3C07**] M-class flares have a moderate effect on HF propagation while X-class flares can cause radio blackouts and disturbances that last for days. Within each class, increasing numeric values such as X1, X2, X3, and so on correspond to increasing intensity. For example, an X3 flare is 50 percent more intense than an X2 flare. [**E3C09**]

GEOMAGNETIC FIELD

E3C02 — What is indicated by a rising A or K index?

E3C04 — What does the value of Bz (B sub Z) represent?

E3C05 — What orientation of Bz (B sub z) increases the likelihood that incoming particles from the sun will cause disturbed conditions?

E3C08 — What does the space weather term "G5" mean?

Once solar energy or charged particles reach the Earth, their energy disturbs the Earth's upper atmosphere and *geomagnetic field* (*GMF*). Disturbances of the GMF and ionosphere also disrupts the refraction of radio waves that travel through it. When the GMF is highly disturbed worldwide, that is called a *geomagnetic storm*. HF propagation through the auroral zones is particularly sensitive to this type of disturbance.

There are several parameters that amateurs use to assess HF propagation:

• B_Z — a measurement of the intensity and orientation of the *interplanetary magnetic field* (*IMF*) generated by the Sun. [**E3C04**] If B_Z is negative, the direction of the IMF is aligned southward (north-to-south). That direction is also aligned with the Earth's GMF so it is easier for charged particles from the Sun to enter and disrupt the GMF. [**E3C05**]

• K index — an evaluation of how disturbed the GMF is at a particular location with a value of 0 (quiet) to 9 (very major storm). The K_p index is an average of reported values worldwide.

• A index — derived from the K index with a wider range of 0 to 400 for the same conditions. Increasing values of the A and K indices indicate increasing disruption of the GMF. [**E3C02**]

• G index— geomagnetic storminess with levels of 0 (none), 1 (minor), 2 (moderate), 3 (strong), 4 (severe), and 5 (extreme), based on the value of the A and K indices. [**E3C08**]

There is a great deal of literature about the effects of solar weather on propagation. Along with the NOAA website, the ARRL's Technical Information Service (**www.arrl.org/tis**) offers numerous articles on propagation and ARRL members can subscribe to a free weekly propagation bulletin with up-to-date information and propagation predictions.

10.3 HF Propagation

In nearly all cases, HF signals make the journey between stations by either traveling along the surface of the Earth (*ground-wave*) or by being returned to Earth after encountering the upper layers of the ionosphere (*sky-wave* or *skip*). The differences in frequency between the lowest current amateur band (1.8 MHz) and the highest HF band (28 MHz) cause the behavior of these modes of propagation to be quite different across the HF spectrum.

GROUND-WAVE PROPAGATION

E3C12 — How does the maximum range of ground-wave propagation change when the signal frequency is increased?

E3C13 — What type of polarization is best for ground-wave propagation?

The direction of waves of all types can be changed by both *diffraction* and *refraction*. Diffraction is created by the construction and reinforcement of wavefronts after the radio wave encounters a reflecting surface's corners or edges. Refraction is a gradual bending of the wave because of changes in the velocity of propagation in the medium through which the wave travels.

There is a special form of diffraction that primarily affects vertically polarized radio waves at HF and lower frequencies. This type of diffraction results from the lower part of the wave losing energy because of currents induced in the ground. This slows the lower portion of the wave, causing the entire wave to tilt forward slightly, following the curvature of the Earth.

Topics in Radio Propagation **10-5**

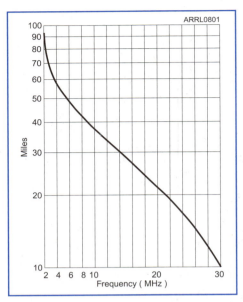

Figure 10.2 — Typical HF ground-wave range as a function of frequency.

This tilting results in *ground-wave propagation*, allowing low-frequency signals to be heard over distances well beyond line of sight. Although the term is often applied to any short-distance communication, the actual mechanism is unique to signals with longer wavelengths. Ground-wave propagation is most noticeable at LF, MF (AM broadcast, 160 meters), and the lower HF bands (80 meters) Practical ground-wave communication distances on these MF and HF bands is in the range of 50 – 100 miles.

Ground-wave propagation is lossy because the vertically polarized portion of the wave's electric field that extends into the ground is mostly absorbed. Over distance, the signal traveling along the ground is increasingly absorbed until the signal is too weak to be received. This loss increases significantly with frequency until at 28 MHz (10 meters), the maximum range of ground-wave is only a few miles. [**E3C12**] **Figure 10.2** shows typical ground-wave range at different frequencies.

Ground-wave propagation is most useful during the day at 1.8 and 3.5 MHz, when losses in the lower ionosphere make sky-wave propagation impossible. Vertically polarized antennas provide the best results. [**E3C13**] Ground-wave losses are reduced considerably over saltwater and are highest over dry and rocky land.

One simple way to observe the effects of ground-wave propagation is to listen to stations on the AM broadcast band. During the day you will regularly hear high-power stations from 100 to 150 miles away. You won't hear stations much farther than 200 miles, however. At night, when sky-wave propagation becomes possible, you will begin to hear stations several hundred miles away. Of course, high-power AM broadcast stations usually have vertical antennas with excellent ground systems to radiate a strong signal!

SKY-WAVE PROPAGATION

E3A06 — What might help to restore contact when DX signals become too weak to copy across an entire HF band a few hours after sunset?
E3B10 — Why is chordal hop propagation desirable?
E3B12 — What is the primary characteristic of chordal hop propagation?

Signals that travel into the ionosphere can be refracted (bent) by ionized gas in the ionosphere's E and F regions, returning to Earth some distance away. This refraction occurs because the region of ionized gases causes the radio wave to slow down, and this bends the wave. Refraction is primarily a propagation mode below VHF. Signals that follow a path away from the surface of the Earth are called *sky waves*. The path of a wave that returns to Earth after being bent by the ionosphere is called a *hop*.

The maximum one-hop skip distance for high-frequency radio signals via the F layer is usually considered to be about 2500 miles. (Skip via the E layer can extend to around 1500 miles.) Most HF communication beyond that distance takes place by means of several ionospheric hops in which the surface of the Earth reflects the signals back into the ionosphere for another hop. It is also possible that signals may reflect between the E and F regions, or even be reflected several times within the F region. When the wave makes two successive reflections from the ionosphere without an intervening reflection from the ground, that is called a *chordal hop*. Avoiding a lossy ground reflections means the signal will be stronger at the receiving end of the path. [**E3B10, E3B12**]

Every day there are big changes as night turns to day and vice versa. Bands open and close quickly, sometimes in minutes. By knowing the "band basics" you can plan your operating periods and react to the conditions you find on the air. For example, if you're making

HF contacts with stations in Europe after sunset, soon the signals all across the band will start to get weaker. When the band is about to close, signals will begin to exhibit the rapid fading that gives them a distinctive fluttery sound. What can you do to keep making those contacts? By learning how HF propagation works, you know that the MUF between your station and Europe is dropping. In response, change to a lower-frequency band. [**E3A06**] This is called "following the bands" and it works in reverse as the MUF moves higher through the morning.

Ordinary and Extraordinary Waves

E3B04 — What is meant by the terms "extraordinary" and "ordinary" waves?

E3B07 — What happens to linearly polarized radio waves that split into ordinary and extraordinary waves in the ionosphere?

An interesting thing happens when a radio wave enters the ionosphere: It divides into two waves polarized at right angles to each other. The *ordinary wave* or *o-wave* E field is parallel to the Earth's magnetic field. The *extraordinary wave* or *x-wave* E field is perpendicular to the Earth's magnetic field. [**E3B04**] The waves travel at slightly different speeds, creating a phase difference between them. The result is that a linearly polarized wave becomes elliptically polarized. [**E3B07**] At high frequencies (10 MHz and higher) both waves travel almost identical paths. On the 7 MHz and lower bands, however, the waves may travel along very different paths and in very different directions.

Propagation studies suggest radio waves may at times propagate for some distance through the uppermost F2 region of the ionosphere. In this type of propagation, a signal radiated at a medium elevation angle sometimes is returned to Earth at a greater distance than a wave radiated at a lower angle. The higher-angle wave, called the *Pedersen ray*, is believed to penetrate to the F2 region, farther than lower-angle rays. In the less-densely ionized F2 region, the amount of refraction is less, nearly equaling the curvature of the region itself as it encircles the Earth. **Figure 10.3** shows how the Pedersen ray could provide propagation beyond the normal single-hop distance. The Pedersen ray theory is further supported by studies of propagation times and signal strengths for signals that travel completely around the Earth. The time required is significantly less than would be necessary to hop between the Earth and the ionosphere 10 or more times while circling the Earth.

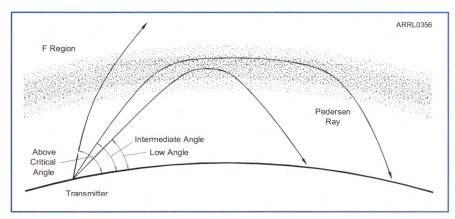

Figure 10.3 — This diagram shows a radio wave entering the F region at an intermediate angle, which penetrates higher than normal into the F region and then follows that region for some distance before being bent enough to return to Earth. A signal that travels for some distance through the F region is called a Pedersen ray.

Predicting Propagation

E3C01 — What does the radio communication term "ray tracing" describe?

E3C11 — What does VOACAP software model?

Models of propagation have been developed based on measurements of the ionosphere throughout the year and over complete sunspot cycles. These have been incorporated into software such as *VOACAP* to predict HF propagation between locations. [**E3C11**] *VOACAP* is available as an online website (**www.voacap.com/prediction.html**) so you can experiment with your own propagation questions at any time. These programs show that there are often multiple paths radio waves can take as they travel. Waves can reflect between layers and in different directions before reaching a destination. Following the possible paths through the ionosphere a wave might take is called *ray tracing*. [**E3C01**] HF propagation is still full of surprises, however, regardless of model sophistication. There is no substitute for actually transmitting a signal to see where it is received!

Absorption

E3C03 — Which of the following signal paths is most likely to experience high levels of absorption when the A index or K index is elevated?

E3C15 — What might be indicated by a sudden rise in radio background noise across a large portion of the HF spectrum?

The lowest of the ionosphere's layers is the D layer, occupying from around 35 to 60 miles above the Earth's surface. The D layer exists in a relatively dense region, compared to the rest of the ionosphere. This means the ionized atoms and molecules are closer together and recombine quickly. As a result, the D layer is present only when illuminated by the Sun. Created at sunrise and reaching its strongest around local noon, the D layer disappears quickly after sunset.

When a passing wave causes D layer electrons to move, they collide with other electrons and ions so frequently that a great deal of the wave's energy is dissipated as heat. This is called *ionospheric absorption*. The longer the wavelength of the radio wave, the farther the electron travels under influence of the wave and the greater the portion of the wave's energy lost as heat. This means absorption eliminates long-distance sky-wave propagation on the 1.8 and 3.5 MHz bands during the day, especially during periods of high solar activity. NVIS (*near vertical incidence sky-wave*) and ground-wave propagation can be used during daylight hours on these bands, however.

Geomagnetic disturbances and solar flares also increase absorption. Disruptions in the ionosphere from a changing magnetic field cause more of a radio wave's energy to be dissipated as heat. As the A and K indices rise, so does absorption, particularly along polar paths that travel through the auroral zones where most charged particles from the Sun enter the Earth's atmosphere. [**E3C03**] The large pulse of X-ray and UV energy generated by a solar flare raises ionization in the entire upper atmosphere, increasing dissipation of lower-frequency radio waves on the Earth's daylight side facing the Sun. If you are on the air when a solar flare occurs, noise levels on the HF bands slowly increase as signals fade. [**E3C15**] The effects last for several hours until ionization levels return to normal.

LONG-PATH AND GRAY-LINE PROPAGATION

E3B05 — Which amateur bands typically support long-path propagation?

E3B06 — Which of the following amateur bands most frequently provides long-path propagation?

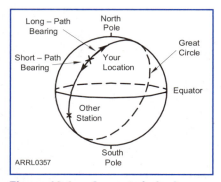

Figure 10.4 — A great circle drawn on the globe between two stations. The short-path and long-path bearings are shown from the Northern Hemisphere station.

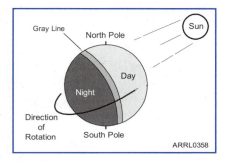

Figure 10.5 — The gray line is the transition region between daylight and darkness. On one side of the Earth the gray line is coming into sunrise, and on the other side just past sunset.

Most of the time, HF signals propagate over a *great circle* path between the transmitter and receiver. The great circle path is illustrated in **Figure 10.4**. A careful inspection shows that there are really two great circle paths, one shorter than the other. The longer of the two paths may also be useful for communications when conditions are favorable. This is called *long-path propagation*. Both stations must have directional antennas, such as beams, that can be pointed in the long path direction to make the best use of this propagation. The long- and short-path directions usually differ by 180°.

Long-path propagation can occur on any band with sky-wave propagation, so you may hear long-path signals on the 160 to 10 meter bands. [**E3B05**] (Six meter long-path has occurred on rare occasions.) Long-path enhancement occurs most often on the 20 meter band. [**E3B06**] All it takes to make long-path QSOs is a modest beam antenna with a relatively high gain compared to a dipole, such as a three-element beam. The antenna should be at a height above ground that allows low takeoff angles.

For paths less than about 6000 miles, the short-path signal will almost always be stronger because of the increased losses caused by multiple-hop ground-reflection losses and ionospheric absorption over the long path. When the short-path is more than 6000 miles, however, long-path propagation usually will be observed either along the *gray line* (the *terminator* between darkness and light that runs completely around the Earth) or over the nighttime side of the Earth.

The gray line is a band along the terminator that extends to either side for a number of miles. **Figure 10.5** illustrates the gray line. Notice that on one side of the Earth, the gray line is coming into daylight (sunrise) and on the other side it is coming into darkness (sunset).

Gray line or *grayline propagation*, often via long path, occurs between two stations that are simultaneously near sunrise and sunset. Gray line propagation can be quite effective because the D layer,

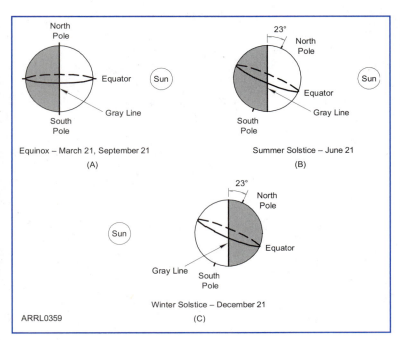

Figure 10.6 — The angle at which the gray line crosses the equator depends on the time of year, and on whether it is at sunset or sunrise. Part A shows that the gray line is perpendicular to the equator twice a year, at the vernal equinox and the autumnal equinox. B shows the North Pole tilted toward the Sun at a 23° angle at the summer solstice, and at C the North Pole tilted away from the Sun at a 23° angle at the winter solstice. This tilt causes the direction of the gray line to change so it is not north-south.

Topics in Radio Propagation   10-9

which absorbs HF signals, disappears rapidly on the sunset side of the gray line before it has had time to build up on the sunrise side. Meanwhile, the E and F layers, being at higher altitudes, are still illuminated and providing propagation. Signals may travel along the terminator but often travel from the gray line region through darkness via a skewed path to the receiving station that may or may not be in its own gray line region. This type of propagation lasts until somewhere along the path D layer absorption or the loss of F layer refraction prevent the signal from reaching the receiving station.

Look for gray-line propagation around sunrise and sunset. If you have a beam, aim it along the terminator. 160 through 20 meters are the most likely to experience gray-line enhancement because they are the most affected by D layer absorption.

Figure 10.6 illustrates the changing the tilt of the gray line. The tilt angle will be between these extremes during the rest of the year. Knowledge of the tilt angle will be helpful in determining what directions are likely to provide gray line propagation on a particular day.

10.4 VHF/UHF/Microwave Propagation

Without regular and dependable sky-wave propagation, VHF and UHF operators utilize alternative modes of propagation to make contacts. There are plenty of options — more than on HF, in fact — and many support communications over very long distances. This section touches on a number of interesting propagation modes that you're likely to encounter, should you decide to try using SSB, CW or one of the digital modes above 30 MHz.

For signals that travel essentially in a straight line between the transmitter and the receiver (also known as *space-wave propagation*), antennas that are low-angle radiators (concentrate signals toward the horizon) are best. Remember that the polarization of both the receiving and transmitting antennas should be the same for VHF and UHF operation because the polarization of a space wave remains constant as it travels.

RADIO HORIZON

E3C06 — By how much does the VHF/UHF radio horizon distance exceed the geometric horizon?

E3C14 — Why does the radio-path horizon distance exceed the geometric horizon?

In the early days of VHF amateur communications, it was generally believed that communications required direct line-of-sight paths between the antennas of the communicating stations. After some experiments with good equipment and antennas, however, it became clear that VHF radio waves are bent or scattered in several ways, making communications possible with stations beyond the *visual* or *geometric horizon*. The farthest point to which radio waves will travel directly via space-wave propagation is called the *radio horizon*.

Under normal conditions, density variations in the atmosphere near the Earth cause radio waves to bend into a curved path that keeps them nearer to the Earth than true straight-line travel would. **Figure 10.7** shows how this bending of the radio waves causes the distance to the radio horizon to exceed the distance to the visual horizon. [**E3C14**] The radio horizon is approximately 15% farther than the geometric horizon as shown graphically in **Figure 10.8**. [**E3C06**]

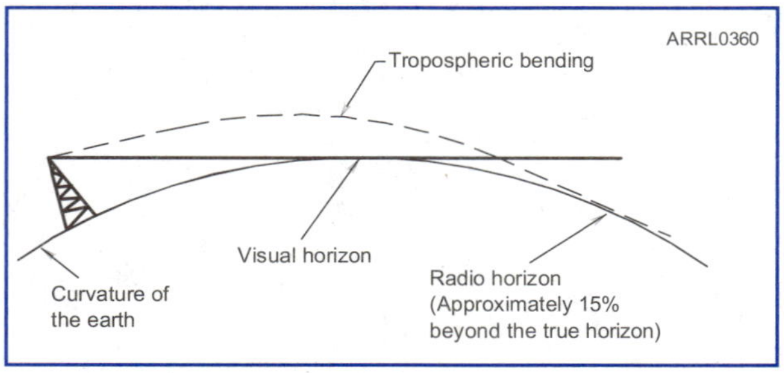

The distance to the radio horizon is assumed to be to a point on the ground. An antenna that is on a high hill or tall building well above any surrounding obstructions has a much farther radio horizon than an antenna located in a valley or shadowed by other obstructions. If the receiving antenna is also elevated, the maximum

Figure 10.7 — Under normal conditions, tropospheric bending causes VHF and UHF radio waves to be returned to Earth beyond the visible horizon.

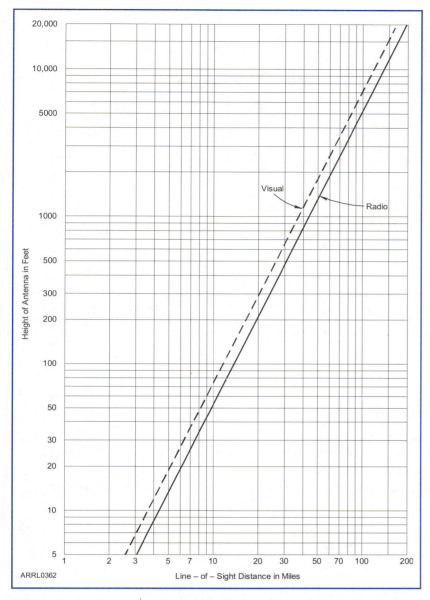

Figure 10.8 — **Distance to the radio horizon from an antenna of given height above average terrain is indicated by the solid line. The broken line indicates the distance to the visual, or geometric, horizon. The radio horizon is approximately 15% farther than the visual horizon.**

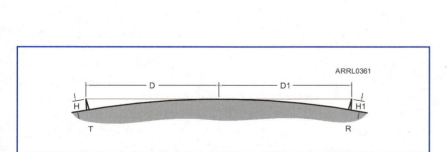

Figure 10.9 — **The distance, D, to the radio horizon is greater from a higher antenna. The maximum distance over which two stations may communicate by space wave is equal to the sum of the distances to their respective radio horizons.**

space-wave distance between the two antennas is equal to the sum of the distance to the radio horizon from the transmitting antenna plus the distance to the radio horizon from the receiving antenna. **Figure 10.9** illustrates this principle. Unless the two stations are identical, each will have a different radio horizon.

Multipath

A common cause of fading is an effect known as *multipath*. Several components of the same transmitted signal may arrive at the receiving antenna from different directions. The phase relationships between the multiple signals may cause them to cancel or reinforce each other. This effect is illustrated in **Figure 10.10**. Multipath effects can occur whenever the transmitted signal follows more than one path to the receiving station. Multipath and the distortion it causes to the received signal is a major challenge to providing high-speed digital service via wireless systems.

Topics in Radio Propagation

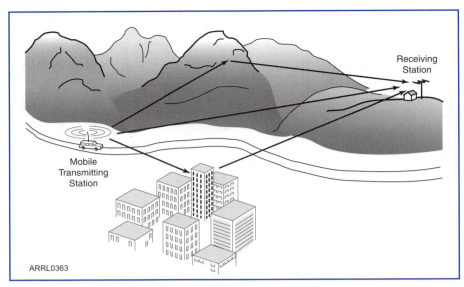

Figure 10.10 — If a signal travels from a transmitter to a receiver over several different paths, the signals may arrive at the receiver slightly out of phase. The out-of-phase signals alternately cancel and reinforce each other, and the result is a fading signal. This effect is known as multipath fading.

TROPOSPHERIC PROPAGATION

E3A04 — What do Hepburn maps predict?
E3A05 — Tropospheric propagation of microwave signals often occurs in association with what phenomenon?
E3A07 — Atmospheric ducts capable of propagating microwave signals often form over what geographic feature?
E3A10 — Which type of atmospheric structure can create a path for microwave propagation?
E3A11 — What is a typical range for tropospheric propagation of microwave signals?

At times, weather conditions such as temperature inversions and warm or cold fronts can create sharp transitions between air layers. These transitions can reflect or guide VHF, UHF, and microwave radio waves, forming *ducts* in the troposphere (lower layers of the atmosphere), similar to propagation in a waveguide. This form of propagation is called *tropospheric ducting*. [**E3A05, E3A10**]

The possibility of propagation via tropospheric ducting increases with frequency. Ducting is rare on 50 MHz, fairly common on 144 MHz and more common on higher frequencies. Because ducts often form over water, Gulf Coast states experience it often, and the Atlantic Seaboard, Great Lakes and Mississippi Valley areas less frequently, usually in September and October. [**E3A07**] Ducting can support microwave propagation over 100 – 300 miles with progressively longer distances achieved at lower frequencies. [**E3A11**]

Because these inversions are invisible to the eye, it can be difficult to tell when propagation is likely to occur. You just have to get on the air and try! The likelihood of tropospheric propagation can be estimated from weather conditions, however. William Hepburn developed a technique for creating a map of locations over which tropospheric propagation was likely. These are called "Hepburn maps" and are available online (**www.dxinfocentre.com/tropo_wam.html**) for amateurs to use. [**E3A04**]

SPORADIC E PROPAGATION

E3B09 — At what time of year is sporadic E propagation most likely to occur?
E3B11 — At what time of day can sporadic E propagation occur?

Sporadic E (Es or E-skip) consists of propagation from thin, highly ionized layers that form temporarily in the E layer of the ionosphere. The process of how these layers form is not entirely clear. Es commonly propagates 28, 50, and 144 MHz radio signals between 500 and 2300 km (300 and 1400 miles). Signals are apt to be exceedingly strong, allowing even modest stations to make Es contacts. Mid-latitude Es events may last only a few minutes or can persist for many hours.

Sporadic E at mid latitudes (roughly 15° to 45°) may occur at any time, but it is most common in the Northern Hemisphere around the summer solstice during May, June and July, with a less-intense season around the winter solstice at the end of December and early January. [**E3B09**] Its appearance is independent of the solar cycle. Sporadic E propagation can occur at any time through the day but is most likely to occur from 9 AM to noon local time and again early in the evening between 5 PM and 8 PM. [**E3B11**]

TRANSEQUATORIAL PROPAGATION

E3B01 — What is transequatorial propagation?
E3B02 — What is the approximate maximum range for signals using transequatorial propagation?
E3B03 — What is the best time of day for transequatorial propagation?

Transequatorial propagation (TE) is a form of F layer ionospheric propagation discovered by amateurs in the late 1940s. Amateurs on all continents reported the phenomenon almost simultaneously on various north-south paths on 50 MHz during the evening hours. At that time, the maximum predicted MUF was around 40 MHz for daylight hours. Research carried out by amateurs has shown that the TE mode works on 144 MHz and even to some degree at 432 MHz. TE occurs between mid-latitude stations approximately the same distance north and south of the Earth's magnetic equator. [**E3B01**] **Figure 10.11** shows the paths of a number of contacts made on 144 MHz using TE propagation.

You might expect the ionization of the ionosphere's upper layers to be at a maximum over the equator around the vernal (spring) and autumnal (fall) equinoxes. In fact, at the equinoxes there is not a single area of maximum ionization, but two. These maxima form in the morning, are well established by noon and last until after midnight. The high-density-ionization regions form approximately between 10° and 15° on either side of the Earth's magnetic equator — not the geographic equator — forming a pair of regions able to reflect VHF and even UHF signals.

As the relative position of the Sun moves away from the equator, the ionization levels in the Northern and Southern Hemispheres become unbalanced, lowering the MUF for TE propagation. So the best time of year to look for transequatorial propagation is around March 21 and September 21. The MUF for TE will also be higher during solar activity peaks. The best conditions for TE exist when the Earth's magnetic field is quiet, as well.

TE also enables very strong signals on the HF bands during the afternoon and early evening, so these are the best times to look for this propa-

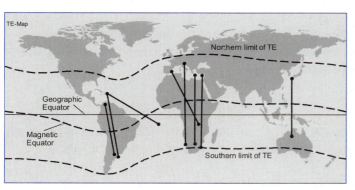

Figure 10.11 — This world map shows contacts made on 144 MHz over paths using transequatorial propagation (TE). Notice the symmetrical distribution of stations with respect to the magnetic equator.

Topics in Radio Propagation **10-13**

gation mode. [E3B03] Later at night, and sometimes in the early morning as well, you will hear weak and watery signals arriving by TE.

As the signal frequency increases, the communication zones become more restricted to those equidistant from, and perpendicular to, the magnetic equator. In addition, the duration of the opening tends to be shorter and closer to 8 PM local time. The rate of flutter fading and the degree of frequency spreading increase with signal frequency. TE range extends to approximately 5000 miles — 2500 miles on each side of the magnetic equator. [E3B02]

AURORAL PROPAGATION

E3A12 — What is the cause of auroral activity?
E3A13 — Which of these emission modes is best for auroral propagation?

Auroral propagation occurs when VHF radio waves are reflected from the ionization created by an auroral curtain. It is a VHF and UHF propagation mode that allows contacts up to about 1400 miles. Auroral propagation occurs for stations near the northern and southern polar regions but the discussion here is limited to auroral propagation in the Northern Hemisphere.

Aurora results from a large-scale interaction between the ionosphere and magnetic field of the Earth and electrically charged particles of the *solar wind*, ejected from the surface of the Sun. Visible aurora, often called the northern lights or *aurora borealis*, is caused by the collision of these solar-wind particles with oxygen and nitrogen molecules in the E layer. [E3A12] These collisions partially ionize the molecules, creating a conductive region capable of reflecting radio waves.

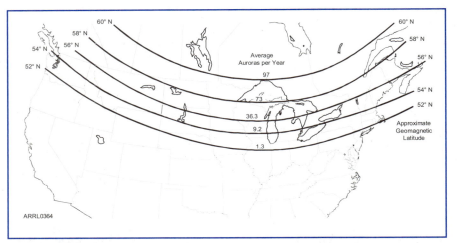

Figure 10.12 — The probability of auroral propagation decreases as distance from the Geomagnetic North Pole increases. The Geomagnetic North Pole is currently near 78° N, 104° W. This map shows the approximate geomagnetic latitude for the northern US.

When the electrons that were knocked loose from the oxygen and nitrogen recombine with the molecules, light is produced. The extent of the ionization determines how bright the aurora will appear. At times, the ionization is so strong that it is able to reflect radio signals with frequencies as low as 20 MHz. This ionization occurs at an altitude of about 70 miles in the E layer of the ionosphere. Not all auroral activity is intense enough to reflect radio signals, so a distinction is made between a visible aurora and a radio aurora. **Figure 10.12** shows the relatively likelihood of auroral propagation at different latitudes in North America.

Using Auroral Propagation

Most common on 10, 6, and 2 meters, auroral contacts have been made on frequencies as high as 222 and 432 MHz. The number and duration of openings decreases rapidly as the operating frequency rises.

The reflecting properties of an aurora vary rapidly, so signals received via this mode are badly distorted, making CW the most effective mode for auroral work. [E3A13] CW signals most often have a buzzing or raspy sound rather than a pure tone as reflection from the aurora makes them appear modulated by white noise. SSB is usable for 6 meter auroral contacts

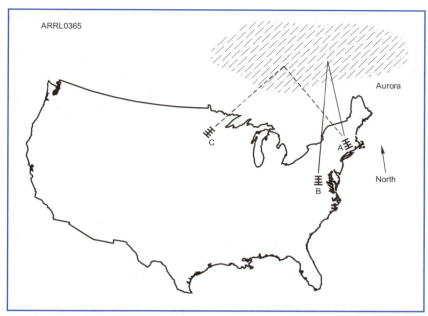

Figure 10.13 — To use auroral propagation, stations point their antennas toward the Geomagnetic North Pole. Station A may have to beam west of north to work station C.

if signals are strong; voices are often intelligible if the operator speaks slowly and distinctly. SSB is rarely usable at 2 meters or higher frequencies.

Stations should point their antennas toward the Geomagnetic North Pole during the aurora and, in effect, "bounce" their signals off the auroral zone as illustrated by **Figure 10.13**. The optimum antenna heading varies with the position of the aurora and may change rapidly, just as the visible aurora does. Constant probing with the antenna is recommended to peak signals, especially for high-gain antennas with narrow beamwidths.

Since auroras are associated with solar disturbances, you often can predict one by monitoring websites that publish geomagnetic data as discussed in the first section of this chapter. In particular, K index values of 3 and rising indicate that conditions associated with auroral propagation may be present. Maximum occurrence of radio aurora is for K index values of 7 to 9. The same websites also publish real-time maps of aurora location and intensity.

METEOR SCATTER COMMUNICATIONS

E2D01 — Which of the following digital modes is designed for meteor scatter communications?

E2D02 — Which of the following is a good technique for making meteor scatter contacts?

E3A08 — When a meteor strikes the Earth's atmosphere, a cylindrical region of free electrons is formed at what layer of the ionosphere?

E3A09 — Which of the following frequency ranges is most suited for meteor-scatter communications?

Meteoroids are particles of mineral or metallic matter that travel in highly elliptical orbits around the Sun. Most of these are microscopic in size. Every day hundreds of millions of these meteoroids enter the Earth's atmosphere. Attracted by the Earth's gravitational field, they attain speeds from 6 to 60 mi/s (22,000 to 220,000 mi/h).

As a meteoroid speeds through the upper atmosphere, it heats up and begins to vaporize

Table 10.1
Major Meteor Showers

Date	Name
January 3-5	Quadrantids
April 19-23	Lyrids
*June 8	Arietids
July 26-31	Aquarids
July 27-August 14	Perseids
October 18-23	Orionids
October 26 – November 16	Taurids
November 14-16	Leonids
December 10-14	Geminids
December 22	Ursids

All showers occur in the evening except those marked (*), which are daytime showers. Evening showers begin at approximately 2300 local standard time; daylight showers at approximately 0500 local standard time.

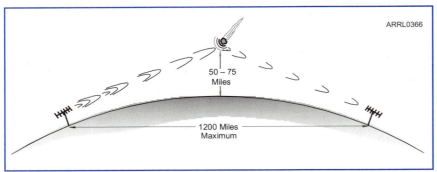

Figure 10.14 — Meteor-scatter communication makes extended-range VHF and UHF communications possible for short periods.

as it collides with air molecules. This action creates heat and light and leaves a trail of free electrons and positively charged ions called a *meteor*. Trail size is directly dependent on particle size and speed. A typical meteoroid the size of a grain of sand creates a trail about 3 feet in diameter and 12 miles or longer, depending on speed.

Meteor showers greatly enhance meteor-scatter communications at VHF. The degree of enhancement depends on the time of day, shower intensity and the frequency in use. The largest meteor showers of the year are the Perseids in August and the Geminids in December. **Table 10.1** is a partial list of meteor showers throughout the year.

Radio waves can be reflected by the ionized trail of a meteor. The ability of a meteor trail to reflect radio signals depends on electron density. Greater density causes greater reflecting ability and reflection at higher frequencies. The electron density in a typical meteor trail will strongly affect radio waves on the upper HF and lower VHF bands. Signal frequencies as low as 20 MHz and as high as 432 MHz will be usable for meteor-scatter communication at times. The best range of frequencies for amateur meteor scatter communications is from 28 to 148 MHz. [**E3A09**]

Meteor trails are formed at approximately the altitude of the ionospheric E layer, 50 to 75 miles above the Earth. [**E3A08**] That means the range for meteor-scatter propagation is about the same as for single-hop E (or sporadic E) skip — a maximum of approximately 1200 miles, as **Figure 10.14** shows.

Meteor Scatter Techniques

The secret to successful meteor scatter communication is short transmissions. An entire QSO with information exchanged and confirmed in both directions may last only a few seconds!

A single meteor may produce a strong enough path to sustain communication long enough to complete a short QSO. At other times multiple bursts are needed to complete the QSO, especially at higher frequencies. In response, amateurs developed an operating protocol based on 15-second transmissions that alternated between station locations. Eventually, a burst of characters would get through. Another convention, once a meteor trail was present and reflecting signals, was to use short transmissions with rapidly repeated call signs and signal reports that could be copied before the trail disappeared.

MSK144, part of the *WSJT-X* software suite, was written specifically for amateur meteor-scatter communications. [**E2D01, E2D02**] In this mode, the stations make repeated short transmissions of specially formatted packets to take advantage of any short meteor "burn" within range of both stations. See the Modulation, Protocols, and Modes chapter for more information on *WSJT-X*.

EARTH-MOON-EARTH COMMUNICATIONS

E2D03 — Which of the following digital modes is especially useful for EME communications?

E2D05 — What is one advantage of the JT65 mode?

E2D06 — Which of the following describes a method of establishing EME contacts?

E3A01 — What is the approximate maximum separation measured along the surface of the Earth between two stations communicating by EME?

E3A02 — What characterizes libration fading of an EME signal?

E3A03 — When scheduling EME contacts, which of these conditions will generally result in the least path loss?

The concept of *Earth-Moon-Earth* (*EME*) communications, popularly known as *moon-bounce*, is straightforward: Stations that can simultaneously see the Moon communicate by reflecting VHF or UHF signals off the lunar surface. Those stations may be separated by nearly half the circumference of the Earth — a distance of nearly 12,000 miles — as long as they can both "see" the Moon. [**E3A01**] This is called their *mutual lunar window* in which the Moon is above the "radio horizon" for both stations at the same time.

Since the Moon's average distance from Earth is 239,000 miles, *path losses* are huge when compared to "local" VHF paths. (Path loss refers to the total signal loss between the transmitting and receiving stations relative to the total radiated signal energy. Path loss is lowest when the Moon is at perigee — closest to the Earth). [**E3A03**] Nevertheless, for any type of amateur communication over a distance of 500 miles or more at 432 MHz, for example, moonbounce comes out the winner over terrestrial propagation paths when all the factors limiting propagation are taken into account.

One of the most troublesome aspects of receiving a moonbounce signal, besides the enormous path loss and Faraday-rotation fading in the ionosphere, is *libration* (pronounced lie-BRAY-shun) *fading*. Librations are short-term oscillations in the apparent motion of the Moon relative to Earth. Libration fading of an EME signal is experienced as fluttery, rapid, irregular fading not unlike that observed in tropospheric-scatter propagation. [**E3A02**] Librations cause fading combined with multipath scattering of the radio waves from the very large (2000-mile diameter) and rough lunar surface.

For analog or digital mode EME contacts, the round-trip time and extremely weak signals make the usual call-and-answer method impractical. In response, amateurs have developed a standard calling procedure for CW and SSB that uses alternating sequences of transmissions. Digital EME uses a special mode, JT65, that also uses alternating, time-synchronized transmissions. [**E2D03, E2D06**] The transmitted messages of JT65 use advanced coding techniques that allow signals to be decoded at extremely low signal-to-noise ratios. [**E2D05**] JT65 is part of the *WSJT-X* software suite that is discussed in the Modulation, Protocols, and Modes chapter.

Topics in Radio Propagation **10-17**

Chapter 11

Safety

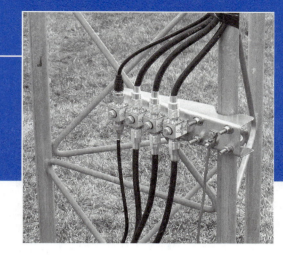

In this chapter, you'll learn about:
- Hazardous substances that may be found in the ham station
- Ionizing and non-ionizing radiation
- Power density, duty cycle and absorption limits
- Controlled and uncontrolled environments
- Evaluating RF exposure
- Steps you can take to minimize RF exposure
- Grounding and bonding in your station

Amateur radio is not a hazardous undertaking! Like driving a car, building and maintaining a station presents opportunities to act in an unsafe manner, but that doesn't mean the hazards are unavoidable. Quite the contrary! Learning about the hazards is the first step and that's why the question pool touches on several topics involving hazardous materials, RF exposure, and electrical safety. *The ARRL Handbook* contains a chapter on safety, as does *The ARRL Antenna Book*. RF exposure is covered thoroughly in the ARRL's *RF Exposure and You*, as well. We recommend having a copy of each in your ham radio library!

11.1 Hazardous Materials

E0A07 — How may dangerous levels of carbon monoxide from an emergency generator be detected?

E0A09 — Which insulating material commonly used as a thermal conductor for some types of electronic devices is extremely toxic if broken or crushed and the particles are accidentally inhaled?

E0A10 — What toxic material may be present in some electronic components such as high voltage capacitors and transformers?

There aren't many materials considered hazardous that are required to communicate using amateur radio. To be realistic, you probably have more hazardous materials in your garage and in your cleaning supplies than any ham station ever will. Nevertheless, here are three that you ought to know about.

PCBs

Not printed-circuit boards, of course, but *polychlorinated biphenyls*, are the PCBs referred to by this section. You may have heard about PCBs contaminating industrial sites or locations where waste oils were dumped. PCBs are an additive to insulating oils once commonly used in electrical equipment. PCBs helped the oil retain its insulating properties without breaking down and so became widely used until the hazard they presented became known. PCBs are known carcinogens — exposure to them, even in small amounts, elevates the risk of certain types of cancer.

Safety   11-1

In the ham station, PCBs may be found in older oil-filled high-voltage capacitors and sometimes utility-style high-voltage transformers referred to as "pole pigs." [**E0A10**] These were used in dc power supplies for tube-type RF amplifiers. Since any component containing PCBs would have to be fairly old, it is a good idea to replace it with a new one. If you are unsure whether a component does or doesn't contain PCBs, you can contact the manufacturer with the model number (if the manufacturer is known and still in business).

If you find such a component or remove one from equipment, wear protective gloves and wipe down the outside of the case with a paper towel. Place the paper towel and the component in a plastic bag or wrap it in plastic and take it to your local electric utility. They have procedures for safely and properly disposing of PCBs and may be able to handle it for you at no charge. Some local governments also have regular toxic-disposal opportunities and you can get rid of the component there, as well.

BERYLLIUM AND BERYLLIUM OXIDE

Beryllium copper is found in spring contacts and other flexible metal items that need to be both conductive and mechanically strong. In this form (or even as pure metal) beryllium is not dangerous. It is the dust and small particles that present a hazard and then only from chronic, extended exposure. To be safe, do not grind, weld, or file metal containing beryllium in an unventilated area.

The oxide of beryllium (BeO) is a tough, durable ceramic that has the rare combination of being an excellent electrical insulator and an excellent thermal conductor (**en.wikipedia.org/wiki/Beryllium_oxide**). It is used inside some power semiconductors to insulate the transistor structure from the case and conduct heat away from the transistor. It is also used in larger vacuum tubes as a thermally conductive insulator. In general, BeO is only found inside the envelopes of tubes and inside transistor packages. Handling BeO in solid form is not dangerous but if a tube or package is broken, BeO pieces could become cracked or produce dust. The dust is toxic if inhaled. [**E0A09**]

It is difficult to tell whether a white ceramic is BeO or a more benign form of ceramic, so treat all such materials with caution. Vacuum any possible dust or small particles with a vacuum cleaner using HEPA-rated bags, place the vacuum cleaner bag in a sealed plastic bag, and contact your local recycling or solid waste disposal service for instructions.

CARBON MONOXIDE

The use of fossil fuel-powered generators and heaters during emergency and portable operation is becoming increasingly common. This presents several hazards of which the amateur should be aware, including electrical, fire, and fuel storage hazards. A particularly worrisome hazard is caused by the carbon monoxide (CO) emitted by these devices.

Carbon monoxide is an odorless and colorless gas, so there is no warning detectable by humans that concentrations of CO have risen to dangerous levels. For that reason, it is important for generators and heaters (including wood-burning stoves) to only be used in well-ventilated areas away from people. The only reliable method of sensing the presence of excessive levels of CO is by using a carbon monoxide detector — a smoke alarm will not respond to CO alone. [**E0A07**] A CO detector should be placed in any area occupied by people in which CO from generator exhaust or heater vent can build up.

11.2 RF Exposure

E0A02 — When evaluating RF exposure levels from your station at a neighbor's home, what must you do?

E0A03 — Over what range of frequencies are the FCC human body RF exposure limits most restrictive?

E0A06 — Why are there separate electric (E) and magnetic (H) field MPE limits?

E0A08 — What does SAR measure?

E0A11 — Which of the following injuries can result from using high-power UHF or microwave transmitters?

You've been exposed (so to speak) to various safety topics regarding exposure to RF in the Technician and General license exams. The Extra class exam reviews familiar topics and introduces a couple of new ones. Although some of this material may be familiar from earlier exam studies (we hope!), it is worth covering again.

Let's start with a reminder that RF exposure at low levels — even continuously — is not hazardous in any way. It is only when the level of exposure is high enough to affect the temperature of the body that hazards occur. Although the power emitted by an antenna is called "radiation" and a hazard from RF exposure may be referred to as a "radiation hazard," it is not the same as radiation from a radioactive source. At RF, radiation does not have sufficient energy to break apart atoms and molecules — it can only cause heating. Radiation from radioactive sources does and that is why it is referred to as *ionizing radiation*. RF radiation is *nonionizing radiation* and is many orders of magnitude weaker than ionizing radiation in this regard.

Exposure to RF at low levels is not hazardous. At high power levels, for some frequencies the amount of energy that the body absorbs can be a problem. For example, exposure to high-power UHF or microwave RF can cause localized heating of the body. [**E0A11**] There are a number of factors to consider along with the power level, including frequency, average exposure, duty cycle of the transmission, and so forth. The two primary factors that determine how much RF the body will absorb are power density and frequency. This section discusses how to take into account the various factors and arrive at a reasonable estimate of what RF exposure results from your transmissions and whether any safety precautions are required.

POWER DENSITY

Heating from exposure to RF signals is caused by the body absorbing RF energy. The intensity of the RF energy is called *power density* and it is measured in mW/cm^2 (milliwatts per square centimeter) which is power per unit of area. For example, if the power density in an RF field is 10 mW/cm^2 and your hand's surface area is 75 cm^2, then your hand is exposed to a total of $10 \times 75 = 750$ mW of RF power in that RF field. Power density is highest near antennas and in the directions in which antennas have the most gain. Increasing transmitter power increases power density around the antenna. Increasing distance from an antenna lowers power density.

While RF exposure is measured in mW/cm^2 for most amateur requirements, the body's response to both E and H fields suggest that the RF exposure can also be measured in V/m (for the E field) and A/m (for the H field). Depending on the source of the RF and the environment, either of these measurements may be more appropriate than power density. For example, around reflecting surfaces or conducting materials, the intensity of the E and H fields can peak in different locations. Under and near antennas, ground reflections and scattering can make the field impedance (the ratio of E field to H field strength) vary with location, as well. [**E0A06**]

Safety 11-3

ABSORPTION AND LIMITS

The rate at which energy is absorbed from the power to which the body is exposed is called the *Specific Absorption Rate* (SAR). [**E0A08**] SAR is the best measure of RF exposure for amateur operators. The SAR varies with frequency, power density, average amount of exposure, and the duty cycle of transmission. Injury is only caused when the combination of frequency and power cause too much energy to be absorbed in too short a time.

SAR depends on the frequency and the size of the body or body part affected and is highest where the body and body parts are naturally resonant. The limbs (arms and legs) and torso have the highest SAR for RF fields in the VHF spectrum from 30 to 300 MHz. The head is most sensitive at UHF frequencies from 300 MHz to 3 GHz and the eyes are most affected by microwave signals above 1 GHz. The frequencies with highest SAR are between 30 and 1500 MHz. At frequencies above and below the ranges of highest absorption, the body responds less and less to the RF energy, just like an antenna responds poorly to signals away from its natural resonant frequency.

Safe levels of SAR based on demonstrated hazards have been established for amateurs by the FCC in the form of *Maximum Permissible Exposure* (MPE) limits that vary with frequency as shown in **Figure 11.1** and **Table 11.1**. These take into account the different sensitivity of the body to RF energy at different frequencies.

As you can see from the graph, safe exposure levels are much lower above 30 MHz. MPE limits are the lowest between 30 and 300 MHz. [**E0A03**] This means extra caution is required around high-power RF sources in the amateur VHF, UHF, and microwave bands. Of particular concern are amplifiers operating at these frequencies. Legal-limit power levels on these bands can create a significant hazard. When testing such equipment, take extra precautions to prevent accidental exposure, either near the transmitter or from an antenna. Radiation leaks above the MPE limits from klystron and magnetron transmitters, in particular, can present a significant hazard because of the power levels they can develop.

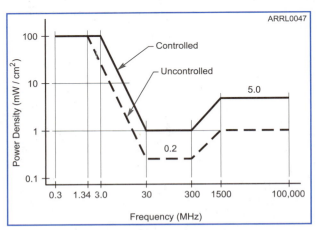

Figure 11.1 — Maximum Permissible Exposure (MPE) limits vary with frequency because the body responds differently to energy at different frequencies. The controlled and uncontrolled limits refer to the environment in which people are exposed to the RF energy.

Table 11.1
Maximum Permissible Exposure (MPE) Limits

Controlled Exposure (6-Minute Average)

Frequency Range (MHz)	Power Density (mW/cm²)
0.3-3.0	(100)*
3.0-30	(900/f²)*
30-300	1.0
300-1500	f/300
1500-100,000	5

Uncontrolled Exposure (30-Minute Average)

Frequency Range (MHz)	Magnetic Field Power Density (mW/cm²)
0.3-1.34	(100)*
1.34-30	(180/f²)*
30-300	0.2
300-1500	f/1500
1500-100,000	1.0

f = frequency in MHz
* = Plane-wave equivalent power density

AVERAGING AND DUTY CYCLE

Exposure to RF energy is averaged over fixed time intervals because the response of the body to heating is different for short duration and long duration exposures. Time-averaging evaluates the total RF exposure over a fixed time interval. In addition, there are two types of environments with different averaging periods: controlled and uncontrolled.

Controlled and Uncontrolled Environments

People in *controlled environments* are considered to be aware of their exposure and are expected to take reasonable steps to minimize exposure. Examples of controlled environments are transmitting facilities (including amateur radio stations) and areas near antennas. In a controlled environment, access is restricted to authorized and informed individuals. The people expected to be in controlled environments would be station employees, licensed amateurs, and the families of licensed amateurs.

Uncontrolled environments are areas where the general public has access, such as public roads and walkways, homes and schools, and even unfenced personal property. The homes of your neighbors are uncontrolled environments. [**E0A02**] People in uncontrolled environments are not aware of their exposure but are much less likely to receive continuous exposure. As a result, RF power density limits are higher for controlled environments and the averaging period is longer for uncontrolled environments. The averaging period is 6 minutes for controlled environments and 30 minutes for uncontrolled environments.

Duty Cycle

Duty cycle is the ratio of transmitter on time to total time during the exposure. Duty cycle has a maximum of 100%. (*Duty factor* is the same as duty cycle expressed as a fraction instead of percent. For example, a duty cycle of 25% is equivalent to a duty factor of 0.25.) The lower the transmission duty cycle (the less the transmitter is on), the lower the average exposure. A lower transmission duty cycle permits greater short-term exposure levels for a given average exposure. This is the *operational duty cycle*. For most amateur operation listening and transmitting time are about the same, so operational duty cycle is rarely higher than 50%.

Along with operational duty cycle, the different modes themselves have different *emission duty cycles* as shown in **Table 11.2**. For example, a normal SSB signal without speech processing to raise average power is considered to have an emission duty cycle of 20%. In contrast, FM is a constant-power mode so its emission duty cycle is 100%. Transmitter PEP multiplied by the emission duty cycle multiplied by the operating duty cycle gives the average power output.

For example, if a station is using SSB without speech processing, transmitting and listening for equal amounts of time and with a PEP of 150 W, then the average power output is 150 W × 20% × 50%, or 15 W.

For the AM entries, note that the table assumes the same PEP for all signals. If PEP is the same, an AM signal with 50% modulation has a higher duty factor (more carrier, less sidebands) than for a signal with 100% modulation. Because the average power of the signal depends on modulation, the duty cycle can range from 25% to 100%.

Table 11.2
Emission Duty Cycle of Modes Commonly Used by Amateurs

Mode	Duty Cycle	Notes
Conversational SSB	20%	1
Conversational SSB	40%	2
SSB AFSK	100%	
SSB SSTV	100%	
Voice AM, 50% modulation	50%	3
Voice AM, 100% modulation	25%	
Voice AM, no modulation	100%	
Voice FM	100%	
Digital FM	100%	
ATV, video portion, image	60%	
ATV, video portion, black screen	80%	
Conversational CW	40%	
Carrier	100%	4

Notes
1) Includes voice characteristics and syllabic duty factor. No speech processing.
2) Includes voice characteristics and syllabic duty factor. Heavy speech processing.
3) Full-carrier, double-sideband modulation, referenced to PEP. Typical for voice speech. Can range from 25% to 100% depending on modulation.
4) A full carrier is commonly used for tune-up purposes.

Safety 11-5

ANTENNA SYSTEM

You must also take into account the amount of gain provided by your antenna and any significant losses from the feed line. High gain antennas increase a signal's average power considerably. For example, let's modify the previous example by using an antenna with 6 dB of gain. If the transmitter PEP is increased from 150 W to 600 W by the 6 dB antenna gain, the average power is 600 W × 20% × 50%, or 60 W.

Including antenna gain in the field strength calculation is required only when the evaluation is being performed in the antenna's *far field*. The far field begins approximately 10 wavelengths or so from the antenna and is generally considered to be the region in which the antenna's radiation pattern has assumed its final shape and does not change with increasing distance from the antenna. If the evaluation is to be performed in the *near field* (anything closer than the far field distance), then a different measure of antenna gain must be used.

ESTIMATING EXPOSURE AND STATION EVALUATION

E0A04 — When evaluating a site with multiple transmitters operating at the same time, the operators and licensees of which transmitters are responsible for mitigating over-exposure situations?

E0A05 — What is one of the potential hazards of operating in the amateur radio microwave bands?

All fixed amateur stations must evaluate their capability to cause RF exposure, no matter whether they use high or low power. (Mobile and handheld transceivers are exempt from having to calculate exposure because they do not stay in one location.) A routine evaluation must then be performed if the transmitter PEP and frequency are within the FCC rule limits. The limits vary with frequency and PEP as shown in **Table 11.3**. You are required to perform the RF exposure evaluation only if your transmitter output power exceeds the levels shown for any band. For example, if your HF transmitter cannot output more than 25 W, you are exempt from having to evaluate exposure caused by it.

You can perform the evaluation by actually measuring the RF field strength with calibrated field strength meters and calibrated antennas. You can also use computer modeling to determine the exposure levels. However, it's easiest for most hams to use the tables provided by the ARRL (**www.arrl.org/rf-exposure**) or an online calculator, such as the one listed on the ARRL website.

If you choose to use the ARRL tables or calculators, you will need to know:

• Power at the antenna, including adjustments for duty cycle and feed line loss

• Antenna type (or gain) and height above ground

• Operating frequency

The ARRL tables are organized by frequency, antenna type, and antenna height. They show the distance required from the antenna to comply with MPE limits for certain levels of transmitter output power.

Exposure can be evaluated in one of two ways. The first way is to determine the power density at a known distance to see if exposure at that distance meets the MPE limit. The second way is to determine the minimum distance from your antenna at which the MPE limit is satisfied. Either way, the goal is to determine if your station meets MPE limits for all controlled and uncontrolled environments present at your station.

If you make changes to your station, such as changing to a higher power transmitter, increasing antenna gain, or changing antenna height, then you must reevaluate the RF exposure from your station. If you reduce output power without making any other changes to a

Table 11.3
Power Thresholds for RF Exposure Evaluation

Band	Power (W)
160 meters	500
80	500
40	500
30	425
20	225
17	125
15	100
12	75
10	50
6	50
2	50
1.25	50
70 cm	70
33	150
23	200
13	250
SHF (all bands)	250
EHF (all bands)	250

11-6   Chapter 11

station already in compliance, you need not make any further changes.

In a multi-transmitter environment, such as at a commercial repeater site, each transmitter operator may be jointly responsible (with all other site operators) for ensuring that the total RF exposure from the site does not exceed the MPE limits. Any transmitter (including the antenna) that produces more than 5% of the total permissible exposure limit for transmissions at that frequency must be included in the site evaluation. (This is 5% of the permitted power density or 5% of the square of the E or H-field MPE limit. It is *not* 5% of the total exposure, which sometimes can be unknown.) [**E0A04**] The situation described by this question is common for amateur repeater installations, which often share a transmitting site.

EXPOSURE SAFETY MEASURES

The measures you can take if your evaluation results exceed MPE limits are summarized in **Figure 11.2**. These are all "good practice" suggestions that can save time and expense if they are followed before doing your evaluation.

• Locate or move antennas away from where people can get close to them and be exposed to excessive RF fields. Raise the antenna or place it away from where people will be. Keep the ends (high voltage) and center (high current) of antennas away from people where people could come in contact with them. Locate the antenna away from property lines and place a fence around the base of ground-mounted antennas. This prevents people from encountering RF in excess of the MPE limits.

• Don't point gain antennas where people are likely to be. Use beam antennas to direct the RF energy away from people. Remember that high-gain antennas have a narrower beam, but exposure in the beam will be more intense. Take special care with high-gain VHF/UHF/microwave antennas (such as Yagis and dishes used for EME) and transmitters — don't transmit when you or other people are close to the antenna or the antenna is pointed close to the horizon. [**E0A05**]

• If you have to use stealth or attic antennas, carefully evaluate whether MPE limits are exceeded in your home's living quarters.

• On VHF and UHF, place mobile antennas on the roof or trunk of the car to maximize shielding of the passengers. Use a remote microphone to hold a handheld transceiver away from your head while transmitting.

• When using microwave signals, take extra care around the high-gain antennas used on these frequencies. Even modest transmitter powers can result in significant RF levels when focused by a dish with more than 20 dB of gain! This is a particular concern when operating in a portable or rover configuration where antennas may not be far off the ground.

• From the transmitter's perspective, use a dummy load or dummy antenna when testing a transmitter. You can also reduce the power and duty cycle of your transmissions. This is often quite effective and has a minimal effect on your signal.

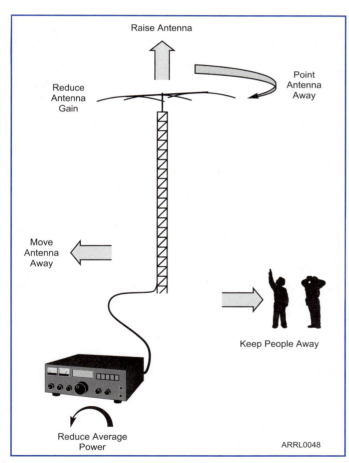

Figure 11.2 — There are many ways to reduce RF exposure to nearby people. Whatever lowers the power density in areas where people are present will work. Raising the antenna will even benefit your signal strength to other stations as it lowers power density on the ground.

11.3 Grounding and Bonding

Hams are concerned with several kinds of things called "ground," even if they really aren't a connection to the Earth. These are easily confused because we call each of them "ground." Three of these ideas are covered here:

1) Electrical safety ground
2) Protection from lightning and transient voltages
3) Common reference potential (chassis ground)

This brief discussion is only intended to introduce three different types of connections referred to under the general idea of grounding. For a more complete discussion and suggestions for building a station, see the *ARRL Handbook* or the ARRL's *Grounding and Bonding for the Radio Amateur*.

ELECTRICAL SAFETY GROUND

An ac power ground is required by building codes to ensure the safety of life and property. The NEC or National Electrical Code also requires that all grounds be bonded together; this is a very important safety feature as well as an NEC requirement. Bonding means connecting two points together so they are at the same electrical potential or voltage.

"Electrical safety ground" is the usual term used for the "third wire" or "green-wire ground." The purpose of the non-load current-carrying wire is to provide a path to ensure that the circuit's overcurrent protection will trip and remove power in the event of a line-to-enclosure short circuit in a piece of equipment. This could either be a fuse or circuit breaker at the main service panel. Fuses inside the equipment offer another layer of protection.

The conductive enclosure of equipment is required to be connected to the bonding system, which is also connected to earth ground at the service entrance. This prevents someone from getting shocked who is connected to "earth" (for example, standing in bare feet on a concrete floor) and touches the exposed enclosure.

An effective safety ground system is necessary for every amateur station, and the NEC requires that all the "grounds" be bonded together to provide a common reference potential for all parts of the ac system. This is not sufficient for lightning protection or RF management. An effective bonding conductor at 60 Hz may present very high impedance at RF because of its length or the inductance may be too high for effective lightning protection.

LIGHTNING DISSIPATION GROUND

E0A01 — What is the primary function of an external earth connection or ground rod?

Lightning dissipation grounding is intended to allow charge from a lightning strike to dissipate in the surrounding earth and to prevent large voltage differences between parts of the ground system. Since the lightning pulse has RF components at and above 1 MHz, it is an RF signal, and low inductance is needed as well as low resistance.

A lightning ground needs to handle a peak current of tens of kiloamperes. Large conductors are used in lightning grounds to reduce inductance, to handle the mechanical forces from high currents, and for ruggedness to prevent inadvertent breakage. A large diameter wire, or even better, a wide flat strap, has the lowest inductance and makes the best ground connection.

The primary function of ground rods is making a suitable earth connection for lightning protection of both ac power systems and antenna systems. [**E0A01**] Vertical antenna radial systems and buried radials connected to a grounded tower also act as earth connections to dissipate charge from a lightning strike.

COMMON REFERENCE POTENTIAL

The typical ham station includes a lot of circuits that are sensitive to interfering signals at millivolt levels, such as audio signals to and from sound cards. To keep from contaminating these small signals with RF, we shouldn't be using the equipment enclosures or shielding conductors as part of the RF circuit.

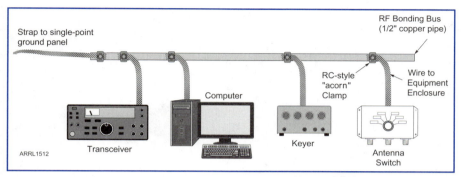

Figure 11.3 — An RF bonding bus connects all of the equipment enclosures together to keep them at the same RF voltage. The bus must be connected to a lightning protection ground using a heavy conductor, fastened securely. The RF bonding bus should also be connected to the ac safety ground.

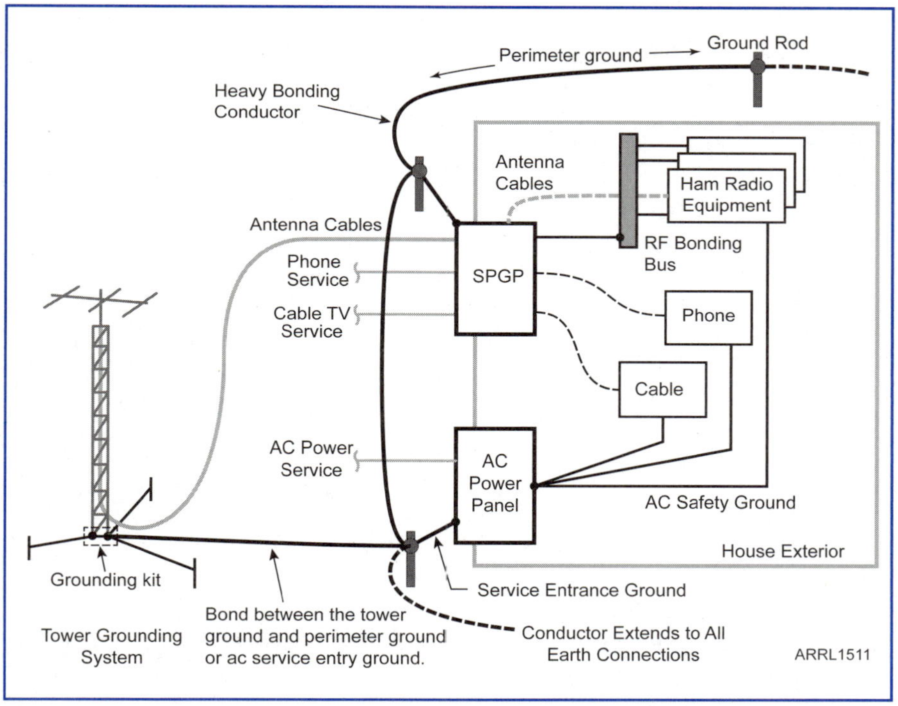

Instead, we design our stations to create a common reference potential, called the "reference plane" or "RF bonding" and we work to keep equipment connected to the reference plane at a common potential. This minimizes RF current that would flow between pieces of equipment. The general idea of an RF bonding bus is shown in **Figure 11.3**.

All grounds, including safety, RF, lightning protection and commercial communications such as cable TV and telephone cables, must be bonded together in order to prevent significant voltages between them. The NEC requires that antenna grounds be bonded to the other grounds in the system. An overview of this type of whole-building ground system is shown in **Figure 11.4**. Remember that the focus of the electrical code bonding requirement is safety in the event of a power distribution fault or transient, such as a direct or nearby lightning strike.

Figure 11.4 — Overview of a station lightning protection system. The station's single-point ground panel (SPGP) is bonded to the building's ac service entrance ground. All external ground rods must be bonded together.

Safety   11-9

Chapter 12

Glossary

The glossary in this edition is primarily focused on words associated with abbreviations or acronyms along with uncommon words. For words and phrases not listed here, please check the index.

A

A index — A geomagnetic-field measurement used to indicate HF propagation conditions. Rising values generally indicate disturbed conditions while falling values indicate improving conditions. See also *K index*.

Admittance (Y) — The reciprocal of impedance.

Alpha (α) — The ratio of transistor collector current to emitter current. It is between 0.92 and 0.98 for a bipolar junction transistor.

Amplitude modulation (AM) — A method of superimposing an information signal on an RF carrier wave in which the amplitude of the RF envelope (carrier and sidebands) is varied in relation to the information signal strength.

Analog-to-digital converter (ADC) — A circuit that converts analog signals to digital values.

Anode — The terminal connected to the positive supply for current to flow through a device.

ASCII — American National Standard Code for Information Interchange

ATV (amateur television) — A fast-scan TV system that can use commercial transmission standards on the 70-cm band and higher frequencies.

Audio FSK (AFSK) — Generating a frequency shift keying (FSK) signal by inputting tones to the transmitter through the audio or microphone input.

Automatic Link Enable (ALE) — A digital mode that uses automatic control to find frequencies at which two stations can communicate.

Automatic Packet Reporting System (APRS) — A system of sending location and other data over packet radio to a common website for tracking and recording purposes.

AX.25 — The amateur implementation of the X.25 communications protocol, used for packet radio.

B

B_z — The B index is a measurement of the Earth's geomagnetic field's strength and orientation with three-dimensional values: B_x, B_y, and B_z. B_z indicates the strength of the north-south field.

Bandwidth (BW) — (1) The frequency range over which a signal or the output of a circuit is within 3 dB of its peak strength within that range. (2) The frequency range over which a circuit or antenna meets a specified performance requirement.

Baseband — (1) The information that modulates a carrier or that is recovered from a modulated signal. (2) A video signal with its lowest frequency component at or near dc.

Glossary **12-1**

Baud — A unit of signaling speed equal to the number of discrete conditions or events per second. (For example, if an FSK signal changes frequency every 3.33 milliseconds, the signaling or baud rate is 300 bauds or the reciprocal of 0.00333 seconds.) Baud is equivalent to *symbol rate*.

Beta (β) — Current gain of a bipolar transistor, the ratio of collector to base current.

BiCMOS — A digital logic family that combines bipolar and CMOS technology in a single integrated circuit.

Bipolar junction transistor (BJT) — A transistor made of three layers of alternating type material (N or P) creating two PN semiconductor junctions between them.

C

Cabrillo format — A standardized digital file format for submitting information in a contest log.

Cardioid (radiation pattern) — A heart-shaped antenna pattern characterized by a single main lobe and a deep, narrow null in the opposite direction.

Cathode — The terminal connected to the negative supply for current to flow through a device.

CEPT (European Conference of Postal and Telecommunications Administrations) agreement — A multilateral operating arrangement that allows US amateurs to operate in many European countries, and amateurs from many European countries to operate in the US.

Certificate of Successful Completion of Examination (CSCE) — A document issued by a Volunteer Examiner Team to certify that a candidate has passed specific exam elements at their test session. If the candidate qualified for a license upgrade at the exam session, the CSCE provides the authority to operate using the newly earned license privileges, with special identification procedures.

Chroma (chrominance) — Information in a composite video signal that carries the color information. A *chroma burst* is a short period of signal used to synchronize color processing circuitry.

Circulator — A passive device with three or more ports that allows radio waves to travel between ports in only one direction.

CMOS — Complementary metal oxide semiconductor (digital logic family)

Code division multiple access (CDMA) — A method of using spread spectrum techniques to share a common frequency range by assigning each signal a different spreading code.

Common — In a transistor circuit (common-emitter/collector/base/source/gate/drain), the transistor electrode that is shared or used as a reference for both input and output circuits.

Complementary metal-oxide semiconductor (CMOS) — A type of construction used to make digital integrated circuits with both N-channel and P-channel MOS devices on the same chip.

Conductance (G) — The reciprocal of resistance. The real part of complex admittance.

Cross-modulation — See *intermodulation distortion*.

Cross-polarized — Antennas or signals that are aligned with their polarization at right angles.

Cutoff frequency (f_c) — The frequency at which (1) the output power of a passive circuit is reduced to half of its input or (2) the power gain of an active circuit is one-half its peak gain.

D

Decibel (dB) — A logarithm of the ratio of two power levels: dB = 10 log (P2/P1). Power gains and losses are expressed in decibels.

Depletion mode — Type of FET in which drain-source current is reduced by reverse bias on the gate.

Dielectric — An insulating material in which energy can be stored by an electric field.

Dielectric constant (k) — Also known as an insulating material's *relative permittivity* compared to that of free space. See also *permittivity*.

Digital-to-analog converter (DAC) — A circuit that converts digital values to analog signals.

Digital Amateur Television (DATV) — Amateur television that uses commercial digital television modulation techniques and equipment.

Digital multimeter (DMM) — An instrument with a digital display that measures voltage, current, and resistance.

Digital Radio Mondiale (DRM) — A digital modulation method used to transfer audio and data on HF bands.

DIP — dual in-line package. A type of integrated circuit package with two parallel rows of pins.

Direct digital conversion (DDC) — In a *software defined radio (SDR)* the conversion of RF directly to and from digital data without an intermediate frequency conversion step.

Direct digital synthesizer (DDS) — The technique of generating a signal from a sequence of digital values stored in a table.

Direct FSK — Generating a frequency shift keying (FSK) signal by shifting the transmitter frequency directly under the control of a digital signal.

Direct sequence (DS) — A spread-spectrum communications system in which a very fast binary bit stream is used to shift the phase of an RF carrier.

DSB-SC — Double-sideband, suppressed carrier

DX — Distance. On HF, often used to describe stations in countries outside your own.

E

E plane — The plane of the electric field of an antenna's radiation.

Effective radiated power (ERP) — A measure of the power radiated from an antenna system. ERP takes into account transmitter output power, feed line losses and other system losses, and antenna gain as compared to a dipole. *Effective isotropic radiated power (EIRP)* is the same as ERP except the reference antenna is an isotropic radiator.

Electromagnetic (EM) waves — Energy moving through space or materials in the form of changing electric and magnetic fields.

EME — Earth-Moon-Earth (see also *moonbounce*)

Enhancement mode — An FET in which drain-source current is increased by forward bias on the gate.

Error correction — see *Forward error correction*

Extraordinary or X wave — The cross-polarized component of a radio wave that splits in two upon encountering the ionosphere. See also *ordinary wave*.

F

Fast-scan TV (FSTV) — See *ATV*.

Field-effect transistor (FET) — A semiconductor device that uses voltage to control output current.

Finite Impulse Response (FIR) filter — A digital filter with a response to an impulse signal that lasts for a finite amount of time. See also *Infinite Impulse Response (IIR) filter*.

Forward error correction (FEC) — The method of adding special codes to a data stream so that a receiving system can detect and correct certain types of transmission errors.

Frequency division multiplexing (FDM) — Combining more than one stream of information in a single transmitted signal by using different modulating frequencies.

Frequency hopping (FH) — A spread-spectrum communications system in which the center frequency of a conventional carrier is altered many times a second in accordance with a pseudorandom list of channels.

Frequency modulation (FM) — A method of superimposing an information signal on an RF carrier wave in which the instantaneous frequency of an RF carrier wave is varied in relation to the information signal strength.

Frequency shift keying (FSK) — A method of digital modulation in which individual bit values are represented by specific frequencies. If two frequencies are used, one is called *mark* and one *space*.

Frequency standard — A circuit or device used to produce a highly accurate reference frequency. The frequency standard may be a crystal oscillator in a marker generator or a radio broadcast, such as from WWV, with a carefully controlled transmit frequency.

Front-to-side (F/S)/back (F/B)/rear (F/R) ratio — The ratio of field strength at the peak of the major lobe to that in the specified direction. Rear implies an average value over a specified angle centered on the back direction.

FT4, FT8 — Digital messaging protocols for use at low signal-to-noise ratios. Part of the *WSJT-X* software suite.

G

G index — An index indicating the impact of a geomagnetic storm on propagation, ranging from minor (G1) to extreme (G5).

G5RV antenna — A multi-band antenna similar to a dipole that is fed in the middle with a specific length of open-wire transmission line to create a low impedance suitable for connecting to a coaxial feed line.

Grid square locator — A 2° longitude by 1° latitude rectangle identified by a four-character label such as "EM48." Grid square locators are exchanged in some contests, and are used as the basis for some VHF/UHF awards.

H

H plane — The plane of the magnetic field of an antenna's radiation.

Height above average terrain (HAAT) — The height of an antenna above an average elevation of the surrounding terrain determined by measurements along several radial lines from the antenna.

I

IARP (International Amateur Radio Permit) — A multilateral operating arrangement that allows US amateurs to operate in many Central and South American countries, and amateurs from many Central and South American countries to operate in the US.

IF — Intermediate frequency.

Impedance (Z) — The general term for opposition to current flow, either ac or dc. Impedance is made up of resistance and reactance.

Infinite impulse response (IIR) filter — A digital filter with a response to an impulse signal that persists forever. See also *Finite impulse response filter*.

Intercept point (IP) — The level of a receiver input signal at which distortion products would be as strong as the desired output.

Intermodulation distortion (IMD) — A type of interference that results from the unwanted mixing of two strong signals, producing a signal on an unintended frequency. Often abbreviated as "intermod".

Isolator — A passive attenuator in which the loss in one direction is much greater than the loss in the other.

Isotropic — The same in all directions.

J

Joule (J) — The unit of energy in the metric system of measure.

Junction field-effect transistor (JFET) — A field-effect transistor in which the gate electrode and channel are in direct contact and made of opposite types of semiconductor materials (N or P).

JT65 — A multi-tone FSK mode used with extremely low signal-to-noise ratios.

K

K index — A geomagnetic-field measurement used to indicate HF propagation conditions. Rising values generally indicate disturbed conditions while falling values indicate improving conditions. See also *A index*.

Keplerian elements — Parameters that describe a satellite's orbit such that it can be located in the sky at any time.

L

Line A — A line parallel to and approximately 50 miles from the Canadian border, north of which US amateurs may not transmit on 420 – 430 MHz because of interference with Canadian stations.

LO — Local oscillator

Low Earth Orbit (LEO) — Orbits from 200 – 500 miles above the Earth. The International Space Station is in LEO.

M

Maximum Permissible Exposure (MPE) — The highest allowed level of exposure to RF.

Metal-oxide semiconductor FET (MOSFET) — A field-effect transistor with the gate insulated from the channel material. Also called an IGFET or *insulated gate FET*.

Minimum discernible signal (MDS) — The input signal level equal to the receiver's internal noise.

Monolithic microwave integrated circuit (MMIC) — An integrated circuit designed for operation at microwave frequencies. MMICs usually provide simple functions such as amplification.

Moonbounce — A common name for Earth-Moon-Earth (EME) communication in which signals are reflected from the Moon before being received.

MSK144 — A digital mode designed for use with meteor scatter propagation. Part of the *WSJT-X* software suite.

N

Noise blanker (NB) — A circuit that removes noise from the receiver output by muting the receiver during a noise pulse.

Noise figure — The ratio in dB of the noise output power to the noise input power with the input termination at a standard temperature of 290 K. It is a measure of the noise generated in the receiver circuitry. *Noise factor* is the same quantity expressed as a linear ratio.

Noise reduction (NR) — A type of adaptive filtering that removes unwanted noise in a signal's passband.

Glossary 12-5

NTSC — National Television Standard Committee. The US analog television standard.

N-type material — Semiconductor material that has been treated with impurities to give it an excess of electrons.

O

OFCD (Off-center fed dipole) — A dipole fed away from its center point to present a similar feed point impedance on different bands.

Ordinary or O wave — The component of a radio wave that retains its original polarization when it splits in two upon encountering the ionosphere. See also *extraordinary wave*.

Orthogonal Frequency Division Multiplexing (OFDM) — The technique of transmitting digital data by modulating multiple carriers separated to minimize interference between them.

P

PCB (hazardous materials) — Polychlorinated biphenyls, carcinogenic hydrocarbons once added to insulating oils

Peak envelope power (PEP) — The maximum average power level in a signal during one cycle during a modulation peak. (Used for modulated RF signals.)

Peak envelope voltage (PEV) — The maximum voltage in a cycle at the peak of a modulated signal envelope.

Peak inverse voltage (PIV) — The maximum instantaneous anode-to-cathode reverse voltage that may be applied to a diode without damage.

Peak-to-peak (P-P) voltage — The difference between the negative and positive peak voltages of a waveform.

Pedersen ray — A high-angle radio wave that penetrates deeper into the F region of the ionosphere so the wave is bent less than a lower-angle wave and thus travels for some distance through the F region, returning to Earth at a distance farther than normally expected for single-hop propagation.

Period (T) — The time it takes to complete one cycle of an ac waveform.

Permeability (μ) — The ability of a material to store energy in a magnetic field.

Permittivity (ε) — The ability of a material to store energy in an electric field.

Phase-locked loop (PLL) — A servo loop consisting of a phase detector, low-pass filter, dc amplifier and voltage-controlled oscillator.

Phase modulation (PM) — A method of superimposing an information signal on an RF carrier wave in which the phase of an RF carrier wave is varied in relation to the information signal strength.

Phase shift keying (PSK) — A method of modulation in which the phase of a carrier signal is varied to represent different digital values.

PIN diode — A diode consisting of a relatively thick layer of nearly pure semiconductor material (intrinsic semiconductor) with a layer of P-type material on one side and a layer of N-type material on the other.

PN junction — The contact area between two layers of opposite-type semiconductor material.

PRB-1 — The FCC regulation requiring local governments to make reasonable accommodations for amateur radio in land-use regulations.

Programmable logic device (PLD) — A digital integrated circuit consisting of individual logic circuit elements and subsystems that can be connected together (programmed) to implement a complex function. If the logic elements are logic gates the device is called a *Programmable Gate Array (PGA)*.

12-6   Chapter 12

Pseudonoise (PN) — A binary sequence designed to appear to be random (contain an approximately equal number of ones and zeros). Pseudonoise is generated by a digital circuit and mixed with digital information to produce a direct-sequence spread-spectrum signal.

P-type material — A semiconductor material that has been treated with impurities to give it an electron shortage. This creates excess positive charge carriers, or "holes."

Q

Q — (circuit or component) A quality factor describing how much energy is lost in a component or circuit due to resistance compared to energy stored in reactance. (frequency response) The ratio of center frequency of a filter to its bandwidth.

Q point — See *operating point*; also called quiescent point.

Q section — A quarter-wave section of transmission line used for impedance matching.

R

RACES — Radio Amateur Civil Emergency Service.

Radians — A unit of angular measurement. There are 2π radians in a circle and 1 radian = $57.3°$

Reactance (X) — The opposition to ac current due to capacitance or inductance. The imaginary component of complex impedance.

Receiving directivity factor (RDF) — The ratio of an antenna's forward gain to its average of gain over all directions, a figure of merit for an antenna's receiving directivity.

Reflection coefficient (ρ or Γ) — The ratio of the reflected voltage at a given point on a transmission line to the incident voltage at the same point. The reflection coefficient is also equal to the ratio of reflected and incident currents.

Root-mean-square (RMS) voltage — A measure of the effective value of an ac voltage. The value of a dc voltage that would produce the same amount of heat in a resistance as the ac voltage.

S

S or scattering parameters — Ratios of incident and reflected voltage waves at and between the signal ports of a circuit. S parameters are used to describe impedance, gain, SWR, and other parameters of circuits at RF.

Signal-to-noise ratio (SNR) — The numeric ratio of signal power to noise power in a given bandwidth. *Signal-to-noise-plus-distortion (SINAD)* adds distortion product power to the noise power.

Single-sideband, suppressed-carrier signal (SSB) — A radio signal in which only one of the two sidebands generated by amplitude modulation is transmitted. The other sideband and the RF carrier wave are removed before the signal is transmitted.

Slow-scan television (SSTV) — A television system used by amateurs to transmit pictures within a voice signal's bandwidth allowed on the HF bands by the FCC. Each frame takes several seconds to transmit.

Software defined radio (SDR) — A receiver and/or transmitter based on DSP techniques and with a modulation/demodulation configuration determined entirely by software.

Specific absorption rate (SAR) — The rate at which the body absorbs electromagnetic energy.

Sporadic E — A type of propagation for upper HF, VHF, and lower UHF signals that occurs when signals are reflected by highly-ionized regions of the E layer.

Glossary 12-7

Spread-spectrum (SS) communication — A communications method in which the RF bandwidth of the transmitted signal is much larger than that needed for traditional modulation schemes, and in which the RF bandwidth is independent of the modulation content. Increasing the bandwidth of the signal by means of a randomizing sequence (*spreading code*) is called *spreading*.

Surface-mount device (SMD) — An electronic component without wire leads, designed to be soldered directly to copper-foil pads on a circuit board.

Surface-mount technology (SMT) — The general term for methods and devices for mounting components directly on printed-circuit boards.

Susceptance (B) — The reciprocal of reactance. The imaginary part of complex admittance.

T

Time division multiplexing (TDM) — Combining more than one stream of information in a single transmitted signal by using different time periods or "slots" for each stream.

Transconductance (g_m) — The ratio of output current to input voltage, primarily used with FETs and vacuum tubes.

Transequatorial propagation (TE) — A form of F-layer ionospheric propagation, in which signals of higher frequency than the expected MUF propagate across the Earth's magnetic equator.

V

Vector network analyzer (VNA) — A test instrument that measures complex impedance, phase, and amplitude in circuits at RF.

Velocity factor (VF, velocity of propagation) — An expression of how fast a radio wave will travel through a material or transmission line. It is usually stated as a fraction of the speed the wave would have in free space (where the wave would have its maximum velocity). Velocity factor is also sometimes specified as a percentage of the speed of a radio wave in free space.

Vertical interval signaling (VIS) — The method of identifying the type of SSTV signal by sending coded information during the vertical synchronization period.

Vestigial sideband (VSB) — A signal-transmission method in which one sideband, the carrier and part of the second sideband are transmitted.

VOACAP — A propagation prediction program.

Volunteer Examiner (VE) — A licensed amateur who is accredited by a Volunteer Examiner Coordinator (VEC) to administer amateur license exams.

Volunteer Examiner Coordinator (VEC) — An organization that has entered into an agreement with the FCC to coordinate amateur license examinations.

Voltage-controlled oscillator (VCO) — An oscillator whose frequency is varied by means of an applied control voltage.

W

WSJT-X — A suite of digital modes developed by K1JT including FT8, FT4, JT65, MSK144, and WSPR which are used at very low signal-to-noise ratios and for scatter or EME propagation.

WSPR — A digital mode that uses extremely low power and advanced coding techniques for evaluating propagation. Part of the *WSJT-X* software suite.

Chapter 13

Question Pool

Extra Class (Element 4) Syllabus

Effective July 1, 2020 to June 30, 2024

SUBELEMENT E1 — COMMISSION'S RULES

[6 Exam Questions — 6 Groups] 75 Questions

E1A — Operating Standards: frequency privileges; automatic message forwarding; stations aboard ships or aircraft; power restriction on 630 and 2200 meter bands

E1B — Station restrictions and special operations: restrictions on station location; general operating restrictions; spurious emissions; antenna structure restrictions; RACES operations

E1C — Definitions and restrictions pertaining to local, automatic and remote control operation; IARP and CEPT licenses; emission and bandwidth standards

E1D — Amateur space and Earth stations; telemetry and telecommand rules; identification of balloon transmissions; one-way communications

E1E — Volunteer examiner program: definitions; qualifications; preparation and administration of exams; accreditation; question pools; documentation requirements

E1F — Miscellaneous rules: external RF power amplifiers; prohibited communications; spread spectrum; auxiliary stations; Canadian amateurs operating in the U.S.; special temporary authority; control operator of an auxiliary station

SUBELEMENT E2 — OPERATING PROCEDURES

[5 Exam Questions — 5 Groups] 61 Questions

E2A — Amateur radio in space: amateur satellites; orbital mechanics; frequencies and modes; satellite hardware; satellite operations

E2B — Television practices: fast scan television standards and techniques; slow scan television standards and techniques

E2C — Operating methods: contest and DX operating; remote operation techniques; Cabrillo format; QSLing; RF network connected systems

E2D — Operating methods: VHF and UHF digital modes and procedures; APRS; EME procedures; meteor scatter procedures

E2E — Operating methods: operating HF digital modes

SUBELEMENT E3 — RADIO WAVE PROPAGATION

[3 Exam Questions — 3 Groups] 40 Questions

E3A — Electromagnetic waves; Earth-Moon-Earth communications; meteor scatter; microwave tropospheric and scatter propagation; aurora propagation; ionospheric propagation changes over the day; circular polarization

E3B — Transequatorial propagation; long-path; ordinary and extraordinary waves; chordal hop; sporadic E mechanisms

E3C — Radio horizon; ground wave; propagation prediction techniques and modeling; effects of space weather parameters on propagation

Valid July 1, 2020 through June 30, 2024

SUBELEMENT E4 — AMATEUR PRACTICES

[5 Exam Questions — 5 Groups] 60 Questions

E4A — Test equipment: analog and digital instruments; spectrum analyzers; antenna analyzers; oscilloscopes; RF measurements; computer-aided measurements

E4B — Measurement technique and limitations: instrument accuracy and performance limitations; probes; techniques to minimize errors; measurement of Q; instrument calibration; S parameters; vector network analyzers

E4C — Receiver performance characteristics: phase noise, noise floor, image rejection, MDS, signal-to-noise ratio, noise figure, reciprocal mixing; selectivity; effects of SDR receiver non-linearity; use of attenuators at low frequencies

E4D — Receiver performance characteristics: blocking dynamic range; intermodulation and cross-modulation interference; third-order intercept; desensitization; preselector

E4E — Noise suppression and interference: system noise; electrical appliance noise; line noise; locating noise sources; DSP noise reduction; noise blankers; grounding for signals; common mode currents

SUBELEMENT E5 — ELECTRICAL PRINCIPLES

[4 Exam Questions — 4 Groups] 55 Questions

E5A — Resonance and Q: characteristics of resonant circuits: series and parallel resonance; definitions and effects of Q; half-power bandwidth; phase relationships in reactive circuits

E5B — Time constants and phase relationships: RL and RC time constants; phase angle in reactive circuits and components; admittance and susceptance

E5C — Coordinate systems and phasors in electronics: rectangular coordinates; polar coordinates; phasors

E5D — AC and RF energy in real circuits: skin effect; electromagnetic fields; reactive power; power factor; electrical length of conductors at UHF and microwave frequencies; microstrip

SUBELEMENT E6 — CIRCUIT COMPONENTS

[6 Exam Questions — 6 Groups] 70 Questions

E6A — Semiconductor materials and devices: semiconductor materials; germanium, silicon, P-type, N-type; transistor types: NPN, PNP, junction, field-effect transistors: enhancement mode; depletion mode; MOS; CMOS; N-channel; P-channel

E6B — Diodes

E6C — Digital ICs: Families of digital ICs; gates; Programmable Logic Devices (PLDs)

E6D — Toroidal and Solenoidal Inductors: permeability, core material, selecting, winding; transformers; piezoelectric devices

E6E — Analog ICs: MMICs, IC packaging characteristics

E6F — Electro-optical technology: photoconductivity; photovoltaic devices; optical sensors and encoders; optical isolation

SUBELEMENT E7 — PRACTICAL CIRCUITS

[8 Exam Questions — 8 Groups] 108 Questions

E7A — Digital circuits: digital circuit principles and logic circuits; classes of logic elements; positive and negative logic; frequency dividers; truth tables

E7B — Amplifiers: Class of operation; vacuum tube and solid-state circuits; distortion and intermodulation; spurious and parasitic suppression; microwave amplifiers; switching-type amplifiers

E7C — Filters and matching networks: types of networks; types of filters; filter applications; filter characteristics; impedance matching; DSP filtering

E7D — Power supplies and voltage regulators; Solar array charge controllers

E7E — Modulation and demodulation: reactance, phase and balanced modulators; detectors; mixer stages

E7F — DSP filtering and other operations; software defined radio fundamentals; DSP modulation and demodulation

E7G — Active filters and op-amp circuits: active audio filters; characteristics; basic circuit design; operational amplifiers

E7H — Oscillators and signal sources: types of oscillators; synthesizers and phase-locked loops; direct digital synthesizers; stabilizing thermal drift; microphonics; high-accuracy oscillators

Valid July 1, 2020 through June 30, 2024

SUBELEMENT E8 — SIGNALS AND EMISSIONS

[4 Exam Questions — 4 Groups] 45 Questions

E8A — AC waveforms: sine, square, and irregular waveforms; AC measurements; average power and PEP of RF signals; Fourier analysis; analog to digital conversion: digital to analog conversion; advantages of digital communications

E8B — Modulation and demodulation: modulation methods; modulation index and deviation ratio; frequency and time division multiplexing; Orthogonal Frequency Division Multiplexing

E8C — Digital signals: digital communication modes; information rate vs. bandwidth; error correction

E8D — Keying defects and overmodulation of digital signals; digital codes; spread spectrum

SUBELEMENT E9 — ANTENNAS AND TRANSMISSION LINES

[8 Exam Questions — 8 Groups] 96 Questions

E9A — Basic Antenna parameters: radiation resistance, gain, beamwidth, efficiency; effective radiated power

E9B — Antenna patterns and designs: E and H plane patterns; gain as a function of pattern; antenna modeling

E9C — Practical wire antennas; folded dipoles; phased arrays; effects of ground near antennas

E9D — Yagi antennas; parabolic reflectors; circular polarization; loading coils; top loading; feed point impedance of electrically short antennas; antenna Q; RF grounding

E9E — Matching: matching antennas to feed lines; phasing lines; power dividers

E9F — Transmission lines: characteristics of open and shorted feed lines; coax versus open-wire; velocity factor; electrical length; coaxial cable dielectrics

E9G — The Smith chart

E9H — Receiving Antennas: radio direction finding antennas; Beverage antennas; specialized receiving antennas; long-wire receiving antennas

SUBELEMENT E0 — SAFETY

[1 exam question — 1 group] 11 Questions

E0A — Safety: RF radiation hazards; hazardous materials; grounding

Element 4 — Extra Class Question Pool
Effective for VEC examinations on July 1, 2020 through June 30, 2024

SUBELEMENT E1 — COMMISSION RULES
[6 Exam Questions — 6 Groups]

E1A — Operating Standards: frequency privileges; automatic message forwarding; stations aboard ships or aircraft; power restriction on 630 and 2200 meter bands

E1A01
(A)
[97.305, 97.307(b)]
Page 3-4

E1A01
Which of the following carrier frequencies is illegal for LSB AFSK emissions on the 17 meter band RTTY and data segment of 18.068 to 18.110 MHz?
A. 18.068 MHz
B. 18.100 MHz
C. 18.107 MHz
D. 18.110 MHz

E1A02
(D)
[97.301, 97.305]
Page 3-4

E1A02
When using a transceiver that displays the carrier frequency of phone signals, which of the following displayed frequencies represents the lowest frequency at which a properly adjusted LSB emission will be totally within the band?
A. The exact lower band edge
B. 300 Hz above the lower band edge
C. 1 kHz above the lower band edge
D. 3 kHz above the lower band edge

E1A03
(C)
[97.305, 97.307(b)]
Page 3-4

E1A03
What is the maximum legal carrier frequency on the 20 meter band for transmitting USB AFSK digital signals having a 1 kHz bandwidth?
A. 14.070 MHz
B. 14.100 MHz
C. 14.149 MHz
D. 14.349 MHz

E1A04
(C)
[97.301, 97.305]
Page 3-4

E1A04
With your transceiver displaying the carrier frequency of phone signals, you hear a DX station calling CQ on 3.601 MHz LSB. Is it legal to return the call using lower sideband on the same frequency?
A. Yes, because the DX station initiated the contact
B. Yes, because the displayed frequency is within the 75 meter phone band segment
C. No, the sideband will extend beyond the edge of the phone band segment
D. No, U.S. stations are not permitted to use phone emissions below 3.610 MHz

E1A05
(C)
[97.313]
Page 3-4

E1A05
What is the maximum power output permitted on the 60 meter band?
A. 50 watts PEP effective radiated power relative to an isotropic radiator
B. 50 watts PEP effective radiated power relative to a dipole
C. 100 watts PEP effective radiated power relative to the gain of a half-wave dipole
D. 100 watts PEP effective radiated power relative to an isotropic radiator

E1A06
Where must the carrier frequency of a CW signal be set to comply with FCC rules for 60 meter operation?
A. At the lowest frequency of the channel
B. At the center frequency of the channel
C. At the highest frequency of the channel
D. On any frequency where the signal's sidebands are within the channel

E1A06
(B)
[97.303(h)(1)]
Page 3-4

E1A07
What is the maximum power permitted on the 2200 meter band?
A. 50 watts PEP
B. 100 watts PEP
C. 1 watt EIRP (Equivalent isotropic radiated power)
D. 5 watts EIRP (Equivalent isotropic radiated power)

E1A07
(C)
[97.313(k)]
Page 3-5

E1A08
If a station in a message forwarding system inadvertently forwards a message that is in violation of FCC rules, who is primarily accountable for the rules violation?
A. The control operator of the packet bulletin board station
B. The control operator of the originating station
C. The control operators of all the stations in the system
D. The control operators of all the stations in the system not authenticating the source from which they accept communications

E1A08
(B)
[97.219]
Page 3-5

E1A09
What action or actions should you take if your digital message forwarding station inadvertently forwards a communication that violates FCC rules?
A. Discontinue forwarding the communication as soon as you become aware of it
B. Notify the originating station that the communication does not comply with FCC rules
C. Notify the nearest FCC Field Engineer's office
D. All these choices are correct

E1A09
(A)
[97.219]
Page 3-6

E1A10
If an amateur station is installed aboard a ship or aircraft, what condition must be met before the station is operated?
A. Its operation must be approved by the master of the ship or the pilot in command of the aircraft
B. The amateur station operator must agree not to transmit when the main radio of the ship or aircraft is in use
C. The amateur station must have a power supply that is completely independent of the main ship or aircraft power supply
D. The amateur operator must have an FCC Marine or Aircraft endorsement on his or her amateur license

E1A10
(A)
[97.11]
Page 3-6

E1A11
Which of the following describes authorization or licensing required when operating an amateur station aboard a U.S.-registered vessel in international waters?
A. Any amateur license with an FCC Marine or Aircraft endorsement
B. Any FCC-issued amateur license
C. Only General Class or higher amateur licenses
D. An unrestricted Radiotelephone Operator Permit

E1A11
(B)
[97.5]
Page 3-6

Question Pool   13-5

E1A12
(C)
Page 8-21

E1A12
What special operating frequency restrictions are imposed on slow scan TV transmissions?
    A. None; they are allowed on all amateur frequencies
    B. They are restricted to 7.245 MHz, 14.245 MHz, 21.345 MHz, and 28.945 MHz
    C. They are restricted to phone band segments
    D. They are not permitted above 54 MHz

E1A13
(B)
[97.5]
Page 3-6

E1A13
Who must be in physical control of the station apparatus of an amateur station aboard any vessel or craft that is documented or registered in the United States?
    A. Only a person with an FCC Marine Radio license grant
    B. Any person holding an FCC issued amateur license or who is authorized for alien reciprocal operation
    C. Only a person named in an amateur station license grant
    D. Any person named in an amateur station license grant or a person holding an unrestricted Radiotelephone Operator Permit

E1A14
(D)
[97.313(l)]
Page 3-6

E1A14
Except in some parts of Alaska, what is the maximum power permitted on the 630 meter band?
    A. 50 watts PEP
    B. 100 watts PEP
    C. 1 watt EIRP
    D. 5 watts EIRP

E1B — Station restrictions and special operations: restrictions on station location; general operating restrictions; spurious emissions; antenna structure restrictions; RACES operations

E1B01
(D)
[97.3]
Page 3-7

E1B01
Which of the following constitutes a spurious emission?
    A. An amateur station transmission made without the proper call sign identification
    B. A signal transmitted to prevent its detection by any station other than the intended recipient
    C. Any transmitted signal that unintentionally interferes with another licensed radio station
    D. An emission outside the signal's necessary bandwidth that can be reduced or eliminated without affecting the information transmitted

E1B02
(A)
[97.307(f)(2)]
Page 8-23

E1B02
Which of the following is an acceptable bandwidth for Digital Radio Mondiale (DRM) based voice or SSTV digital transmissions made on the HF amateur bands?
    A. 3 kHz
    B. 10 kHz
    C. 15 kHz
    D. 20 kHz

E1B03
(A)
[97.13]
Page 3-8

E1B03
Within what distance must an amateur station protect an FCC monitoring facility from harmful interference?
    A. 1 mile
    B. 3 miles
    C. 10 miles
    D. 30 miles

13-6   Chapter 13

E1B04

What must be done before placing an amateur station within an officially designated wilderness area or wildlife preserve, or an area listed in the National Register of Historic Places?

A. A proposal must be submitted to the National Park Service
B. A letter of intent must be filed with the Environmental Protection Agency
C. An Environmental Assessment must be submitted to the FCC
D. A form FSD-15 must be submitted to the Department of the Interior

E1B04
(C)
[97.13, 1.1305-1.1319]
Page 3-8

E1B05

What is the National Radio Quiet Zone?

A. An area in Puerto Rico surrounding the Arecibo Radio Telescope
B. An area in New Mexico surrounding the White Sands Test Area
C. An area surrounding the National Radio Astronomy Observatory
D. An area in Florida surrounding Cape Canaveral

E1B05
(C)
[97.3]
Page 3-20

E1B06

Which of the following additional rules apply if you are installing an amateur station antenna at a site at or near a public use airport?

A. You may have to notify the Federal Aviation Administration and register it with the FCC as required by Part 17 of the FCC rules
B. You must submit engineering drawings to the FAA
C. You must file an Environmental Impact Statement with the EPA before construction begins
D. You must obtain a construction permit from the airport zoning authority

E1B06
(A)
[97.15]
Page 3-9

E1B07

To what type of regulations does PRB-1 apply?

A. Homeowners associations
B. FAA tower height limits
C. State and local zoning
D. Use of wireless devices in vehicles

E1B07
(C)
[97.15]
Page 3-8

E1B08

What limitations may the FCC place on an amateur station if its signal causes interference to domestic broadcast reception, assuming that the receivers involved are of good engineering design?

A. The amateur station must cease operation
B. The amateur station must cease operation on all frequencies below 30 MHz
C. The amateur station must cease operation on all frequencies above 30 MHz
D. The amateur station must avoid transmitting during certain hours on frequencies that cause the interference

E1B08
(D)
[97.121]
Page 3-7

E1B09

Which amateur stations may be operated under RACES rules?

A. Only those club stations licensed to Amateur Extra Class operators
B. Any FCC-licensed amateur station except a Technician Class
C. Any FCC-licensed amateur station certified by the responsible civil defense organization for the area served
D. Any FCC-licensed amateur station participating in the Military Auxiliary Radio System (MARS)

E1B09
(C)
[97.407]
Page 3-6

E1B10
(A)
[97.407]
Page 3-6

E1B10
What frequencies are authorized to an amateur station operating under RACES rules?
A. All amateur service frequencies authorized to the control operator
B. Specific segments in the amateur service MF, HF, VHF and UHF bands
C. Specific local government channels
D. Military Auxiliary Radio System (MARS) channels

E1B11
(B)
[97.15]
Page 3-8

E1B11
What does PRB-1 require of regulations affecting amateur radio?
A. No limitations may be placed on antenna size or placement
B. Reasonable accommodations of amateur radio must be made
C. Amateur radio operations must be permitted in any private residence
D. Use of wireless devices in a vehicle is exempt from regulation

E1B12
(A)
[97.303(b)]
Page 3-19

E1B12
What must the control operator of a repeater operating in the 70 cm band do if a radiolocation system experiences interference from that repeater?
A. Cease operation or make changes to the repeater to mitigate the interference
B. File an FAA NOTAM (Notice to Airmen) with the repeater system's ERP, call sign, and six-character grid locator
C. Reduce the repeater antenna HAAT (Height Above Average Terrain)
D. All these choices are correct

E1C — Rules pertaining to automatic and remote control; band-specific regulations; operating in, and communicating with foreign countries; spurious emission standards; HF modulation index limit; bandwidth definition

E1C01
(D)
[97.303]
Page 3-4

E1C01
What is the maximum bandwidth for a data emission on 60 meters?
A. 60 Hz
B. 170 Hz
C. 1.5 kHz
D. 2.8 kHz

E1C02
(C)
[97.117]
Page 3-20

E1C02
Which of the following types of communications may be transmitted to amateur stations in foreign countries?
A. Business-related messages for non-profit organizations
B. Messages intended for users of the maritime satellite service
C. Communications incidental to the purpose of the amateur service and remarks of a personal nature
D. All these choices are correct

E1C03
(B)
[97.109(d)]
Page 3-12

E1C03
How do the control operator responsibilities of a station under automatic control differ from one under local control?
A. Under local control there is no control operator
B. Under automatic control the control operator is not required to be present at the control point
C. Under automatic control there is no control operator
D. Under local control a control operator is not required to be present at a control point

13-8   Chapter 13

E1C04

What is meant by IARP?

A. An international amateur radio permit that allows U.S. amateurs to operate in certain countries of the Americas

B. The internal amateur radio practices policy of the FCC

C. An indication of increased antenna reflected power

D. A forecast of intermittent aurora radio propagation

E1C04
(A)
Page 3-21

E1C05

When may an automatically controlled station originate third party communications?

A. Never

B. Only when transmitting RTTY or data emissions

C. When agreed upon by the sending or receiving station

D. When approved by the National Telecommunication and Information Administration

E1C05
(A)
[97.221(c)(1), 97.115(c)]
Page 3-12

E1C06

Which of the following is required in order to operate in accordance with CEPT rules in foreign countries where permitted?

A. You must identify in the official language of the country in which you are operating

B. The U.S. embassy must approve of your operation

C. You must bring a copy of FCC Public Notice DA 16-1048

D. You must append "/CEPT" to your call sign

E1C06
(C)
Page 3-21

E1C07

At what level below a signal's mean power level is its bandwidth determined according to FCC rules?

A. 3 dB

B. 6 dB

C. 23 dB

D. 26 dB

E1C07
(D)
[97.3(a)(8)]
Page 3-4

E1C08

What is the maximum permissible duration of a remotely controlled station's transmissions if its control link malfunctions?

A. 30 seconds

B. 3 minutes

C. 5 minutes

D. 10 minutes

E1C08
(B)
[97.213]
Page 3-10

E1C09

What is the highest modulation index permitted at the highest modulation frequency for angle modulation below 29.0 MHz?

A. 0.5

B. 1.0

C. 2.0

D. 3.0

E1C09
(B)
[97.307]
Page 8-4

E1C10
(A)
[97.307]
Page 3-7

E1C10
What is the permitted mean power of any spurious emission relative to the mean power of the fundamental emission from a station transmitter or external RF amplifier installed after January 1, 2003 and transmitting on a frequency below 30 MHz?
   A. At least 43 dB below
   B. At least 53 dB below
   C. At least 63 dB below
   D. At least 73 dB below

E1C11
(A)
[97.5]
Page 3-21

E1C11
Which of the following operating arrangements allows an FCC-licensed U.S. citizen to operate in many European countries, and alien amateurs from many European countries to operate in the U.S.?
   A. CEPT agreement
   B. IARP agreement
   C. ITU reciprocal license
   D. All these choices are correct

E1C12
(D)
[97.305(c)]
Page 3-5

E1C12
On what portion of the 630 meter band are phone emissions permitted?
   A. None
   B. Only the top 3 kHz
   C. Only the bottom 3 kHz
   D. The entire band

E1C13
(C)
[97.303(g)]
Page 3-5

E1C13
What notifications must be given before transmitting on the 630 meter or 2200 meter bands?
   A. A special endorsement must be requested from the FCC
   B. An environmental impact statement must be filed with the Department of the Interior
   C. Operators must inform the Utilities Technology Council (UTC) of their call sign and coordinates of the station
   D. Operators must inform the FAA of their intent to operate, giving their call sign and distance to the nearest runway

E1C14
(B)
[97.303(g)]
Page 3-5

E1C14
How long must an operator wait after filing a notification with the Utilities Technology Council (UTC) before operating on the 2200 meter or 630 meter band?
   A. Operators must not operate until approval is received
   B. Operators may operate after 30 days, providing they have not been told that their station is within 1 km of PLC systems using those frequencies
   C. Operators may not operate until a test signal has been transmitted in coordination with the local power company
   D. Operations may commence immediately, and may continue unless interference is reported by the UTC

E1D — Amateur space and Earth stations; telemetry and telecommand rules; identification of balloon transmissions; one-way communications

E1D01
What is the definition of telemetry?
A. One-way transmission of measurements at a distance from the measuring instrument
B. Two-way transmissions in excess of 1000 feet
C. Two-way transmissions of data
D. One-way transmission that initiates, modifies, or terminates the functions of a device at a distance

E1D01
(A)
[97.3]
Page 3-12

E1D02
Which of the following may transmit special codes intended to obscure the meaning of messages?
A. Telecommand signals from a space telecommand station
B. Data containing personal information
C. Auxiliary relay links carrying repeater audio
D. Binary control characters

E1D02
(A)
[97.211(b)]
Page 3-13

E1D03
What is a space telecommand station?
A. An amateur station located on the surface of the Earth for communication with other Earth stations by means of Earth satellites
B. An amateur station that transmits communications to initiate, modify or terminate functions of a space station
C. An amateur station located in a satellite or a balloon more than 50 kilometers above the surface of the Earth
D. An amateur station that receives telemetry from a satellite or balloon more than 50 kilometers above the surface of the Earth

E1D03
(B)
[97.3(a)(45)]
Page 3-13

E1D04
Which of the following is required in the identification transmissions from a balloon-borne telemetry station?
A. Call sign
B. The output power of the balloon transmitter
C. The station's six-character Maidenhead grid locator
D. All these choices are correct

E1D04
(A)
[97.119(a)]
Page 3-12

E1D05
What must be posted at the station location of a station being operated by telecommand on or within 50 km of the earth's surface?
A. A photocopy of the station license
B. A label with the name, address, and telephone number of the station licensee
C. A label with the name, address, and telephone number of the control operator
D. All these choices are correct

E1D05
(D)
[97.213(d)]
Page 3-10

E1D06
What is the maximum permitted transmitter output power when operating a model craft by telecommand?
A. 1 watt
B. 2 watts
C. 5 watts
D. 100 watts

E1D06
(A)
[97.215(c)]
Page 3-10

Question Pool   13-11

E1D07
(A)
[97.207]
Page 3-13

E1D07
Which HF amateur bands have frequencies authorized for space stations?
 A. Only the 40, 20, 17, 15, 12, and 10 meter bands
 B. Only the 40, 20, 17, 15, and 10 meter bands
 C. Only the 40, 30, 20, 15, 12, and 10 meter bands
 D. All HF bands

E1D08
(D)
[97.207]
Page 3-13

E1D08
Which VHF amateur bands have frequencies authorized for space stations?
 A. 6 meters and 2 meters
 B. 6 meters, 2 meters, and 1.25 meters
 C. 2 meters and 1.25 meters
 D. 2 meters

E1D09
(B)
[97.207]
Page 3-13

E1D09
Which UHF amateur bands have frequencies authorized for space stations?
 A. 70 cm only
 B. 70 cm and 13 cm
 C. 70 cm and 33 cm
 D. 33 cm and 13 cm

E1D10
(B)
[97.211]
Page 3-13

E1D10
Which amateur stations are eligible to be telecommand stations of space stations (subject to the privileges of the class of operator license held by the control operator of the station)?
 A. Any amateur station designated by NASA
 B. Any amateur station so designated by the space station licensee
 C. Any amateur station so designated by the ITU
 D. All these choices are correct

E1D11
(D)
[97.209]
Page 3-13

E1D11
Which amateur stations are eligible to operate as Earth stations?
 A. Any amateur station whose licensee has filed a pre-space notification with the FCC's International Bureau
 B. Only those of General, Advanced or Amateur Extra Class operators
 C. Only those of Amateur Extra Class operators
 D. Any amateur station, subject to the privileges of the class of operator license held by the control operator

E1D12
(A)
[97.207(e),
97.203(g)]
Page 3-10

E1D12
Which of the following amateur stations may transmit one-way communications?
 A. A space station, beacon station, or telecommand station
 B. A local repeater or linked repeater station
 C. A message forwarding station or automatically controlled digital station
 D. All these choices are correct

13-12   Chapter 13

E1E — Volunteer examiner program: definitions; qualifications; preparation and administration of exams; accreditation; question pools; documentation requirements

E1E01

For which types of out-of-pocket expenses do the Part 97 rules state that VEs and VECs may be reimbursed?

A. Preparing, processing, administering, and coordinating an examination for an amateur radio operator license
B. Teaching an amateur operator license examination preparation course
C. No expenses are authorized for reimbursement
D. Providing amateur operator license examination preparation training materials

E1E02

Who does Part 97 task with maintaining the pools of questions for all U.S. amateur license examinations?

A. The VEs
B. The FCC
C. The VECs
D. The ARRL

E1E03

What is a Volunteer Examiner Coordinator?

A. A person who has volunteered to administer amateur operator license examinations
B. A person who has volunteered to prepare amateur operator license examinations
C. An organization that has entered into an agreement with the FCC to coordinate, prepare, and administer amateur operator license examinations
D. The person who has entered into an agreement with the FCC to be the VE session manager

E1E04

Which of the following best describes the Volunteer Examiner accreditation process?

A. Each General, Advanced and Amateur Extra Class operator is automatically accredited as a VE when the license is granted
B. The amateur operator applying must pass a VE examination administered by the FCC Enforcement Bureau
C. The prospective VE obtains accreditation from the FCC
D. The procedure by which a VEC confirms that the VE applicant meets FCC requirements to serve as an examiner

E1E05

What is the minimum passing score on all amateur operator license examinations?

A. Minimum passing score of 70%
B. Minimum passing score of 74%
C. Minimum passing score of 80%
D. Minimum passing score of 77%

E1E06

Who is responsible for the proper conduct and necessary supervision during an amateur operator license examination session?

A. The VEC coordinating the session
B. The FCC
C. Each administering VE
D. The VE session manager

E1E01
(A)
[97.527]
Page 3-17

E1E02
(C)
[97.523]
Page 3-16

E1E03
(C)
[97.521]
Page 3-14

E1E04
(D)
[97.509, 97.525]
Page 3-16

E1E05
(B)
[97.503]
Page 3-18

E1E06
(C)
[97.509]
Page 3-17

E1E07
(B)
[97.509]
Page 3-17

E1E07
What should a VE do if a candidate fails to comply with the examiner's instructions during an amateur operator license examination?
A. Warn the candidate that continued failure to comply will result in termination of the examination
B. Immediately terminate the candidate's examination
C. Allow the candidate to complete the examination, but invalidate the results
D. Immediately terminate everyone's examination and close the session

E1E08
(C)
[97.509]
Page 3-16

E1E08
To which of the following examinees may a VE not administer an examination?
A. Employees of the VE
B. Friends of the VE
C. Relatives of the VE as listed in the FCC rules
D. All these choices are correct

E1E09
(A)
[97.509]
Page 3-18

E1E09
What may be the penalty for a VE who fraudulently administers or certifies an examination?
A. Revocation of the VE's amateur station license grant and the suspension of the VE's amateur operator license grant
B. A fine of up to $1000 per occurrence
C. A sentence of up to one year in prison
D. All these choices are correct

E1E10
(C)
[97.509(h)]
Page 3-18

E1E10
What must the administering VEs do after the administration of a successful examination for an amateur operator license?
A. They must collect and send the documents to the NCVEC for grading
B. They must collect and submit the documents to the coordinating VEC for grading
C. They must submit the application document to the coordinating VEC according to the coordinating VEC instructions
D. They must collect and send the documents to the FCC according to instructions

E1E11
(B)
[97.509(m)]
Page 3-18

E1E11
What must the VE team do if an examinee scores a passing grade on all examination elements needed for an upgrade or new license?
A. Photocopy all examination documents and forward them to the FCC for processing
B. Three VEs must certify that the examinee is qualified for the license grant and that they have complied with the administering VE requirements
C. Issue the examinee the new or upgrade license
D. All these choices are correct

E1E12
(A)
[97.509(j)]
Page 3-18

E1E12
What must the VE team do with the application form if the examinee does not pass the exam?
A. Return the application document to the examinee
B. Maintain the application form with the VEC's records
C. Send the application form to the FCC and inform the FCC of the grade
D. Destroy the application form

13-14   Chapter 13

E1F — Miscellaneous rules: external RF power amplifiers; prohibited communications; spread spectrum; auxiliary stations; Canadian amateurs operating in the U.S.; special temporary authority; control operator of an auxiliary station

E1F01
On what frequencies are spread spectrum transmissions permitted?
A. Only on amateur frequencies above 50 MHz
B. Only on amateur frequencies above 222 MHz
C. Only on amateur frequencies above 420 MHz
D. Only on amateur frequencies above 144 MHz

E1F01
(B)
[97.305]
Page 3-21

E1F02
What privileges are authorized in the U.S. to persons holding an amateur service license granted by the government of Canada?
A. None, they must obtain a U.S. license
B. All privileges of the Amateur Extra Class license
C. The operating terms and conditions of the Canadian amateur service license, not to exceed U.S. Amateur Extra Class license privileges
D. Full privileges, up to and including those of the Amateur Extra Class license, on the 80, 40, 20, 15, and 10 meter bands

E1F02
(C)
[97.107]
Page 3-21

E1F03
Under what circumstances may a dealer sell an external RF power amplifier capable of operation below 144 MHz if it has not been granted FCC certification?
A. It was purchased in used condition from an amateur operator and is sold to another amateur operator for use at that operator's station
B. The equipment dealer assembled it from a kit
C. It was imported from a manufacturer in a country that does not require certification of RF power amplifiers
D. It was imported from a manufacturer in another country and was certificated by that country's government

E1F03
(A)
[97.315]
Page 3-19

E1F04
Which of the following geographic descriptions approximately describes "Line A"?
A. A line roughly parallel to and south of the border between the U.S. and Canada
B. A line roughly parallel to and west of the U.S. Atlantic coastline
C. A line roughly parallel to and north of the border between the U.S. and Mexico
D. A line roughly parallel to and east of the U.S. Pacific coastline

E1F04
(A)
[97.3]
Page 3-19

E1F05
Amateur stations may not transmit in which of the following frequency segments if they are located in the contiguous 48 states and north of Line A?
A. 440 MHz - 450 MHz
B. 53 MHz - 54 MHz
C. 222 MHz - 223 MHz
D. 420 MHz - 430 MHz

E1F05
(D)
[97.303]
Page 3-19

Question Pool   13-15

E1F06
(A)
[1.931]
Page 3-21

E1F06
Under what circumstances might the FCC issue a Special Temporary Authority (STA) to an amateur station?
A. To provide for experimental amateur communications
B. To allow regular operation on Land Mobile channels
C. To provide additional spectrum for personal use
D. To provide temporary operation while awaiting normal licensing

E1F07
(D)
[97.113]
Page 3-20

E1F07
When may an amateur station send a message to a business?
A. When the total money involved does not exceed $25
B. When the control operator is employed by the FCC or another government agency
C. When transmitting international third-party communications
D. When neither the amateur nor his or her employer has a pecuniary interest in the communications

E1F08
(A)
[97.113(c)]
Page 3-20

E1F08
Which of the following types of amateur station communications are prohibited?
A. Communications transmitted for hire or material compensation, except as otherwise provided in the rules
B. Communications that have political content, except as allowed by the Fairness Doctrine
C. Communications that have religious content
D. Communications in a language other than English

E1F09
(D)
[97.311]
Page 3-21

E1F09
Which of the following conditions apply when transmitting spread spectrum emissions?
A. A station transmitting SS emission must not cause harmful interference to other stations employing other authorized emissions
B. The transmitting station must be in an area regulated by the FCC or in a country that permits SS emissions
C. The transmission must not be used to obscure the meaning of any communication
D. All these choices are correct

E1F10
(B)
[97.201]
Page 3-18

E1F10
Who may be the control operator of an auxiliary station?
A. Any licensed amateur operator
B. Only Technician, General, Advanced or Amateur Extra Class operators
C. Only General, Advanced or Amateur Extra Class operators
D. Only Amateur Extra Class operators

E1F11
(D)
[97.317]
Page 3-19

E1F11
Which of the following best describes one of the standards that must be met by an external RF power amplifier if it is to qualify for a grant of FCC certification?
A. It must produce full legal output when driven by not more than 5 watts of mean RF input power
B. It must be capable of external RF switching between its input and output networks
C. It must exhibit a gain of 0 dB or less over its full output range
D. It must satisfy the FCC's spurious emission standards when operated at the lesser of 1500 watts or its full output power

13-16   Chapter 13

SUBELEMENT E2 — OPERATING PROCEDURES
[5 Exam Questions — 5 Groups]

E2A — Amateur radio in space: amateur satellites; orbital mechanics; frequencies and modes; satellite hardware; satellite operations

E2A01
What is the direction of an ascending pass for an amateur satellite?
- A. From west to east
- B. From east to west
- C. From south to north
- D. From north to south

E2A01
(C)
Page 2-9

E2A02
Which of the following occurs when a satellite is using an inverting linear transponder?
- A. Doppler shift is reduced because the uplink and downlink shifts are in opposite directions
- B. Signal position in the band is reversed
- C. Upper sideband on the uplink becomes lower sideband on the downlink, and vice versa
- D. All these choices are correct

E2A02
(D)
Page 2-11

E2A03
How is the signal inverted by an inverting linear transponder?
- A. The signal is detected and remodulated on the reverse sideband
- B. The signal is passed through a non-linear filter
- C. The signal is reduced to I and Q components and the Q component is filtered out
- D. The signal is passed through a mixer and the difference rather than the sum is transmitted

E2A03
(D)
Page 2-11

E2A04
What is meant by the term "mode" as applied to an amateur radio satellite?
- A. Whether the satellite is in a low earth or geostationary orbit
- B. The satellite's uplink and downlink frequency bands
- C. The satellite's orientation with respect to the Earth
- D. Whether the satellite is in a polar or equatorial orbit

E2A04
(B)
Page 2-12

E2A05
What do the letters in a satellite's mode designator specify?
- A. Power limits for uplink and downlink transmissions
- B. The location of the ground control station
- C. The polarization of uplink and downlink signals
- D. The uplink and downlink frequency ranges

E2A05
(D)
Page 2-12

E2A06
What are Keplerian elements?
- A. Parameters that define the orbit of a satellite
- B. Phase reversing elements in a Yagi antenna
- C. High-emission heater filaments used in magnetron tubes
- D. Encrypting codes used for spread spectrum modulation

E2A06
(A)
Page 2-8

Question Pool   13-17

E2A07
(D)
Page 2-10

E2A07
Which of the following types of signals can be relayed through a linear transponder?
A. FM and CW
B. SSB and SSTV
C. PSK and packet
D. All these choices are correct

E2A08
(B)
Page 2-11

E2A08
Why should effective radiated power to a satellite that uses a linear transponder be limited?
A. To prevent creating errors in the satellite telemetry
B. To avoid reducing the downlink power to all other users
C. To prevent the satellite from emitting out-of-band signals
D. To avoid interfering with terrestrial QSOs

E2A09
(A)
Page 2-12

E2A09
What do the terms "L band" and "S band" specify regarding satellite communications?
A. The 23 centimeter and 13 centimeter bands
B. The 2 meter and 70 centimeter bands
C. FM and Digital Store-and-Forward systems
D. Which sideband to use

E2A10
(B)
Page 2-8

E2A10
What type of satellite appears to stay in one position in the sky?
A. HEO
B. Geostationary
C. Geomagnetic
D. LEO

E2A11
(B)
Page 2-9

E2A11
What type of antenna can be used to minimize the effects of spin modulation and Faraday rotation?
A. A linearly polarized antenna
B. A circularly polarized antenna
C. An isotropic antenna
D. A log-periodic dipole array

E2A12
(C)
Page 2-12

E2A12
What is the purpose of digital store-and-forward functions on an amateur radio satellite?
A. To upload operational software for the transponder
B. To delay download of telemetry between satellites
C. To store digital messages in the satellite for later download by other stations
D. To relay messages between satellites

E2A13
(B)
Page 2-12

E2A13
Which of the following techniques is normally used by low Earth orbiting digital satellites to relay messages around the world?
A. Digipeating
B. Store-and-forward
C. Multi-satellite relaying
D. Node hopping

13-18   Chapter 13

E2B — Television practices: fast scan television standards and techniques; slow scan television standards and techniques

E2B01
How many times per second is a new frame transmitted in a fast-scan (NTSC) television system?
  A. 30
  B. 60
  C. 90
  D. 120

E2B02
How many horizontal lines make up a fast-scan (NTSC) television frame?
  A. 30
  B. 60
  C. 525
  D. 1080

E2B03
How is an interlaced scanning pattern generated in a fast-scan (NTSC) television system?
  A. By scanning two fields simultaneously
  B. By scanning each field from bottom to top
  C. By scanning lines from left to right in one field and right to left in the next
  D. By scanning odd numbered lines in one field and even numbered lines in the next

E2B04
How is color information sent in analog SSTV?
  A. Color lines are sent sequentially
  B. Color information is sent on a 2.8 kHz subcarrier
  C. Color is sent in a color burst at the end of each line
  D. Color is amplitude modulated on the frequency modulated intensity signal

E2B05
Which of the following describes the use of vestigial sideband in analog fast-scan TV transmissions?
  A. The vestigial sideband carries the audio information
  B. The vestigial sideband contains chroma information
  C. Vestigial sideband reduces bandwidth while allowing for simple video detector circuitry
  D. Vestigial sideband provides high frequency emphasis to sharpen the picture

E2B06
What is vestigial sideband modulation?
  A. Amplitude modulation in which one complete sideband and a portion of the other are transmitted
  B. A type of modulation in which one sideband is inverted
  C. Narrow-band FM modulation achieved by filtering one sideband from the audio before frequency modulating the carrier
  D. Spread spectrum modulation achieved by applying FM modulation following single sideband amplitude modulation

E2B01
(A)
Page 8-19

E2B02
(C)
Page 8-19

E2B03
(D)
Page 8-20

E2B04
(A)
Page 8-23

E2B05
(C)
Page 8-21

E2B06
(A)
Page 8-21

E2B07
(B)
Page 8-20

E2B07
What is the name of the signal component that carries color information in NTSC video?
A. Luminance
B. Chroma
C. Hue
D. Spectral intensity

E2B08
(A)
Page 8-19

E2B08
What technique allows commercial analog TV receivers to be used for fast-scan TV operations on the 70 cm band?
A. Transmitting on channels shared with cable TV
B. Using converted satellite TV dishes
C. Transmitting on the abandoned TV channel 2
D. Using USB and demodulating the signal with a computer sound card

E2B09
(D)
Page 8-23

E2B09
What hardware, other than a receiver with SSB capability and a suitable computer, is needed to decode SSTV using Digital Radio Mondiale (DRM)?
A. A special IF converter
B. A special front end limiter
C. A special notch filter to remove synchronization pulses
D. No other hardware is needed

E2B10
(A)
Page 8-23

E2B10
What aspect of an analog slow-scan television signal encodes the brightness of the picture?
A. Tone frequency
B. Tone amplitude
C. Sync amplitude
D. Sync frequency

E2B11
(B)
Page 8-23

E2B11
What is the function of the Vertical Interval Signaling (VIS) code sent as part of an SSTV transmission?
A. To lock the color burst oscillator in color SSTV images
B. To identify the SSTV mode being used
C. To provide vertical synchronization
D. To identify the call sign of the station transmitting

E2B12
(A)
Page 8-23

E2B12
What signals SSTV receiving software to begin a new picture line?
A. Specific tone frequencies
B. Elapsed time
C. Specific tone amplitudes
D. A two-tone signal

13-20   Chapter 13

E2C — Operating methods: contest and DX operating; remote operation techniques; Cabrillo format; QSLing; RF network connected systems

E2C01
What indicator is required to be used by U.S.-licensed operators when operating a station via remote control and the remote transmitter is located in the U.S.?
  A. / followed by the USPS two-letter abbreviation for the state in which the remote station is located
  B. /R# where # is the district of the remote station
  C. / followed by the ARRL Section of the remote station
  D. No additional indicator is required

E2C01
(D)
Page 2-7

E2C02
Which of the following best describes the term "self-spotting" in connection with HF contest operating?
  A. The often-prohibited practice of posting one's own call sign and frequency on a spotting network
  B. The acceptable practice of manually posting the call signs of stations on a spotting network
  C. A manual technique for rapidly zero beating or tuning to a station's frequency before calling that station
  D. An automatic method for rapidly zero beating or tuning to a station's frequency before calling that station

E2C02
(A)
Page 2-7

E2C03
From which of the following bands is amateur radio contesting generally excluded?
  A. 30 meters
  B. 6 meters
  C. 2 meters
  D. 33 centimeters

E2C03
(A)
Page 2-6

E2C04
Which of the following frequencies are sometimes used for amateur radio mesh networks?
  A. HF frequencies where digital communications are permitted
  B. Frequencies shared with various unlicensed wireless data services
  C. Cable TV channels 41 through 43
  D. The 60 meter band channel centered on 5373 kHz

E2C04
(B)
Page 8-16

E2C05
What is the function of a DX QSL Manager?
  A. To allocate frequencies for DXpeditions
  B. To handle the receiving and sending of confirmation cards for a DX station
  C. To run a net to allow many stations to contact a rare DX station
  D. To relay calls to and from a DX station

E2C05
(B)
Page 2-3

E2C06
During a VHF/UHF contest, in which band segment would you expect to find the highest level of SSB or CW activity?
  A. At the top of each band, usually in a segment reserved for contests
  B. In the middle of each band, usually on the national calling frequency
  C. In the weak signal segment of the band, with most of the activity near the calling frequency
  D. In the middle of the band, usually 25 kHz above the national calling frequency

E2C06
(C)
Page 2-6

Question Pool   13-21

E2C07
(A)
Page 2-6

E2C07
What is the Cabrillo format?
A. A standard for submission of electronic contest logs
B. A method of exchanging information during a contest QSO
C. The most common set of contest rules
D. The rules of order for meetings between contest sponsors

E2C08
(B)
Page 2-3

E2C08
Which of the following contacts may be confirmed through the U.S. QSL bureau system?
A. Special event contacts between stations in the U.S.
B. Contacts between a U.S. station and a non-U.S. station
C. Repeater contacts between U.S. club members
D. Contacts using tactical call signs

E2C09
(C)
Page 8-16

E2C09
What type of equipment is commonly used to implement an amateur radio mesh network?
A. A 2 meter VHF transceiver with a 1200 baud modem
B. An optical cable connection between the USB ports of 2 separate computers
C. A wireless router running custom firmware
D. A 440 MHz transceiver with a 9600 baud modem

E2C10
(D)
Page 2-4

E2C10
Why might a DX station state that they are listening on another frequency?
A. Because the DX station may be transmitting on a frequency that is prohibited to some responding stations
B. To separate the calling stations from the DX station
C. To improve operating efficiency by reducing interference
D. All these choices are correct

E2C11
(A)
Page 2-4

E2C11
How should you generally identify your station when attempting to contact a DX station during a contest or in a pileup?
A. Send your full call sign once or twice
B. Send only the last two letters of your call sign until you make contact
C. Send your full call sign and grid square
D. Send the call sign of the DX station three times, the words "this is," then your call sign three times

E2C12
(C)
Page 8-16

E2C12
What technique do individual nodes use to form a mesh network?
A. Forward error correction and Viterbi codes
B. Acting as store-and-forward digipeaters
C. Discovery and link establishment protocols
D. Custom code plugs for the local trunking systems

E2D — Operating methods: VHF and UHF digital modes and procedures; APRS; EME procedures; meteor scatter procedures

E2D01
Which of the following digital modes is designed for meteor scatter communications?
  A. WSPR
  B. MSK144
  C. Hellschreiber
  D. APRS

E2D01
(B)
Page 10-16

E2D02
Which of the following is a good technique for making meteor scatter contacts?
  A. 15-second timed transmission sequences with stations alternating based on location
  B. Use of special digital modes
  C. Short transmissions with rapidly repeated call signs and signal reports
  D. All these choices are correct

E2D02
(D)
Page 10-16

E2D03
Which of the following digital modes is especially useful for EME communications?
  A. MSK144
  B. PACTOR III
  C. Olivia
  D. JT65

E2D03
(D)
Page 10-17

E2D04
What technology is used to track, in real time, balloons carrying amateur radio transmitters?
  A. Ultrasonics
  B. Bandwidth compressed LORAN
  C. APRS
  D. Doppler shift of beacon signals

E2D04
(C)
Page 8-12

E2D05
What is one advantage of the JT65 mode?
  A. Uses only a 65 Hz bandwidth
  B. The ability to decode signals which have a very low signal-to-noise ratio
  C. Easily copied by ear if necessary
  D. Permits fast-scan TV transmissions over narrow bandwidth

E2D05
(B)
Page 10-17

E2D06
Which of the following describes a method of establishing EME contacts?
  A. Time synchronous transmissions alternately from each station
  B. Storing and forwarding digital messages
  C. Judging optimum transmission times by monitoring beacons reflected from the moon
  D. High-speed CW identification to avoid fading

E2D06
(A)
Page 10-17

E2D07
What digital protocol is used by APRS?
  A. PACTOR
  B. 802.11
  C. AX.25
  D. AMTOR

E2D07
(C)
Page 8-11

Question Pool   13-23

E2D08
(A)
Page 8-12

E2D08
What type of packet frame is used to transmit APRS beacon data?
A. Unnumbered Information
B. Disconnect
C. Acknowledgement
D. Connect

E2D09
(A)
Page 8-13

E2D09
What type of modulation is used for JT65 contacts?
A. Multi-tone AFSK
B. PSK
C. RTTY
D. IEEE 802.11

E2D10
(C)
Page 8-12

E2D10
How can an APRS station be used to help support a public service communications activity?
A. An APRS station with an emergency medical technician can automatically transmit medical data to the nearest hospital
B. APRS stations with General Personnel Scanners can automatically relay the participant numbers and time as they pass the check points
C. An APRS station with a Global Positioning System unit can automatically transmit information to show a mobile station's position during the event
D. All these choices are correct

E2D11
(D)
Page 8-12

E2D11
Which of the following data are used by the APRS network to communicate station location?
A. Polar coordinates
B. Time and frequency
C. Radio direction finding spectrum analysis
D. Latitude and longitude

E2E — Operating methods: operating HF digital modes

E2E01
(B)
Page 8-10

E2E01
Which of the following types of modulation is common for data emissions below 30 MHz?
A. DTMF tones modulating an FM signal
B. FSK
C. Pulse modulation
D. Spread spectrum

E2E02
(A)
Page 8-17

E2E02
What do the letters FEC mean as they relate to digital operation?
A. Forward Error Correction
B. First Error Correction
C. Fatal Error Correction
D. Final Error Correction

E2E03
(C)
Page 8-13

E2E03
How is the timing of FT4 contacts organized?
A. By exchanging ACK/NAK packets
B. Stations take turns on alternate days
C. Alternating transmissions at 7.5 second intervals
D. It depends on the lunar phase

13-24   Chapter 13

E2E04
What is indicated when one of the ellipses in an FSK crossed-ellipse display suddenly disappears?
  A. Selective fading has occurred
  B. One of the signal filters is saturated
  C. The receiver has drifted 5 kHz from the desired receive frequency
  D. The mark and space signal have been inverted

E2E04
(A)
Page 8-10

E2E05
Which of these digital modes does not support keyboard-to-keyboard operation?
  A. PACTOR
  B. RTTY
  C. PSK31
  D. MFSK

E2E05
(A)
Page 8-12

E2E06
What is the most common data rate used for HF packet?
  A. 48 baud
  B. 110 baud
  C. 300 baud
  D. 1200 baud

E2E06
(C)
Page 8-12

E2E07
Which of the following is a possible reason that attempts to initiate contact with a digital station on a clear frequency are unsuccessful?
  A. Your transmit frequency is incorrect
  B. The protocol version you are using is not supported by the digital station
  C. Another station you are unable to hear is using the frequency
  D. All these choices are correct

E2E07
(D)
Page 8-14

E2E08
Which of the following HF digital modes can be used to transfer binary files?
  A. Hellschreiber
  B. PACTOR
  C. RTTY
  D. AMTOR

E2E08
(B)
Page 8-12

E2E09
Which of the following HF digital modes uses variable-length coding for bandwidth efficiency?
  A. RTTY
  B. PACTOR
  C. MT63
  D. PSK31

E2E09
(D)
Page 8-7

E2E10
Which of these digital modes has the narrowest bandwidth?
  A. MFSK16
  B. 170 Hz shift, 45-baud RTTY
  C. PSK31
  D. 300-baud packet

E2E10
(C)
Page 8-11

Question Pool   13-25

E2E11
(A)
Page 8-10

E2E11
What is the difference between direct FSK and audio FSK?
  A. Direct FSK applies the data signal to the transmitter VFO, while AFSK transmits tones via phone
  B. Direct FSK occupies less bandwidth
  C. Direct FSK can transmit faster baud rates
  D. Only direct FSK can be decoded by computer

E2E12
(A)
Page 8-6

E2E12
How do ALE stations establish contact?
  A. ALE constantly scans a list of frequencies, activating the radio when the designated call sign is received
  B. ALE radios monitor an internet site for the frequency they are being paged on
  C. ALE radios send a constant tone code to establish a frequency for future use
  D. ALE radios activate when they hear their signal echoed by back scatter

E2E13
(D)
Page 8-12

E2E13
Which of these digital modes has the fastest data throughput under clear communication conditions?
  A. AMTOR
  B. 170 Hz shift, 45 baud RTTY
  C. PSK31
  D. 300 baud packet

SUBELEMENT E3 — RADIO WAVE PROPAGATION
[3 Exam Questions — 3 Groups]

E3A — Electromagnetic waves; Earth-Moon-Earth communications; meteor scatter; microwave tropospheric and scatter propagation; aurora propagation; ionospheric propagation changes over the day; circular polarization

E3A01
(D)
Page 10-17

E3A01
What is the approximate maximum separation measured along the surface of the Earth between two stations communicating by EME?
  A. 500 miles, if the moon is at perigee
  B. 2000 miles, if the moon is at apogee
  C. 5000 miles, if the moon is at perigee
  D. 12,000 miles, if the moon is visible by both stations

E3A02
(B)
Page 10-17

E3A02
What characterizes libration fading of an EME signal?
  A. A slow change in the pitch of the CW signal
  B. A fluttery irregular fading
  C. A gradual loss of signal as the sun rises
  D. The returning echo is several hertz lower in frequency than the transmitted signal

E3A03
When scheduling EME contacts, which of these conditions will generally result in the least path loss?
A. When the moon is at perigee
B. When the moon is full
C. When the moon is at apogee
D. When the MUF is above 30 MHz

E3A03
(A)
Page 10-17

E3A04
What do Hepburn maps predict?
A. Sporadic E propagation
B. Locations of auroral reflecting zones
C. Likelihood of rain scatter along cold or warm fronts
D. Probability of tropospheric propagation

E3A04
(D)
Page 10-12

E3A05
Tropospheric propagation of microwave signals often occurs in association with what phenomenon?
A. Grayline
B. Lightning discharges
C. Warm and cold fronts
D. Sprites and jets

E3A05
(C)
Page 10-12

E3A06
What might help to restore contact when DX signals become too weak to copy across an entire HF band a few hours after sunset?
A. Switch to a higher frequency HF band
B. Switch to a lower frequency HF band
C. Wait 90 minutes or so for the signal degradation to pass
D. Wait 24 hours before attempting another communication on the band

E3A06
(B)
Page 10-7

E3A07
Atmospheric ducts capable of propagating microwave signals often form over what geographic feature?
A. Mountain ranges
B. Forests
C. Bodies of water
D. Urban areas

E3A07
(C)
Page 10-12

E3A08
When a meteor strikes the Earth's atmosphere, a cylindrical region of free electrons is formed at what layer of the ionosphere?
A. The E layer
B. The F1 layer
C. The F2 layer
D. The D layer

E3A08
(A)
Page 10-16

E3A09
Which of the following frequency ranges is most suited for meteor scatter communications?
A. 1.8 MHz - 1.9 MHz
B. 10 MHz - 14 MHz
C. 28 MHz - 148 MHz
D. 220 MHz - 450 MHz

E3A09
(C)
Page 10-16

E3A10
(B)
Page 10-12

E3A10
Which type of atmospheric structure can create a path for microwave propagation?
A. The jet stream
B. Temperature inversion
C. Wind shear
D. Dust devil

E3A11
(B)
Page 10-12

E3A11
What is a typical range for tropospheric propagation of microwave signals?
A. 10 miles to 50 miles
B. 100 miles to 300 miles
C. 1200 miles
D. 2500 miles

E3A12
(C)
Page 10-14

E3A12
What is the cause of auroral activity?
A. The interaction in the F2 layer between the solar wind and the Van Allen belt
B. An extreme low-pressure area in the polar regions
C. The interaction in the E layer of charged particles from the Sun with the Earth's magnetic field
D. Meteor showers concentrated in the extreme northern and southern latitudes

E3A13
(A)
Page 10-14

E3A13
Which of these emission modes is best for auroral propagation?
A. CW
B. SSB
C. FM
D. RTTY

E3A14
(B)
Page 10-3

E3A14
What is meant by circularly polarized electromagnetic waves?
A. Waves with an electric field bent into a circular shape
B. Waves with a rotating electric field
C. Waves that circle the Earth
D. Waves produced by a loop antenna

E3B — Transequatorial propagation; long-path; grayline; ordinary and extraordinary waves; chordal hop; sporadic E mechanisms

E3B01
(A)
Page 10-13

E3B01
What is transequatorial propagation?
A. Propagation between two mid-latitude points at approximately the same distance north and south of the magnetic equator
B. Propagation between points located on the magnetic equator
C. Propagation between a point on the equator and its antipodal point
D. Propagation between points at the same latitude

E3B02
(C)
Page 10-14

E3B02
What is the approximate maximum range for signals using transequatorial propagation?
A. 1000 miles
B. 2500 miles
C. 5000 miles
D. 7500 miles

E3B03
What is the best time of day for transequatorial propagation?
A. Morning
B. Noon
C. Afternoon or early evening
D. Late at night

E3B03
(C)
Page 10-14

E3B04
What is meant by the terms "extraordinary" and "ordinary" waves?
A. Extraordinary waves describe rare long-skip propagation compared to ordinary waves, which travel shorter distances
B. Independent waves created in the ionosphere that are elliptically polarized
C. Long-path and short-path waves
D. Refracted rays and reflected waves

E3B04
(B)
Page 10-7

E3B05
Which amateur bands typically support long-path propagation?
A. Only 160 meters to 40 meters
B. Only 30 meters to 10 meters
C. 160 meters to 10 meters
D. 6 meters to 2 meters

E3B05
(C)
Page 10-9

E3B06
Which of the following amateur bands most frequently provides long-path propagation?
A. 80 meters
B. 20 meters
C. 10 meters
D. 6 meters

E3B06
(B)
Page 10-9

E3B07
What happens to linearly polarized radio waves that split into ordinary and extraordinary waves in the ionosphere?
A. They are bent toward the magnetic poles
B. They become depolarized
C. They become elliptically polarized
D. They become phase locked

E3B07
(C)
Page 10-7

E3B08 — Question has been withdrawn

E3B09
At what time of year is sporadic E propagation most likely to occur?
A. Around the solstices, especially the summer solstice
B. Around the solstices, especially the winter solstice
C. Around the equinoxes, especially the spring equinox
D. Around the equinoxes, especially the fall equinox

E3B09
(A)
Page 10-13

Question Pool   13-29

E3B10
(A)
Page 10-6

E3B11
(D)
Page 10-13

E3B12
(B)
Page 10-6

E3B10
Why is chordal hop propagation desirable?
   A. The signal experiences less loss compared to multi-hop using Earth as a reflector
   B. The MUF for chordal hop propagation is much lower than for normal skip propagation
   C. Atmospheric noise is lower in the direction of chordal hop propagation
   D. Signals travel faster along ionospheric chords

E3B11
At what time of day can sporadic E propagation occur?
   A. Only around sunset
   B. Only around sunset and sunrise
   C. Only in hours of darkness
   D. Any time

E3B12
What is the primary characteristic of chordal hop propagation?
   A. Propagation away from the great circle bearing between stations
   B. Successive ionospheric refractions without an intermediate reflection from the ground
   C. Propagation across the geomagnetic equator
   D. Signals reflected back toward the transmitting station

E3C — Radio horizon; ground wave; propagation prediction techniques and modeling; effects of space weather parameters on propagation

E3C01
(B)
Page 10-8

E3C02
(A)
Page 10-5

E3C03
(B)
Page 10-8

E3C04
(C)
Page 10-5

E3C01
What does the radio communication term "ray tracing" describe?
   A. The process in which an electronic display presents a pattern
   B. Modeling a radio wave's path through the ionosphere
   C. Determining the radiation pattern from an array of antennas
   D. Evaluating high voltage sources for x-rays

E3C02
What is indicated by a rising A or K index?
   A. Increasing disruption of the geomagnetic field
   B. Decreasing disruption of the geomagnetic field
   C. Higher levels of solar UV radiation
   D. An increase in the critical frequency

E3C03
Which of the following signal paths is most likely to experience high levels of absorption when the A index or K index is elevated?
   A. Transequatorial
   B. Polar
   C. Sporadic E
   D. NVIS

E3C04
What does the value of Bz (B sub Z) represent?
   A. Geomagnetic field stability
   B. Critical frequency for vertical transmissions
   C. Direction and strength of the interplanetary magnetic field
   D. Duration of long-delayed echoes

E3C05
What orientation of Bz (B sub z) increases the likelihood that incoming particles from the sun will cause disturbed conditions?
  A. Southward
  B. Northward
  C. Eastward
  D. Westward

E3C06
By how much does the VHF/UHF radio horizon distance exceed the geometric horizon?
  A. By approximately 15 percent of the distance
  B. By approximately twice the distance
  C. By approximately 50 percent of the distance
  D. By approximately four times the distance

E3C07
Which of the following descriptors indicates the greatest solar flare intensity?
  A. Class A
  B. Class B
  C. Class M
  D. Class X

E3C08
What does the space weather term "G5" mean?
  A. An extreme geomagnetic storm
  B. Very low solar activity
  C. Moderate solar wind
  D. Waning sunspot numbers

E3C09
How does the intensity of an X3 flare compare to that of an X2 flare?
  A. 10 percent greater
  B. 50 percent greater
  C. Twice as great
  D. Four times as great

E3C10
What does the 304A solar parameter measure?
  A. The ratio of x-ray flux to radio flux, correlated to sunspot number
  B. UV emissions at 304 angstroms, correlated to the solar flux index
  C. The solar wind velocity at 304 degrees from the solar equator, correlated to solar activity
  D. The solar emission at 304 GHz, correlated to x-ray flare levels

E3C11
What does VOACAP software model?
  A. AC voltage and impedance
  B. VHF radio propagation
  C. HF propagation
  D. AC current and impedance

E3C05
(A)
Page 10-5

E3C06
(A)
Page 10-10

E3C07
(D)
Page 10-4

E3C08
(A)
Page 10-5

E3C09
(B)
Page 10-4

E3C10
(B)
Page 10-4

E3C11
(C)
Page 10-8

E3C12
(C)
Page 10-6

E3C12
How does the maximum range of ground-wave propagation change when the signal frequency is increased?
    A. It stays the same
    B. It increases
    C. It decreases
    D. It peaks at roughly 14 MHz

E3C13
(A)
Page 10-6

E3C13
What type of polarization is best for ground-wave propagation?
    A. Vertical
    B. Horizontal
    C. Circular
    D. Elliptical

E3C14
(D)
Page 10-10

E3C14
Why does the radio-path horizon distance exceed the geometric horizon?
    A. E-region skip
    B. D-region skip
    C. Due to the Doppler effect
    D. Downward bending due to density variations in the atmosphere

E3C15
(B)
Page 10-8

E3C15
What might be indicated by a sudden rise in radio background noise across a large portion of the HF spectrum?
    A. A temperature inversion has occurred
    B. A solar flare has occurred
    C. Increased transequatorial propagation is likely
    D. Long-path propagation is likely

SUBELEMENT E4 — AMATEUR PRACTICES
[5 Exam Questions — 5 Groups]

E4A — Test equipment: analog and digital instruments; spectrum analyzers; antenna analyzers; oscilloscopes; RF measurements; computer-aided measurements

E4A01
(A)
Page 7-7

E4A01
Which of the following limits the highest frequency signal that can be accurately displayed on a digital oscilloscope?
    A. Sampling rate of the analog-to-digital converter
    B. Amount of memory
    C. Q of the circuit
    D. All these choices are correct

E4A02
(B)
Page 7-10

E4A02
Which of the following parameters does a spectrum analyzer display on the vertical and horizontal axes?
    A. RF amplitude and time
    B. RF amplitude and frequency
    C. SWR and frequency
    D. SWR and time

13-32   Chapter 13

E4A03

Which of the following test instruments is used to display spurious signals and/or intermodulation distortion products generated by an SSB transmitter?

A. A wattmeter
B. A spectrum analyzer
C. A logic analyzer
D. A time-domain reflectometer

E4A03
(B)
Page 7-10

E4A04

How is the compensation of an oscilloscope probe typically adjusted?

A. A square wave is displayed and the probe is adjusted until the horizontal portions of the displayed wave are as nearly flat as possible
B. A high frequency sine wave is displayed and the probe is adjusted for maximum amplitude
C. A frequency standard is displayed and the probe is adjusted until the deflection time is accurate
D. A DC voltage standard is displayed and the probe is adjusted until the displayed voltage is accurate

E4A04
(A)
Page 7-7

E4A05

What is the purpose of the prescaler function on a frequency counter?

A. It amplifies low-level signals for more accurate counting
B. It multiplies a higher frequency signal so a low-frequency counter can display the operating frequency
C. It prevents oscillation in a low-frequency counter circuit
D. It divides a higher frequency signal so low-frequency counter can display the input frequency

E4A05
(D)
Page 7-4

E4A06

What is the effect of aliasing on a digital oscilloscope caused by setting the time base too slow?

A. A false, jittery low-frequency version of the signal is displayed
B. All signals will have a DC offset
C. Calibration of the vertical scale is no longer valid
D. Excessive blanking occurs, which prevents display of the signal

E4A06
(A)
Page 7-7

E4A07

Which of the following is an advantage of using an antenna analyzer compared to an SWR bridge to measure antenna SWR?

A. Antenna analyzers automatically tune your antenna for resonance
B. Antenna analyzers do not need an external RF source
C. Antenna analyzers display a time-varying representation of the modulation envelope
D. All these choices are correct

E4A07
(B)
Page 9-40

E4A08

Which of the following measures SWR?

A. A spectrum analyzer
B. A Q meter
C. An ohmmeter
D. An antenna analyzer

E4A08
(D)
Page 9-39

Question Pool   13-33

E4A09
(A)
Page 7-6

E4A09
Which of the following is good practice when using an oscilloscope probe?
A. Keep the signal ground connection of the probe as short as possible
B. Never use a high-impedance probe to measure a low-impedance circuit
C. Never use a DC-coupled probe to measure an AC circuit
D. All these choices are correct

E4A10
(D)
Page 7-7

E4A10
Which of the following displays multiple digital signal states simultaneously?
A. Network analyzer
B. Bit error rate tester
C. Modulation monitor
D. Logic analyzer

E4A11
(D)
Page 9-39

E4A11
How should an antenna analyzer be connected when measuring antenna resonance and feed point impedance?
A. Loosely couple the analyzer near the antenna base
B. Connect the analyzer via a high-impedance transformer to the antenna
C. Loosely couple the antenna and a dummy load to the analyzer
D. Connect the antenna feed line directly to the analyzer's connector

E4B — Measurement technique and limitations: instrument accuracy and performance limitations; probes; techniques to minimize errors; measurement of Q; instrument calibration; S parameters; vector network analyzers

E4B01
(B)
Page 7-4

E4B01
Which of the following factors most affects the accuracy of a frequency counter?
A. Input attenuator accuracy
B. Time base accuracy
C. Decade divider accuracy
D. Temperature coefficient of the logic

E4B02
(A)
Page 7-2

E4B02
What is the significance of voltmeter sensitivity expressed in ohms per volt?
A. The full scale reading of the voltmeter multiplied by its ohms per volt rating will indicate the input impedance of the voltmeter
B. When used as a galvanometer, the reading in volts multiplied by the ohms per volt rating will determine the power drawn by the device under test
C. When used as an ohmmeter, the reading in ohms divided by the ohms per volt rating will determine the voltage applied to the circuit
D. When used as an ammeter, the full scale reading in amps divided by ohms per volt rating will determine the size of shunt needed

E4B03
(C)
Page 9-39

E4B03
Which S parameter is equivalent to forward gain?
A. S11
B. S12
C. S21
D. S22

13-34   Chapter 13

E4B04
Which S parameter represents input port return loss or reflection coefficient (equivalent to VSWR)?
  A. S11
  B. S12
  C. S21
  D. S22

E4B04
(A)
Page 9-38

E4B05
What three test loads are used to calibrate an RF vector network analyzer?
  A. 50 ohms, 75 ohms, and 90 ohms
  B. Short circuit, open circuit, and 50 ohms
  C. Short circuit, open circuit, and resonant circuit
  D. 50 ohms through ⅛ wavelength, ¼ wavelength, and ½ wavelength of coaxial cable

E4B05
(B)
Page 9-40

E4B06
How much power is being absorbed by the load when a directional power meter connected between a transmitter and a terminating load reads 100 watts forward power and 25 watts reflected power?
  A. 100 watts
  B. 125 watts
  C. 25 watts
  D. 75 watts

E4B06
(D)
Page 9-32

E4B07
What do the subscripts of S parameters represent?
  A. The port or ports at which measurements are made
  B. The relative time between measurements
  C. Relative quality of the data
  D. Frequency order of the measurements

E4B07
(A)
Page 9-38

E4B08
Which of the following can be used to measure the Q of a series-tuned circuit?
  A. The inductance to capacitance ratio
  B. The frequency shift
  C. The bandwidth of the circuit's frequency response
  D. The resonant frequency of the circuit

E4B08
(C)
Page 4-32

E4B09
What is indicated if the current reading on an RF ammeter placed in series with the antenna feed line of a transmitter increases as the transmitter is tuned to resonance?
  A. There is possibly a short to ground in the feed line
  B. The transmitter is not properly neutralized
  C. There is an impedance mismatch between the antenna and feed line
  D. There is more power going into the antenna

E4B09
(D)
Page 9-32

Question Pool   13-35

E4B10
(B)
Page 7-11

E4B10
Which of the following methods measures intermodulation distortion in an SSB transmitter?
    A. Modulate the transmitter using two RF signals having non-harmonically related frequencies and observe the RF output with a spectrum analyzer
    B. Modulate the transmitter using two AF signals having non-harmonically related frequencies and observe the RF output with a spectrum analyzer
    C. Modulate the transmitter using two AF signals having harmonically related frequencies and observe the RF output with a peak reading wattmeter
    D. Modulate the transmitter using two RF signals having harmonically related frequencies and observe the RF output with a logic analyzer

E4B11
(D)
Page 9-40

E4B11
Which of the following can be measured with a vector network analyzer?
    A. Input impedance
    B. Output impedance
    C. Reflection coefficient
    D. All these choices are correct

E4C — Receiver performance characteristics: phase noise, noise floor, image rejection, MDS, signal-to-noise ratio, noise figure, reciprocal mixing; selectivity; effects of SDR receiver non-linearity; use of attenuators at low frequencies

E4C01
(D)
Page 7-20

E4C01
What is an effect of excessive phase noise in a receiver's local oscillator?
    A. It limits the receiver's ability to receive strong signals
    B. It can affect the receiver's frequency calibration
    C. It decreases receiver third-order intercept point
    D. It can combine with strong signals on nearby frequencies to generate interference

E4C02
(A)
Page 7-14

E4C02
Which of the following receiver circuits can be effective in eliminating interference from strong out-of-band signals?
    A. A front-end filter or pre-selector
    B. A narrow IF filter
    C. A notch filter
    D. A properly adjusted product detector

E4C03
(C)
Page 7-21

E4C03
What is the term for the suppression in an FM receiver of one signal by another stronger signal on the same frequency?
    A. Desensitization
    B. Cross-modulation interference
    C. Capture effect
    D. Frequency discrimination

E4C04
(D)
Page 7-13

E4C04
What is the noise figure of a receiver?
    A. The ratio of atmospheric noise to phase noise
    B. The ratio of the noise bandwidth in hertz to the theoretical bandwidth of a resistive network
    C. The ratio of thermal noise to atmospheric noise
    D. The ratio in dB of the noise generated by the receiver to the theoretical minimum noise

E4C05

What does a receiver noise floor of −174 dBm represent?

A. The minimum detectable signal as a function of receive frequency

B. The theoretical noise in a 1 Hz bandwidth at the input of a perfect receiver at room temperature

C. The noise figure of a 1 Hz bandwidth receiver

D. The galactic noise contribution to minimum detectable signal

E4C06

A CW receiver with the AGC off has an equivalent input noise power density of −174 dBm/Hz. What would be the level of an unmodulated carrier input to this receiver that would yield an audio output SNR of 0 dB in a 400 Hz noise bandwidth?

A. −174 dBm

B. −164 dBm

C. −155 dBm

D. −148 dBm

E4C07

What does the MDS of a receiver represent?

A. The meter display sensitivity

B. The minimum discernible signal

C. The multiplex distortion stability

D. The maximum detectable spectrum

E4C08

An SDR receiver is overloaded when input signals exceed what level?

A. One-half the maximum sample rate

B. One-half the maximum sampling buffer size

C. The maximum count value of the analog-to-digital converter

D. The reference voltage of the analog-to-digital converter

E4C09

Which of the following choices is a good reason for selecting a high frequency for the design of the IF in a superheterodyne HF or VHF communications receiver?

A. Fewer components in the receiver

B. Reduced drift

C. Easier for front-end circuitry to eliminate image responses

D. Improved receiver noise figure

E4C10

What is an advantage of having a variety of receiver IF bandwidths from which to select?

A. The noise figure of the RF amplifier can be adjusted to match the modulation type, thus increasing receiver sensitivity

B. Receiver power consumption can be reduced when wider bandwidth is not required

C. Receive bandwidth can be set to match the modulation bandwidth, maximizing signal-to-noise ratio and minimizing interference

D. Multiple frequencies can be received simultaneously if desired

---

E4C05
(B)
Page 7-12

E4C06
(D)
Page 7-13

E4C07
(B)
Page 7-12

E4C08
(D)
Page 7-16

E4C09
(C)
Page 7-14

E4C10
(C)
Page 7-15

E4C11
(D)
Page 7-12

E4C11
Why can an attenuator be used to reduce receiver overload on the lower frequency HF bands with little or no impact on signal-to-noise ratio?
    A. The attenuator has a low-pass filter to increase the strength of lower frequency signals
    B. The attenuator has a noise filter to suppress interference
    C. Signals are attenuated separately from the noise
    D. Atmospheric noise is generally greater than internally generated noise even after attenuation

E4C12
(D)
Page 7-16

E4C12
Which of the following has the largest effect on an SDR receiver's dynamic range?
    A. CPU register width in bits
    B. Anti-aliasing input filter bandwidth
    C. RAM speed used for data storage
    D. Analog-to-digital converter sample width in bits

E4C13
(C)
Page 7-15

E4C13
How does a narrow-band roofing filter affect receiver performance?
    A. It improves sensitivity by reducing front end noise
    B. It improves intelligibility by using low Q circuitry to reduce ringing
    C. It improves dynamic range by attenuating strong signals near the receive frequency
    D. All these choices are correct

E4C14
(D)
Page 7-15

E4C14
What transmit frequency might generate an image response signal in a receiver tuned to 14.300 MHz and that uses a 455 kHz IF frequency?
    A. 13.845 MHz
    B. 14.755 MHz
    C. 14.445 MHz
    D. 15.210 MHz

E4C15
(D)
Page 7-20

E4C15
What is reciprocal mixing?
    A. Two out-of-band signals mixing to generate an in-band spurious signal
    B. In-phase signals cancelling in a mixer resulting in loss of receiver sensitivity
    C. Two digital signals combining from alternate time slots
    D. Local oscillator phase noise mixing with adjacent strong signals to create interference to desired signals

E4D — Receiver performance characteristics: blocking dynamic range; intermodulation and cross-modulation interference; third-order intercept; desensitization; preselector

E4D01
(A)
Page 7-17

E4D01
What is meant by the blocking dynamic range of a receiver?
    A. The difference in dB between the noise floor and the level of an incoming signal that will cause 1 dB of gain compression
    B. The minimum difference in dB between the levels of two FM signals that will cause one signal to block the other
    C. The difference in dB between the noise floor and the third-order intercept point
    D. The minimum difference in dB between two signals which produce third-order intermodulation products greater than the noise floor

E4D02

Which of the following describes problems caused by poor dynamic range in a receiver?

A. Spurious signals caused by cross-modulation and desensitization from strong adjacent signals
B. Oscillator instability requiring frequent retuning and loss of ability to recover the opposite sideband
C. Cross-modulation of the desired signal and insufficient audio power to operate the speaker
D. Oscillator instability and severe audio distortion of all but the strongest received signals

E4D03

How can intermodulation interference between two repeaters occur?

A. When the repeaters are in close proximity and the signals cause feedback in the final amplifier of one or both transmitters
B. When the repeaters are in close proximity and the signals mix in the final amplifier of one or both transmitters
C. When the signals from the transmitters are reflected out of phase from airplanes passing overhead
D. When the signals from the transmitters are reflected in phase from airplanes passing overhead

E4D04

Which of the following may reduce or eliminate intermodulation interference in a repeater caused by another transmitter operating in close proximity?

A. A band-pass filter in the feed line between the transmitter and receiver
B. A properly terminated circulator at the output of the repeater's transmitter
C. Utilizing a Class C final amplifier
D. Utilizing a Class D final amplifier

E4D05

What transmitter frequencies would cause an intermodulation-product signal in a receiver tuned to 146.70 MHz when a nearby station transmits on 146.52 MHz?

A. 146.34 MHz and 146.61 MHz
B. 146.88 MHz and 146.34 MHz
C. 146.10 MHz and 147.30 MHz
D. 173.35 MHz and 139.40 MHz

E4D06

What is the term for spurious signals generated by the combination of two or more signals in a non-linear device or circuit?

A. Amplifier desensitization
B. Neutralization
C. Adjacent channel interference
D. Intermodulation

E4D07

Which of the following reduces the likelihood of receiver desensitization?

A. Decrease the RF bandwidth of the receiver
B. Raise the receiver IF frequency
C. Increase the receiver front end gain
D. Switch from fast AGC to slow AGC

E4D02
(A)
Page 7-20

E4D03
(B)
Page 7-22

E4D04
(B)
Page 7-22

E4D05
(A)
Page 7-18

E4D06
(D)
Page 7-22

E4D07
(A)
Page 7-17

E4D08
(C)
Page 7-22

E4D08
What causes intermodulation in an electronic circuit?
  A. Too little gain
  B. Lack of neutralization
  C. Nonlinear circuits or devices
  D. Positive feedback

E4D09
(C)
Page 7-14

E4D09
What is the purpose of the preselector in a communications receiver?
  A. To store often-used frequencies
  B. To provide a range of AGC time constants
  C. To increase rejection of signals outside the desired band
  D. To allow selection of the optimum RF amplifier device

E4D10
(C)
Page 7-19

E4D10
What does a third-order intercept level of 40 dBm mean with respect to receiver performance?
  A. Signals less than 40 dBm will not generate audible third-order intermodulation products
  B. The receiver can tolerate signals up to 40 dB above the noise floor without producing third-order intermodulation products
  C. A pair of 40 dBm input signals will theoretically generate a third-order intermodulation product that has the same output amplitude as either of the input signals
  D. A pair of 1 mW input signals will produce a third-order intermodulation product that is 40 dB stronger than the input signal

E4D11
(A)
Page 7-18

E4D11
Why are odd-order intermodulation products, created within a receiver, of particular interest compared to other products?
  A. Odd-order products of two signals in the band of interest are also likely to be within the band
  B. Odd-order products overload the IF filters
  C. Odd-order products are an indication of poor image rejection
  D. Odd-order intermodulation produces three products for every input signal within the band of interest

E4D12
(A)
Page 7-16

E4D12
What is the term for the reduction in receiver sensitivity caused by a strong signal near the received frequency?
  A. Desensitization
  B. Quieting
  C. Cross-modulation interference
  D. Squelch gain rollback

E4E — Noise suppression and interference: system noise; electrical appliance noise; line noise; locating noise sources; DSP noise reduction; noise blankers; grounding for signals; common mode currents

E4E01
(A)
Page 7-27

E4E01
What problem can occur when using an automatic notch filter (ANF) to remove interfering carriers while receiving CW signals?
  A. Removal of the CW signal as well as the interfering carrier
  B. Any nearby signal passing through the DSP system will overwhelm the desired signal
  C. Received CW signals will appear to be modulated at the DSP clock frequency
  D. Ringing in the DSP filter will completely remove the spaces between the CW characters

13-40   Chapter 13

E4E02

Which of the following types of noise can often be reduced with a digital signal processing noise filter?
    A. Broadband white noise
    B. Ignition noise
    C. Power line noise
    D. All these choices are correct

E4E02
(D)
Page 7-27

E4E03

Which of the following signals might a receiver noise blanker be able to remove from desired signals?
    A. Signals that are constant at all IF levels
    B. Signals that appear across a wide bandwidth
    C. Signals that appear at one IF but not another
    D. Signals that have a sharply peaked frequency distribution

E4E03
(B)
Page 7-26

E4E04

How can conducted and radiated noise caused by an automobile alternator be suppressed?
    A. By installing filter capacitors in series with the DC power lead and a blocking capacitor in the field lead
    B. By installing a noise suppression resistor and a blocking capacitor in both leads
    C. By installing a high-pass filter in series with the radio's power lead and a low-pass filter in parallel with the field lead
    D. By connecting the radio's power leads directly to the battery and by installing coaxial capacitors in line with the alternator leads

E4E04
(D)
Page 7-26

E4E05

How can radio frequency interference from an AC motor be suppressed?
    A. By installing a high-pass filter in series with the motor's power leads
    B. By installing a brute-force AC-line filter in series with the motor leads
    C. By installing a bypass capacitor in series with the motor leads
    D. By using a ground-fault current interrupter in the circuit used to power the motor

E4E05
(B)
Page 7-25

E4E06

What is one type of electrical interference that might be caused by a nearby personal computer?
    A. A loud AC hum in the audio output of your station receiver
    B. A clicking noise at intervals of a few seconds
    C. The appearance of unstable modulated or unmodulated signals at specific frequencies
    D. A whining type noise that continually pulses off and on

E4E06
(C)
Page 7-25

E4E07

Which of the following can cause shielded cables to radiate or receive interference?
    A. Low inductance ground connections at both ends of the shield
    B. Common-mode currents on the shield and conductors
    C. Use of braided shielding material
    D. Tying all ground connections to a common point resulting in differential-mode currents in the shield

E4E07
(B)
Page 7-25

Question Pool   13-41

E4E08
(B)
Page 7-25

E4E08
What current flows equally on all conductors of an unshielded multi-conductor cable?
   A. Differential-mode current
   B. Common-mode current
   C. Reactive current only
   D. Return current

E4E09
(C)
Page 7-26

E4E09
What undesirable effect can occur when using an IF noise blanker?
   A. Received audio in the speech range might have an echo effect
   B. The audio frequency bandwidth of the received signal might be compressed
   C. Nearby signals may appear to be excessively wide even if they meet emission standards
   D. FM signals can no longer be demodulated

E4E10
(D)
Page 7-25

E4E10
What might be the cause of a loud roaring or buzzing AC line interference that comes and goes at intervals?
   A. Arcing contacts in a thermostatically controlled device
   B. A defective doorbell or doorbell transformer inside a nearby residence
   C. A malfunctioning illuminated advertising display
   D. All these choices are correct

E4E11
(B)
Page 7-23

E4E11
What could cause local AM broadcast band signals to combine to generate spurious signals in the MF or HF bands?
   A. One or more of the broadcast stations is transmitting an over-modulated signal
   B. Nearby corroded metal joints are mixing and re-radiating the broadcast signals
   C. You are receiving skywave signals from a distant station
   D. Your station receiver IF amplifier stage is defective

SUBELEMENT E5 — ELECTRICAL PRINCIPLES
[4 Exam Questions — 4 Groups]

E5A — Resonance and Q: characteristics of resonant circuits: series and parallel resonance; definitions and effects of Q; half-power bandwidth; phase relationships in reactive circuits

E5A01
(A)
Page 4-30

E5A01
What can cause the voltage across reactances in a series RLC circuit to be higher than the voltage applied to the entire circuit?
   A. Resonance
   B. Capacitance
   C. Conductance
   D. Resistance

E5A02
(C)
Page 4-27

E5A02
What is resonance in an LC or RLC circuit?
   A. The highest frequency that will pass current
   B. The lowest frequency that will pass current
   C. The frequency at which the capacitive reactance equals the inductive reactance
   D. The frequency at which the reactive impedance equals the resistive impedance

13-42   **Chapter 13**

E5A03
What is the magnitude of the impedance of a series RLC circuit at resonance?
A. High, as compared to the circuit resistance
B. Approximately equal to capacitive reactance
C. Approximately equal to inductive reactance
D. Approximately equal to circuit resistance

E5A03
(D)
Page 4-30

E5A04
What is the magnitude of the impedance of a parallel RLC circuit at resonance?
A. Approximately equal to circuit resistance
B. Approximately equal to inductive reactance
C. Low compared to the circuit resistance
D. High compared to the circuit resistance

E5A04
(A)
Page 4-30

E5A05
What is the result of increasing the Q of an impedance-matching circuit?
A. Matching bandwidth is decreased
B. Matching bandwidth is increased
C. Matching range is increased
D. It has no effect on impedance matching

E5A05
(A)
Page 4-33

E5A06
What is the magnitude of the circulating current within the components of a parallel LC circuit at resonance?
A. It is at a minimum
B. It is at a maximum
C. It equals 1 divided by the quantity 2 times pi, multiplied by the square root of inductance L multiplied by capacitance C
D. It equals 2 multiplied by pi, multiplied by frequency, multiplied by inductance

E5A06
(B)
Page 4-30

E5A07
What is the magnitude of the current at the input of a parallel RLC circuit at resonance?
A. Minimum
B. Maximum
C. R/L
D. L/R

E5A07
(A)
Page 4-30

E5A08
What is the phase relationship between the current through and the voltage across a series resonant circuit at resonance?
A. The voltage leads the current by 90 degrees
B. The current leads the voltage by 90 degrees
C. The voltage and current are in phase
D. The voltage and current are 180 degrees out of phase

E5A08
(C)
Page 4-31

E5A09
How is the Q of an RLC parallel resonant circuit calculated?
A. Reactance of either the inductance or capacitance divided by the resistance
B. Reactance of either the inductance or capacitance multiplied by the resistance
C. Resistance divided by the reactance of either the inductance or capacitance
D. Reactance of the inductance multiplied by the reactance of the capacitance

E5A09
(C)
Page 4-32

Question Pool   13-43

E5A10
(A)
Page 4-32

E5A10
How is the Q of an RLC series resonant circuit calculated?
  A. Reactance of either the inductance or capacitance divided by the resistance
  B. Reactance of either the inductance or capacitance multiplied by the resistance
  C. Resistance divided by the reactance of either the inductance or capacitance
  D. Reactance of the inductance multiplied by the reactance of the capacitance

E5A11
(C)
Page 4-33

E5A11
What is the half-power bandwidth of a resonant circuit that has a resonant frequency of 7.1 MHz and a Q of 150?
  A. 157.8 Hz
  B. 315.6 Hz
  C. 47.3 kHz
  D. 23.67 kHz

E5A12
(C)
Page 4-33

E5A12
What is the half-power bandwidth of a resonant circuit that has a resonant frequency of 3.7 MHz and a Q of 118?
  A. 436.6 kHz
  B. 218.3 kHz
  C. 31.4 kHz
  D. 15.7 kHz

E5A13
(C)
Page 4-32

E5A13
What is an effect of increasing Q in a series resonant circuit?
  A. Fewer components are needed for the same performance
  B. Parasitic effects are minimized
  C. Internal voltages increase
  D. Phase shift can become uncontrolled

E5A14
(C)
Page 4-28

E5A14
What is the resonant frequency of an RLC circuit if R is 22 ohms, L is 50 microhenries and C is 40 picofarads?
  A. 44.72 MHz
  B. 22.36 MHz
  C. 3.56 MHz
  D. 1.78 MHz

E5A15
(A)
Page 4-31

E5A15
Which of the following increases Q for inductors and capacitors?
  A. Lower losses
  B. Lower reactance
  C. Lower self-resonant frequency
  D. Higher self-resonant frequency

E5A16
(D)
Page 4-29

E5A16
What is the resonant frequency of an RLC circuit if R is 33 ohms, L is 50 microhenries and C is 10 picofarads?
  A. 23.5 MHz
  B. 23.5 kHz
  C. 7.12 kHz
  D. 7.12 MHz

E5B — Time constants and phase relationships: RL and RC time constants; phase angle in reactive circuits and components; admittance and susceptance

E5B01
What is the term for the time required for the capacitor in an RC circuit to be charged to 63.2% of the applied voltage or to discharge to 36.8% of its initial voltage?
  A. An exponential rate of one
  B. One time constant
  C. One exponential period
  D. A time factor of one

E5B01
(B)
Page 4-9

E5B02
What letter is commonly used to represent susceptance?
  A. G
  B. X
  C. Y
  D. B

E5B02
(D)
Page 4-19

E5B03
How is impedance in polar form converted to an equivalent admittance?
  A. Take the reciprocal of the angle and change the sign of the magnitude
  B. Take the reciprocal of the magnitude and change the sign of the angle
  C. Take the square root of the magnitude and add 180 degrees to the angle
  D. Square the magnitude and subtract 90 degrees from the angle

E5B03
(B)
Page 4-20

E5B04
What is the time constant of a circuit having two 220-microfarad capacitors and two 1-megohm resistors, all in parallel?
  A. 55 seconds
  B. 110 seconds
  C. 440 seconds
  D. 220 seconds

E5B04
(D)
Page 4-11

E5B05
What happens to the magnitude of a pure reactance when it is converted to a susceptance?
  A. It is unchanged
  B. The sign is reversed
  C. It is shifted by 90 degrees
  D. It becomes the reciprocal

E5B05
(D)
Page 4-19

E5B06
What is susceptance?
  A. The magnetic impedance of a circuit
  B. The ratio of magnetic field to electric field
  C. The imaginary part of admittance
  D. A measure of the efficiency of a transformer

E5B06
(C)
Page 4-19

Question Pool   13-45

E5B07
(C)
Page 4-22

E5B07
What is the phase angle between the voltage across and the current through a series RLC circuit if XC is 500 ohms, R is 1 kilohm, and XL is 250 ohms?
    A. 68.2 degrees with the voltage leading the current
    B. 14.0 degrees with the voltage leading the current
    C. 14.0 degrees with the voltage lagging the current
    D. 68.2 degrees with the voltage lagging the current

E5B08
(A)
Page 4-22

E5B08
What is the phase angle between the voltage across and the current through a series RLC circuit if XC is 100 ohms, R is 100 ohms, and XL is 75 ohms?
    A. 14 degrees with the voltage lagging the current
    B. 14 degrees with the voltage leading the current
    C. 76 degrees with the voltage leading the current
    D. 76 degrees with the voltage lagging the current

E5B09
(D)
Page 4-14

E5B09
What is the relationship between the AC current through a capacitor and the voltage across a capacitor?
    A. Voltage and current are in phase
    B. Voltage and current are 180 degrees out of phase
    C. Voltage leads current by 90 degrees
    D. Current leads voltage by 90 degrees

E5B10
(A)
Page 4-15

E5B10
What is the relationship between the AC current through an inductor and the voltage across an inductor?
    A. Voltage leads current by 90 degrees
    B. Current leads voltage by 90 degrees
    C. Voltage and current are 180 degrees out of phase
    D. Voltage and current are in phase

E5B11
(B)
Page 4-23

E5B11
What is the phase angle between the voltage across and the current through a series RLC circuit if XC is 25 ohms, R is 100 ohms, and XL is 50 ohms?
    A. 14 degrees with the voltage lagging the current
    B. 14 degrees with the voltage leading the current
    C. 76 degrees with the voltage lagging the current
    D. 76 degrees with the voltage leading the current

E5B12
(A)
Page 4-19

E5B12
What is admittance?
    A. The inverse of impedance
    B. The term for the gain of a field effect transistor
    C. The turns ratio of a transformer
    D. The inverse of Q factor

E5C — Coordinate systems and phasors in electronics: rectangular coordinates; polar coordinates; phasors

E5C01
Which of the following represents capacitive reactance in rectangular notation?
A. –jX
B. +jX
C. Delta
D. Omega

E5C01
(A)
Page 4-16

E5C02
How are impedances described in polar coordinates?
A. By X and R values
B. By real and imaginary parts
C. By phase angle and magnitude
D. By Y and G values

E5C02
(C)
Page 4-16

E5C03
Which of the following represents an inductive reactance in polar coordinates?
A. A positive magnitude
B. A negative magnitude
C. A positive phase angle
D. A negative phase angle

E5C03
(C)
Page 4-16

E5C04
What coordinate system is often used to display the resistive, inductive, and/or capacitive reactance components of impedance?
A. Maidenhead grid
B. Faraday grid
C. Elliptical coordinates
D. Rectangular coordinates

E5C04
(D)
Page 4-16

E5C05
What is the name of the diagram used to show the phase relationship between impedances at a given frequency?
A. Venn diagram
B. Near field diagram
C. Phasor diagram
D. Far field diagram

E5C05
(C)
Page 4-16

E5C06
What does the impedance 50–j25 represent?
A. 50 ohms resistance in series with 25 ohms inductive reactance
B. 50 ohms resistance in series with 25 ohms capacitive reactance
C. 25 ohms resistance in series with 50 ohms inductive reactance
D. 25 ohms resistance in series with 50 ohms capacitive reactance

E5C06
(B)
Page 4-16

E5C07
(D)
Page 4-16

E5C07
Where is the impedance of a pure resistance plotted on rectangular coordinates?
A. On the vertical axis
B. On a line through the origin, slanted at 45 degrees
C. On a horizontal line, offset vertically above the horizontal axis
D. On the horizontal axis

E5C08
(D)
Page 4-16

E5C08
What coordinate system is often used to display the phase angle of a circuit containing resistance, inductive and/or capacitive reactance?
A. Maidenhead grid
B. Faraday grid
C. Elliptical coordinates
D. Polar coordinates

E5C09
(A)
Page 4-16

E5C09
When using rectangular coordinates to graph the impedance of a circuit, what do the axes represent?
A. The X axis represents the resistive component and the Y axis represents the reactive component
B. The X axis represents the reactive component and the Y axis represents the resistive component
C. The X axis represents the phase angle and the Y axis represents the magnitude
D. The X axis represents the magnitude and the Y axis represents the phase angle

13-48   **Chapter 13**

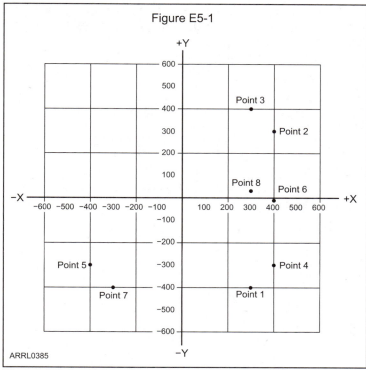

Figure E5-1 — This figure is used for questions E5C10 through E5C12.

E5C10
Which point on Figure E5-1 best represents the impedance of a series circuit consisting of a 400-ohm resistor and a 38-picofarad capacitor at 14 MHz?
    A. Point 2
    B. Point 4
    C. Point 5
    D. Point 6

E5C11
Which point in Figure E5-1 best represents the impedance of a series circuit consisting of a 300-ohm resistor and an 18-microhenry inductor at 3.505 MHz?
    A. Point 1
    B. Point 3
    C. Point 7
    D. Point 8

E5C12
Which point on Figure E5-1 best represents the impedance of a series circuit consisting of a 300-ohm resistor and a 19-picofarad capacitor at 21.200 MHz?
    A. Point 1
    B. Point 3
    C. Point 7
    D. Point 8

E5C10
(B)
Page 4-21

E5C11
(B)
Page 4-20

E5C12
(A)
Page 4-22

E5D — AC and RF energy in real circuits: skin effect; electromagnetic fields; reactive power; power factor; electrical length of conductors at UHF and microwave frequencies; microstrip

E5D01
(A)
Page 4-33

E5D01
What is the result of skin effect?
   A. As frequency increases, RF current flows in a thinner layer of the conductor, closer to the surface
   B. As frequency decreases, RF current flows in a thinner layer of the conductor, closer to the surface
   C. Thermal effects on the surface of the conductor increase the impedance
   D. Thermal effects on the surface of the conductor decrease the impedance

E5D02
(B)
Page 4-35

E5D02
Why is it important to keep lead lengths short for components used in circuits for VHF and above?
   A. To increase the thermal time constant
   B. To avoid unwanted inductive reactance
   C. To maintain component lifetime
   D. All these choices are correct

E5D03
(D)
Page 5-13

E5D03
What is microstrip?
   A. Lightweight transmission line made of common zip cord
   B. Miniature coax used for low power applications
   C. Short lengths of coax mounted on printed circuit boards to minimize time delay between microwave circuits
   D. Precision printed circuit conductors above a ground plane that provide constant impedance interconnects at microwave frequencies

E5D04
(B)
Page 4-35

E5D04
Why are short connections used at microwave frequencies?
   A. To increase neutralizing resistance
   B. To reduce phase shift along the connection
   C. To increase compensating capacitance
   D. To reduce noise figure

E5D05
(C)
Page 4-26

E5D05
What is the power factor of an RL circuit having a 30-degree phase angle between the voltage and the current?
   A. 1.73
   B. 0.5
   C. 0.866
   D. 0.577

E5D06
(D)
Page 4-7

E5D06
In what direction is the magnetic field oriented about a conductor in relation to the direction of electron flow?
   A. In the same direction as the current
   B. In a direction opposite to the current
   C. In all directions; omni-directional
   D. In a circle around the conductor

E5D07
How many watts are consumed in a circuit having a power factor of 0.71 if the apparent power is 500 VA?
A. 704 W
B. 355 W
C. 252 W
D. 1.42 mW

E5D08
How many watts are consumed in a circuit having a power factor of 0.6 if the input is 200 VAC at 5 amperes?
A. 200 watts
B. 1000 watts
C. 1600 watts
D. 600 watts

E5D09
What happens to reactive power in an AC circuit that has both ideal inductors and ideal capacitors?
A. It is dissipated as heat in the circuit
B. It is repeatedly exchanged between the associated magnetic and electric fields, but is not dissipated
C. It is dissipated as kinetic energy in the circuit
D. It is dissipated in the formation of inductive and capacitive fields

E5D10
How can the true power be determined in an AC circuit where the voltage and current are out of phase?
A. By multiplying the apparent power by the power factor
B. By dividing the reactive power by the power factor
C. By dividing the apparent power by the power factor
D. By multiplying the reactive power by the power factor

E5D11
What is the power factor of an RL circuit having a 60-degree phase angle between the voltage and the current?
A. 1.414
B. 0.866
C. 0.5
D. 1.73

E5D12
How many watts are consumed in a circuit having a power factor of 0.2 if the input is 100 VAC at 4 amperes?
A. 400 watts
B. 80 watts
C. 2000 watts
D. 50 watts

E5D07
(B)
Page 4-26

E5D08
(D)
Page 4-26

E5D09
(B)
Page 4-24

E5D10
(A)
Page 4-25

E5D11
(C)
Page 4-26

E5D12
(B)
Page 4-26

Question Pool   13-51

E5D13
(B)
Page 4-26

E5D13
How many watts are consumed in a circuit consisting of a 100-ohm resistor in series with a 100-ohm inductive reactance drawing 1 ampere?
  A. 70.7 watts
  B. 100 watts
  C. 141.4 watts
  D. 200 watts

E5D14
(A)
Page 4-24

E5D14
What is reactive power?
  A. Wattless, nonproductive power
  B. Power consumed in wire resistance in an inductor
  C. Power lost because of capacitor leakage
  D. Power consumed in circuit Q

E5D15
(D)
Page 4-26

E5D15
What is the power factor of an RL circuit having a 45-degree phase angle between the voltage and the current?
  A. 0.866
  B. 1.0
  C. 0.5
  D. 0.707

SUBELEMENT E6 — CIRCUIT COMPONENTS
[6 Exam Questions — 6 Groups]

E6A — Semiconductor materials and devices: semiconductor materials; germanium, silicon, P-type, N-type; transistor types: NPN, PNP, junction, field-effect transistors: enhancement mode; depletion mode; MOS; CMOS; N-channel; P-channel

E6A01
(C)
Page 5-13

E6A01
In what application is gallium arsenide used as a semiconductor material?
  A. In high-current rectifier circuits
  B. In high-power audio circuits
  C. In microwave circuits
  D. In very low-frequency RF circuits

E6A02
(A)
Page 5-2

E6A02
Which of the following semiconductor materials contains excess free electrons?
  A. N-type
  B. P-type
  C. Bipolar
  D. Insulated gate

E6A03
(C)
Page 5-3

E6A03
Why does a PN-junction diode not conduct current when reverse biased?
  A. Only P-type semiconductor material can conduct current
  B. Only N-type semiconductor material can conduct current
  C. Holes in P-type material and electrons in the N-type material are separated by the applied voltage, widening the depletion region
  D. Excess holes in P-type material combine with the electrons in N-type material, converting the entire diode into an insulator

13-52   Chapter 13

E6A04

What is the name given to an impurity atom that adds holes to a semiconductor crystal structure?

A. Insulator impurity
B. N-type impurity
C. Acceptor impurity
D. Donor impurity

E6A04
(C)
Page 5-2

E6A05

How does DC input impedance at the gate of a field-effect transistor compare with the DC input impedance of a bipolar transistor?

A. They are both low impedance
B. An FET has lower input impedance
C. An FET has higher input impedance
D. They are both high impedance

E6A05
(C)
Page 5-10

E6A06

What is the beta of a bipolar junction transistor?

A. The frequency at which the current gain is reduced to 0.707
B. The change in collector current with respect to base current
C. The breakdown voltage of the base to collector junction
D. The switching speed

E6A06
(B)
Page 5-9

E6A07

Which of the following indicates that a silicon NPN junction transistor is biased on?

A. Base-to-emitter resistance of approximately 6 to 7 ohms
B. Base-to-emitter resistance of approximately 0.6 to 0.7 ohms
C. Base-to-emitter voltage of approximately 6 to 7 volts
D. Base-to-emitter voltage of approximately 0.6 to 0.7 volts

E6A07
(D)
Page 5-8

E6A08

What term indicates the frequency at which the grounded-base current gain of a transistor has decreased to 0.7 of the gain obtainable at 1 kHz?

A. Corner frequency
B. Alpha rejection frequency
C. Beta cutoff frequency
D. Alpha cutoff frequency

E6A08
(D)
Page 5-9

E6A09

What is a depletion-mode FET?

A. An FET that exhibits a current flow between source and drain when no gate voltage is applied
B. An FET that has no current flow between source and drain when no gate voltage is applied
C. Any FET without a channel
D. Any FET for which holes are the majority carriers

E6A09
(A)
Page 5-11

Question Pool   13-53

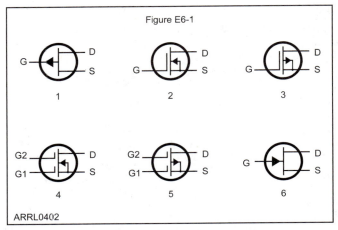

Figure E6-1 — This figure is used for questions E6A10 and E6A11.

E6A10
(B)
Page 5-11

E6A10
In Figure E6-1, what is the schematic symbol for an N-channel dual-gate MOSFET?
  A. 2
  B. 4
  C. 5
  D. 6

E6A11
(A)
Page 5-10

E6A11
In Figure E6-1, what is the schematic symbol for a P-channel junction FET?
  A. 1
  B. 2
  C. 3
  D. 6

E6A12
(D)
Page 5-11

E6A12
Why do many MOSFET devices have internally connected Zener diodes on the gates?
  A. To provide a voltage reference for the correct amount of reverse-bias gate voltage
  B. To protect the substrate from excessive voltages
  C. To keep the gate voltage within specifications and prevent the device from overheating
  D. To reduce the chance of static damage to the gate

E6B — Diodes

E6B01
What is the most useful characteristic of a Zener diode?
  A. A constant current drop under conditions of varying voltage
  B. A constant voltage drop under conditions of varying current
  C. A negative resistance region
  D. An internal capacitance that varies with the applied voltage

E6B01
(B)
Page 5-5

E6B02
What is an important characteristic of a Schottky diode as compared to an ordinary silicon diode when used as a power supply rectifier?
  A. Much higher reverse voltage breakdown
  B. More constant reverse avalanche voltage
  C. Longer carrier retention time
  D. Less forward voltage drop

E6B02
(D)
Page 5-4

E6B03
What type of bias is required for an LED to emit light?
  A. Reverse bias
  B. Forward bias
  C. Zero bias
  D. Inductive bias

E6B03
(B)
Page 5-7

E6B04
What type of semiconductor device is designed for use as a voltage-controlled capacitor?
  A. Varactor diode
  B. Tunnel diode
  C. Silicon-controlled rectifier
  D. Zener diode

E6B04
(A)
Page 5-6

E6B05
What characteristic of a PIN diode makes it useful as an RF switch?
  A. Extremely high reverse breakdown voltage
  B. Ability to dissipate large amounts of power
  C. Reverse bias controls its forward voltage drop
  D. Low junction capacitance

E6B05
(D)
Page 5-7

E6B06
Which of the following is a common use of a Schottky diode?
  A. As a rectifier in high current power supplies
  B. As a variable capacitance in an automatic frequency control circuit
  C. As a constant voltage reference in a power supply
  D. As a VHF/UHF mixer or detector

E6B06
(D)
Page 5-5

E6B07
What is the failure mechanism when a junction diode fails due to excessive current?
  A. Excessive inverse voltage
  B. Excessive junction temperature
  C. Insufficient forward voltage
  D. Charge carrier depletion

E6B07
(B)
Page 5-4

E6B08
(A)
Page 5-4

E6B08
Which of the following is a Schottky barrier diode?
   A. Metal-semiconductor junction
   B. Electrolytic rectifier
   C. PIN junction
   D. Thermionic emission diode

E6B09
(C)
Page 5-5

E6B09
What is a common use for point-contact diodes?
   A. As a constant current source
   B. As a constant voltage source
   C. As an RF detector
   D. As a high-voltage rectifier

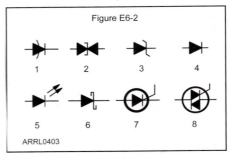

Figure E6-2 — This figure is used for question E6B10.

E6B10
(B)
Page 5-7

E6B10
In Figure E6-2, what is the schematic symbol for a light-emitting diode?
   A. 1
   B. 5
   C. 6
   D. 7

E6B11
(A)
Page 5-7

E6B11
What is used to control the attenuation of RF signals by a PIN diode?
   A. Forward DC bias current
   B. A sub-harmonic pump signal
   C. Reverse voltage larger than the RF signal
   D. Capacitance of an RF coupling capacitor

E6C — Digital ICs: Families of digital ICs; gates; Programmable Logic Devices (PLDs)

E6C01
What is the function of hysteresis in a comparator?
A. To prevent input noise from causing unstable output signals
B. To allow the comparator to be used with AC input signals
C. To cause the output to change states continually
D. To increase the sensitivity

E6C01
(A)
Page 6-10

E6C02
What happens when the level of a comparator's input signal crosses the threshold?
A. The IC input can be damaged
B. The comparator changes its output state
C. The comparator enters latch-up
D. The feedback loop becomes unstable

E6C02
(B)
Page 6-10

E6C03
What is tri-state logic?
A. Logic devices with 0, 1, and high-impedance output states
B. Logic devices that utilize ternary math
C. Low-power logic devices designed to operate at 3 volts
D. Proprietary logic devices manufactured by Tri-State Devices

E6C03
(A)
Page 5-21

E6C04
Which of the following is an advantage of BiCMOS logic?
A. Its simplicity results in much less expensive devices than standard CMOS
B. It is immune to electrostatic damage
C. It has the high input impedance of CMOS and the low output impedance of bipolar transistors
D. All these choices are correct

E6C04
(C)
Page 5-26

E6C05
What is an advantage of CMOS logic devices over TTL devices?
A. Differential output capability
B. Lower distortion
C. Immune to damage from static discharge
D. Lower power consumption

E6C05
(D)
Page 5-26

E6C06
Why do CMOS digital integrated circuits have high immunity to noise on the input signal or power supply?
A. Large bypass capacitance is inherent
B. The input switching threshold is about two times the power supply voltage
C. The input switching threshold is about one-half the power supply voltage
D. Bandwidth is very limited

E6C06
(C)
Page 5-26

Question Pool   13-57

E6C07
(B)
Page 5-25

E6C07
What best describes a pull-up or pull-down resistor?
A. A resistor in a keying circuit used to reduce key clicks
B. A resistor connected to the positive or negative supply line used to establish a voltage when an input or output is an open circuit
C. A resistor that ensures that an oscillator frequency does not drift
D. A resistor connected to an op-amp output that prevents signals from exceeding the power supply voltage

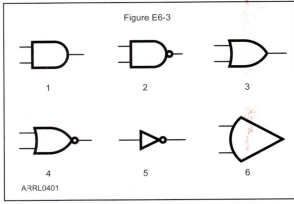

Figure E6-3 — This figure is used for question E6C08, E6C10, and E6C11.

E6C08
(B)
Page 5-20

E6C08
In Figure E6-3, what is the schematic symbol for a NAND gate?
A. 1
B. 2
C. 3
D. 4

E6C09
(B)
Page 5-26

E6C09
What is a Programmable Logic Device (PLD)?
A. A logic circuit that can be modified during use
B. A programmable collection of logic gates and circuits in a single integrated circuit
C. Programmable equipment used for testing digital logic integrated circuits
D. A type of transistor whose gain can be changed by digital logic circuits

E6C10
(D)
Page 5-20

E6C10
In Figure E6-3, what is the schematic symbol for a NOR gate?
A. 1
B. 2
C. 3
D. 4

E6C11
(C)
Page 5-19

E6C11
In Figure E6-3, what is the schematic symbol for the NOT operation (inverter)?
A. 2
B. 4
C. 5
D. 6

E6D — Toroidal and Solenoidal Inductors: permeability, core material, selecting, winding; transformers; piezoelectric devices

E6D01
Why should core saturation of an impedance matching transformer be avoided?
A. Harmonics and distortion could result
B. Magnetic flux would increase with frequency
C. RF susceptance would increase
D. Temporary changes of the core permeability could result

E6D01
(A)
Page 4-36

E6D02
What is the equivalent circuit of a quartz crystal?
A. Motional capacitance, motional inductance, and loss resistance in series, all in parallel with a shunt capacitor representing electrode and stray capacitance
B. Motional capacitance, motional inductance, loss resistance, and a capacitor representing electrode and stray capacitance all in parallel
C. Motional capacitance, motional inductance, loss resistance, and a capacitor representing electrode and stray capacitance all in series
D. Motional inductance and loss resistance in series, paralleled with motional capacitance and a capacitor representing electrode and stray capacitance

E6D02
(A)
Page 6-15

E6D03
Which of the following is an aspect of the piezoelectric effect?
A. Mechanical deformation of material by the application of a voltage
B. Mechanical deformation of material by the application of a magnetic field
C. Generation of electrical energy in the presence of light
D. Increased conductivity in the presence of light

E6D03
(A)
Page 6-15

E6D04
Which materials are commonly used as a core in an inductor?
A. Polystyrene and polyethylene
B. Ferrite and brass
C. Teflon and Delrin
D. Cobalt and aluminum

E6D04
(B)
Page 4-36

E6D05
What is one reason for using ferrite cores rather than powdered iron in an inductor?
A. Ferrite toroids generally have lower initial permeability
B. Ferrite toroids generally have better temperature stability
C. Ferrite toroids generally require fewer turns to produce a given inductance value
D. Ferrite toroids are easier to use with surface mount technology

E6D05
(C)
Page 4-36

E6D06
What core material property determines the inductance of an inductor?
A. Thermal impedance
B. Resistance
C. Reactivity
D. Permeability

E6D06
(D)
Page 4-36

Question Pool 13-59

E6D07
(A)
Page 4-36

E6D07
What is current in the primary winding of a transformer called if no load is attached to the secondary?
A. Magnetizing current
B. Direct current
C. Excitation current
D. Stabilizing current

E6D08
(B)
Page 4-36

E6D08
What is one reason for using powdered-iron cores rather than ferrite cores in an inductor?
A. Powdered-iron cores generally have greater initial permeability
B. Powdered-iron cores generally maintain their characteristics at higher currents
C. Powdered-iron cores generally require fewer turns to produce a given inductance
D. Powdered-iron cores use smaller diameter wire for the same inductance

E6D09
(C)
Page 4-37

E6D09
What devices are commonly used as VHF and UHF parasitic suppressors at the input and output terminals of a transistor HF amplifier?
A. Electrolytic capacitors
B. Butterworth filters
C. Ferrite beads
D. Steel-core toroids

E6D10
(A)
Page 4-36

E6D10
What is a primary advantage of using a toroidal core instead of a solenoidal core in an inductor?
A. Toroidal cores confine most of the magnetic field within the core material
B. Toroidal cores make it easier to couple the magnetic energy into other components
C. Toroidal cores exhibit greater hysteresis
D. Toroidal cores have lower Q characteristics

E6D11
(B)
Page 4-36

E6D11
Which type of core material decreases inductance when inserted into a coil?
A. Ceramic
B. Brass
C. Ferrite
D. Powdered iron

E6D12
(C)
Page 4-36

E6D12
What is inductor saturation?
A. The inductor windings are over-coupled
B. The inductor's voltage rating is exceeded causing a flashover
C. The ability of the inductor's core to store magnetic energy has been exceeded
D. Adjacent inductors become over-coupled

E6D13
(A)
Page 4-34

E6D13
What is the primary cause of inductor self-resonance?
A. Inter-turn capacitance
B. The skin effect
C. Inductive kickback
D. Non-linear core hysteresis

E6E — Analog ICs: MMICs, IC packaging characteristics

E6E01
Why is gallium arsenide (GaAs) useful for semiconductor devices operating at UHF and higher frequencies?
   A. Higher noise figures
   B. Higher electron mobility
   C. Lower junction voltage drop
   D. Lower transconductance

E6E01
(B)
Page 5-13

E6E02
Which of the following device packages is a through-hole type?
   A. DIP
   B. PLCC
   C. Ball grid array
   D. SOT

E6E02
(A)
Page 4-35

E6E03
Which of the following materials is likely to provide the highest frequency of operation when used in MMICs?
   A. Silicon
   B. Silicon nitride
   C. Silicon dioxide
   D. Gallium nitride

E6E03
(D)
Page 5-13

E6E04
Which is the most common input and output impedance of circuits that use MMICs?
   A. 50 ohms
   B. 300 ohms
   C. 450 ohms
   D. 10 ohms

E6E04
(A)
Page 5-12

E6E05
Which of the following noise figure values is typical of a low-noise UHF preamplifier?
   A. 2 dB
   B. –10 dB
   C. 44 dBm
   D. –20 dBm

E6E05
(A)
Page 5-13

E6E06
What characteristics of the MMIC make it a popular choice for VHF through microwave circuits?
   A. The ability to retrieve information from a single signal even in the presence of other strong signals
   B. Plate current that is controlled by a control grid
   C. Nearly infinite gain, very high input impedance, and very low output impedance
   D. Controlled gain, low noise figure, and constant input and output impedance over the specified frequency range

E6E06
(D)
Page 5-12

Question Pool   13-61

E6E07
(D)
Page 5-13

E6E07
What type of transmission line is used for connections to MMICs?
  A. Miniature coax
  B. Circular waveguide
  C. Parallel wire
  D. Microstrip

E6E08
(A)
Page 5-12

E6E08
How is power supplied to the most common type of MMIC?
  A. Through a resistor and/or RF choke connected to the amplifier output lead
  B. MMICs require no operating bias
  C. Through a capacitor and RF choke connected to the amplifier input lead
  D. Directly to the bias voltage (VCC IN) lead

E6E09
(D)
Page 4-35

E6E09
Which of the following component package types would be most suitable for use at frequencies above the HF range?
  A. TO-220
  B. Axial lead
  C. Radial lead
  D. Surface mount

E6E10
(D)
Page 4-35

E6E10
What advantage does surface-mount technology offer at RF compared to using through-hole components?
  A. Smaller circuit area
  B. Shorter circuit-board traces
  C. Components have less parasitic inductance and capacitance
  D. All these choices are correct

E6E11
(D)
Page 4-35

E6E11
What is a characteristic of DIP packaging used for integrated circuits?
  A. Package mounts in a direct inverted position
  B. Low leakage doubly insulated package
  C. Two chips in each package (Dual In Package)
  D. A total of two rows of connecting pins placed on opposite sides of the package (Dual In-line Package)

E6E12
(C)
Page 4-33

E6E12
Why are DIP through-hole package ICs not typically used at UHF and higher frequencies?
  A. Too many pins
  B. Epoxy coating is conductive above 300 MHz
  C. Excessive lead length
  D. Unsuitable for combining analog and digital signals

E6F — Electro-optical technology: photoconductivity; photovoltaic devices; optical sensors and encoders; optical isolation

E6F01
What absorbs the energy from light falling on a photovoltaic cell?
A. Protons
B. Photons
C. Electrons
D. Holes

E6F01
(C)
Page 5-17

E6F02
What happens to the conductivity of a photoconductive material when light shines on it?
A. It increases
B. It decreases
C. It stays the same
D. It becomes unstable

E6F02
(A)
Page 5-15

E6F03
What is the most common configuration of an optoisolator or optocoupler?
A. A lens and a photomultiplier
B. A frequency modulated helium-neon laser
C. An amplitude modulated helium-neon laser
D. An LED and a phototransistor

E6F03
(D)
Page 5-16

E6F04
What is the photovoltaic effect?
A. The conversion of voltage to current when exposed to light
B. The conversion of light to electrical energy
C. The conversion of electrical energy to mechanical energy
D. The tendency of a battery to discharge when exposed to light

E6F04
(B)
Page 5-17

E6F05
Which describes an optical shaft encoder?
A. A device that detects rotation of a control by interrupting a light source with a patterned wheel
B. A device that measures the strength of a beam of light using analog to digital conversion
C. A digital encryption device often used to encrypt spacecraft control signals
D. A device for generating RTTY signals by means of a rotating light source

E6F05
(A)
Page 5-16

E6F06
Which of these materials is most commonly used to create photoconductive devices?
A. A crystalline semiconductor
B. An ordinary metal
C. A heavy metal
D. A liquid semiconductor

E6F06
(A)
Page 5-15

Question Pool   13-63

E6F07
(B)
Page 5-16

E6F07
What is a solid-state relay?
  A. A relay using transistors to drive the relay coil
  B. A device that uses semiconductors to implement the functions of an electromechanical relay
  C. A mechanical relay that latches in the on or off state each time it is pulsed
  D. A semiconductor passive delay line

E6F08
(C)
Page 5-16

E6F08
Why are optoisolators often used in conjunction with solid-state circuits when switching 120 VAC?
  A. Optoisolators provide a low impedance link between a control circuit and a power circuit
  B. Optoisolators provide impedance matching between the control circuit and power circuit
  C. Optoisolators provide a very high degree of electrical isolation between a control circuit and the circuit being switched
  D. Optoisolators eliminate the effects of reflected light in the control circuit

E6F09
(D)
Page 5-18

E6F09
What is the efficiency of a photovoltaic cell?
  A. The output RF power divided by the input DC power
  B. Cost per kilowatt-hour generated
  C. The open-circuit voltage divided by the short-circuit current under full illumination
  D. The relative fraction of light that is converted to current

E6F10
(B)
Page 5-17

E6F10
What is the most common type of photovoltaic cell used for electrical power generation?
  A. Selenium
  B. Silicon
  C. Cadmium Sulfide
  D. Copper oxide

E6F11
(B)
Page 5-17

E6F11
What is the approximate open-circuit voltage produced by a fully illuminated silicon photovoltaic cell?
  A. 0.1 V
  B. 0.5 V
  C. 1.5 V
  D. 12 V

SUBELEMENT E7 — PRACTICAL CIRCUITS
[8 Exam Questions — 8 Groups]

E7A — Digital circuits: digital circuit principles and logic circuits; classes of logic elements; positive and negative logic; frequency dividers; truth tables

E7A01
(C)
Page 5-22

E7A01
Which circuit is bistable?
  A. An AND gate
  B. An OR gate
  C. A flip-flop
  D. A bipolar amplifier

13-64   Chapter 13

E7A02
What is the function of a decade counter?
A. It produces one output pulse for every 10 input pulses
B. It decodes a decimal number for display on a seven-segment LED display
C. It produces 10 output pulses for every input pulse
D. It decodes a binary number for display on a seven-segment LED display

E7A02
(A)
Page 5-24

E7A03
Which of the following can divide the frequency of a pulse train by 2?
A. An XOR gate
B. A flip-flop
C. An OR gate
D. A multiplexer

E7A03
(B)
Page 5-23

E7A04
How many flip-flops are required to divide a signal frequency by 4?
A. 1
B. 2
C. 4
D. 8

E7A04
(B)
Page 5-23

E7A05
Which of the following is a circuit that continuously alternates between two states without an external clock?
A. Monostable multivibrator
B. J-K flip-flop
C. T flip-flop
D. Astable multivibrator

E7A05
(D)
Page 5-23

E7A06
What is a characteristic of a monostable multivibrator?
A. It switches momentarily to the opposite binary state and then returns to its original state after a set time
B. It produces a continuous square wave oscillating between 1 and 0
C. It stores one bit of data in either a 0 or 1 state
D. It maintains a constant output voltage, regardless of variations in the input voltage

E7A06
(A)
Page 5-23

E7A07
What logical operation does a NAND gate perform?
A. It produces logic 0 at its output only when all inputs are logic 0
B. It produces logic 1 at its output only when all inputs are logic 1
C. It produces logic 0 at its output if some but not all inputs are logic 1
D. It produces logic 0 at its output only when all inputs are logic 1

E7A07
(D)
Page 5-20

E7A08
What logical operation does an OR gate perform?
A. It produces logic 1 at its output if any or all inputs are logic 1
B. It produces logic 0 at its output if all inputs are logic 1
C. It only produces logic 0 at its output when all inputs are logic 1
D. It produces logic 1 at its output if all inputs are logic 0

E7A08
(A)
Page 5-20

Question Pool   13-65

E7A09
(C)
Page 5-20

E7A09
What logical operation is performed by an exclusive NOR gate?
A. It produces logic 0 at its output only if all inputs are logic 0
B. It produces logic 1 at its output only if all inputs are logic 1
C. It produces logic 0 at its output if only one input is logic 1
D. It produces logic 1 at its output if only one input is logic 1

E7A10
(C)
Page 5-19

E7A10
What is a truth table?
A. A table of logic symbols that indicate the high logic states of an op-amp
B. A diagram showing logic states when the digital device output is true
C. A list of inputs and corresponding outputs for a digital device
D. A table of logic symbols that indicate the logic states of an op-amp

E7A11
(D)
Page 5-20

E7A11
What type of logic defines "1" as a high voltage?
A. Reverse Logic
B. Assertive Logic
C. Negative logic
D. Positive Logic

E7B — Amplifiers: Class of operation; vacuum tube and solid-state circuits; distortion and intermodulation; spurious and parasitic suppression; microwave amplifiers; switching-type amplifiers

E7B01
(A)
Page 6-11

E7B01
For what portion of the signal cycle does each active element in a push-pull Class AB amplifier conduct?
A. More than 180 degrees but less than 360 degrees
B. Exactly 180 degrees
C. The entire cycle
D. Less than 180 degrees

E7B02
(A)
Page 6-12

E7B02
What is a Class D amplifier?
A. A type of amplifier that uses switching technology to achieve high efficiency
B. A low power amplifier that uses a differential amplifier for improved linearity
C. An amplifier that uses drift-mode FETs for high efficiency
D. A frequency doubling amplifier

E7B03
(A)
Page 6-12

E7B03
Which of the following components form the output of a class D amplifier circuit?
A. A low-pass filter to remove switching signal components
B. A high-pass filter to compensate for low gain at low frequencies
C. A matched load resistor to prevent damage by switching transients
D. A temperature compensating load resistor to improve linearity

E7B04
(A)
Page 6-11

E7B04
Where on the load line of a Class A common emitter amplifier would bias normally be set?
A. Approximately halfway between saturation and cutoff
B. Where the load line intersects the voltage axis
C. At a point where the bias resistor equals the load resistor
D. At a point where the load line intersects the zero bias current curve

13-66   Chapter 13

E7B05
What can be done to prevent unwanted oscillations in an RF power amplifier?
A. Tune the stage for maximum SWR
B. Tune both the input and output for maximum power
C. Install parasitic suppressors and/or neutralize the stage
D. Use a phase inverter in the output filter

E7B05
(C)
Page 6-13

E7B06
Which of the following amplifier types reduces even-order harmonics?
A. Push-push
B. Push-pull
C. Class C
D. Class AB

E7B06
(B)
Page 6-11

E7B07
Which of the following is a likely result when a Class C amplifier is used to amplify a single-sideband phone signal?
A. Reduced intermodulation products
B. Increased overall intelligibility
C. Signal inversion
D. Signal distortion and excessive bandwidth

E7B07
(D)
Page 6-11

E7B08
How can an RF power amplifier be neutralized?
A. By increasing the driving power
B. By reducing the driving power
C. By feeding a 180-degree out-of-phase portion of the output back to the input
D. By feeding an in-phase component of the output back to the input

E7B08
(C)
Page 6-14

E7B09
Which of the following describes how the loading and tuning capacitors are to be adjusted when tuning a vacuum tube RF power amplifier that employs a Pi-network output circuit?
A. The loading capacitor is set to maximum capacitance and the tuning capacitor is adjusted for minimum allowable plate current
B. The tuning capacitor is set to maximum capacitance and the loading capacitor is adjusted for minimum plate permissible current
C. The loading capacitor is adjusted to minimum plate current while alternately adjusting the tuning capacitor for maximum allowable plate current
D. The tuning capacitor is adjusted for minimum plate current, and the loading capacitor is adjusted for maximum permissible plate current

E7B09
(D)
Page 6-40

Question Pool   13-67

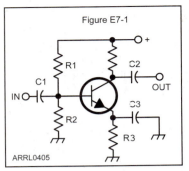

Figure E7-1 — This figure is used for questions E7B10 through E7B12.

E7B10
(B)
Page 6-3

E7B10
In Figure E7-1, what is the purpose of R1 and R2?
A. Load resistors
B. Voltage divider bias
C. Self bias
D. Feedback

E7B11
(D)
Page 6-3

E7B11
In Figure E7-1, what is the purpose of R3?
A. Fixed bias
B. Emitter bypass
C. Output load resistor
D. Self bias

E7B12
(C)
Page 6-3

E7B12
What type of amplifier circuit is shown in Figure E7-1?
A. Common base
B. Common collector
C. Common emitter
D. Emitter follower

E7B13
(D)
Page 6-4

E7B13
Which of the following describes an emitter follower (or common collector) amplifier?
A. A two-transistor amplifier with the emitters sharing a common bias resistor
B. A differential amplifier with both inputs fed to the emitter of the input transistor
C. An OR circuit with only one emitter used for output
D. An amplifier with a low impedance output that follows the base input voltage

E7B14
(B)
Page 6-12

E7B14
Why are switching amplifiers more efficient than linear amplifiers?
A. Switching amplifiers operate at higher voltages
B. The power transistor is at saturation or cutoff most of the time
C. Linear amplifiers have high gain resulting in higher harmonic content
D. Switching amplifiers use push-pull circuits

13-68   Chapter 13

E7B15
What is one way to prevent thermal runaway in a bipolar transistor amplifier?
A. Neutralization
B. Select transistors with high beta
C. Use a resistor in series with the emitter
D. All these choices are correct

E7B15
(C)
Page 6-3

E7B16
What is the effect of intermodulation products in a linear power amplifier?
A. Transmission of spurious signals
B. Creation of parasitic oscillations
C. Low efficiency
D. All these choices are correct

E7B16
(A)
Page 6-12

E7B17
Why are odd-order rather than even-order intermodulation distortion products of concern in linear power amplifiers?
A. Because they are relatively close in frequency to the desired signal
B. Because they are relatively far in frequency from the desired signal
C. Because they invert the sidebands causing distortion
D. Because they maintain the sidebands, thus causing multiple duplicate signals

E7B17
(A)
Page 6-12

E7B18
What is a characteristic of a grounded-grid amplifier?
A. High power gain
B. High filament voltage
C. Low input impedance
D. Low bandwidth

E7B18
(C)
Page 6-4

E7C — Filters and matching networks: types of networks; types of filters; filter applications; filter characteristics; impedance matching; DSP filtering

E7C01
How are the capacitors and inductors of a low-pass filter Pi-network arranged between the network's input and output?
A. Two inductors are in series between the input and output, and a capacitor is connected between the two inductors and ground
B. Two capacitors are in series between the input and output, and an inductor is connected between the two capacitors and ground
C. An inductor is connected between the input and ground, another inductor is connected between the output and ground, and a capacitor is connected between the input and output
D. A capacitor is connected between the input and ground, another capacitor is connected between the output and ground, and an inductor is connected between input and output

E7C01
(D)
Page 6-39

E7C02
Which of the following is a property of a T-network with series capacitors and a parallel shunt inductor?
A. It is a low-pass filter
B. It is a band-pass filter
C. It is a high-pass filter
D. It is a notch filter

E7C02
(C)
Page 6-40

Question Pool   13-69

E7C03
(A)
Page 6-40

E7C03
What advantage does a series-L Pi-L-network have over a series-L Pi-network for impedance matching between the final amplifier of a vacuum-tube transmitter and an antenna?
A. Greater harmonic suppression
B. Higher efficiency
C. Does not require a capacitor
D. Greater transformation range

E7C04
(C)
Page 6-39

E7C04
How does an impedance-matching circuit transform a complex impedance to a resistive impedance?
A. It introduces negative resistance to cancel the resistive part of impedance
B. It introduces transconductance to cancel the reactive part of impedance
C. It cancels the reactive part of the impedance and changes the resistive part to a desired value
D. Reactive currents are dissipated in matched resistances

E7C05
(D)
Page 6-34

E7C05
Which filter type is described as having ripple in the passband and a sharp cutoff?
A. A Butterworth filter
B. An active LC filter
C. A passive op-amp filter
D. A Chebyshev filter

E7C06
(C)
Page 6-34

E7C06
What are the distinguishing features of an elliptical filter?
A. Gradual passband rolloff with minimal stop band ripple
B. Extremely flat response over its pass band with gradually rounded stop band corners
C. Extremely sharp cutoff with one or more notches in the stop band
D. Gradual passband rolloff with extreme stop band ripple

E7C07
(B)
Page 6-40

E7C07
Which describes a Pi-L-network used for matching a vacuum tube final amplifier to a 50-ohm unbalanced output?
A. A Phase Inverter Load network
B. A Pi-network with an additional series inductor on the output
C. A network with only three discrete parts
D. A matching network in which all components are isolated from ground

E7C08
(A)
Page 6-35

E7C08
Which of the following factors has the greatest effect on the bandwidth and response shape of a crystal ladder filter?
A. The relative frequencies of the individual crystals
B. The DC voltage applied to the quartz crystal
C. The gain of the RF stage preceding the filter
D. The amplitude of the signals passing through the filter

E7C09
(D)
Page 6-35

E7C09
What is a crystal lattice filter?
A. A power supply filter made with interlaced quartz crystals
B. An audio filter made with four quartz crystals that resonate at 1 kHz intervals
C. A filter using lattice-shaped quartz crystals for high-Q performance
D. A filter with narrow bandwidth and steep skirts made using quartz crystals

13-70   Chapter 13

E7C10

Which of the following filters would be the best choice for use in a 2 meter band repeater duplexer?

A. A crystal filter
B. A cavity filter
C. A DSP filter
D. An L-C filter

E7C10
(B)
Page 6-33

E7C11

Which of the following describes a receiving filter's ability to reject signals occupying an adjacent channel?

A. Passband ripple
B. Phase response
C. Shape factor
D. Noise factor

E7C11
(C)
Page 6-34

E7C12

What is one advantage of a Pi-matching network over an L-matching network consisting of a single inductor and a single capacitor?

A. The Q of Pi-networks can be controlled
B. L-networks cannot perform impedance transformation
C. Pi-networks are more stable
D. Pi-networks provide balanced input and output

E7C12
(A)
Page 6-40

E7D — Power supplies and voltage regulators; Solar array charge controllers

E7D01

How does a linear electronic voltage regulator work?

A. It has a ramp voltage as its output
B. It eliminates the need for a pass transistor
C. The control element duty cycle is proportional to the line or load conditions
D. The conduction of a control element is varied to maintain a constant output voltage

E7D01
(D)
Page 6-41

E7D02

What is a characteristic of a switching electronic voltage regulator?

A. The resistance of a control element is varied in direct proportion to the line voltage or load current
B. It is generally less efficient than a linear regulator
C. The controlled device's duty cycle is changed to produce a constant average output voltage
D. It gives a ramp voltage at its output

E7D02
(C)
Page 6-43

E7D03

What device is typically used as a stable voltage reference in a linear voltage regulator?

A. A Zener diode
B. A tunnel diode
C. An SCR
D. A varactor diode

E7D03
(A)
Page 6-42

E7D04
(B)
Page 6-42

E7D04
Which of the following types of linear voltage regulator usually make the most efficient use of the primary power source?
A. A series current source
B. A series regulator
C. A shunt regulator
D. A shunt current source

E7D05
(D)
Page 6-41

E7D05
Which of the following types of linear voltage regulator places a constant load on the unregulated voltage source?
A. A constant current source
B. A series regulator
C. A shunt current source
D. A shunt regulator

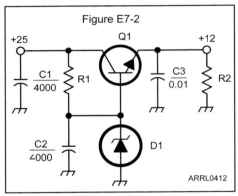

Figure E7-2 — This figure is used for questions E7D06 through E7D08.

E7D06
(C)
Page 6-42

E7D06
What is the purpose of Q1 in the circuit shown in Figure E7-2?
A. It provides negative feedback to improve regulation
B. It provides a constant load for the voltage source
C. It controls the current supplied to the load
D. It provides D1 with current

E7D07
(A)
Page 6-42

E7D07
What is the purpose of C2 in the circuit shown in Figure E7-2?
A. It bypasses rectifier output ripple around D1
B. It is a brute force filter for the output
C. To self-resonate at the hum frequency
D. To provide fixed DC bias for Q1

E7D08
(C)
Page 6-42

E7D08
What type of circuit is shown in Figure E7-2?
A. Switching voltage regulator
B. Grounded emitter amplifier
C. Linear voltage regulator
D. Monostable multivibrator

13-72 Chapter 13

E7D09
What is the main reason to use a charge controller with a solar power system?
A. Prevention of battery undercharge
B. Control of electrolyte levels during battery discharge
C. Prevention of battery damage due to overcharge
D. Matching of day and night charge rates

E7D09
(C)
Page 6-42

E7D10
What is the primary reason that a high-frequency switching type high-voltage power supply can be both less expensive and lighter in weight than a conventional power supply?
A. The inverter design does not require any output filtering
B. It uses a diode bridge rectifier for increased output
C. The high frequency inverter design uses much smaller transformers and filter components for an equivalent power output
D. It uses a large power factor compensation capacitor to recover power from the unused portion of the AC cycle

E7D10
(C)
Page 6-43

E7D11
What is the function of the pass transistor in a linear voltage regulator circuit?
A. Permits a wide range of output voltage settings
B. Provides a stable input impedance over a wide range of source voltage
C. Maintains nearly constant output impedance over a wide range of load current
D. Maintains nearly constant output voltage over a wide range of load current

E7D11
(D)
Page 6-42

E7D12
What is the dropout voltage of an analog voltage regulator?
A. Minimum input voltage for rated power dissipation
B. Maximum output voltage drops when the input voltage is varied over its specified range
C. Minimum input-to-output voltage required to maintain regulation
D. Maximum that the output voltage may decrease at rated load

E7D12
(C)
Page 6-42

E7D13
What is the equation for calculating power dissipated by a series linear voltage regulator?
A. Input voltage multiplied by input current
B. Input voltage divided by output current
C. Voltage difference from input to output multiplied by output current
D. Output voltage multiplied by output current

E7D13
(C)
Page 6-42

E7D14
What is the purpose of connecting equal-value resistors across power supply filter capacitors connected in series?
A. Equalize the voltage across each capacitor
B. Discharge the capacitors when voltage is removed
C. Provide a minimum load on the supply
D. All these choices are correct

E7D14
(D)
Page 6-43

E7D15
What is the purpose of a step-start circuit in a high-voltage power supply?
A. To provide a dual-voltage output for reduced power applications
B. To compensate for variations of the incoming line voltage
C. To allow for remote control of the power supply
D. To allow the filter capacitors to charge gradually

E7D15
(D)
Page 6-43

Question Pool   13-73

E7E — Modulation and demodulation: reactance, phase and balanced modulators; detectors; mixer stages

E7E01
(B)
Page 6-23

E7E01
Which of the following can be used to generate FM phone emissions?
A. A balanced modulator on the audio amplifier
B. A reactance modulator on the oscillator
C. A reactance modulator on the final amplifier
D. A balanced modulator on the oscillator

E7E02
(D)
Page 6-22

E7E02
What is the function of a reactance modulator?
A. To produce PM signals by using an electrically variable resistance
B. To produce AM signals by using an electrically variable inductance or capacitance
C. To produce AM signals by using an electrically variable resistance
D. To produce PM or FM signals by using an electrically variable inductance or capacitance

E7E03
(D)
Page 6-24

E7E03
What is a frequency discriminator stage in a FM receiver?
A. An FM generator circuit
B. A circuit for filtering two closely adjacent signals
C. An automatic band-switching circuit
D. A circuit for detecting FM signals

E7E04
(A)
Page 6-21

E7E04
What is one way a single-sideband phone signal can be generated?
A. By using a balanced modulator followed by a filter
B. By using a reactance modulator followed by a mixer
C. By using a loop modulator followed by a mixer
D. By driving a product detector with a DSB signal

E7E05
(D)
Page 6-23

E7E05
What circuit is added to an FM transmitter to boost the higher audio frequencies?
A. A de-emphasis network
B. A heterodyne suppressor
C. A heterodyne enhancer
D. A pre-emphasis network

E7E06
(A)
Page 6-23

E7E06
Why is de-emphasis commonly used in FM communications receivers?
A. For compatibility with transmitters using phase modulation
B. To reduce impulse noise reception
C. For higher efficiency
D. To remove third-order distortion products

E7E07
(B)
Page 6-20

E7E07
What is meant by the term "baseband" in radio communications?
A. The lowest frequency band that the transmitter or receiver covers
B. The frequency range occupied by a message signal prior to modulation
C. The unmodulated bandwidth of the transmitted signal
D. The basic oscillator frequency in an FM transmitter that is multiplied to increase the deviation and carrier frequency

13-74   Chapter 13

E7E08

What are the principal frequencies that appear at the output of a mixer circuit?
A. Two and four times the original frequency
B. The square root of the product of input frequencies
C. The two input frequencies along with their sum and difference frequencies
D. 1.414 and 0.707 times the input frequency

E7E08
(C)
Page 6-19

E7E09

What occurs when an excessive amount of signal energy reaches a mixer circuit?
A. Spurious mixer products are generated
B. Mixer blanking occurs
C. Automatic limiting occurs
D. A beat frequency is generated

E7E09
(A)
Page 6-20

E7E10

How does a diode envelope detector function?
A. By rectification and filtering of RF signals
B. By breakdown of the Zener voltage
C. By mixing signals with noise in the transition region of the diode
D. By sensing the change of reactance in the diode with respect to frequency

E7E10
(A)
Page 6-23

E7E11

Which type of detector circuit is used for demodulating SSB signals?
A. Discriminator
B. Phase detector
C. Product detector
D. Phase comparator

E7E11
(C)
Page 6-24

E7F — DSP filtering and other operations; software defined radio fundamentals; DSP modulation and demodulation

E7F01

What is meant by direct digital conversion as applied to software defined radios?
A. Software is converted from source code to object code during operation of the receiver
B. Incoming RF is converted to a control voltage for a voltage controlled oscillator
C. Incoming RF is digitized by an analog-to-digital converter without being mixed with a local oscillator signal
D. A switching mixer is used to generate I and Q signals directly from the RF input

E7F01
(C)
Page 6-29

E7F02

What kind of digital signal processing audio filter is used to remove unwanted noise from a received SSB signal?
A. An adaptive filter
B. A crystal-lattice filter
C. A Hilbert-transform filter
D. A phase-inverting filter

E7F02
(A)
Page 6-36

E7F03

What type of digital signal processing filter is used to generate an SSB signal?
A. An adaptive filter
B. A notch filter
C. A Hilbert-transform filter
D. An elliptical filter

E7F03
(C)
Page 6-32

Question Pool   13-75

E7F04
(D)
Page 6-32

E7F04
What is a common method of generating an SSB signal using digital signal processing?
 A. Mixing products are converted to voltages and subtracted by adder circuits
 B. A frequency synthesizer removes the unwanted sidebands
 C. Varying quartz crystal characteristics emulated in digital form
 D. Signals are combined in quadrature phase relationship

E7F05
(B)
Page 6-26

E7F05
How frequently must an analog signal be sampled by an analog-to-digital converter so that the signal can be accurately reproduced?
 A. At least half the rate of the highest frequency component of the signal
 B. At least twice the rate of the highest frequency component of the signal
 C. At the same rate as the highest frequency component of the signal
 D. At four times the rate of the highest frequency component of the signal

E7F06
(D)
Page 6-28

E7F06
What is the minimum number of bits required for an analog-to-digital converter to sample a signal with a range of 1 volt at a resolution of 1 millivolt?
 A. 4 bits
 B. 6 bits
 C. 8 bits
 D. 10 bits

E7F07
(C)
Page 6-28

E7F07
What function is performed by a Fast Fourier Transform?
 A. Converting analog signals to digital form
 B. Converting digital signals to analog form
 C. Converting digital signals from the time domain to the frequency domain
 D. Converting 8-bit data to 16-bit data

E7F08
(B)
Page 6-28

E7F08
What is the function of decimation?
 A. Converting data to binary code decimal form
 B. Reducing the effective sample rate by removing samples
 C. Attenuating the signal
 D. Removing unnecessary significant digits

E7F09
(A)
Page 6-28

E7F09
Why is an anti-aliasing digital filter required in a digital decimator?
 A. It removes high-frequency signal components that would otherwise be reproduced as lower frequency components
 B. It peaks the response of the decimator, improving bandwidth
 C. It removes low-frequency signal components to eliminate the need for DC restoration
 D. It notches out the sampling frequency to avoid sampling errors

E7F10
(A)
Page 6-30

E7F10
What aspect of receiver analog-to-digital conversion determines the maximum receive bandwidth of a Direct Digital Conversion SDR?
 A. Sample rate
 B. Sample width in bits
 C. Sample clock phase noise
 D. Processor latency

13-76   Chapter 13

E7F11

What sets the minimum detectable signal level for a direct-sampling SDR receiver in the absence of atmospheric or thermal noise?
A. Sample clock phase noise
B. Reference voltage level and sample width in bits
C. Data storage transfer rate
D. Missing codes and jitter

E7F11
(B)
Page 6-28

E7F12

Which of the following is an advantage of a Finite Impulse Response (FIR) filter vs an Infinite Impulse Response (IIR) digital filter?
A. FIR filters can delay all frequency components of the signal by the same amount
B. FIR filters are easier to implement for a given set of passband rolloff requirements
C. FIR filters can respond faster to impulses
D. All these choices are correct

E7F12
(A)
Page 6-38

E7F13

What is the function of taps in a digital signal processing filter?
A. To reduce excess signal pressure levels
B. Provide access for debugging software
C. Select the point at which baseband signals are generated
D. Provide incremental signal delays for filter algorithms

E7F13
(D)
Page 6-37

E7F14

Which of the following would allow a digital signal processing filter to create a sharper filter response?
A. Higher data rate
B. More taps
C. Complex phasor representations
D. Double-precision math routines

E7F14
(B)
Page 6-37

E7G — Active filters and op-amp circuits: active audio filters; characteristics; basic circuit design; operational amplifiers

E7G01

What is the typical output impedance of an op-amp?
A. Very low
B. Very high
C. 100 ohms
D. 1000 ohms

E7G01
(A)
Page 6-7

E7G02

What is ringing in a filter?
A. An echo caused by a long time delay
B. A reduction in high frequency response
C. Partial cancellation of the signal over a range of frequencies
D. Undesired oscillations added to the desired signal

E7G02
(D)
Page 6-36

Question Pool   13-77

E7G03
(D)
Page 6-7

E7G03
What is the typical input impedance of an op-amp?
A. 100 ohms
B. 1000 ohms
C. Very low
D. Very high

E7G04
(C)
Page 6-8

E7G04
What is meant by the term "op-amp input offset voltage"?
A. The output voltage of the op-amp minus its input voltage
B. The difference between the output voltage of the op-amp and the input voltage required in the immediately following stage
C. The differential input voltage needed to bring the open loop output voltage to zero
D. The potential between the amplifier input terminals of the op-amp in an open loop condition

E7G05
(A)
Page 6-36

E7G05
How can unwanted ringing and audio instability be prevented in an op-amp RC audio filter circuit?
A. Restrict both gain and Q
B. Restrict gain but increase Q
C. Restrict Q but increase gain
D. Increase both gain and Q

E7G06
(B)
Page 6-7

E7G06
What is the gain-bandwidth of an operational amplifier?
A. The maximum frequency for a filter circuit using that type of amplifier
B. The frequency at which the open-loop gain of the amplifier equals one
C. The gain of the amplifier at a filter's cutoff frequency
D. The frequency at which the amplifier's offset voltage is zero

13-78   Chapter 13

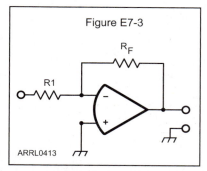

Figure E7-3 — This figure is used for questions E7G07 and E7G09 through E7G11.

E7G07
What magnitude of voltage gain can be expected from the circuit in Figure E7-3 when R1 is 10 ohms and RF is 470 ohms?
  A. 0.21
  B. 94
  C. 47
  D. 24

E7G08
How does the gain of an ideal operational amplifier vary with frequency?
  A. It increases linearly with increasing frequency
  B. It decreases linearly with increasing frequency
  C. It decreases logarithmically with increasing frequency
  D. It does not vary with frequency

E7G09
What will be the output voltage of the circuit shown in Figure E7-3 if R1 is 1000 ohms, RF is 10,000 ohms, and 0.23 volts DC is applied to the input?
  A. 0.23 volts
  B. 2.3 volts
  C. –0.23 volts
  D. –2.3 volts

E7G10
What absolute voltage gain can be expected from the circuit in Figure E7-3 when R1 is 1800 ohms and RF is 68 kilohms?
  A. 1
  B. 0.03
  C. 38
  D. 76

E7G11
What absolute voltage gain can be expected from the circuit in Figure E7-3 when R1 is 3300 ohms and RF is 47 kilohms?
  A. 28
  B. 14
  C. 7
  D. 0.07

E7G07
(C)
Page 6-9

E7G08
(D)
Page 6-7

E7G09
(D)
Page 6-9

E7G10
(C)
Page 6-9

E7G11
(B)
Page 6-7

E7G12
(A)
Page 6-7

E7G12
What is an operational amplifier?
    A. A high-gain, direct-coupled differential amplifier with very high input impedance and very low output impedance
    B. A digital audio amplifier whose characteristics are determined by components external to the amplifier
    C. An amplifier used to increase the average output of frequency modulated amateur signals to the legal limit
    D. A RF amplifier used in the UHF and microwave regions

E7H — Oscillators and signal sources: types of oscillators; synthesizers and phase-locked loops; direct digital synthesizers; stabilizing thermal drift; microphonics; high-accuracy oscillators

E7H01
(D)
Page 6-15

E7H01
What are three oscillator circuits used in amateur radio equipment?
    A. Taft, Pierce and negative feedback
    B. Pierce, Fenner and Beane
    C. Taft, Hartley and Pierce
    D. Colpitts, Hartley and Pierce

E7H02
(C)
Page 6-16

E7H02
What is a microphonic?
    A. An IC used for amplifying microphone signals
    B. Distortion caused by RF pickup on the microphone cable
    C. Changes in oscillator frequency due to mechanical vibration
    D. Excess loading of the microphone by an oscillator

E7H03
(A)
Page 6-14

E7H03
How is positive feedback supplied in a Hartley oscillator?
    A. Through a tapped coil
    B. Through a capacitive divider
    C. Through link coupling
    D. Through a neutralizing capacitor

E7H04
(C)
Page 6-14

E7H04
How is positive feedback supplied in a Colpitts oscillator?
    A. Through a tapped coil
    B. Through link coupling
    C. Through a capacitive divider
    D. Through a neutralizing capacitor

E7H05
(D)
Page 6-15

E7H05
How is positive feedback supplied in a Pierce oscillator?
    A. Through a tapped coil
    B. Through link coupling
    C. Through a neutralizing capacitor
    D. Through a quartz crystal

E7H06
Which of the following oscillator circuits are commonly used in VFOs?
A. Pierce and Zener
B. Colpitts and Hartley
C. Armstrong and deForest
D. Negative feedback and balanced feedback

E7H06
(B)
Page 6-15

E7H07
How can an oscillator's microphonic responses be reduced?
A. Use NP0 capacitors
B. Reduce noise on the oscillator's power supply
C. Increase the bias voltage
D. Mechanically isolate the oscillator circuitry from its enclosure

E7H07
(D)
Page 6-16

E7H08
Which of the following components can be used to reduce thermal drift in crystal oscillators?
A. NP0 capacitors
B. Toroidal inductors
C. Wirewound resistors
D. Non-inductive resistors

E7H08
(A)
Page 6-16

E7H09
What type of frequency synthesizer circuit uses a phase accumulator, lookup table, digital to analog converter, and a low-pass anti-alias filter?
A. A direct digital synthesizer
B. A hybrid synthesizer
C. A phase-locked loop synthesizer
D. A diode-switching matrix synthesizer

E7H09
(A)
Page 6-17

E7H10
What information is contained in the lookup table of a direct digital synthesizer (DDS)?
A. The phase relationship between a reference oscillator and the output waveform
B. Amplitude values that represent the desired waveform
C. The phase relationship between a voltage-controlled oscillator and the output waveform
D. Frequently used receiver and transmitter frequencies

E7H10
(B)
Page 6-17

E7H11
What are the major spectral impurity components of direct digital synthesizers?
A. Broadband noise
B. Digital conversion noise
C. Spurious signals at discrete frequencies
D. Nyquist limit noise

E7H11
(C)
Page 6-17

E7H12
Which of the following must be done to ensure that a crystal oscillator provides the frequency specified by the crystal manufacturer?
A. Provide the crystal with a specified parallel inductance
B. Provide the crystal with a specified parallel capacitance
C. Bias the crystal at a specified voltage
D. Bias the crystal at a specified current

E7H12
(B)
Page 6-16

Question Pool   13-81

E7H13
(D)
Page 6-15

E7H13

Which of the following is a technique for providing highly accurate and stable oscillators needed for microwave transmission and reception?

A. Use a GPS signal reference
B. Use a rubidium stabilized reference oscillator
C. Use a temperature-controlled high Q dielectric resonator
D. All these choices are correct

E7H14
(C)
Page 6-18

E7H14

What is a phase-locked loop circuit?

A. An electronic servo loop consisting of a ratio detector, reactance modulator, and voltage-controlled oscillator
B. An electronic circuit also known as a monostable multivibrator
C. An electronic servo loop consisting of a phase detector, a low-pass filter, a voltage-controlled oscillator, and a stable reference oscillator
D. An electronic circuit consisting of a precision push-pull amplifier with a differential input

E7H15
(D)
Page 6-19

E7H15

Which of these functions can be performed by a phase-locked loop?

A. Wide-band AF and RF power amplification
B. Comparison of two digital input signals, digital pulse counter
C. Photovoltaic conversion, optical coupling
D. Frequency synthesis, FM demodulation

SUBELEMENT E8 — SIGNALS AND EMISSIONS
[4 Exam Questions — 4 Groups]

E8A — AC waveforms: sine, square, and irregular waveforms; AC measurements; average power and PEP of RF signals; Fourier analysis; analog to digital conversion: digital to analog conversion; advantages of digital communications

E8A01
(A)
Page 7-8

E8A01

What is the name of the process that shows that a square wave is made up of a sine wave plus all its odd harmonics?

A. Fourier analysis
B. Vector analysis
C. Numerical analysis
D. Differential analysis

E8A02
(A)
Page 6-30

E8A02

Which of the following is a type of analog-to-digital conversion?

A. Successive approximation
B. Harmonic regeneration
C. Level shifting
D. Phase reversal

E8A03

What type of wave does a Fourier analysis show to be made up of sine waves of a given fundamental frequency plus all its harmonics?
A. A sawtooth wave
B. A square wave
C. A sine wave
D. A cosine wave

E8A03
(A)
Page 7-9

E8A04

What is "dither" with respect to analog-to-digital converters?
A. An abnormal condition where the converter cannot settle on a value to represent the signal
B. A small amount of noise added to the input signal to allow more precise representation of a signal over time
C. An error caused by irregular quantization step size
D. A method of decimation by randomly skipping samples

E8A04
(B)
Page 6-28

E8A05

What of the following instruments would be the most accurate for measuring the RMS voltage of a complex waveform?
A. A grid dip meter
B. A D'Arsonval meter
C. An absorption wave meter
D. A true-RMS calculating meter

E8A05
(D)
Page 7-2

E8A06

What is the approximate ratio of PEP-to-average power in a typical single-sideband phone signal?
A. 2.5 to 1
B. 25 to 1
C. 1 to 1
D. 100 to 1

E8A06
(A)
Page 7-3

E8A07

What determines the PEP-to-average power ratio of a single-sideband phone signal?
A. The frequency of the modulating signal
B. Speech characteristics
C. The degree of carrier suppression
D. Amplifier gain

E8A07
(B)
Page 7-3

E8A08

Why would a direct or flash conversion analog-to-digital converter be useful for a software defined radio?
A. Very low power consumption decreases frequency drift
B. Immunity to out-of-sequence coding reduces spurious responses
C. Very high speed allows digitizing high frequencies
D. All these choices are correct

E8A08
(C)
Page 6-30

E8A09

How many different input levels can be encoded by an analog-to-digital converter with 8-bit resolution?
A. 8
B. 8 multiplied by the gain of the input amplifier
C. 256 divided by the gain of the input amplifier
D. 256

E8A09
(D)
Page 6-27

Question Pool   13-83

E8A10
(C)
Page 6-28

E8A10
What is the purpose of a low-pass filter used in conjunction with a digital-to-analog converter?
    A. Lower the input bandwidth to increase the effective resolution
    B. Improve accuracy by removing out-of-sequence codes from the input
    C. Remove harmonics from the output caused by the discrete analog levels generated
    D. All these choices are correct

E8A11
(A)
Page 6-28

E8A11
Which of the following is a measure of the quality of an analog-to-digital converter?
    A. Total harmonic distortion
    B. Peak envelope power
    C. Reciprocal mixing
    D. Power factor

E8B — Modulation and demodulation: modulation methods; modulation index and deviation ratio; frequency and time division multiplexing; Orthogonal Frequency Division Multiplexing

E8B01
(A)
Page 8-3

E8B01
What is the modulation index of an FM signal?
    A. The ratio of frequency deviation to modulating signal frequency
    B. The ratio of modulating signal amplitude to frequency deviation
    C. The type of modulation used by the transmitter
    D. The bandwidth of the transmitted signal divided by the modulating signal frequency

E8B02
(D)
Page 8-4

E8B02
How does the modulation index of a phase-modulated emission vary with RF carrier frequency?
    A. It increases as the RF carrier frequency increases
    B. It decreases as the RF carrier frequency increases
    C. It varies with the square root of the RF carrier frequency
    D. It does not depend on the RF carrier frequency

E8B03
(A)
Page 8-4

E8B03
What is the modulation index of an FM-phone signal having a maximum frequency deviation of 3000 Hz either side of the carrier frequency when the modulating frequency is 1000 Hz?
    A. 3
    B. 0.3
    C. 3000
    D. 1000

E8B04
(B)
Page 8-4

E8B04
What is the modulation index of an FM-phone signal having a maximum carrier deviation of plus or minus 6 kHz when modulated with a 2 kHz modulating frequency?
    A. 6000
    B. 3
    C. 2000
    D. $^1/_3$

E8B05

What is the deviation ratio of an FM-phone signal having a maximum frequency swing of plus-or-minus 5 kHz when the maximum modulation frequency is 3 kHz?

A. 60
B. 0.167
C. 0.6
D. 1.67

E8B05
(D)
Page 8-3

E8B06

What is the deviation ratio of an FM-phone signal having a maximum frequency swing of plus or minus 7.5 kHz when the maximum modulation frequency is 3.5 kHz?

A. 2.14
B. 0.214
C. 0.47
D. 47

E8B06
(A)
Page 8-3

E8B07

Orthogonal Frequency Division Multiplexing is a technique used for which type of amateur communication?

A. High-speed digital modes
B. Extremely low-power contacts
C. EME
D. OFDM signals are not allowed on amateur bands

E8B07
(A)
Page 8-14

E8B08

What describes Orthogonal Frequency Division Multiplexing?

A. A frequency modulation technique that uses non-harmonically related frequencies
B. A bandwidth compression technique using Fourier transforms
C. A digital mode for narrow-band, slow-speed transmissions
D. A digital modulation technique using subcarriers at frequencies chosen to avoid intersymbol interference

E8B08
(D)
Page 8-14

E8B09

What is deviation ratio?

A. The ratio of the audio modulating frequency to the center carrier frequency
B. The ratio of the maximum carrier frequency deviation to the highest audio modulating frequency
C. The ratio of the carrier center frequency to the audio modulating frequency
D. The ratio of the highest audio modulating frequency to the average audio modulating frequency

E8B09
(B)
Page 8-3

E8B10

What is frequency division multiplexing?

A. The transmitted signal jumps from band to band at a predetermined rate
B. Two or more information streams are merged into a baseband, which then modulates the transmitter
C. The transmitted signal is divided into packets of information
D. Two or more information streams are merged into a digital combiner, which then pulse position modulates the transmitter

E8B10
(B)
Page 8-5

Question Pool 13-85

E8B11
(B)
Page 8-5

E8B11
What is digital time division multiplexing?
A. Two or more data streams are assigned to discrete sub-carriers on an FM transmitter
B. Two or more signals are arranged to share discrete time slots of a data transmission
C. Two or more data streams share the same channel by transmitting time of transmission as the sub-carrier
D. Two or more signals are quadrature modulated to increase bandwidth efficiency

E8C — Digital signals: digital communication modes; information rate vs. bandwidth; error correction

E8C01
(C)
Page 8-17

E8C01
How is Forward Error Correction implemented?
A. By the receiving station repeating each block of three data characters
B. By transmitting a special algorithm to the receiving station along with the data characters
C. By transmitting extra data that may be used to detect and correct transmission errors
D. By varying the frequency shift of the transmitted signal according to a predefined algorithm

E8C02
(C)
Page 8-5

E8C02
What is the definition of symbol rate in a digital transmission?
A. The number of control characters in a message packet
B. The duration of each bit in a message sent over the air
C. The rate at which the waveform changes to convey information
D. The number of characters carried per second by the station-to-station link

E8C03
(A)
Page 8-11

E8C03
Why should phase-shifting of a PSK signal be done at the zero crossing of the RF signal?
A. To minimize bandwidth
B. To simplify modulation
C. To improve carrier suppression
D. All these choices are correct

E8C04
(C)
Page 8-11

E8C04
What technique minimizes the bandwidth of a PSK31 signal?
A. Zero-sum character encoding
B. Reed-Solomon character encoding
C. Use of sinusoidal data pulses
D. Use of trapezoidal data pulses

E8C05
(C)
Page 8-9

E8C05
What is the approximate bandwidth of a 13-WPM International Morse Code transmission?
A. 13 Hz
B. 26 Hz
C. 52 Hz
D. 104 Hz

E8C06
(C)
Page 8-10

E8C06
What is the bandwidth of a 170-hertz shift, 300-baud ASCII transmission?
A. 0.1 Hz
B. 0.3 kHz
C. 0.5 kHz
D. 1.0 kHz

13-86   Chapter 13

E8C07
What is the bandwidth of a 4800-Hz frequency shift, 9600-baud ASCII FM transmission?
A. 15.36 kHz
B. 9.6 kHz
C. 4.8 kHz
D. 5.76 kHz

E8C07
(A)
Page 8-10

E8C08
How does ARQ accomplish error correction?
A. Special binary codes provide automatic correction
B. Special polynomial codes provide automatic correction
C. If errors are detected, redundant data is substituted
D. If errors are detected, a retransmission is requested

E8C08
(D)
Page 8-17

E8C09
Which digital code allows only one bit to change between sequential code values?
A. Binary Coded Decimal Code
B. Extended Binary Coded Decimal Interchange Code
C. Excess 3 code
D. Gray code

E8C09
(D)
Page 8-8

E8C10
How may data rate be increased without increasing bandwidth?
A. It is impossible
B. Increasing analog-to-digital conversion resolution
C. Using a more efficient digital code
D. Using forward error correction

E8C10
(C)
Page 8-6

E8C11
What is the relationship between symbol rate and baud?
A. They are the same
B. Baud is twice the symbol rate
C. Symbol rate is only used for packet-based modes
D. Baud is only used for RTTY

E8C11
(A)
Page 8-5

E8C12
What factors affect the bandwidth of a transmitted CW signal?
A. IF bandwidth and Q
B. Modulation index and output power
C. Keying speed and shape factor (rise and fall time)
D. All these choices are correct

E8C12
(C)
Page 8-9

E8D — Keying defects and overmodulation of digital signals; digital codes; spread spectrum

E8D01
Why are received spread spectrum signals resistant to interference?
A. Signals not using the spread spectrum algorithm are suppressed in the receiver
B. The high power used by a spread spectrum transmitter keeps its signal from being easily overpowered
C. The receiver is always equipped with a digital blanker
D. If interference is detected by the receiver it will signal the transmitter to change frequencies

E8D01
(A)
Page 8-15

E8D02
(B)
Page 8-16

E8D02
What spread spectrum communications technique uses a high-speed binary bit stream to shift the phase of an RF carrier?
A. Frequency hopping
B. Direct sequence
C. Binary phase-shift keying
D. Phase compandored spread spectrum

E8D03
(D)
Page 8-16

E8D03
How does the spread spectrum technique of frequency hopping work?
A. If interference is detected by the receiver it will signal the transmitter to change frequencies
B. If interference is detected by the receiver it will signal the transmitter to wait until the frequency is clear
C. A binary bit stream is used to shift the phase of an RF carrier very rapidly in a pseudorandom sequence
D. The frequency of the transmitted signal is changed very rapidly according to a pseudorandom sequence also used by the receiving station

E8D04
(C)
Page 8-9

E8D04
What is the primary effect of extremely short rise or fall time on a CW signal?
A. More difficult to copy
B. The generation of RF harmonics
C. The generation of key clicks
D. Limits data speed

E8D05
(A)
Page 8-9

E8D05
What is the most common method of reducing key clicks?
A. Increase keying waveform rise and fall times
B. Low-pass filters at the transmitter output
C. Reduce keying waveform rise and fall times
D. High-pass filters at the transmitter output

E8D06
(D)
Page 8-8

E8D06
What is the advantage of including parity bits in ASCII characters?
A. Faster transmission rate
B. The signal can overpower interfering signals
C. Foreign language characters can be sent
D. Some types of errors can be detected

E8D07
(D)
Page 8-14

E8D07
What is a common cause of overmodulation of AFSK signals?
A. Excessive numbers of retries
B. Ground loops
C. Bit errors in the modem
D. Excessive transmit audio levels

E8D08
(D)
Page 8-14

E8D08
What parameter evaluates distortion of an AFSK signal caused by excessive input audio levels?
A. Signal-to-noise ratio
B. Baud rate
C. Repeat Request Rate (RRR)
D. Intermodulation Distortion (IMD)

E8D09

What is considered an acceptable maximum IMD level for an idling PSK signal?

A. +10 dB
B. +15 dB
C. –20 dB
D. –30 dB

E8D09
(D)
Page 8-14

E8D10

What are some of the differences between the Baudot digital code and ASCII?

A. Baudot uses 4 data bits per character, ASCII uses 7 or 8; Baudot uses 1 character as a letters/figures shift code, ASCII has no letters/figures code
B. Baudot uses 5 data bits per character, ASCII uses 7 or 8; Baudot uses 2 characters as letters/figures shift codes, ASCII has no letters/figures shift code
C. Baudot uses 6 data bits per character, ASCII uses 7 or 8; Baudot has no letters/figures shift code, ASCII uses 2 letters/figures shift codes
D. Baudot uses 7 data bits per character, ASCII uses 8; Baudot has no letters/figures shift code, ASCII uses 2 letters/figures shift codes

E8D10
(B)
Page 8-7

E8D11

What is one advantage of using ASCII code for data communications?

A. It includes built-in error correction features
B. It contains fewer information bits per character than any other code
C. It is possible to transmit both upper and lower case text
D. It uses one character as a shift code to send numeric and special characters

E8D11
(C)
Page 8-7

SUBELEMENT E9 — ANTENNAS AND TRANSMISSION LINES
[8 Exam Questions — 8 Groups]

E9A — Basic Antenna parameters: radiation resistance, gain, beamwidth, efficiency; effective radiated power

E9A01

What is an isotropic antenna?

A. A grounded antenna used to measure Earth conductivity
B. A horizontally polarized antenna used to compare Yagi antennas
C. A theoretical, omnidirectional antenna used as a reference for antenna gain
D. A spacecraft antenna used to direct signals toward Earth

E9A01
(C)
Page 9-3

E9A02

What is the effective radiated power relative to a dipole of a repeater station with 150 watts transmitter power output, 2 dB feed line loss, 2.2 dB duplexer loss, and 7 dBd antenna gain?

A. 1977 watts
B. 78.7 watts
C. 420 watts
D. 286 watts

E9A02
(D)
Page 9-24

Question Pool   13-89

E9A03
(C)
Page 9-5

E9A03
What is the radiation resistance of an antenna?
    A. The combined losses of the antenna elements and feed line
    B. The specific impedance of the antenna
    C. The value of a resistance that would dissipate the same amount of power as that radiated from an antenna
    D. The resistance in the atmosphere that an antenna must overcome to be able to radiate a signal

E9A04
(B)
Page 9-6

E9A04
Which of the following factors affect the feed point impedance of an antenna?
    A. Transmission line length
    B. Antenna height
    C. The settings of an antenna tuner at the transmitter
    D. The input power level

E9A05
(D)
Page 9-5

E9A05
What is included in the total resistance of an antenna system?
    A. Radiation resistance plus space impedance
    B. Radiation resistance plus transmission resistance
    C. Transmission-line resistance plus radiation resistance
    D. Radiation resistance plus loss resistance

E9A06
(A)
Page 9-24

E9A06
What is the effective radiated power relative to a dipole of a repeater station with 200 watts transmitter power output, 4 dB feed line loss, 3.2 dB duplexer loss, 0.8 dB circulator loss, and 10 dBd antenna gain?
    A. 317 watts
    B. 2000 watts
    C. 126 watts
    D. 300 watts

E9A07
(B)
Page 9-24

E9A07
What is the effective isotropic radiated power of a repeater station with 200 watts transmitter power output, 2 dB feed line loss, 2.8 dB duplexer loss, 1.2 dB circulator loss, and 7 dBi antenna gain?
    A. 159 watts
    B. 252 watts
    C. 632 watts
    D. 63.2 watts

E9A08
(B)
Page 9-8

E9A08
What is antenna bandwidth?
    A. Antenna length divided by the number of elements
    B. The frequency range over which an antenna satisfies a performance requirement
    C. The angle between the half-power radiation points
    D. The angle formed between two imaginary lines drawn through the element ends

E9A09
(B)
Page 9-6

E9A09
What is antenna efficiency?
    A. Radiation resistance divided by transmission resistance
    B. Radiation resistance divided by total resistance
    C. Total resistance divided by radiation resistance
    D. Effective radiated power divided by transmitter output

13-90   **Chapter 13**

E9A10
Which of the following improves the efficiency of a ground-mounted quarter-wave vertical antenna?
A. Installing a radial system
B. Isolating the coax shield from ground
C. Shortening the radiating element
D. All these choices are correct

E9A10
(A)
Page 9-8

E9A11
Which of the following factors determines ground losses for a ground-mounted vertical antenna operating in the 3 MHz to 30 MHz range?
A. The standing wave ratio
B. Distance from the transmitter
C. Soil conductivity
D. Take-off angle

E9A11
(C)
Page 9-8

E9A12
How much gain does an antenna have compared to a ½-wavelength dipole when it has 6 dB gain over an isotropic antenna?
A. 3.85 dB
B. 6.0 dB
C. 8.15 dB
D. 2.79 dB

E9A12
(A)
Page 9-3

E9A13
What term describes station output, taking into account all gains and losses?
A. Power factor
B. Half-power bandwidth
C. Effective radiated power
D. Apparent power

E9A13
(C)
Page 9-23

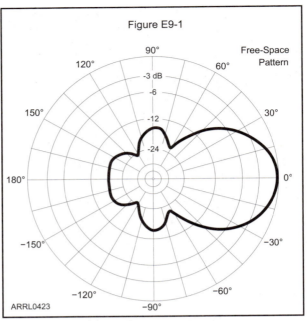

Figure E9-1 — This figure is used for questions E9B01 through E9B03.

E9B — Antenna patterns and designs: E and H plane patterns; gain as a function of pattern; antenna modeling

E9B01
In the antenna radiation pattern shown in Figure E9-1, what is the beamwidth?
A. 75 degrees
B. 50 degrees
C. 25 degrees
D. 30 degrees

E9B02
In the antenna radiation pattern shown in Figure E9-1, what is the front-to-back ratio?
A. 36 dB
B. 18 dB
C. 24 dB
D. 14 dB

E9B03
In the antenna radiation pattern shown in Figure E9-1, what is the front-to-side ratio?
A. 12 dB
B. 14 dB
C. 18 dB
D. 24 dB

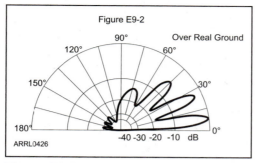

Figure E9-2 — This figure is used for questions E9B04 through E9B06.

E9B04
What is the front-to-back ratio of the radiation pattern shown in Figure E9-2?
A. 15 dB
B. 28 dB
C. 3 dB
D. 38 dB

E9B05
What type of antenna pattern is shown in Figure E9-2?
A. Elevation
B. Azimuth
C. Radiation resistance
D. Polarization

E9B06
What is the elevation angle of peak response in the antenna radiation pattern shown in Figure E9-2?
A. 45 degrees
B. 75 degrees
C. 7.5 degrees
D. 25 degrees

E9B07
How does the total amount of radiation emitted by a directional gain antenna compare with the total amount of radiation emitted from a theoretical isotropic antenna, assuming each is driven by the same amount of power?
A. The total amount of radiation from the directional antenna is increased by the gain of the antenna
B. The total amount of radiation from the directional antenna is stronger by its front-to-back ratio
C. They are the same
D. The radiation from the isotropic antenna is 2.15 dB stronger than that from the directional antenna

E9B04
(B)
Page 9-7

E9B05
(A)
Page 9-7

E9B06
(C)
Page 9-7

E9B07
(C)
Page 9-3

E9B08
(D)
Page 9-2

E9B08
What is the far field of an antenna?
    A. The region of the ionosphere where radiated power is not refracted
    B. The region where radiated power dissipates over a specified time period
    C. The region where radiated field strengths are constant
    D. The region where the shape of the antenna pattern is independent of distance

E9B09
(B)
Page 9-40

E9B09
What type of computer program technique is commonly used for modeling antennas?
    A. Graphical analysis
    B. Method of Moments
    C. Mutual impedance analysis
    D. Calculus differentiation with respect to physical properties

E9B10
(A)
Page 9-40

E9B10
What is the principle of a Method of Moments analysis?
    A. A wire is modeled as a series of segments, each having a uniform value of current
    B. A wire is modeled as a single sine-wave current generator
    C. A wire is modeled as a single sine-wave voltage source
    D. A wire is modeled as a series of segments, each having a distinct value of voltage across it

E9B11
(C)
Page 9-41

E9B11
What is a disadvantage of decreasing the number of wire segments in an antenna model below 10 segments per half-wavelength?
    A. Ground conductivity will not be accurately modeled
    B. The resulting design will favor radiation of harmonic energy
    C. The computed feed point impedance may be incorrect
    D. The antenna will become mechanically unstable

E9C — Practical wire antennas; folded dipoles; phased arrays; effects of ground near antennas

E9C01
(D)
Page 9-18

E9C01
What is the radiation pattern of two ¼-wavelength vertical antennas spaced ½-wavelength apart and fed 180 degrees out of phase?
    A. Cardioid
    B. Omni-directional
    C. A figure-8 broadside to the axis of the array
    D. A figure-8 oriented along the axis of the array

E9C02
(A)
Page 9-18

E9C02
What is the radiation pattern of two ¼-wavelength vertical antennas spaced ¼-wavelength apart and fed 90 degrees out of phase?
    A. Cardioid
    B. A figure-8 end-fire along the axis of the array
    C. A figure-8 broadside to the axis of the array
    D. Omni-directional

E9C03

What is the radiation pattern of two ¼-wavelength vertical antennas spaced ½-wavelength apart and fed in phase?

A. Omni-directional
B. Cardioid
C. A Figure-8 broadside to the axis of the array
D. A Figure-8 end-fire along the axis of the array

E9C03
(C)
Page 9-18

E9C04

What happens to the radiation pattern of an unterminated long wire antenna as the wire length is increased?

A. The lobes become more perpendicular to the wire
B. The lobes align more in the direction of the wire
C. The vertical angle increases
D. The front-to-back ratio decreases

E9C04
(B)
Page 9-14

E9C05

Which of the following is a type of OCFD antenna?

A. A dipole fed approximately ⅓ the way from one end with a 4:1 balun to provide multiband operation
B. A remotely tunable dipole antenna using orthogonally controlled frequency diversity
C. A folded dipole center-fed with 300-ohm transmission line
D. A multiband dipole antenna using one-way circular polarization for frequency diversity

E9C05
(A)
Page 9-12

E9C06

What is the effect of adding a terminating resistor to a rhombic antenna?

A. It reflects the standing waves on the antenna elements back to the transmitter
B. It changes the radiation pattern from bidirectional to unidirectional
C. It changes the radiation pattern from horizontal to vertical polarization
D. It decreases the ground loss

E9C06
(B)
Page 9-15

E9C07

What is the approximate feed point impedance at the center of a two-wire folded dipole antenna?

A. 300 ohms
B. 72 ohms
C. 50 ohms
D. 450 ohms

E9C07
(A)
Page 9-10

E9C08

What is a folded dipole antenna?

A. A dipole one-quarter wavelength long
B. A type of ground-plane antenna
C. A half-wave dipole with an additional parallel wire connecting its two ends
D. A dipole configured to provide forward gain

E9C08
(C)
Page 9-10

E9C09

Which of the following describes a G5RV antenna?

A. A multi-band dipole antenna fed with coax and a balun through a selected length of open wire transmission line
B. A multi-band trap antenna
C. A phased array antenna consisting of multiple loops
D. A wide band dipole using shorted coaxial cable for the radiating elements and fed with a 4:1 balun

E9C09
(A)
Page 9-11

Question Pool   13-95

E9C10
(B)
Page 9-10

E9C10
Which of the following describes a Zepp antenna?
  A. A dipole constructed from zip cord
  B. An end-fed dipole antenna
  C. An omni-directional antenna commonly used for satellite communications
  D. A vertical array capable of quickly changing the direction of maximum radiation by changing phasing lines

E9C11
(D)
Page 9-8

E9C11
How is the far-field elevation pattern of a vertically polarized antenna affected by being mounted over seawater versus soil?
  A. The low-angle radiation decreases
  B. Additional higher vertical angle lobes will appear
  C. Fewer vertical angle lobes will be present
  D. The low-angle radiation increases

E9C12
(C)
Page 9-11

E9C12
Which of the following describes an Extended Double Zepp antenna?
  A. A wideband vertical antenna constructed from precisely tapered aluminum tubing
  B. A portable antenna erected using two push support poles
  C. A center-fed 1.25-wavelength antenna (two ⅝-wave elements in phase)
  D. An end-fed folded dipole antenna

E9C13
(B)
Page 9-9

E9C13
How does the radiation pattern of a horizontally polarized 3-element beam antenna vary with increasing height above ground?
  A. The takeoff angle of the lowest elevation lobe increases
  B. The takeoff angle of the lowest elevation lobe decreases
  C. The horizontal beamwidth increases
  D. The horizontal beamwidth decreases

E9C14
(B)
Page 9-9

E9C14
How does the performance of a horizontally polarized antenna mounted on the side of a hill compare with the same antenna mounted on flat ground?
  A. The main lobe takeoff angle increases in the downhill direction
  B. The main lobe takeoff angle decreases in the downhill direction
  C. The horizontal beamwidth decreases in the downhill direction
  D. The horizontal beamwidth increases in the uphill direction

E9D —Yagi antennas; parabolic reflectors; circular polarization; loading coils; top loading; feed point impedance of electrically short antennas; antenna Q; RF grounding

E9D01
(D)
Page 9-19

E9D01
How much does the gain of an ideal parabolic dish antenna change when the operating frequency is doubled?
  A. 2 dB
  B. 3 dB
  C. 4 dB
  D. 6 dB

13-96   Chapter 13

E9D02

How can linearly polarized Yagi antennas be used to produce circular polarization?
A. Stack two Yagis fed 90 degrees out of phase to form an array with the respective elements in parallel planes
B. Stack two Yagis fed in phase to form an array with the respective elements in parallel planes
C. Arrange two Yagis perpendicular to each other with the driven elements at the same point on the boom fed 90 degrees out of phase
D. Arrange two Yagis collinear to each other with the driven elements fed 180 degrees out of phase

E9D02
(C)
Page 9-19

E9D03

Where should a high Q loading coil be placed to minimize losses in a shortened vertical antenna?
A. Near the center of the vertical radiator
B. As low as possible on the vertical radiator
C. As close to the transmitter as possible
D. At a voltage node

E9D03
(A)
Page 9-13

E9D04

Why should an HF mobile antenna loading coil have a high ratio of reactance to resistance?
A. To swamp out harmonics
B. To lower the radiation angle
C. To minimize losses
D. To minimize the Q

E9D04
(C)
Page 9-13

E9D05

What usually occurs if a Yagi antenna is designed solely for maximum forward gain?
A. The front-to-back ratio increases
B. The front-to-back ratio decreases
C. The frequency response is widened over the whole frequency band
D. The SWR is reduced

E9D05
(B)
Page 9-42

E9D06

What happens to the SWR bandwidth when one or more loading coils are used to resonate an electrically short antenna?
A. It is increased
B. It is decreased
C. It is unchanged if the loading coil is located at the feed point
D. It is unchanged if the loading coil is located at a voltage maximum point

E9D06
(B)
Page 9-13

E9D07

What is an advantage of using top loading in a shortened HF vertical antenna?
A. Lower Q
B. Greater structural strength
C. Higher losses
D. Improved radiation efficiency

E9D07
(D)
Page 9-13

E9D08

What happens as the Q of an antenna increases?
A. SWR bandwidth increases
B. SWR bandwidth decreases
C. Gain is reduced
D. More common-mode current is present on the feed line

E9D08
(B)
Page 9-8

Question Pool   **13-97**

E9D09
(D)
Page 9-12

E9D09
What is the function of a loading coil used as part of an HF mobile antenna?
    A. To increase the SWR bandwidth
    B. To lower the losses
    C. To lower the Q
    D. To cancel capacitive reactance

E9D10
(B)
Page 9-12

E9D10
What happens to feed-point impedance at the base of a fixed length HF mobile antenna when operated below its resonant frequency?
    A. The radiation resistance decreases and the capacitive reactance decreases
    B. The radiation resistance decreases and the capacitive reactance increases
    C. The radiation resistance increases and the capacitive reactance decreases
    D. The radiation resistance increases and the capacitive reactance increases

E9D11
(B)
Page 9-9

E9D11
Which of the following conductors would be best for minimizing losses in a station's RF ground system?
    A. Resistive wire, such as spark plug wire
    B. Wide flat copper strap
    C. Stranded wire
    D. Solid wire

E9D12
(C)
Page 9-9

E9D12
Which of the following would provide the best RF ground for your station?
    A. A 50-ohm resistor connected to ground
    B. An electrically short connection to a metal water pipe
    C. An electrically short connection to 3 or 4 interconnected ground rods driven into the Earth
    D. An electrically short connection to 3 or 4 interconnected ground rods via a series RF choke

E9E — Matching: matching antennas to feed lines; phasing lines; power dividers

E9E01
(B)
Page 9-25

E9E01
What system matches a higher-impedance transmission line to a lower-impedance antenna by connecting the line to the driven element in two places spaced a fraction of a wavelength each side of element center?
    A. The gamma matching system
    B. The delta matching system
    C. The omega matching system
    D. The stub matching system

E9E02
(A)
Page 9-25

E9E02
What is the name of an antenna matching system that matches an unbalanced feed line to an antenna by feeding the driven element both at the center of the element and at a fraction of a wavelength to one side of center?
    A. The gamma match
    B. The delta match
    C. The epsilon match
    D. The stub match

13-98   **Chapter 13**

E9E03

What is the name of the matching system that uses a section of transmission line connected in parallel with the feed line at or near the feed point?

A. The gamma match
B. The delta match
C. The omega match
D. The stub match

E9E03
(D)
Page 9-26

E9E04

What is the purpose of the series capacitor in a gamma-type antenna matching network?

A. To provide DC isolation between the feed line and the antenna
B. To cancel the inductive reactance of the matching network
C. To provide a rejection notch that prevents the radiation of harmonics
D. To transform the antenna impedance to a higher value

E9E04
(B)
Page 9-26

E9E05

How must an antenna's driven element be tuned to use a hairpin matching system?

A. The driven element reactance must be capacitive
B. The driven element reactance must be inductive
C. The driven element resonance must be lower than the operating frequency
D. The driven element radiation resistance must be higher than the characteristic impedance of the transmission line

E9E05
(A)
Page 9-26

E9E06

Which of these feed line impedances would be suitable for constructing a quarter-wave Q-section for matching a 100-ohm loop to 50-ohm feed line?

A. 50 ohms
B. 62 ohms
C. 75 ohms
D. 450 ohms

E9E06
(C)
Page 9-38

E9E07

What parameter describes the interactions at the load end of a mismatched transmission line?

A. Characteristic impedance
B. Reflection coefficient
C. Velocity factor
D. Dielectric constant

E9E07
(B)
Page 9-31

E9E08

What is a use for a Wilkinson divider?

A. It divides the operating frequency of a transmitter signal so it can be used on a lower frequency band
B. It is used to feed high-impedance antennas from a low-impedance source
C. It is used to divide power equally between two 50-ohm loads while maintaining 50-ohm input impedance
D. It is used to feed low-impedance loads from a high-impedance source

E9E08
(C)
Page 9-18

E9E09

Which of the following is used to shunt-feed a grounded tower at its base?

A. Double-bazooka match
B. Hairpin match
C. Gamma match
D. All these choices are correct

E9E09
(C)
Page 9-26

Question Pool   13-99

E9E10
(C)
Page 9-38

E9E10
Which of these choices is an effective way to match an antenna with a 100-ohm feed point impedance to a 50-ohm coaxial cable feed line?
    A. Connect a ¼-wavelength open stub of 300-ohm twinlead in parallel with the coaxial feed line where it connects to the antenna
    B. Insert a ½ wavelength piece of 300-ohm twinlead in series between the antenna terminals and the 50-ohm feed cable
    C. Insert a ¼-wavelength piece of 75-ohm coaxial cable transmission line in series between the antenna terminals and the 50-ohm feed cable
    D. Connect a ½ wavelength shorted stub of 75-ohm cable in parallel with the 50-ohm cable where it attaches to the antenna

E9E11
(A)
Page 9-18

E9E11
What is the primary purpose of phasing lines when used with an antenna having multiple driven elements?
    A. It ensures that each driven element operates in concert with the others to create the desired antenna pattern
    B. It prevents reflected power from traveling back down the feed line and causing harmonic radiation from the transmitter
    C. It allows single-band antennas to operate on other bands
    D. It creates a low-angle radiation pattern

E9F — Transmission lines: characteristics of open and shorted feed lines; coax versus open-wire; velocity factor; electrical length; coaxial cable dielectrics

E9F01
(D)
Page 9-28

E9F01
What is the velocity factor of a transmission line?
    A. The ratio of the characteristic impedance of the line to the terminating impedance
    B. The index of shielding for coaxial cable
    C. The velocity of the wave in the transmission line multiplied by the velocity of light in a vacuum
    D. The velocity of the wave in the transmission line divided by the velocity of light in a vacuum

E9F02
(C)
Page 9-28

E9F02
Which of the following has the biggest effect on the velocity factor of a transmission line?
    A. The termination impedance
    B. The line length
    C. Dielectric materials used in the line
    D. The center conductor resistivity

E9F03
(D)
Page 9-29

E9F03
Why is the physical length of a coaxial cable transmission line shorter than its electrical length?
    A. Skin effect is less pronounced in the coaxial cable
    B. The characteristic impedance is higher in a parallel feed line
    C. The surge impedance is higher in a parallel feed line
    D. Electrical signals move more slowly in a coaxial cable than in air

E9F04
What impedance does a ½-wavelength transmission line present to a generator when the line is shorted at the far end?
  A. Very high impedance
  B. Very low impedance
  C. The same as the characteristic impedance of the line
  D. The same as the output impedance of the generator

E9F04
(B)
Page 9-36

E9F05
What is the approximate physical length of a solid polyethylene dielectric coaxial transmission line that is electrically ¼ wavelength long at 14.1 MHz?
  A. 10.6 meters
  B. 5.3 meters
  C. 4.3 meters
  D. 3.5 meters

E9F05
(D)
Page 9-29

E9F06
What is the approximate physical length of an air-insulated, parallel conductor transmission line that is electrically ½ wavelength long at 14.10 MHz?
  A. 7.0 meters
  B. 8.5 meters
  C. 10.6 meters
  D. 13.3 meters

E9F06
(C)
Page 9-30

E9F07
How does ladder line compare to small-diameter coaxial cable such as RG-58 at 50 MHz?
  A. Lower loss
  B. Higher SWR
  C. Smaller reflection coefficient
  D. Lower velocity factor

E9F07
(A)
Page 9-30

E9F08
Which of the following is a significant difference between foam dielectric coaxial cable and solid dielectric cable, assuming all other parameters are the same?
  A. Foam dielectric has lower safe operating voltage limits
  B. Foam dielectric has lower loss per unit of length
  C. Foam dielectric has higher velocity factor
  D. All these choices are correct

E9F08
(D)
Page 9-31

E9F09
What is the approximate physical length of a foam polyethylene dielectric coaxial transmission line that is electrically ¼ wavelength long at 7.2 MHz?
  A. 10.4 meters
  B. 8.3 meters
  C. 6.9 meters
  D. 5.2 meters

E9F09
(B)
Page 9-29

Question Pool   13-101

E9F10
(C)
Page 9-37

E9F10
What impedance does a ⅛-wavelength transmission line present to a generator when the line is shorted at the far end?
  A. A capacitive reactance
  B. The same as the characteristic impedance of the line
  C. An inductive reactance
  D. Zero

E9F11
(C)
Page 9-36

E9F11
What impedance does a ⅛-wavelength transmission line present to a generator when the line is open at the far end?
  A. The same as the characteristic impedance of the line
  B. An inductive reactance
  C. A capacitive reactance
  D. Infinite

E9F12
(D)
Page 9-36

E9F12
What impedance does a ¼-wavelength transmission line present to a generator when the line is open at the far end?
  A. The same as the characteristic impedance of the line
  B. The same as the input impedance to the generator
  C. Very high impedance
  D. Very low impedance

E9F13
(A)
Page 9-36

E9F13
What impedance does a ¼-wavelength transmission line present to a generator when the line is shorted at the far end?
  A. Very high impedance
  B. Very low impedance
  C. The same as the characteristic impedance of the transmission line
  D. The same as the generator output impedance

E9G — The Smith chart

E9G01
(A)
Page 9-33

E9G01
Which of the following can be calculated using a Smith chart?
  A. Impedance along transmission lines
  B. Radiation resistance
  C. Antenna radiation pattern
  D. Radio propagation

E9G02
(B)
Page 9-35

E9G02
What type of coordinate system is used in a Smith chart?
  A. Voltage circles and current arcs
  B. Resistance circles and reactance arcs
  C. Voltage lines and current chords
  D. Resistance lines and reactance chords

E9G03
(C)
Page 9-33

E9G03
Which of the following is often determined using a Smith chart?
  A. Beam headings and radiation patterns
  B. Satellite azimuth and elevation bearings
  C. Impedance and SWR values in transmission lines
  D. Trigonometric functions

E9G04
What are the two families of circles and arcs that make up a Smith chart?
  A. Resistance and voltage
  B. Reactance and voltage
  C. Resistance and reactance
  D. Voltage and impedance

E9G05
Which of the following is a common use for a Smith chart?
  A. Determine the length and position of an impedance matching stub
  B. Determine the impedance of a transmission line, given the physical dimensions
  C. Determine the gain of an antenna given the physical and electrical parameters
  D. Determine the loss/100 feet of a transmission line, given the velocity factor and conductor materials

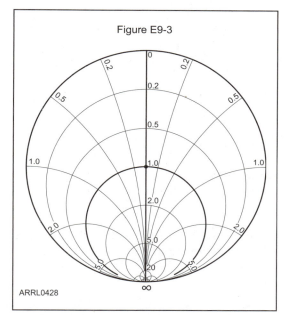

Figure E9-3 — This figure is used for questions E9G06 and E9G07.

E9G06
On the Smith chart shown in Figure E9-3, what is the name for the large outer circle on which the reactance arcs terminate?
  A. Prime axis
  B. Reactance axis
  C. Impedance axis
  D. Polar axis

E9G07
On the Smith chart shown in Figure E9-3, what is the only straight line shown?
  A. The reactance axis
  B. The current axis
  C. The voltage axis
  D. The resistance axis

E9G04
(C)
Page 9-35

E9G05
(A)
Page 9-27

E9G06
(B)
Page 9-35

E9G07
(D)
Page 9-33

Question Pool   13-103

E9G08
(C)
Page 9-35

E9G08
What is the process of normalization with regard to a Smith chart?
  A. Reassigning resistance values with regard to the reactance axis
  B. Reassigning reactance values with regard to the resistance axis
  C. Reassigning impedance values with regard to the prime center
  D. Reassigning prime center with regard to the reactance axis

E9G09
(A)
Page 9-35

E9G09
What third family of circles is often added to a Smith chart during the process of solving problems?
  A. Standing wave ratio circles
  B. Antenna-length circles
  C. Coaxial-length circles
  D. Radiation-pattern circles

E9G10
(D)
Page 9-35

E9G10
What do the arcs on a Smith chart represent?
  A. Frequency
  B. SWR
  C. Points with constant resistance
  D. Points with constant reactance

E9G11
(B)
Page 9-35

E9G11
How are the wavelength scales on a Smith chart calibrated?
  A. In fractions of transmission line electrical frequency
  B. In fractions of transmission line electrical wavelength
  C. In fractions of antenna electrical wavelength
  D. In fractions of antenna electrical frequency

E9H — Receiving Antennas: radio direction finding antennas; Beverage antennas; specialized receiving antennas; long-wire receiving antennas

E9H01
(D)
Page 9-15

E9H01
When constructing a Beverage antenna, which of the following factors should be included in the design to achieve good performance at the desired frequency?
  A. Its overall length must not exceed ¼ wavelength
  B. It must be mounted more than 1 wavelength above ground
  C. It should be configured as a four-sided loop
  D. It should be one or more wavelengths long

E9H02
(A)
Page 9-15

E9H02
Which is generally true for low band (160 meter and 80 meter) receiving antennas?
  A. Atmospheric noise is so high that gain over a dipole is not important
  B. They must be erected at least ½ wavelength above the ground to attain good directivity
  C. Low loss coax transmission line is essential for good performance
  D. All these choices are correct

E9H03
(D)
Page 9-21

E9H03
What is Receiving Directivity Factor (RDF)?
  A. Forward gain compared to the gain in the reverse direction
  B. Relative directivity compared to isotropic
  C. Relative directivity compared to a dipole
  D. Forward gain compared to average gain over the entire hemisphere

E9H04
What is an advantage of placing a grounded electrostatic shield around a small loop direction-finding antenna?
   A. It adds capacitive loading, increasing the bandwidth of the antenna
   B. It eliminates unbalanced capacitive coupling to the surroundings, improving the nulls
   C. It eliminates tracking errors caused by strong out-of-band signals
   D. It increases signal strength by providing a better match to the feed line

E9H04
(B)
Page 9-20

E9H05
What is the main drawback of a small wire-loop antenna for direction finding?
   A. It has a bidirectional pattern
   B. It has no clearly defined null
   C. It is practical for use only on VHF and higher bands
   D. All these choices are correct

E9H05
(A)
Page 9-20

E9H06
What is the triangulation method of direction finding?
   A. The geometric angles of sky waves from the source are used to determine its position
   B. A fixed receiving station plots three headings to the signal source
   C. Antenna headings from several different receiving locations are used to locate the signal source
   D. A fixed receiving station uses three different antennas to plot the location of the signal source

E9H06
(C)
Page 9-22

E9H07
Why is RF attenuation used when direction-finding?
   A. To narrow the receiver bandwidth
   B. To compensate for isotropic directivity and the antenna effect of feed lines
   C. To increase receiver sensitivity
   D. To prevent receiver overload which reduces pattern nulls

E9H07
(D)
Page 9-21

E9H08
What is the function of a sense antenna?
   A. It modifies the pattern of a DF antenna array to provide a null in one direction
   B. It increases the sensitivity of a DF antenna array
   C. It allows DF antennas to receive signals at different vertical angles
   D. It provides diversity reception that cancels multipath signals

E9H08
(A)
Page 9-20

E9H09
What is a Pennant antenna?
   A. A four-element, high-gain vertical array invented by George Pennant
   B. A small, vertically oriented receiving antenna consisting of a triangular loop terminated in approximately 900 ohms
   C. A form of rhombic antenna terminated in a variable capacitor to provide frequency diversity
   D. A stealth antenna built to look like a flagpole

E9H09
(B)
Page 9-21

E9H10
How can the output voltage of a multiple-turn receiving loop antenna be increased?
   A. By reducing the permeability of the loop shield
   B. By utilizing high impedance wire for the coupling loop
   C. By winding adjacent turns in opposing directions
   D. By increasing the number of turns and/or the area

E9H10
(D)
Page 9-21

E9H11
(B)
Page 9-20

E9H11
What feature of a cardioid pattern antenna makes it useful for direction finding?
A. A very sharp peak
B. A very sharp single null
C. Broadband response
D. High radiation angle

SUBELEMENT E0 — SAFETY
[1 exam question — 1 group]

E0A — Safety: RF radiation hazards; hazardous materials; grounding

E0A01
(B)
Page 11-8

E0A01
What is the primary function of an external earth connection or ground rod?
A. Reduce received noise
B. Lightning protection
C. Reduce RF current flow between pieces of equipment
D. Reduce RFI to telephones and home entertainment systems

E0A02
(B)
Page 11-5

E0A02
When evaluating RF exposure levels from your station at a neighbor's home, what must you do?
A. Ensure signals from your station are less than the controlled Maximum Permitted Exposure (MPE) limits
B. Ensure signals from your station are less than the uncontrolled Maximum Permitted Exposure (MPE) limits
C. Ensure signals from your station are less than the controlled Maximum Permitted Emission (MPE) limits
D. Ensure signals from your station are less than the uncontrolled Maximum Permitted Emission (MPE) limits

E0A03
(C)
Page 11-4

E0A03
Over what range of frequencies are the FCC human body RF exposure limits most restrictive?
A. 300 kHz to 3 MHz
B. 3 to 30 MHz
C. 30 to 300 MHz
D. 300 to 3000 MHz

E0A04
(C)
Page 11-7

E0A04
When evaluating a site with multiple transmitters operating at the same time, the operators and licensees of which transmitters are responsible for mitigating over-exposure situations?
A. Only the most powerful transmitter
B. Only commercial transmitters
C. Each transmitter that produces 5 percent or more of its MPE limit in areas where the total MPE limit is exceeded
D. Each transmitter operating with a duty cycle greater than 50 percent

E0A05
(B)
Page 11-7

E0A05
What is one of the potential hazards of operating in the amateur radio microwave bands?
A. Microwaves are ionizing radiation
B. The high gain antennas commonly used can result in high exposure levels
C. Microwaves often travel long distances by ionospheric reflection
D. The extremely high frequency energy can damage the joints of antenna structures

13-106   Chapter 13

E0A06

Why are there separate electric (E) and magnetic (H) field MPE limits?
A. The body reacts to electromagnetic radiation from both the E and H fields
B. Ground reflections and scattering make the field strength vary with location
C. E field and H field radiation intensity peaks can occur at different locations
D. All these choices are correct

E0A06
(D)
Page 11-3

E0A07

How may dangerous levels of carbon monoxide from an emergency generator be detected?
A. By the odor
B. Only with a carbon monoxide detector
C. Any ordinary smoke detector can be used
D. By the yellowish appearance of the gas

E0A07
(B)
Page 11-2

E0A08

What does SAR measure?
A. Synthetic Aperture Ratio of the human body
B. Signal Amplification Rating
C. The rate at which RF energy is absorbed by the body
D. The rate of RF energy reflected from stationary terrain

E0A08
(C)
Page 11-4

E0A09

Which insulating material commonly used as a thermal conductor for some types of electronic devices is extremely toxic if broken or crushed and the particles are accidentally inhaled?
A. Mica
B. Zinc oxide
C. Beryllium Oxide
D. Uranium Hexafluoride

E0A09
(C)
Page 11-2

E0A10

What toxic material may be present in some electronic components such as high voltage capacitors and transformers?
A. Polychlorinated biphenyls
B. Polyethylene
C. Polytetrafluoroethylene
D. Polymorphic silicon

E0A10
(A)
Page 11-2

E0A11

Which of the following injuries can result from using high-power UHF or microwave transmitters?
A. Hearing loss caused by high voltage corona discharge
B. Blood clotting from the intense magnetic field
C. Localized heating of the body from RF exposure in excess of the MPE limits
D. Ingestion of ozone gas from the cooling system

E0A11
(C)
Page 11-3

Question Pool   13-107

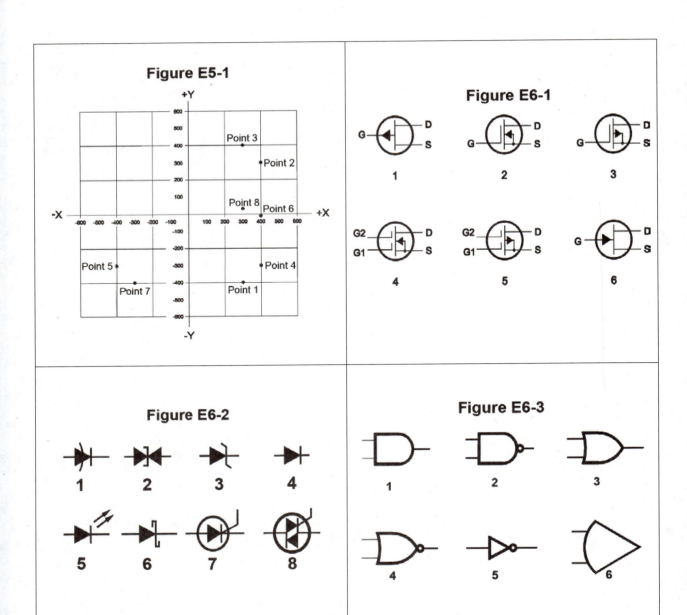

13-108 Chapter 13

Figure E7-1

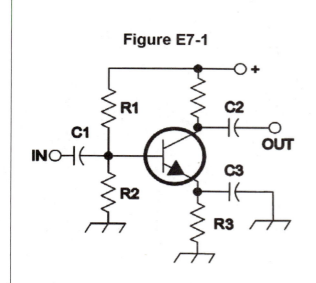

Figure E7-2

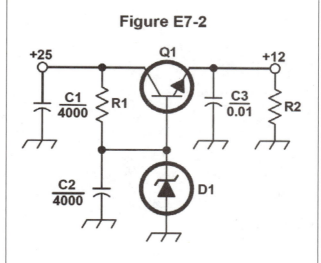

Figure E7-3

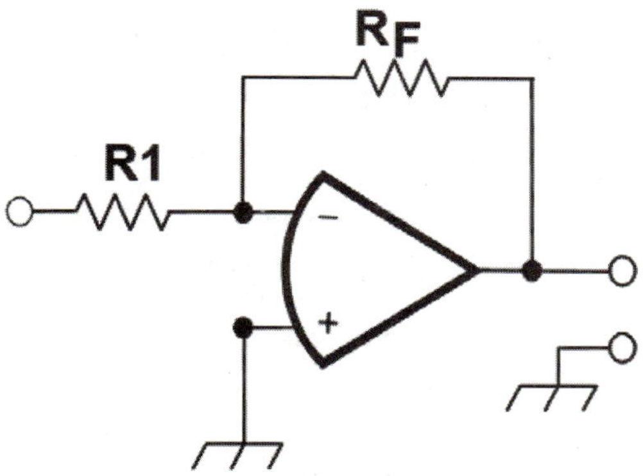

Figure E9-1

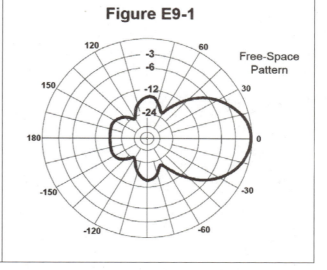

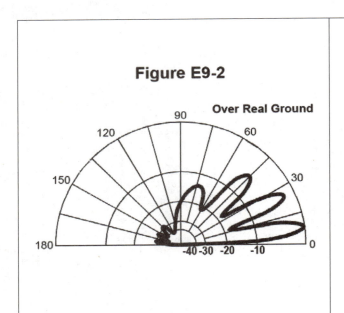

Figure E9-2

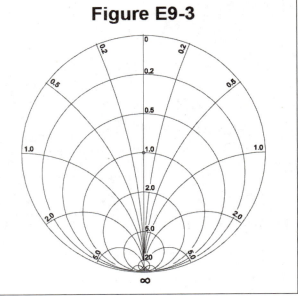

Figure E9-3

iCOM

FIRST IN TECHNOLOGY

Your New Partner for Field Operations

IC-705
HF/6m/2m/70cm Portable SDR* Transceiver

Be Active! In the field with the IC-705

IC-9700
2m/70cm/23cm SDR* Transceiver

IC-7300
HF/6m SDR Transceiver

IC-7610
HF/6m SDR Transceiver

IC-R8600
Wideband SDR* Receiver

*refer to specifications for details of the Direct Sampling SDR

©2020 Icom America Inc. The Icom logo is a registered trademark of Icom Inc. All other trademarks remain the property of their respective owners. All specifications are subject to change without notice or obligation. 31374

YAESU

FT-991A

The **Yaesu FT-991A** does it all. HF to 70cm with 100W on HF/6m. Multimode including C4FM Digital, HF/6m tuner, stunning 3.5" color TFT touch panel with spectrum display. Enhancements in this new "A" version include realtime spectrum scope and three year limited mfg. warranty.

FT-891

Looking for a *truly* compact 100 watt HF +6m transceiver? Consider the new **Yaesu FT-891.** With 32 bit DSP, detachable head, optional tuner, backlit keys, 99 memories, APF, auto notch USB/CAT ports, and more. Receive is 30 kHz - 56 MHz.

FTM-400XDR

The **Yaesu FTM-400XDR** is as capable as it is attractive. Enjoy sophisticated 2 meter-440 operation in FM or C4FM digital *System Fusion* modes. Features include: auto-mode select, 3.5" TFT color display, band scope, altitude display, APRS, GPS, clock, V+U/V+V/U+U, 50/20/5W RF, microSD port, plus 108-470 & 800-999 MHz receive.

FT3DR

The **FT3DR** 2meter/440 HT offers cutting edge features including FM and C4FM digital, built in GPS, Bluetooth®, band scope, color TFT touch-screen, V+V/U+U/ V+U/U+V, voice record and receive from 100 kHz to 999.999 MHz (less cellular).

Visit **www.universal-radio.com** for full information *and* videos on these radios!

Universal Radio
651-B Lakeview Plaza Blvd.
Worthington, OH 43085
♦ Phone: 614 866-4267
♦ dx@universal-radio.com
www.universal-radio.com

The ARRL Extra Class License Manual

Index of Advertisers

ARRL
Page A-4
www.arrl.org

bhi Ltd
Cover 3-
www.bhi-ltd.com

Dave's Hobby Shop
Page A-4
www.w5swl.com

FlexRadio Systems
Cover 2
www.flex-radio.com

HamTest Online
Page A-2
www.hamtestonline.com

ICOM America
Page A-1
www.icomamerica.com

Mosley
Page A-2
www.mosley-electronics.com

Palstar
Page A-3
www.palstar.com

Universal Radio
Page A-2
www.universal-radio.com

HamTestOnline™ students are 50 times more likely to give us 5 stars than request a refund because they failed an exam!

eHam.net reviews
★★★★★ 780
5.0 out of 5 stars

5 star	759
4 star	18
3 star	2
2 star	1
1 star	0

Best study method, study materials, customer support, and guarantee in the industry!
www.hamtestonline.com

MOSLEY ANTENNAS
Mosley
AIRCRAFT GRADE ALUMINUM ELEMENTS & BOOMS and STAINLESS STEEL HARDWARE
STRONG!
AVERAGE IN USE LIFE OF 27+ YEARS WITHOUT MAINTENANCE
..."a better Antenna"!
Call **800-325-4016**
REQUEST A CATALOG
...built to last!   www.mosley-electronics.com

www.palstar.com
800-773-7931

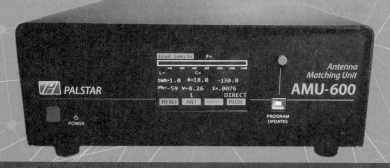

NEW IN 2020! AMU-600 • 600W Antenna Matching Unit

- High-Speed Single Solution Technology*
- Fully Automatic Tuning
- 600W PEP, 160 - 6m Coverage
- Touch Screen Color Display
- Supports 3 Antennas
- 10"W x 4"H x 13"D

Coming Soon: AMU-1K

*Protected By US Patent

LA-1K • 1000W HF Amplifier

HF-AUTO • Automatic Antenna Tuner

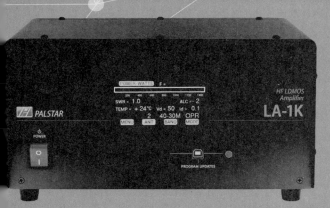

- RF Sensing Dual HF LDMOS
- 1000W (CW ICAS), 500W (FM), 275W (AM)
- 1.8 - 54MHz, All MARS Bands
- 13 dB + or -1 dB gain (nominal)
- Touch screen color display
- Output: 3 x RF SO-239 or Type N

- RF Sensing
- Automatic tuning with memory
- 1800W PEP
- 160 to 6 meters
- 3 Antenna outputs
- Stepper motor controlled L & C

PALSTAR INC. • 9676 N. LOONEY RD, PIQUA, OH 45356 • TEL: 800-773-7931 • www.palstar.com • info@palstar.com

RF Connectors and Adapters

**DIN – BNC
C – FME
Low Pim
MC – MCX
MUHF
N – QMA
SMA – SMB
TNC
UHF & More**

....

Attenuators

Loads & Terminations

Component Parts

Hardware

Mic & Headset Jacks

Mounts

Feet – Knobs

Speakers & Surge Protectors

Test Gear Parts

Gadgets – Tools

www.W5SWL.com

The ARRL Antenna Book for Radio Communications

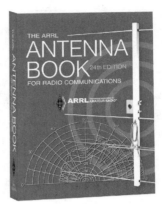

Showcasing 80 Years of Antenna Know How

The ARRL Antenna Book is a single resource covering antenna design, practical treatments, and projects. It contains everything you need to understand how radio signals propagate, how antennas work, and how to construct your own antenna system.

- Learn more about those highly popular HF transmitting loops.
- See the all-new chapter on VHF/UHF antennas.
- Try building some new MF, HF, or 6-meter antennas.
- Discover the importance of ground and bonding.
- Find out how to troubleshoot your antenna tuner quickly and easily.
- Get the most out of your antenna analyzer.
- Design a high-performance Band-Optimized Log Periodic Dipole Array.

Plus, download the fully searchable digital edition of the printed book, as well as utility programs and supplemental content including expanded technical papers, in depth construction guides, and referenced articles.

Softcover Edition
Only $49.95

Order online at
www.arrl.org/shop

INDEX

2200 meter band
    Mode and power restrictions: 3-5
30 meter band
    Mode and power restrictions: 3-4
60 meter band
    Mode and power restrictions: 3-4
630 meter band
    Mode and power restrictions: 3-5

A

Absorption (ionospheric): 10-8
Absorption (RF exposure): 11-3
AC line noise: 7-23
Active filter: 6-35
Admittance: 4-19
Air link: 8-5
Aircraft (stations on board): 3-6
Alias, aliasing: 6-26
Alpha (transistor): 5-9
Amateur bands (chart): 2-2
Amateur radio (overview): 1-2
Amateur Television (ATV): 8-17
    Audio channel: 8-21
    Equipment: 8-18
    FM: 8-21
    RF ATV signals: 8-21
    Signal definitions: 8-19
Amateur-satellite service: 3-13
    Definitions: 3-13
Amplifier
    Basic circuits: 6-2
    Class of operation: 6-10
    Common-base: 6-6
    Common-collector: 6-3
    Common-emitter: 6-3
    Cutoff: 6-10
    Distortion: 6-12
    Emitter follower: 6-3
    Gain: 6-2
    Gain calculation: 6-5
    Impedance, input and output: 6-2
    Intermodulation: 6-12
    Neutralization: 6-13
    Op amp: 6-7
    Parasitic oscillation: 6-14
    Saturation: 6-10
    Self-, fixed-voltage bias: 6-3
    Stability: 6-13
    Switching (switchmode): 6-11
    Vacuum tube: 6-4
Analog-to-digital converter (ADC): 6-27, 6-30
    Harmonic distortion: 6-28

AND gate operation: 5-19
Antenna
    Analyzer: 9-39
    Bandwidth: 9-7
    Basics: 9-1
    Beamwidth: 9-4
    Beverage: 9-15
    Design tradeoffs: 9-41
    Direction finding (DFing): 9-19
    Directional: 9-2
    Efficiency: 9-6
    Far-field: 9-2
    Feed point impedance: 9-5
    Folded dipole: 9-10
    Front-to-back ratio (F/B): 9-5
    Front-to-side ratio (F/S): 9-5
    G5RV antenna: 9-11
    Gain: 9-2
    Ground effects: 9-8
    Ground systems: 9-8
    Height above ground: 9-9
    Impedance matching: 9-24
    Isotropic: 9-3
    Loaded whips: 9-12
    Lobes: 9-3
    Loop (receiving): 9-20
    Modeling: 9-40
    Off-center fed dipole (OFCD): 9-12
    Optimization: 9-41
    Pattern ratio: 9-5
    Phased array: 9-16
    Polarization: 9-7
    Radiation patterns: 9-2
    Radiation resistance: 9-6
    Reference dipole: 9-3
    Restrictions: 3-8
    Rhombic: 9-14
    Satellite: 9-18
    Sense: 9-20
    Shortened: 9-12
    Terrain effects: 9-9
    Traveling wave: 9-14
    Zepp and Extended Zepp: 9-10
Apparent power: 4-24
Arcing: 7-23
ARRL
    Amateur Radio Contesting for Beginners: 2-5
    Antenna Book: 1-4
    Basic Electronics: 1-4
    Basic Radio: 1-4
    Contest Update (newsletter): 2-5
    DX Bulletin: 2-3

Handbook: 1-4
Operating Manual: 1-4, 2-3, 2-5
Understanding Basic Electronics: 1-4
Propagation bulletin: 2-3
ASCII: 8-7
Atmospheric static: 7-21
Audio Frequency Shift Keying (AFSK): 8-10
Auroral propagation: 10-14
Automatic control: 3-10
Automatic Link Establishment (ALE): 8-6
Automatic Packet Reporting System (APRS): 8-11
Automatic repeat request (ARQ): 8-17
Auxiliary stations: 3-12, 3-18
Avalanche current: 5-5
Average forward current: 5-4
AX.25 standard: 8-11
Azimuthal patterns: 9-7

B

Back EMF: 4-7
Balanced modulator: 6-21
Band plans: 2-3
Band-pass filter: 6-33
Bandwidth
    Antenna: 9-7
    CW: 8-9
    Digital signal: 8-8
    FCC definition: 3-4
    FSK/AFSK: 8-10
    PSK: 8-11
    Resonant circuit: 4-32
Base-loaded antenna: 9-12
Battery charging: 6-42
Baud: 8-5
Baudot: 8-7
Beacon stations: 3-12
Beamwidth: 9-4
Beta (transistor): 5-9
Beverage antenna: 9-15
Bias
    Diodes: 5-3
    Transistors: 5-8
BiCMOS (logic): 5-25
Bipolar junction transistor (BJT): 5-8
Bleeder resistor: 6-43
Blocking dynamic range (BDR): 7-16
Bonding (RF and safety): 11-8
Boolean algebra: 5-18
    AND operation: 5-18
    Exclusive OR (NOR) operation: 5-20
    Inversion (NOT) operation: 5-19
    NAND operation: 5-20
    NOR operation: 5-20
    OR operation: 5-20
    XOR operation: 5-20
Business communications: 3-20
Butterworth filter: 6-33

C

Cabrillo (contest log format): 2-6
Call signs: 1-4
Capacitive hat (antenna): 9-13
Capacitors
    Voltage-current relationship: 4-13
Capture effect: 7-21
Carbon monoxide: 11-2
CEPT license: 3-21

Certificate of Successful Completion of Examination
(CSCE): 1-8, 3-17
Chebyshev filter: 6-33
Circular polarization: 10-3
Class of operation (amplifier)
    Class A, B, AB, C, D: 6-11
Club locator service: 1-12
CMOS (logic): 5-25
Code-division, multiple access (CDMA): 8-15
Colpitts oscillator: 6-15
Common-mode signal: 7-25
Comparators: 6-9
    Hysteresis: 6-10
Complete DXer (*The*): 2-6
Complex
    Coordinates: 4-4
    Impedance: 4-16
    Number: 4-2
Component packaging
    Effects at RF: 4-34
    Surface-mount: 4-35
    Through-hole: 4-35
Compression (speech): 7-3
Contesting: 2-5
    Cabrillo log format: 2-6
    Calendars: 2-5
    Contest calendars: 2-1
    Email reflectors: 2-5
    Log submission: 2-6
    Operating techniques: 2-5
    Resources: 2-5
    Self-spotting: 2-7
    Spotting networks: 2-7
Control (station)
    Automatic: 3-10
    Local: 3-10
    Remote: 3-10
Controlled environment: 11-5
Controller, charge: 6-42
Conventional current: 4-7, 5-3
Coordinates
    Complex: 4-4
    Polar: 4-1
    Rectangular: 4-1
Cross-modulation: 7-21
Crossed-ellipse display: 8-10
Crystal
    Controlled oscillator: 6-15
    Filter: 6-35
CubeSat: 2-11
Current
    Circulating: 4-30
    Conventional: 4-7
    Electronic: 4-7
Current gain (transistor): 5-9
Cutoff (amplifier): 6-10
CW bandwidth: 8-9

D

Daily DX (newsletter): 2-3
Data rate: 8-5
dBd: 9-3
dBi: 9-3
Decimation (DSP): 6-28
Delta match: 9-25
Depletion mode (FET): 5-11
Desensitization (desense): 7-16

Detector
FM and PM: 6-24
Product: 6-24
Receiver: 6-23
Deviation ratio: 8-3
Digital logic: 5-18
Digital modes
AFSK: 8-10
ALC: 8-14
ASCII code: 8-7
Automatic message forwarding: 3-5
Automatic repeat request (ARQ): 8-17
Bandwidth: 8-8
Baudot code: 8-7
Codes: 8-6
CW: 8-9
Data (symbol) rate: 8-5
Error detection and correction: 8-17
Forward error correction (FEC): 8-17
FSK: 8-10
Gray code: 8-8
HF packet: 8-12
Hidden transmitter: 8-14
Morse code: 8-7
Orthogonal Frequency Division Multiplexing (OFDM): .
8-13
Overmodulation: 8-14
Packet radio: 8-11
PACTOR: 8-12
Protocols: 8-5
PSK: 8-11
Transmitting properly: 8-14
Varicode: 8-7
WSJT-X: 8-13
Digital signal processing (DSP): 6-25
Adaptive filter: 6-36
Aliasing: 6-26
Decimation (DSP): 6-28
Fast Fourier Transform (FFT): 6-28
Filter: 6-36
Finite Impulse Response (FIR) filter: 6-37
Infinite Impulse Response (IIR) filter: 6-37
Interpolation (DSP): 6-28
Modulation: 6-30
Noise reduction: 7-26
Nyquist sampling theorem: 6-26
Quadrature method of SSB modulation: ...6-33
Sequential sampling: 6-25
Tap, filter: 6-37
Digital-to-analog converter (DAC): 6-28, 6-30
Diodes
Average forward current: 5-4
Hot-carrier: 5-5
Junction: 5-3
Light-emitting (LED): 5-7, 5-14
PIN: 5-6
Point-contact: 5-5
Schottky barrier: 5-4
Varactor: 5-6
Zener: 5-5, 6-41
Dipole
Folded: 9-10
Reference: 9-3
Direct digital conversion (DDC): 6-29
Direct digital synthesizer (DDS): 6-17
Direct sequence spread spectrum: 8-16
Direction finding (RDF): 9-19

Dual in-line package (DIP): 4-35
Duty cycle: 11-5
DX windows: 2-3
DXing
Logbook of the World: 2-3
NCDXF Beacons: 2-4
Overview: 2-3
Pileups: 2-4
Propagation: 2-5
Resources: 2-3
Reverse Beacon Network: 2-4
Split frequency operation: 2-4
Dynamic range
Blocking: 7-16
Intermodulation (IMD): 7-15

E

E-plane: 9-7
Earth station: 3-12
Earth-Moon-Earth communications (EME): ...10-17
Libration fading: 10-17
Operating: 10-17
Effective radiated power (ERP, EIRP): 9-22
Efficiency
Antenna: 9-5
Voltage regulator: 6-42
Electric
Energy storage: 4-6
Field: 4-4, 10-2
Electrical safety ground: 11-9
Electromagnetic (EM) waves: 10-1
Wavefront: 10-2
Electronic current: 4-7, 5-3
Elevation patterns: 9-7
Elliptical filter: 6-33
Emission (spurious): 3-7
Emission designators and types: 8-1
Enhancement mode (FET): 5-11
Environment, controlled and uncontrolled: ...11-5
Error detection and correction: 8-17
Evaluation (RF exposure): 11-6
Examinations, preparation of: 3-16
Exclusive OR (NOR) gate operation: 5-20
Extra class
Element 4 syllabus: 1-13
Exam elements required: 1-1
Exam review software: 1-11
Frequencies: 3-2
HF frequencies: 2-1
License manual website: 1-10
License overview: 1-3
Licensing resources: 1-8
Practice exams: 1-10
Question pool: 1-10
Study topics: 1-5
Volunteer testing process: 1-5

F

FAA antenna restrictions: 3-9
Fading: 10-11
Libration: 10-17
Far-field (antenna): 9-2
Faraday rotation: 2-9
Fast Fourier Transform (FFT): 6-28
Fast-scan television: 8-17

FCC
- Monitoring stations: ...3-8
- Part 97 rules: ...3-1

Federal Registration Number (FRN): ...1-6

Ferrite: ...4-36
- Bead: ...4-36
- Toroid core: ...4-36

Field
- Electric: ...4-4
- Magnetic: ...4-4

Field Effect Transistor (FET)
- Depletion mode: ...5-11
- Enhancement: ...5-11
- JFET: ...5-10
- MOSFET: ...5-10

Filter
- Active: ...6-35
- Band-pass: ...6-33
- Butterworth: ...6-33
- Chebyshev: ...6-33
- Crystal: ...6-35
- DSP: ...6-36
- Elliptical: ...6-33
- Families: ...6-33
- Front-end: ...7-14
- High-pass: ...6-33
- Low-pass: ...6-33
- Noise reduction (DSP): ...7-26
- Notch: ...6-33
- Receiver IF: ...7-14
- Response: ...6-33
- Ripple: ...6-33
- Roofing: ...7-15
- Shape factor: ...6-34

Filter method of SSB modulation: ...6-20

Finite Impulse Response (FIR) filter: ...6-37

Flag antenna: ...9-21

Flip-flop: ...5-22

Folded dipole: ...9-10

Forward error correction (FEC): ...8-17

Forward voltage: ...5-3

Fourier analysis: ...6-28, 7-8
- Sawtooth wave: ...7-8
- Square wave: ...7-8

Frequency
- 30 and 60 meter band restrictions: ...3-4
- Amateur bands chart: ...2-2
- Band edges: ...3-4
- Band plans: ...2-4
- Emission types: ...3-2
- Extra class bands: ...2-1, 3-2
- Mixer: ...6-19
- Satellite: ...2-12
- Selection: ...2-3
- Synthesis: ...6-16

Frequency allocations...2-2
- Repeaters: ...3-12

Frequency counter
- Accuracy: ...7-4
- Frequency: ...7-3
- Prescaler: ...7-3
- Time-base: ...7-3

Frequency discriminator: ...6-24

Frequency domain: ...7-7

Frequency hopping: ...8-16

Frequency modulation (FM): ...8-2

Frequency Shift Keying (FSK): ...8-10

Frequency-division multiplexing (FDM): ...8-4

Front-to-back ratio (F/B): ...9-5

Front-to-side ratio (F/S): ...9-5

FT8 (FT4): ...8-13

G

G5RV antenna: ...9-11

Gain
- Amplifier: ...6-2
- Antenna: ...9-2
- Operational amplifier (op amp): ...6-7

Gain compression: ...7-16

Gamma match: ...9-25

Gray code: ...8-8

Gray-line propagation: ...10-9

Ground effects (antennas): ...9-8

Ground systems (antennas): ...9-9

Ground-wave propagation: ...10-5

Grounded-grid amplifier: ...6-4

Grounding: ...11-8
- Reference potential: ...11-9

H

H-plane: ...9-7

Hairpin match: ...9-26

Ham Radio for Dummies: ...1-4

Hartley oscillator: ...6-15

Height above average terrain (HAAT): ...9-23

Hepburn maps: ...10-12

HF operating: ...1-5

HF packet: ...8-12

Hidden transmitter (digital modes): ...8-14

High-pass filter: ...6-33

Hilbert transform: ...6-32

Horizon (radio): ...10-10

Hot-carrier diode: ...5-5

Hysteresis: ...6-10

I

I/Q modulation: ...6-31

Imaginary number (j): ...4-2

Impedance
- Antenna feed point: ...9-5
- Calculation: ...4-16
- Complex: ...4-16
- Resonant circuit: ...4-30

Impedance matching: ...6-38, 9-24
- L-network: ...6-39
- Pi-, Pi-L-network: ...6-39
- Synchronous transformer: ...9-37
- T-network: ...6-40

Inductors
- Calculations: ...4-37
- Core shape: ...4-36
- Magnetic cores: ...4-35
- Voltage-current relationship: ...4-14

Infinite Impulse Response (IIR) filter: ...6-37

Insertion loss (filter response): ...6-32

Integrated circuit
- Microwave (MMIC): ...5-12
- RF: ...5-12

Intercept point: ...7-18

Interference
- AC line noise: ...7-23
- Arcing: ...7-23
- Atmospheric static: ...7-21

Common-mode signals: 7-25
Computer: 7-25
Intermodulation: 7-21
Locating source of: 7-24
Noise blanker: 7-26
Noise reduction (DSP): 7-26
Spurious emissions: 3-7
Vehicle noise: 7-25
Interlaced video: 8-19
Intermodulation distortion (IMD)
Amplifier: 6-12
Receiver: 7-15
Transmitter: 7-21
Transmitter testing: 7-10
Intermodulation dynamic range (IDR): 7-16
Intermodulation products: 7-17
International Amateur Radio Permit (IARP): 3-21
Interpolation (DSP): 6-28
Inversion (NOT) operation: 5-19
Isotropic radiator: 9-3

J
JFET: 5-10
JT65: 8-13

K
Kepler's Laws: 2-8

L
L-network: 6-39
Left-hand (right-hand) rule: 4-7
Libration fading: 10-17
License exams: 1-6
Certificate of Successful Completion of Examination
(CSCE): 1-8
Identification requirements: 1-6
Locator service: 1-12
Light-emitting diode (LED): 5-7, 5-14
Lightning dissipation ground: 11-9
Line A: 3-19
Line noise (ac): 7-23
Linear voltage regulator: 6-41
Loading
Base: 9-12
Coils: 9-12
Top: 9-13
Whip antenna: 9-12
Lobes (antenna): 9-3
Local control: 3-10
Logic
Basic functions: 5-19
BiCMOS: 5-25
CMOS: 5-25
Counter, decade counter: 5-24
Divider, divide-by-n: 5-24
Families: 5-24
Flip-flop: 5-22
Positive and negative-true: 5-20
Programmable (PLD, PGA): 5-26
Synchronous: 5-21
Tri-state logic: 5-21
TTL: 5-24
Logic analyzer: 7-7
Long-path propagation: 10-9
Loop antenna (receiving): 9-20
Low-pass filter: 6-33

M
Magnetic
Cores: 4-35
Energy storage: 4-4, 4-7
Field: 4-4, 10-2
Matched line loss: 9-30
Maximum Permissible Exposure (MPE): 11-4
Measurements
Accuracy: 7-2
Antenna analyzer: 9-39
Oscilloscope: 7-5
Precision: 7-2
Resolution: 7-2
RF power: 9-32
Root-mean-square (RMS): 7-2
Spectrum analyzer: 7-7
Test equipment: 7-1
Wattmeter (RF): 7-3
Message forwarding (automatic): 3-5, 3-11
Meteor scatter propagation: 10-15
Microphonics: 6-16
Microstrip construction: 5-13
Minimum discernible signal (MDS): 7-12
Mixer (frequency): 6-19
Active: 6-19
Passive: 6-19
Mixing product: 6-19
Modulation index: 8-3
Modulator (modulation)
AM and DSB: 6-20
Balanced: 6-21
Deviation ratio: 8-3
Direct FM: 6-22
Frequency: 8-2
Hilbert transform: 6-32
Indirect FM: 6-22
Phase: 6-22, 8-2
Pre-emphasis and de-emphasis: 6-23
Quadrature: 6-21
Reactance: 6-22
SSB: 6-20
Monolithic microwave integrated circuit (MMIC): 5-12
Moonbounce: 10-17
Morse code: 1-3, 8-7
MOSFET: 5-10
Multimeter (DMM): 7-1
Multipath propagation: 10-11
Multiplexing: 8-4
Multivibrator
Astable (free-running): 5-23
Monostable: 5-23
One-shot: 5-23

N
N-type semiconductor: 5-2
NAND gate operation: 5-20
National Contest Journal (*NCJ*): 2-5
National Radio Quiet Zone: 3-20
Neutralization (amplifier): 6-13
Noise figure: 7-13
Noise locating: 7-24
Noise reduction: 7-26
NOR gate operation: 5-20
Notch filter: 6-33
NPN transistor: 5-8
Nyquist sampling theorem: 6-26

O

Off-center-fed dipole (OFCD): ...9-12
OPDX (newsletter): ...2-3
Operating techniques
    Aurora: ...10-14
    Digital modes: ...8-10
    DX pileups: ...2-4
    DXing: ...2-4
    Earth-Moon-Earth communications (EME): ...10-17
    Meteor scatter: ...10-16
    Satellite communications: ...2-10
    Split frequency operation: ...2-4
Operational amplifier (op amp): ...6-7
    Characteristics: ...6-7
    Circuits: ...6-8
    Gain: ...6-7
Optical shaft encoder: ...5-16
Optocoupler, optoisolator: ...5-16
Optoelectronics: ...5-13
OR gate operation: ...5-20
Orbit (satellite): ...2-8
    Terminology: ...2-9
Ordinary (O) and Extraordinary (X) waves: ...10-7
Orthogonal Frequency Division Multiplexing
    (OFDM): ...8-13
Oscillator
    Circuits: ...6-14
    Colpitts: ...6-15
    Crystal controlled: ...6-15
    Hartley: ...6-15
    Microphonics: ...6-16
    Pierce: ...6-15
    Thermal drift: ...6-16
    Variable frequency (VFO): ...6-15
Oscilloscope: ...7-5
    Digital: ...7-7
    Probes: ...7-6
Overmodulation of digital signals: ...8-14

P

P-type semiconductor: ...5-2
Packet radio: ...8-11
    Automatic message forwarding: ...3-5
    HF: ...8-12
    VHF: ...8-10
PACSAT: ...2-11
PACTOR: ...8-12
Parasitic
    Capacitance: ...4-34
    Inductance: ...4-34
Parasitic oscillation: ...6-14
Pattern ratio: ...9-5
Payment for communication services: ...3-20
PCBs (polychlorinated biphenyls): ...11-1
Peak envelope power (PEP): ...7-3
Pedersen ray: ...10-7
Pennant antenna: ...9-21
Permeability: ...4-36
Phase angle: ...4-12
    Calculation: ...4-16
Phase locked loop (PLL): ...6-18
Phase modulation (PM): ...8-2
Phase noise: ...7-20
Phased array: ...9-16
Photoconductive effect: ...5-15

Photoconductivity: ...5-14
Photoelectric effect: ...5-15
Phototransistor: ...5-16
Photovoltaic cells: ...5-17
Photovoltaic effect: ...5-17
Pi-, Pi-L-network: ...6-39
Pierce oscillator: ...6-15
Piezoelectric effect: ...6-15
PIN diode: ...5-6
PN junction: ...5-3
PNP transistor: ...5-8
Point-contact diode: ...5-4
Polar coordinates: ...4-1
Polarization
    Circular: ...9-19, 10-3
    Electromagnetic wave: ...10-3
Polarization (antenna): ...9-7
Polychlorinated biphenyls (PCBs): ...11-1
Positive and negative-true logic: ...5-20
Powdered iron core: ...4-36
Power
    Apparent: ...4-24
    Effective radiated (ERP): ...9-22
    Peak envelope power (PEP): ...7-3
    Reactive: ...4-24
    Real: ...4-24
    RF signals: ...7-3
Power density (RF exposure): ...11-3
Power factor: ...4-23
Power supplies: ...6-41
    High-voltage: ...6-43
PRB-1 (FCC regulation): ...3-8
Pre-emphasis and de-emphasis: ...6-22
Prescaler: ...7-3
Preselector: ...7-14
Product (mixing): ...6-19
Product detector: ...6-24
Programmable logic (PLD, PGA): ...5-26
Propagation
    A, Bz, K, G indexes: ...10-5
    Absorption (ionospheric): ...10-8
    Auroral: ...10-14
    Chordal hop: ...10-7
    Ducting: ...10-12
    DXing: ...2-4
    Earth-Moon-Earth communications (EME): ...10-17
    Geomagnetic field: ...10-5
    Gray-line: ...10-9
    Ground-wave: ...10-5
    HF: ...10-5
    Long-path: ...10-9
    Meteor scatter: ...10-15
    Multipath: ...10-11
    Ordinary (O) and Extraordinary (X) waves: ...10-7
    Pedersen ray: ...10-7
    Prediction software: ...10-8
    Sky-wave: ...10-6
    Sporadic E: ...10-11113
    Transequatorial (TE): ...10-13
    Tropospheric: ...10-12
    VHF/UHF/microwave: ...10-10
Protocols (digital): ...8-5
PSK: ...8-11

Q

Q (quality factor): 4-31
   Skin effect: 4-33
Quadrature method of SSB modulation: 6-21
Question pool: 1-10
   Committee: 3-16
Quiet hours: 3-7

R

Radians: 4-2
Radiation
   Patterns: 9-2
   Resistance: 9-5
Radio Amateur Civil Emergency Service (RACES): ...3-6
Radio horizon: 10-10
Ratings, diode: 5-4
Reactance: 4-15
Reactance modulator: 6-22
Reactive power: 4-24
Real number: 4-2
Real power: 4-24
Receive directivity factor (RDF): 9-21
Receiver
   Blocking dynamic range (BDR): 7-16
   Capture effect: 7-21
   Detector: 6-23
   Dynamic range: 7-15
   Filters: 7-14
   Frequency discriminator: 6-24
   Gain compression: 7-16
   Intercept point: 7-18
   Intermodulation dynamic range (IDR): 7-15
   Minimum discernible signal (MDS): 7-12
   Noise figure: 7-13
   Noise floor: 7-12
   Performance measurements: 7-12
   Phase noise: 7-20
   Preselector: 7-14
   SDR dynamic range: 7-16
   Selectivity: 7-14
   Sensitivity: 7-12
   Signal-to-noise ratio (SNR): 7-12
   Signal-to-noise-and-distortion (SINAD): 7-12
Reciprocal operating: 3-21
   CEPT license: 3-21
   IARP license: 3-21
Rectangular coordinates: 4-1
Reference dipole: 9-3
Reflection coefficient: 9-31
Remote control: 3-10
Remote stations: 2-7
   Contesting and DXing: 2-7
   Identification requirements: 2-7
Repeaters
   Frequency allocations: 3-12
   Satellites: 2-10
Resistor
   Bleeder: 6-43
   Equalizing: 6-43
Resonant circuit: 4-27
   Bandwidth: 4-30, 4-32
   Frequency calculations: 4-27
   Impedance: 4-30
   Q: 4-31
Return loss (RL): 9-38

RF exposure
   Absorption and limits: 11-4
   Controlled environments: 11-5
   Duty cycle: 11-5
   Estimating: 11-6
   Maximum Permissible Exposure (MPE): 11-4
   Power density: 11-3
   Safety measures: 11-7
   Specific Absorption Rate (SAR): 11-4
   Uncontrolled environments: 11-5
RF integrated circuit: 5-12
RF power amplifier
   Certification of: 3-18
   Class of operation: 6-10
RF power measurement: 9-32
Rhombic antenna: 9-14
Ripple (filter response): 6-33
Roofing filter: 7-15
Root-mean-square (RMS): 7-2

S

Safety
   Beryllium oxide: 11-2
   Carbon monoxide: 11-2
   Hazardous materials: 11-1
Satellites
   Amateur: 2-7
   Amateur-satellite service: 3-12
   Antennas: 9-18
   Ascending and descending nodes: 2-9
   Communication techniques: 2-10
   Doppler shift: 2-11
   Effects of motion: 2-11
   Faraday rotation: 2-9
   Frequencies: 2-12
   Licensing procedures: 3-13
   Modes of operation: 2-12
   Orbits: 2-8, 2-9
   Packet radio: 2-11
   Repeaters: 2-10
   Rules: 3-13
   Spin modulation: 2-9
   Telecommand: 3-13
   Telemetry: 3-12
   Transponders: 2-10
Saturation
   Amplifier: 6-10
Sawtooth wave: 7-8
Scattering (S) parameters: 9-38
Schottky barrier diode: 5-4
Selectivity: 7-14
Self-resonance: 4-34
Semiconductors
   Bipolar junction transistor (BJT): 5-8
   Junction diodes: 5-3
   Materials: 5-1
   N-type: 5-2
   P-type: 5-2
Sense antenna: 9-20
Sequential logic: 5-21
Sequential sampling: 6-25
Series and shunt regulators: 6-41
Shaft encoder, optical: 5-16
Shannon, Claude
   Information theorem: 8-8
Shape factor (filter): 6-34
Ships (stations on board): 3-6

Signal-to-noise ratio (SNR): ...7-12
Signal-to-noise-and-distortion (SINAD): ...7-12
Skin effect: ...4-33
Sky-wave propagation: ...10-6
Slow-Scan Television (SSTV): ...8-21
  Analog basics: ...8-22
  Color: ...8-23
  Digital: ...8-23
  Equipment: ...8-22
  Sync signals: ...8-23
Smith chart: ...9-32
Software-defined radio (SDR): ...6-28
  Direct digital conversion (DDC): ...6-29
Solar cells: ...5-18
Solar effects on propagation: ...10-4
  Solar flares: ...10-4
  Solar flux: ...10-4
Special events: ...2-1
Special Temporary Authority (STA): ...3-22
Specific Absorption Rate (SAR): ...11-4
Spectrum analyzer: ...7-7
  Transmitter testing: ...7-10
Split frequency operation: ...2-4
Sporadic E propagation: ...10-13
Spotting networks: ...2-7
Spread spectrum communications
  Direct sequence: ...8-16
  Frequency hopping: ...8-16
  Mesh network: ...8-16
  Rules: ...3-20
  Techniques: ...8-15
Square wave: ...7-8
Stability (amplifier): ...6-13
Standing wave ratio (SWR): ...9-31
Static, atmospheric (QRN): ...7-21
Station
  Evaluation (RF exposure): ...11-6
  Location restrictions: ...3-8
Station control: ...3-9
Stub
  Matching: ...9-26
  Transmission line: ...9-36
Susceptance: ...4-19
Switching (switchmode) amplifier: ...6-11
Switching (switchmode) voltage regulator: ...6-43
Syllabus (Element 4, Extra class): ...1-13, 13-1
Symbol rate: ...8-5
Synchronous
  Logic: ...5-21
  Transformer: ...9-37

T

T-network: ...6-40
Tables
  AC Measurements for Sine and Square Waves: ...7-2
  Allocations for Ground-Based Repeater Stations: 3-12
  Amateur Extra Class HF Bands and Sharing
    Requirements: ...3-2
  Amateur License Class Examinations: ...1-6
  Amateur Satellite Service Definitions: ...3-12
  Black and White SSTV Standards: ...8-23
  Characteristics of Commonly Used
    Transmission Lines: ...9-30
  DX Windows and Watering Holes: ...2-3
  Emission Designators: ...8-2

Emission Duty Cycle of Modes Commonly Used by
  Amateurs: ...11-5
Exam Elements Needed for Extra Class License: ...1-1
Exclusive HF Privileges for Amateur Extra Class
  Operators: ...3-2
How Mode Affects the Frequency of Transceiver
  Displays: ...3-3
IF Filter Bandwidths by Signal Type: ...7-15
Major Meteor Showers: ...10-16
Maximum Permissible Exposure (MPE): ...11-4
NTSC Standards for ATV: ...8-19
Power Thresholds for RF Exposure Evaluation: ...11-6
Protected FCC locations: ...3-9
Satellite Operating Modes: ...2-12
Technical Information Service (TIS): ...1-12
Telecommand: ...3-10
Telemetry: ...3-12
Television
  Fast-scan: ...8-17
  Slow-scan: ...8-21
Thermal drift: ...6-16
Time constant: ...4-8
  Calculation, RC and RL circuits: ...4-8
  Circuit examples: ...4-10
Time domain: ...7-7
Time-division multiplexing (TDM): ...8-4
Top-loaded antenna: ...9-13
Toroid
  Calculations: ...4-37
  Core shape: ...4-36
  Ferrite: ...4-36
  Powdered iron: ...4-36
  Turns count: ...4-37
Transconductance: ...5-10
Transequatorial (TE) propagation: ...10-13
Transistors
  Alpha: ...5-9
  Beta: ...5-9
  Bipolar: ...5-8
  Characteristics: ...5-9
  Cutoff, saturation: ...5-9
  Field-effect (FET): ...5-10
  Forward bias: ...5-8
  NPN: ...5-8
  PNP: ...5-8
Transmission Line
  Electrical length: ...9-28
  Loss: ...9-30
  Reflection coefficient: ...9-31
  Standing wave ratio (SWR): ...9-31
  Stubs: ...9-36
  Transformers: ...9-37
  Velocity factor: ...9-28
  Velocity of propagation: ...9-36
Transponders: ...2-10
  Modes of operation: ...2-12
Traveling wave antenna: ...9-14
Tri-state logic: ...5-21
Triangulation: ...9-22
Tropospheric propagation: ...10-12
  Ducting: ...10-12
  Hepburn maps: ...10-12
TTL logic: ...5-24

U

Uncontrolled environment: 11-5

V

Vacuum tube amplifier: 6-4
Varactor diode: 5-6
Variable frequency oscillator (VFO): 6-15
Varicode: 8-7
Vehicle noise: 7-25
Velocity factor: 9-28
Velocity of propagation: 9-36
Video
    Composite, RS-170: 8-20
    Interlacing: 8-19
    RGB: 8-20
    Signal types: 8-19
Volt-amperes reactive (VAR): 4-24
Voltage regulator: 6-41
    Battery charging: 6-42
    Drop-out: 6-42
    Efficiency: 6-42
    Linear: 6-41
    Switching (switchmode): 6-45
Volunteer Examiner (VE): 3-14
    Accreditation: 3-16
    Application form: 3-15
    Exam preparation: 3-16

    Exam session administration: 3-16
    Testing process: 1-5
Volunteer Examiner Coordinator (VEC): 3-14
    CSCE: 3-17
    Expenses: 3-17
    Form 605: 1-7
    Special testing procedures: 1-8

W

Watering holes (DX): 2-3
Wattmeter (RF): 7-3, 9-32
Waveforms
    AC: 4-12
    Root-mean-square (RMS): 7-2
Wavefront: 10-2
Website (*Extra Class License Manual*): 1-10
Whip antenna: 9-12
Winlink: 3-11
WSJT-X: 8-13

X

XOR gate operation: 5-20

Z

Zener diode: 5-5, 6-41
Zepp and Extended Zepp antenna: 9-10

Notes

Notes

Notes

Notes

Notes

Notes

Notes

Notes

Notes

Notes